T0176895

13. Section 4.1 (page 94; line 21) Inserted 'the total length of Γ_0 is strictly positive and' after 'where'

14. Subsection 5.1.5 (page 176; Equation 5.70) After $\bar{C}$ amended $\| H(u) \|_{0,2}$ into $(\| H(u) \|_{0,2} + \| \textbf{grad } f \|_{0,\infty})$ so that Equation (5.70) becomes,

$$\| u_h - u \|_{0,2} \leq \left[\sqrt{area(\Omega)} \, \| \textbf{grad } u \|_{0,\infty} + \bar{C} \, (\| H(u) \|_{0,2} + \| \textbf{grad } f \|_{0,\infty}) \right].$$

15. Subsection 5.1.6 (page 188; line 4) Modified the complete sentence into 'Therefore $\forall \vec{q} \in \textbf{Q}_h$, $(\vec{q}|\vec{v}_i^T)$ is also constant along any mesh edge e_i^T.'

16. Subsection 5.2.2 (page 203; line 8) $\vec{a} - \vec{b} \leq |\vec{a}| + |\vec{b}|$ has been changed to '$|\vec{a} - \vec{b}|^2 \leq 2(|\vec{a}|^2 + |\vec{b}|^2)$'

17. Subsection 6.1.6 (page 227; after line 10) Inserted the paragraph "The first term vanishes if f is extended by zero in $\Omega_h \setminus \Omega$ (please check!). However for pedagogic purposes we will assume that f is well defined beyond Ω, and $\forall h \, |f|$ is bounded away from zero in $\Omega_h \setminus \Omega$."

The publisher apologizes for these errors and any confusion they may have caused.

Errata

NUMERICAL METHODS FOR PARTIAL DIFFERENTIAL EQUATIONS
Vitoriano Ruas
9781119111351

The publisher wishes to bring to the reader's attention that errors were inadvertently included in text, equations and references. The following corrections replace those in the printed book. In addition, some page numbers in the index were incorrectly listed – a corrected Index is supplied in a separate file on the companion website page.

1. Section 1.1 (page 4) Reference [110] has been changed to [111]

2. Remark 3.3 (page 90) Reference [195] has been changed to [194]

3. Section 7.3 (page 295) Reference [111] has been changed to [110]

4. Section 7.4 (page 301) Reference [111] has been changed to [110]

5. Appendix A.1 (page 312) Reference [111] has been changed to [110]

6. Appendix A.1 (page 313) Reference [111] has been changed to [110]

7. Appendix A.3 (page 318) Reference [194] has been changed to [195]

8. Subsection 3.4.2 (page 84; line 10) '/2' has been added at the end of the equation, which becomes,

$$\{u([j-1]h, [k-1]\tau) + u([j+1]h, [k-1]\tau)\}/2 = u(jh, [k-1]\tau) + h^2\partial_{xx}u(\sigma_j^k, [k-1]\tau)/2$$

9. Subsection 3.4.2 (page 84; line 14) The following sentence "(3.52) and using the fact that $h < L$, after ..." has been replaced with "(3.52) and taking $h = \lambda\tau$ for a certain $\lambda \geq |o|$, after ...".

10. Subsection 3.4.2 (page 84; Equation 3.63) L has been replaced by λ and inserted $/2$ at the end of the first line. Equation(3.63) becomes:

$$\begin{cases} H_j^{k-1}(u) = \tau\partial_{tt}u(jh, \underline{\xi}_j^k)/2 + h\lambda\partial_{xx}u(\sigma_j^k, [k-1]\tau)/2 \\ +oh[\partial_{xx}u(\overline{\zeta}_j^k, [k-1]\tau) - \partial_{xx}u(\underline{\zeta}_j^k, [k-1]\tau)]/4. \end{cases}$$

11. Subsection 3.4.2 (page 84; Equation 3.64) L has been replaced by λ, inserted [after Q and add] at the end of the equation. Equation (3.64) becomes:

$$|H_j^k(u)| \leq Q\left[\tau \max_{0\leq x\leq L; 0\leq t\leq \Theta}|\partial_{tt}u(x,t)| + (|o|+\lambda)h \max_{0\leq x\leq L; 0\leq t\leq \Theta}|\partial_{xx}u(x,t)|\right].$$

12. Subsection 3.4.2 (page 85; line 4) L has been replaced by λ in the expression of $C^*(u)$.

NUMERICAL METHODS FOR PARTIAL DIFFERENTIAL EQUATIONS

NUMERICAL METHODS FOR PARTIAL DIFFERENTIAL EQUATIONS

AN INTRODUCTION FINITE DIFFERENCES, FINITE ELEMENTS AND FINITE VOLUMES

Vitoriano Ruas

Sorbonne Universités, UPMC - Université Paris 6, France

WILEY

This edition first published 2016

© 2016 John Wiley & Sons Ltd

Registered office
John Wiley & Sons Ltd, The Atrium, Southern Gate, Chichester, West Sussex, PO19 8SQ, United Kingdom

For details of our global editorial offices, for customer services and for information about how to apply for permission to reuse the copyright material in this book please see our website at www.wiley.com.

Library of Congress Cataloging-in-Publication Data

Names: Ruas, Vitoriano, 1946- author.
Title: Numerical methods for partial differential equations : an introduction
  / Vitoriano Ruas.
Description: Chichester, West Sussex, United Kingdom : John Wiley & Sons,
  Inc., [2016] | Includes bibliographical references and index.
Identifiers: LCCN 2016002914 (print) | LCCN 2016007091 (ebook) | ISBN
  9781119111351 (cloth) | ISBN 9781119111375 (pdf) | ISBN 9781119111368
  (epub)
Subjects: LCSH: Differential equations, Partial. | Boundary value problems.
Classification: LCC QA374 .R83 2016 (print) | LCC QA374 (ebook) | DDC
  515/.353–dc23
LC record available at http://lccn.loc.gov/2016002914

A catalogue record for this book is available from the British Library.

Set in 10/12pt, TimesLTStd by SPi Global, Chennai, India.
Printed and bound in Singapore by Markono Print Media Pte Ltd

1 2016

A Alex, Léo & Romy

En hommage à mes maîtres de l'école parisienne d'analyse numérique

Contents

Preface by Eugenio Oñate   xi

Preface by Larisa Beilina   xiii

Acknowledgements   xv

About the Companion Website   xvii

Introduction   xix

Key Reminders on Linear Algebra   xxvii

1   Getting Started in One Space Variable   **1**
1.1   A Model Two-point Boundary Value Problem   2
1.2   The Basic FDM   7
1.3   The Piecewise Linear FEM ($\mathcal{P}_1$ FEM)   12
1.4   The Basic FVM   17
       1.4.1   The Vertex-centred FVM   17
       1.4.2   The Cell-centred FVM   20
       1.4.3   Connections to the Other Methods   22
1.5   Handling Nonzero Boundary Conditions   24
1.6   Effective Resolution   25
       1.6.1   Solving SLAEs for one-dimensional problems   26
       1.6.2   Example 1.1: Numerical Experiments with the Cell-centred FVM   27
1.7   Exercises   28

2   Qualitative Reliability Analysis   **30**
2.1   Norms and Inner Products   31
       2.1.1   Normed Vector Spaces   32
       2.1.2   Inner Product Spaces   33
2.2   Stability of a Numerical Method   35
       2.2.1   Stability in the Maximum Norm   35
       2.2.2   Stability in the Mean-square Sense   39

2.3	Scheme Consistency	42
	2.3.1 *Consistency of the Three-point FD Scheme*	42
	2.3.2 *Consistency of the $\mathcal{P}_1$ FE Scheme*	44
2.4	Convergence of the Discretisation Methods	48
	2.4.1 *Convergence of the Three-point FDM*	49
	2.4.2 *Convergence of the $\mathcal{P}_1$ FEM*	50
	2.4.3 *Remarks on the Convergence of the FVM*	52
	2.4.4 *Example 2.1: Sensitivity Study of Three Equivalent Methods*	54
2.5	Exercises	59
3	**Time-dependent Boundary Value Problems**	**61**
3.1	Numerical Solution of the Heat Equation	64
	3.1.1 *Implicit Time Discretisation*	65
	3.1.2 *Explicit Time Discretisation*	66
	3.1.3 *Example 3.1: Numerical Behaviour of the Forward Euler Scheme*	68
3.2	Numerical Solution of the Transport Equation	70
	3.2.1 *Natural Schemes*	70
	3.2.2 *The Lax Scheme*	72
	3.2.3 *Upwind Schemes*	72
	3.2.4 *Extensions to the FVM and the FEM*	73
3.3	Stability of the Numerical Models	76
	3.3.1 *Schemes for the Heat Equation*	77
	3.3.2 *The Lax Scheme for the Transport Equation*	79
3.4	Consistency and Convergence Results	81
	3.4.1 *Euler Schemes for the Heat Equation*	81
	3.4.2 *Schemes for the Transport Equation*	84
3.5	Complements on the Equation of the Vibrating String (VSE)	85
	3.5.1 *The Lax Scheme to Solve the VS First-order System*	85
	3.5.2 *Example 3.2: Numerical Study of Schemes for the VS First-order System*	86
	3.5.3 *A Natural Explicit Scheme for the VSE*	87
3.6	Exercises	90
4	**Methods for Two-dimensional Problems**	**92**
4.1	The Poisson Equation	93
4.2	The Five-point FDM	95
	4.2.1 *Framework and Method Description*	95
	4.2.2 *A Few Words on Possible Extensions*	98
4.3	The $\mathcal{P}_1$ FEM	100
	4.3.1 *Green's Identities*	100
	4.3.2 *The Standard Galerkin Variational Formulation*	103
	4.3.3 *Method Description*	104
	4.3.4 *Implementation Aspects*	110
	4.3.5 *The Master Element Technique*	115
	4.3.6 *Application to Linear Elasticity*	117
4.4	Basic FVM	121
	4.4.1 *The Vertex-centred FVM: Equivalence with the $\mathcal{P}_1$ FEM*	122
	4.4.2 *The Cell-centred FVM: Focus on Flux Computations*	126

4.5	SLAE Resolution	138
	4.5.1 Example 4.1: A Crout Solver for Banded Matrices	140
	4.5.2 Example 4.2: Iterative Solution of Equivalent FD–FE–FV SLAEs	143
4.6	Exercises	147
5	**Analyses in Two Space Variables**	**149**
5.1	Methods for the Poisson Equation	150
	5.1.1 Convergence of the Five-point FDM	150
	5.1.2 Convergence of the $\mathcal{P}_1$ FEM	153
	5.1.3 Example 5.1: Solving the Poisson Equation with Neumann Boundary Conditions	164
	5.1.4 Example 5.2: Convergence of the $\mathcal{P}_1$ FEM to Non-smooth Solutions	165
	5.1.5 Convergence of the FVM	168
	5.1.6 Example 5.3: Triangle-centred FVM versus RT_0 Mixed FEM	187
5.2	Time Integration Schemes for the Heat Equation	192
	5.2.1 Pointwise Convergence of Five-point FD Schemes	193
	5.2.2 Convergence of $\mathcal{P}_1$ FE Schemes in the Mean-square Sense	196
	5.2.3 Pointwise Behaviour of FE and FV Schemes: An Overview	204
5.3	Exercises	205
6	**Extensions**	**210**
6.1	Lagrange FEM of Degree Greater than One	211
	6.1.1 The $\mathcal{P}_k$ FEM in One-dimension Space for $k > 1$	211
	6.1.2 A FEM for Quadrilateral Meshes	217
	6.1.3 Piecewise Quadratic FEs in Two Space Variables	223
	6.1.4 The Case of Curved Domains	225
	6.1.5 Example 6.1: $\mathcal{P}_2$-FE Solution of the Equation $u - \Delta u = f$	231
	6.1.6 More about Implementation in Two-dimensional Space	234
6.2	Extensions to the Three-dimensional Case	240
	6.2.1 Methods for Rectangular Domains	241
	6.2.2 Tetrahedron-based Methods	245
	6.2.3 Implementation Aspects	249
	6.2.4 Example 6.2: A MATLAB Code for Three-dimensional FE Computations	252
6.3	Exercises	258
7	**Miscellaneous Complements**	**261**
7.1	Numerical Solution of Biharmonic Equations in Rectangles	261
	7.1.1 Model Fourth-order Elliptic PDEs	262
	7.1.2 The 13-point FD Scheme	263
	7.1.3 Hermite FEM in Intervals and Rectangles	265
7.2	The Advection–Diffusion Equation	272
	7.2.1 A Model One-Dimensional Equation	272
	7.2.2 Overcoming the Main Difficulties with the FDM	274
	7.2.3 Example 7.1: Numerical Study of the Upwind FD Scheme	277
	7.2.4 The SUPG Formulation	278
	7.2.5 Example 7.2: Numerics of the SUPG Formulation for the $\mathcal{P}_1$ FEM	281
	7.2.6 An Upwind FV Scheme	282

      7.2.7   *A FE Scheme for the Time-Dependent Problem*   286

      7.2.8   *Example 7.3: Numerical Study of the Weighted Mass FE Scheme*   292

7.3   Basics of a Posteriori Error Estimates and Adaptivity   294

      7.3.1   *A Posteriori Error Estimates*   295

      7.3.2   *Mesh Adaptivity: h, p and h–p Methods*   298

7.4   A Word about Non-linear PDEs   300

      7.4.1   *Example 7.4: Solving Non-linear Two-point Boundary Value Problems*   301

      7.4.2   *Example 7.5: A Quasi-explicit Method for the Navier–Stokes Equations*   305

7.5   Exercises   309

Appendix   **311**

References   **320**

Index   **331**

Preface by Eugenio Oñate

Numerical methods have become nowadays indispensable tools for the quantitative solution of the differential equations that express the behaviour of any system in the universe. Examples are found in the study of engineering systems such as mechanical devices, structures and vehicles, and of biological and medical systems such as human organs, prosthesis and cells, to name a few. Numerical methods are applicable to the solution of differential equations that represent mathematical models of an underlying real system. These models are conceptual representation of reality using mathematics. It is interesting that in mechanics (which governs the behaviour of all engineering systems and many problems in physics) these models are obtained by simple balance (or equilibrium) laws applied to infinitesimal parts of the continuous system and its boundary. The beauty of the modelling process is that many times the resulting differential equations, even if they have the same mathematical form, are applicable to many different real life problems. A typical example is the Poisson equation that is used to represent the behaviour of such different problems as the heat conduction in a body, the seepage flow in a porous media, the flow of an incompressible fluid and the torsion of a bar, among others. Clearly a mathematical model represents a simplification of reality using geometrical data and physical properties of the constituent materials that are close enough to those of the real system under study and at the same time help to express its behaviour in mathematical form. On the other hand, the numerical solution of the resulting governing differential equations of the mathematical model introduces a series of simplifications. Numerical methods work on discretized forms of the model geometry which is split into simple geometrical entities, such as triangles and tetrahedra in two (2D) or three (3D) dimensions, respectively, as is typical in finite element or finite volume procedures or Cartesian grids, as is done in finite difference methods. The balance equations for each discretization unit (an element, a volume or a cell) are obtained in terms of a finite set of parameters and then used to obtain (assemble) the behaviour of the whole system by expressing the satisfaction of the balance laws at all the discrete points of the discrete system. The so-called discretization process leads to a system of algebraic equations that typically involve several millions of unknown parameters for real life problems. These equations are solved in modern computers using state of the art of computer science technology. This explains why numerical methods are many times referred to as computational methods. The two terms are in fact equivalent. The simplifications made along the modelling and numerical solution path lead to a final quantitative solution that is only approximate. The quantification of the modelling and numerical solution errors are topics of much current research. The timely book of Professor Ruas fits precisely in the context described above as it provides a comprehensive insight on the derivation and best use of numerical methods for solving a number of differential equations that govern a wide class of practical problems. The emphasis is in providing the reader with the knowledge and tools to assess the performance and reliability of a particular numerical method via the study of its stability, consistency, convergence and approximation order. The book uses simple one-dimensional (1D) problems for introducing the three basic numerical methods addressed in the book: finite differences, finite elements and finite volumes. A quantitative reliability analysis of the three families of methods is presented as well as the extension of the methods to transient situations and 2D problems. A convergence analysis for a well know multidimensional scalar problem (the Poisson equation) is also included. The book devotes a chapter to presenting higher order Lagrange finite elements and the extension of the finite element method to 3D problems. A final chapter describes a number of interesting topics such as the numerical solution of bi-harmonic equations in rectangles, the solution of the advection- diffusion equation, an outline of a posteriori error estimation and adaptive numerical solution procedures and a brief introduction to the numerical solution of nonlinear partial differential equations. Other numerical techniques are briefly presented in the Appendix.

I consider the book of Professor Ruas a valuable contribution to the extensive literature on numerical methods for partial differential equations. The book addresses modern and practical issues that are of paramount importance for the use of numerical methods with enough confidence. It is written in a clear form with many examples that help to understand the different concepts. I believe that it will be very useful to students in mathematics, physics, engineering and general sciences, as well as to the practitioners of numerical methods.

Barcelona, October 12, 2015
Eugenio Oñate
Professor of Continuum and Structural Mechanics
Technical University of Catalunya, Barcelona, Spain
Director of International Center for Numerical Methods in Engineering (CIMNE)

Preface by Larisa Beilina

This book presents the basics of the mathematical theory for three main numerical methods applied to the solution of Partial Differential Equations: the Finite Element Method (FEM), the Finite Difference Method (FDM) and the Finite Volume Method (FVM).

Author's proposal contains a lot of new material, such us several optimal results, which have not been considered before. An outstanding advantage of Professor Ruas' book is the fact that the basic FEM, FDM and FVM, are studied in a simple and popular fashion, although their reliability is addressed with all rigor by exploiting the concepts of method's stability, consistency and convergence. Main attention is given to low order methods as the most employed ones in real-life applications.

The first two chapters in the book describe the three discretization methods and present in a simplified framework the mathematical analysis applied to the one-dimensional boundary value problem in space. The next chapter considers discretization methods for time-dependent boundary value problems, as well as corresponding stability, consistency and convergence results. Then further chapters of the book gradually take the reader to the discretization and analysis of spatial and time-dependent PDEs in higher dimensions. More precisely, in a very easy form the FEM with piecewise-linear functions as well as the Vertex and Cell-centred FVM are explained, and convergence and reliability analysis of these methods are presented. Several important extensions of the FEM with Lagrange elements of degree greater than or equal to one applied to the solution of elliptic PDE in two and three dimensions are consistently shown. Finally, some basic numerical methods for the solution of non-linear PDEs are also described.

Several numerical examples appended to every chapter, run with codes written in MATLAB or in FORTRAN 95, illustrate approximation properties of numerical methods. At the end of every chapter a number of useful exercises are proposed in order to help the reader to consolidate understanding of the material presented therein, for a number of applied problems.

This book is undoubtedly of great interest for undergraduate and graduate students in computational and applied mathematics, as well as for researchers and engineers working in the field of numerical methods for PDEs.

Gothenburg, October 30, 2015
L. Beilina, PhD
Department of Mathematical Sciences
Chalmers University of Technology and Gothenburg University
Gothenburg, Sweden

Acknowledgements

The author is most thankful to his colleague and friend Marco Antonio S. Ramos for a scrupulous and diligent help with coding and numerical examples.

The author is most grateful to his colleague and friend Marco Antonio S. Ramos for a scrupulous and diligent help with coding and numerical examples.

He is also pleased to acknowledge that the skills in computer science and engineering he acquired at PUC-Rio were fundamental for the accomplishment of this work.

Many thanks are due to author's colleagues of Institut Jean Le Rond d'Alembert at Université Pierre et Marie Curie-Sorbonne Universités, and in particular to its Director, Stéphane Zaleski, for their support and encouragements during the preparation of the manuscript.

About the Companion Website

This book is accompanied by a companion website:

www.wiley.com/go/ruas/numericalmethodsforpartial

The website includes:

- Host codes
- Solutions for the exercises

Introduction

> In the realm of spirit, seek
> clarity; in the material world,
> seek utility.
> Gottfried Wilhelm Leibniz

Since Leonhard Euler, **numerical methods** gradually became the currently widespread techniques to solve real-life problems governed by **differential equations**. It was, however, the invention of modern computers in the middle of the 20th century that significantly pushed this branch of applied science to play such a prominent role in contemporary technological development. Supercomputers among other high-performance machines became available since the 1980s, and this has favoured an even more spectacular evolution of numerical methods for differential equations, as tools capable of producing exploitable responses out of mathematical models of the kind.

In case a system of differential equations is expressed in terms of more than one independent variable, a system member is referred to as a **partial differential equation** (i.e. a **PDE**) and otherwise as an **ordinary differential equation**, (i.e. an **ODE**). About 100 years ago, numerical methods became practitioners' preferred alternative to solve PDEs, whose analytical solution is out of reach. This book is intended for presentation, to specialists acting in various technological and scientific fields, of basic elements on numerical methods to solve PDEs. Although the equations can be posed in any kind of spatial domain, here we confine ourselves to the case of **boundary value problems**, supplemented with **initial conditions** in case they are also time-dependent. This implies assuming that an equation's definition domain is bounded. Moreover, except for a few cases, the presentation is restricted to equations in terms of real independent variables, whose solution range is a subset of the real line.

It is well-known that PDEs model the behaviour of relevant unknown quantities in a large amount of situations of practical interest. These cover domains as diverse as engineering, physics, geo- and biomedical sciences, chemistry and economics, among many others. For example, in aircraft design the knowledge of the way air flows about fuselages is of paramount importance, as much as mastering the propagation of acoustic waves in vehicle interiors is a must in modern automobile design. Also in recent decades, more and more such models are being employed in the search for better understanding of human body systems. This is surely helping to prevent highly lethal diseases such as cardiovascular ones.

Of course, whenever possible, analytical methods should be employed to solve certain types of PDEs. Among these, the method of separation of variables is an outstanding example. However, for different reasons, including model complexity, irregular geometries or inaccurate field data, it is no point trying to determine exact analytical solutions to PDEs in most cases. Instead, numerical methods, naturally designed for use in a computational environment, provide a valid alternative to mathematical expressions representing solutions to the theoretical model. These can be as diverse as fluid velocity, blood pressure, electromagnetic fields, structural stresses, species fractions in biological evolution or chemical reactions, among many others having their behavior modelled by a PDE. That is why running a computer code, in which a numerical solution procedure is implemented, is called a **numerical simulation**. Indeed, the thus generated numerical values replace, in a way, a physical response to input data characterising a specific application.

Aim

The purpose of this book is to study numerical methods for solving PDEs, designed to possibly generate accurate substitutes of unknown fields, in terms of which the equations are expressed. Generally speaking, instead of values provided by a solution's analytical expression at every point of the physical domain in which a given phenomenon or process is being modelled by a PDE, in the numerical approach only solution's approximate values or related quantities at a finite number of points are determined. Owing to this feature, the underlying numerical method is also known as a **discretisation method**. Otherwise stated, the term **discrete** qualifies numerical solution techniques, as much as the terms **analytical** and **continuous** do for procedures aimed at finding exact mathematical expressions for a solution, when they exist.

To a large extent PDEs, being used in mathematical modelling, are of the second order, which means that the highest partial derivative order of the unknown fields appearing in the equations is two. For this reason, a particular emphasis is given to this class of PDEs throughout the text. However, for the sake of conciseness and clarity, we will confine the whole presentation of the numerical methods to the case of **linear differential equations**. Nevertheless, the types of linear PDEs to be studied are the most frequently encountered in practical applications, namely, **elliptic, parabolic and hyperbolic equations**. We assume throughout the text that the reader is familiar with basic concepts of linear PDEs of these representative types. Nevertheless, it would not be superfluous to recall the criteria that characterise them, by restricting the definitions to the case where the solution w of a linear second-order PDE is a function of two independent variables r and s, and moreover to the case of constant coefficients, that is, an equation of the form

$$A\frac{\partial w^2}{\partial r^2} + 2B\frac{\partial w^2}{\partial r \partial s} + C\frac{\partial w^2}{\partial s^2} + D\frac{\partial w}{\partial r} + E\frac{\partial w}{\partial s} + Fw = G,$$

where G is a given function and $A^2 + B^2 + C^2 > 0$. Letting $\Delta = B^2 - AC$, we have:

- If $\Delta > 0$, the equation is hyperbolic;
- If $\Delta = 0$, the equation is parabolic;
- If $\Delta < 0$, the equation is elliptic.

One of the main differences between the three types of equations relies on the boundary and/or initial conditions that must be prescribed, in order to ensure existence and uniqueness of a solution. This issue will be clarified in Chapter 3, as far as the first two types of PDEs are concerned, and in Chapter 4, whose purpose is the study of the Poisson equation, the most typical elliptic PDE.

As many authors believe, in starting from linear PDEs, it is easier to take on otherwise challenging and complicated problems in more advanced studies. Furthermore, this linear approach has an undeniable virtue: if a numerical method is unreliable to find solutions to a simplified linear (i.e. a linearised form), of a true nonlinear model, let alone its application to the latter.

Scope

First of all, we should emphasise that in contemporary numerical simulations of complex physical events, powerful computational tools such as graphics processing units (GPUs) are at practitioners' disposal. Moreover, high-performance techniques to optimise simulation codes, in order to save RAM and storage in general and make them run faster, have been in current use for the past few decades. Here one might think of **vectorisation**, a technique aimed at speeding up matrix and vector arithmetics, featuring more recent scientific computing-oriented programming languages such as FORTRAN 95 and MATLAB. **Parallel computing** based on distributed systems consisting of a computer network or several processors running concurrently in parallel, in order to accomplish different tasks of large-scale numerical processing, has been a facility in use in research centers and industries around the world for a few decades now. Although we are convinced that the reader should be aware of these possibilities, we do not address them at all because our book is an introductory one. In other words, its scope is limited to the study of numerical methods in the framework of rather simple model problems, whose solutions do not require sophisticated tools employed in intensive computer simulations.

The subject this book deals with is continuously evolving. New proposals for the numerical solution of PDEs of particular types are being published in specialised journals practically every week. However, a glance at the present state of the art suffices to show out that a milestone was reached about 50 years ago. At that time, the concepts lying behind three big families of discretisation methods to solve PDEs became well accepted in the worldwide scientific and industrial communities. More precisely, we mean the **finite difference method** (FDM), the **finite element method** (FEM) and the **finite volume method** (FVM), which we chose to study in detail in this book, as techniques playing central roles among several ingredients, in a recipe for the computational determination of **numerical solutions** to PDEs.

The FDM is the oldest and the simplest numerical method to solve differential equations. It has been known since Euler's work in the 18th century, and countless specialists in numerical mathematics contributed to its development up to now. Its routine use among specialists in the numerical solution of PDEs dates back to the beginning of the 20th century. Pioneering work of **Courant**, **Friedrichs** and **Levy** in Europe and in the United States (cf. [55]) set the bases to justify the method's effectiveness and reliability. Almost in parallel, **Gerschgorin** [85] derived decisive results in the framework of numerical analysis as applied to PDEs. Much later, other prominent members of the Russian school such as **Godunov** [91] and **Marchuk** [133] gave relevant contributions in this direction. The latter also collaborated hand in hand with **Lions** (cf. [127]), the respected founder of the prolific Paris school of analysis and numerical mathematics for PDEs in the late 1960s. However as Lions himself and his disciples realised very soon, the development of the FDM was considerably slowing down in the last third of the 20th century, because of its limitation to handle efficiently complex geometries and/or boundary conditions. Nevertheless, the author should stress that, owing to its simplicity and easy implementation, as a rule the FDM should be preferred to other methods, as long as severe applicability restrictions do not come into play. This certainly explains why the method is still very popular today.

In contrast, from the mid-1950s on, the FEM gradually emerged as a flexible tool capable of giving appropriate responses to most geometric and mathematical challenges encountered in practical applications of PDEs, more particularly in engineering. At that time, work on this new numerical simulation technique was carried out simultaneously in several places, in connection with big-aircraft industry among other major building or manufacturing enterprises. The method was tailored to its present form by people like **Clough**, **Turner** [196] and co-workers in the United States; **Argyris** [8] in Germany and in England; **Zienkiewicz** [210] in Wales; **Fraeijs de Veubeke** in Belgium [75] and **Arantes e Oliveira** in Portugal [7], among many other names.

A little later the FVM appeared as a new tool, aimed at overcoming the aforementioned limitation of the FDM. Scientists like **Samarskii** and **Tikhonov** [193] from the the Russian school can be credited as outstanding developers of the FVM, as much as **Spalding** and **Patankar** in the United Kingdom (cf. [151] and [152]). A remarkable growth of the method's user community followed, and to date several commercial simulation codes widely in use in industry are FVM-based, more particularly in fluid dynamics. Such a great success seems to be due to the fact that the principles the method is based upon are close to those of physics. Indeed, by using this method, local mass or heat fluxes, discharges among many other fundamental physical quantities, can be easily controlled in a direct and transparent manner. This is certainly very appealing to industry designers who ordinarily are not available to undertake too many technical interventions in order to run a simulation code, or to post-process numerical results.

On the other hand, the FEM is by far the richest among these three numerical methods, in terms of both versions and applicability. The number of efficient FEM packages available, as applied to a wide spectrum of technological domains, is very large, and this is particularly true of structural analysis. Actually, the range of the FEM and its multiple variants to solve countless classes of problems governed by PDEs is steadily growing, and is practically unlimited. This has certainly something to do with the method's great versatility. However, it is true that in some domains such as fluid dynamics, good use and understanding of the FEM require some more mathematical insight, which is perhaps the reason why the FVM is often preferred in such applications.

These preliminary considerations on the strengths and weaknesses of the three methods will be enriched by additional comments to be made as they are presented and studied along the chapters.

Approach

It is important to recall that values provided by the numerical solution procedure are only **approximations** of those of the theoretical solution, at least in an overwhelming majority of cases. But then a fundamental question arises: How reliable are such approximations as compared to the corresponding values the model would provide if it could be solved analytically? An attempt is made to give precise answers to this question without resorting to functional analysis, in contrast to what many authors do.

Actually, in a book aimed at being used in the framework of an introductory and intensive course it would be advisable to avoid as much as possible going into details on mathematical aspects of the numerical methods being presented. Addressing rather their functional features, such as limits and merits in practical terms, or implementation, might be considered more appropriate. However, many authors are persuaded that, for a better comprehension of numerical methods, mastering underlying basic mathematical concepts is of paramount importance. As a former assistant of the late Professor **Daniel Euvrard** at the Division of Mechanics of the University of Paris 6, the author believes to have learned well with him the right approach to motivate into the subject students in fields other than mathematics, and to convince them why a numerical method is reliable or not. Nevertheless, in his book, he declined to go into ultimate considerations on such aspects of numerical methods. This is perhaps because he felt this would require mathematical concepts supposedly not mastered by his students. Within a certain time after Daniel Euvrard passed away, the author endeavoured to find means to render complete and rigorous reliability studies of numerical methods for PDEs accessible to final-term undergraduate or graduate students in any scientific domain. In a sense, this book is the outcome of what the author believes to be the ideal pedagogic approach, after a rather long maturation period.

More precisely, we shall introduce and thoroughly exploit the concepts of **stability, consistency, convergence** and **order** of a numerical method to solve PDEs, as fundamental tools to understand why they work or not, or how well they work. This is because both the stability and the consistency of a numerical method imply that it is reliable. More precisely, convergence to an equation's exact solution is guaranteed at a certain rate related to the method's order. However, it must be emphasised again that this book utilises neither high-level nor abstract mathematics to qualify numerical methods. Indeed, only the mathematical knowledge a student in scientific or technical fields is supposed to master, when she or he reaches the stage of studying numerical methods for PDEs is taken for granted. Nevertheless in some cases, optimality in the mathematical sense is sacrificed, so as to make possible such an approach.

In short, in many respects, this book can be viewed as original. The following issues should definitively be underlined:

- The pre-requisites for full understanding of this book's material are differential & integral calculus in multiple variables, besides elementary analysis, tensor calculus, numerical analysis and linear algebra. Of course, basics of differential equations including linear PDEs are supposed to be known.
- The reliability analyses for the three methods are carried out in a unified framework, by exploiting in a structured and visible manner the equivalence between convergence and the pair of properties stability and consistency, that is to say, the **Lax–Richtmyer equivalence theorem** [123].
- In addition to this unified treatment, here and there new techniques are employed to derive known results, thereby simplifying their proof.
- Emphasis is given to low-order methods, as practitioners' overwhelming default options for everyday use.
- In the chapters, any time a particular PDE's feature is important for better comprehension of a numerical solution method, the corresponding property is duly presented, recalled or cited from the literature.

- The book is placed halfway between texts addressed to students in mathematics and those for students in other sciences. This means that a balanced emphasis is given to both practical considerations and a rigorous mathematical treatment.
- To the best of the author's knowledge several optimal results rigorously established in the text cannot be found elsewhere, not even in more advanced books. In this sense, although this one is basically education oriented, marginally it can also be a valid reference for research work on numerical methods for PDEs.

More commonplace features are lists of exercises proposed at the end of each chapter. Most of them are just complements to theoretical studies, in order to help the student to consolidate her or his understanding of a particular issue. In some cases, exercises are simply a way to abridge the text itself. In addition to this, several numerical examples are supplied using codes programmed either in **FORTRAN 95** language or in a **MATLAB** environment. Almost all the given examples are academic in the sense that their exact solution is known beforehand. This is because they are essentially aimed at either illustrating or assessing methods' approximation capabilities. By adding this kind of stuff to all chapters, our approach turns out to be similar to the one adopted in some text books on the subject such as **Hughes'** [100]. However, in contrast to this one, the latter are either devoted to a single type of method or addressed to a specific scientific community. In short, coding was just aimed at supporting the material addressed in the book and in no case a goal itself. This is the reason why we did not care about using techniques such as vectorisation, parallelism or computational tools more powerful than an ordinary laptop for running our examples. The reader will certainly realise it by examining some of the programs incorporated into the text.

The general subject organisation throughout the book is as follows. For the sake of clarity and objectiveness, the basic concepts inherent to numerical methods are first presented in the framework of differential equations whose spatial definition domain is one-dimensional, more specifically a bounded interval of the real line. Nevertheless, this presentation is enriched by the treatment of the **time-dependent case**. This means equations having the time as an independent variable, besides the space variable, which model the so-called **transient problems**. Once these concepts are introduced in such a simplified framework, their application to problems defined in domains of higher dimensions can be assimilated more smoothly as we believe. However, we only apply to two-dimensional problems (surely closer to real-life ones!) the conclusions in terms of reliability that the previous studies allow for. A point of view shared by many authors is that this is good pedagogy, since properties that hold for a multidimensional problem can often be regarded as mere extrapolations of valid ones for lower dimensional counterparts.

From the beginning, we should point out that in a significant majority of cases, a numerical method to solve boundary value problems leads to the solution of **systems of linear algebraic equations**. Since such systems play a central role in the framework of numerical methods for PDEs, throughout the text the acronym **SLAE** will be used instead of the expression in full. The SLAE unknowns are usually approximations of the solution function at a set of points of a discretisation lattice. As pointed out above, for this reason numerical methods are often referred to as discretisation methods. Notice, however, that SLAE unknowns can also be other expressions, depending on the solution function, such as derivatives, integrals or combinations of all these.

Owing to the topic's relevance, we supply right after this Introduction some reminders of linear algebra as related to SLAEs, which will be particularly useful throughout the text.

Whenever necessary, these will be complemented with further details on a method to solve SLAEs particularly pertinent to a subject addressed in a given chapter.

Book's outline

Following the introductory section on linear algebra, we describe in Chapter 1 the three discretisation methods named above, as applied to a one-dimensional model problem. Chapter 2 is devoted to mathematical analyses of these methods in such a simplified framework; more precisely, the basic concepts that characterise the reliability of a numerical method for solving a PDE are introduced. In Chapter 3, we apply the same concepts in order to examine schemes for time-dependent counterparts of the model problem. In Chapter 4, we describe the use of the three approaches to solve problems posed in two spatial dimensions without time dependence. The aim of Chapter 5 is the rigorous mathematical study of numerical methods in two space variables. Besides the problems addressed in Chapter 4, their time-dependent counterparts are also considered. This chapter is by far the heaviest part of the book in terms of calculations, which could be left aside in a short or intensive course. In this case, the student could just concentrate on the final results. Chapter 6 brings about several important extensions of the material studied in all the previous chapters, in terms of practical applications, including three-dimensional modelling. Chapter 7 is a complementary one; specific numerical methods are considered for a few scattered but relevant classes of linear PDEs that are not studied in the previous chapters. In the final section of Chapter 7, some basics of the numerical solution of nonlinear PDEs close the book. This outline is completed by those supplied at the beginning of each chapter, specifying its organisation and providing a brief summary of its different sections.

To conclude, we must underline that in contemporary numerical simulations, users can count on numerous methodological options. At an introductory level, however, it would be out of purpose to attempt to address them all. That is what motivated our choice of three methods out of several other techniques for solving PDEs. Nevertheless, for the reader's information and better guidance, we added an Appendix, where main lines are highlighted of several numerical methods of current use not covered by the book. More specifically, brief descriptions are given of the boundary element method, discontinuous Galerkin methods, meshless methods, spectral methods, hybrid finite elements, iso-geometric analysis, the virtual element method, domain decomposition techniques and multigrid methods.

WARNING

Throughout the text, the range of any subscript or superscript is assumed to be sufficiently large, for the surrounding statements or formulae involving it to make sense.

Incidentally it should be stressed that, to a great extent, numerical analysis is the art of handling concepts closely related to integer symbols standing for array subscripts, grid point positions, time stages or iteration numbers in recursive procedures, among many other abstract things of unspecified dimensions or sizes. For this reason, the reader should be aware that following such notations requires much attention. Moreover, symbol-correct typing forces authors on the subject to make an iterative improvement of their texts, a process that usually converges very slowly, and eventually never converges …

Key Reminders on Linear Algebra

> Everything that is written
> merely to please the author is
> worthless.
>
> Blaise Pascal

The aim of these preliminaries is to recall some basics of linear algebra that are particularly useful in the framework of the study of numerical methods for PDEs. For the sake of objectiveness, we focus on aspects pertinent to SLAEs, assuming that the reader is well acquainted with elementary matrix analysis. We also take it for granted that the concept of linear vector space has been mastered by the reader, together with underlying operations, definitions and entities such as sum, product by a scalar, null vector, dimension, linear combination and basis. If a refreshment is found necessary, the reader is advised to resort to classical books that treat linear vector spaces as a tool, and thus in abridged form. In this case, the author recommends the reader to consult, for instance, the two introductory chapters of reference [132]. For more details on matrix analysis or on the methods for solving SLAEs to be addressed hereafter, the author refers to countless textbooks on the subject, such as references [83], [46] and [183].

Generalities

First of all, we recall that an eigenvalue of a real $n \times n$ matrix A with entries $a_{i,j}, 1 \leq i, j \leq n$, is a real or complex number λ such that $A\vec{e} = \lambda\vec{e}$ for some (column) vector $\vec{e} \neq \vec{0}$ with n real or complex components $e_1, e_2, \ldots, e_n$, where $\vec{0}$ is the null vector, (i.e. the vector having only zero components). $\vec{e}$ is called an eigenvector of A. The eigenvalues of an $n \times n$ matrix A are the roots, distinct or not, of the characteristic polynomial $p(A) := det(A - \lambda I)$ of degree n, where I is the identity $n \times n$ matrix and $det(C)$ denotes the determinant of a square matrix C. From this definition follows that $det(A) = 0$ (i.e. A is singular), if A has an eigenvalue equal to zero. Conversely, if $det(A) = 0$, there exists a nonzero $\vec{e} \in \Re^n$ such that $A\vec{e} = \vec{0}$, which is an eigenvector of A associated with a zero eigenvalue.

A real $n \times n$ matrix A is symmetric if $a_{j,i} = a_{i,j}$ for every pair $(i; j)$. Let $\vec{x}^T$ be the n-component row vector having the same ordered components $x_1, x_2, \ldots, x_n$ as the real column vector $\vec{x}$. A symmetric matrix A is said to be **positive-definite** if $\vec{x}^T A\vec{x} \geq 0$ for every n-component real (column) vector $\vec{x}$, and the zero value is attained only if $\vec{x} = \vec{0}$. A symmetric positive-definite matrix has only real strictly positive eigenvalues.

Let B be a rectangular real array with mn real entries $b_{i,j}$, and $\vec{c}$ be a (column) vector of $\Re^n$, that is, an $n \times 1$ array $\vec{c} = [c_1, c_2, \ldots, c_n]^T$ with real components. B^T stands for the transpose of B, that is, the $n \times m$ array $D = \{d_{i,j}\}$ such that $d_{i,j} = b_{j,i}$, for $i = 1, \ldots, n$ and $j = 1, 2, \ldots, m$.

For any vector $\vec{c} \in \Re^n$ its norm $\|\vec{c}\|$ is a real number satisfying

1. $\|\vec{c}\| \geq 0$ and if $\|\vec{c}\| = 0$ then $\vec{c} = \vec{0}$;
2. $\|\alpha\vec{c}\| = |\alpha| \|\vec{c}\| \ \forall \alpha \in \Re$;
3. $\|\vec{c} + \vec{d}\| \leq \|\vec{c}\| + \|\vec{d}\| \ \forall \vec{c}, \vec{d} \in \Re^n$.

The three most used norms of n-component vectors $\vec{c} = [c_1, c_2, \ldots, c_n]^T$ are defined as follows:

$$\|\vec{c}\|_\infty := \max_{1 \leq i \leq n} |c_i| \ \text{(maximum norm)}; \quad \|\vec{c}\|_1 := \sum_{i=1}^{n} |c_i|; \quad \text{and} \quad \|\vec{c}\|_2 := \sqrt{\sum_{i=1}^{n} |c_i|^2}$$

(euclidean norm).

The norm $\|B\|$ of any $m \times n$ matrix B subordinate to a given vector norm $\|(\cdot)\|$ is the smallest real number β satisfying $\|B\vec{c}\| \leq \beta \|\vec{c}\|$ for every $\vec{c} \in \Re^n$. For any subordinate matrix norm, properties analogous to 1, 2 and 3 hold. In particular, if $\|B\| = 0$, then B is the null matrix (i.e. the matrix having only zero entries), for this means that $B\vec{c} = \vec{0}$ for every $\vec{c} \in \Re^n$. If $\|\cdot\|$ is a matrix norm subordinate to a given vector norm, we necessarily have $\|I\| = 1$. Moreover, it satisfies $\|BD\| \leq \|B\| \|D\|$, for all $m \times n$ matrix B and $n \times p$ matrix D.

The matrix norm subordinate to the $\|\cdot\|_\infty$ vector norm is given by $\|B\|_\infty = \max_{1 \leq i \leq m} \sum_{j=1}^{n} |b_{i,j}|$.

The matrix norm subordinate to the vector norm $\|\cdot\|_1$ is given by $\|B\|_1 = \max_{1 \leq j \leq n} \sum_{i=1}^{m} |b_{i,j}|$.

The matrix norm subordinate to the euclidean vector norm $\|\cdot\|_2$ is given by $\|B\|_2 = \sqrt{\mu}$, where μ is the largest eigenvalue of the $m \times m$ matrix $B^T B$, a necessarily nonnegative number. This matrix norm is called the **spectral norm**.

Let A be an $n \times n$ matrix and $\vec{b}$ be a given vector of $\Re^n$. The SLAE $A\vec{x} = \vec{b}$ has a unique solution $\vec{x} \in \Re^n$ if and only if $det(A) \neq 0$. This means that it is possible to determine the inverse A^{-1} of matrix A, which satisfies $A^{-1}A = I$. In this case, $\vec{x} = A^{-1}\vec{b}$.

Assume that for some reason such as measurement errors, vector $\vec{b}$ is replaced by $\vec{b}^* \in \Re^n$. Setting $\Delta\vec{b} = \vec{b}^* - \vec{b}$, let $\vec{x}^*$ be the solution of the SLAE $A\vec{x}^* = \vec{b}^*$. By linearity, it is clear that $\Delta\vec{x} : \vec{x}^* - \vec{x}$ satisfies $A[\Delta\vec{x}] = \Delta\vec{b}$. Then, since $\Delta\vec{x} = A^{-1}\Delta\vec{b}$, we have $\|\Delta\vec{x}\| \leq \|A^{-1}\| \|\Delta\vec{b}\|$. On the other hand, $\|A\| \|\vec{x}\| \geq \|\vec{b}\|$. Therefore, dividing each side of the first inequality by the respective side of the second one, we obtain $\dfrac{\|\Delta\vec{x}\|}{\|\vec{x}\|} \leq \|A^{-1}\| \|A\| \dfrac{\|\Delta\vec{b}\|}{\|\vec{b}\|}$. This inequality tells us that the number $\|A^{-1}\| \|A\|$ relates, in some sense, the relative error of the solution to the relative error of the right side measured in a certain norm. It is called the **condition number** of matrix A, denoted by $cond(A)$. Notice that $cond(A) \geq 1$ since $1 = \| I \| = \| A^{-1}A \| \leq cond(A)$. However, it may happen that $cond(A) >> 1$, and in this case we say that matrix A is **ill-conditioned**. This happens, for instance, if the absolute value of the determinant of A is very small, for in this case A^{-1} will have large entries in absolute value. Anyhow, with this definition, we definitively have

$$\frac{\|\Delta\vec{x}\|}{\|\vec{x}\|} \leq cond(A) \frac{\|\Delta\vec{b}\|}{\|\vec{b}\|}.$$

It is clear that the larger $cond(A)$, the greater will be the amplification of the error of the exact solution resulting from an inaccurate right side.

Assume now that the right side is exact, but instead of $A\vec{x} = \vec{b}$, the SLAE being actually solved is $(A + \Delta A)(\vec{x} + \Delta\vec{x}) = \vec{b}$, where ΔA is the matrix whose entries are the errors of the corresponding entries of A. Clearly enough, $(A + \Delta A)(\vec{x} + \Delta\vec{x}) = A\vec{x}$ and hence $\Delta\vec{x} = -A^{-1}[\Delta A(\vec{x} + \Delta\vec{x})]$. Then, discarding the (highly unlikely) case where $\Delta\vec{x} = -\vec{x}$, we have

$$\frac{\|\Delta\vec{x}\|}{\|\vec{x} + \Delta\vec{x}\|} \le cond(A)\frac{\|\Delta A\|}{\|A\|}.$$

This indicates that the relative error of the true system matrix amplifies the solution relative error with respect to the effectively computed solution $\vec{x}^* := \vec{x} + \Delta\vec{x}$ by a factor of $cond(A)$. Similar expressions can be derived for the case where the SLAE $A\vec{x} = \vec{b}$ is solved with perturbed A and $\vec{b}$ at a time. In any event the condition number is seen to play a crucial role in SLAE solving error propagation.

More particularly, if A is an $n \times n$ invertible matrix, the eigenvalues of A^{-1} are λ_i^{-1}, $i = 1, \ldots, n$, and if A is symmetric, λ_i is necessarily a nonzero real number $\forall i$ (cf. [46]). Hence if $\mu_{\max}$ and $\mu_{\min}$ are defined as $\max_{1 \le i \le n} |\lambda_i|$ and $\min_{1 \le i \le n} |\lambda_i|$, respectively, we have $\|A\|_2 = \mu_{\max}$ and $\|A^{-1}\|_2 = [\mu_{\min}]^{-1}$. It follows that $cond(A)$ in terms of the spectral norm equals the quotient between the largest absolute value and the smallest absolute value of the eigenvalues of A. We recall that the **spectral radius** of a square matrix A denoted by $\rho(A)$ is the largest modulus of its eigenvalues. Hence, in case A is a symmetric matrix, $\|A\|_2 = \rho(A)$.

Solving SLAEs

We next highlight the basics of methods for solving a SLAE of the form $A\vec{x} = \vec{b}$, where A is an invertible $n \times n$ matrix. They can be grouped into two big families of methods, namely, iterative and direct methods.

Iterative methods

Let us first rewrite the SLAE to solve in the equivalent form $\vec{x} = B\vec{x} + \vec{c}$, where B is an $n \times n$ matrix and $\vec{c} \in \Re^n$. Given an initial guess $\vec{x}^0$, we determine for $k = 1, 2, \ldots$ successive approximations $\vec{x}^k$ of $\vec{x}$ by $\vec{x}^k = B\vec{x}^{k-1} + \vec{c}$. Particular choices of B and $\vec{c}$ give rise to classical iterative methods, such as those listed in this chapter. The convergence of an iterative method is characterised by the fact that $\lim_{k \to \infty} \|\vec{x}^k - \vec{x}\| = 0$ occurs, for any choice of $\vec{x}^0$, if and only if $\rho(B) < 1$. However, a subordinate norm of B being strictly less than one is a sufficient condition for convergence of an iterative method. Indeed, setting $\vec{e}^k := \vec{x} - \vec{x}^k$, we easily see that $\vec{e}^k = B\vec{e}^{k-1} \ \forall k > 0$. Hence, $\|\vec{e}^k\| \le \|B\| \|\vec{e}^{k-1}\|$. Thus, by mathematical induction $\|\vec{e}^k\| \le \|B\|^k \|\vec{e}^0\|$, and hence $\lim_{k \to \infty} \vec{x}^k = \vec{x}$ if $\|B\| < 1$. Incidentally, we observe that $\|A\| \ge \rho(A)$ for all square matrices A, $\|\cdot\|$ being any subordinate matrix norm.

In practice, the iterations go on until the norm of the difference between two successive approximations $\vec{x}^{k-1}$ and $\vec{x}^k$ becomes smaller than a given tolerance.

For all the following iterative methods, we assume that $a_{i,i} \ne 0$ for every i. Notice that under this condition, the diagonal matrix $D = \{d_{i,j}\} := diag(A)$ is invertible.

1. The Jacobi method: Clearly, $\vec{x} = D^{-1}\vec{b} + (I - D^{-1}A)\vec{x}$, and this observation gives rise to the Jacobi method. More precisely, it corresponds to the choice $\vec{c} = D^{-1}\vec{b}$ and $B = B_J := -D^{-1}(E + F)$, where E (resp. F) is the lower (resp. upper) triangular matrix with zero diagonal, whose lower (resp. upper) triangle coincides with the lower (resp. upper) triangle of A. In practical terms, setting $\vec{x}^k = [x_1^k, x_2^k, \ldots, x_n^k]^T$ for every k, we have

$$x_i^k = a_{i,i}^{-1}\left[b_i - \sum_{j=1, j\neq i}^{n} a_{i,j}x_j^{k-1}\right], \quad \text{for } i = 1, 2, \ldots, n.$$

2. The Gauss–Seidel method: The principle of this method is to take advantage of the fact that new values of the $i - 1$ first unknowns at the k-th iteration are already available when computing x_i^k. Formally we proceed as follows. Since D is invertible so is $D + E$, and hence we may rewrite the SLAE in the form $\vec{x} = (D + E)^{-1}\vec{b} - (D + E)^{-1}F\vec{x}$. Then the Gauss-Seidel method corresponds to the choice $\vec{c} = (D + E)^{-1}\vec{b}$ and $B = B_1 := -(D + E)^{-1}F$. In practical terms we determine successively,

$$x_1^k = a_{1,1}^{-1}\left[b_1 - \sum_{j=2}^{n} a_{1,j}x_j^{k-1}\right];$$

$$x_i^k = a_{i,i}^{-1}\left[b_i - \sum_{j=1}^{i-1} a_{i,j}x_j^k - \sum_{j=i+1}^{n} a_{i,j}x_j^{k-1}\right], \quad \text{for } i = 2, \ldots, n-1;$$

$$x_n^k = a_{n,n}^{-1}\left[b_n - \sum_{j=1}^{n-1} a_{n,j}x_j^k\right].$$

3. Successive over-relaxation (SOR) methods: Let $\omega \in (0, 2)$ be a relaxation parameter. Noting that the lower triangular matrix $D + \omega E$ is invertible, the SOR method corresponds to the choice $B = B_\omega := (D + \omega E)^{-1}[(1 - \omega)D - \omega F]$ and $\vec{c} = \omega(D + \omega E)^{-1}\vec{b}$. Notice that if $\omega = 1$, the SOR method reduces to the Gauss–Seidel method. The values of the iterates x_i^k are determined by

$$x_1^k = (1 - \omega)x_1^{k-1} + \omega a_{1,1}^{-1}\left[b_1 - \sum_{j=2}^{n} a_{1,j}x_j^{k-1}\right];$$

$$x_i^k = (1 - \omega)x_i^{k-1} + \omega a_{i,i}^{-1}\left[b_i - \sum_{j=1}^{i-1} a_{i,j}x_j^k - \sum_{j=i+1}^{n} a_{i,j}x_j^{k-1}\right], \quad \text{for } i = 2, \ldots, n-1;$$

$$x_n^k = (1 - \omega)x_n^{k-1} + \omega a_{n,n}^{-1}\left[b_n - \sum_{j=1}^{n-1} a_{n,j}x_j^k\right].$$

The restriction that ω must belong to the interval $(0, 2)$ is related to the fact that the product of the eigenvalues of a square matrix equals its determinant. Notice that $det(B_\omega) = (1 - \omega)^n det(D - \omega F/[1 - \omega])/det(D + \omega E)$. Since both matrices involved in this quotient are triangular matrices with the same diagonal, $det(B_\omega) = (1 - \omega)^n$, and hence $\rho(B_\omega) \geq |1 - \omega|$. As a consequence, $\rho(B_\omega)$ can only be strictly less than one if $\omega \in (0, 2)$.

Among other situations of practical interest, all the above methods converge if A is symmetric and positive-definite. It is also so for nonsymmetric matrices which are **strictly diagonally dominant**. The latter concept means that $\forall i, 2a_{i,i} > \sum_{j=1}^{n} |a_{i,j}|$. In both cases, the Gauss–Seidel method converges faster than the Jacobi method, that is, $\rho(B_1) < \rho(B_J)$. On the other hand, the value of ω minimising $\rho(B_\omega)$, and hence optimising the convergence speed, strongly depends on the type of matrix A. The interested reader can find relevant information on these issues in Varga's classic book [199].

More recent successful iterative methods for solving SLAEs are those based on **Krylov spaces**, which are spanned by the vectors $\vec{y}, A\vec{y}, A^2\vec{y}, A^{k-1}\vec{y}$ for a certain k and a suitable $\vec{y}$ (cf. [176]). A fundamental iterative method of this type is the **conjugate gradient method**. It applies to the case where A is symmetric positive-definite, which often occurs in numerical solution of PDEs. The procedure goes as follows: Set $\vec{r}^0 = \vec{b} - A\vec{x}^0$, $\vec{p}^0 = \vec{r}^0$. If $\vec{r}^0 = \vec{0}$, then $\vec{x}^0$ is the solution. Otherwise, for $k > 0$ determine approximations $\vec{x}^k$ of $\vec{x}$ by $\vec{x}^k = \vec{x}^{k-1} + \alpha^{k-1}\vec{p}^{k-1}$, where $\alpha^{k-1} = [\vec{r}^{k-1}]^T \vec{r}^{k-1} / [\vec{p}^{k-1}]^T A\vec{p}^{k-1}$. If the norm of $\vec{r}^k = \vec{r}^{k-1} - \alpha^{k-1} A\vec{p}^{k-1}$ is sufficiently small, then $\vec{x}^k$ is the final approximation of $\vec{x}$. Otherwise, compute $\beta^k = [\vec{r}^k]^T \vec{r}^k / [\vec{r}^{k-1}]^T \vec{r}^{k-1}$, set $\vec{p}^k = \beta^k \vec{p}^{k-1} + \vec{r}^k$ and determine $\vec{x}^{k+1}$ as above, by replacing k with $k + 1$.

There are efficient versions of this method for nonsymmetric and non-positive-definite matrices (see e.g. [197]).

It is important to point out that the conjugate gradient method for symmetric positive-definite matrices yields the theoretical exact solution, at most within n iterations [132]. Here, the term 'theoretical' means that the latter property is being considered by disregarding round-off errors. Notice that, in this sense, the conjugate gradient method can be viewed as belonging to the class of direct methods, which we review next.

Direct methods

Direct methods for solving SLAEs are those that yield the exact solution up to round-off errors, within a finite number of operations. The most important methods of this class are **Gaussian elimination** and its variant, known as Crout's method, both applying to any invertible matrix; and **Cholesky's method** which is restricted to symmetric positive-definite matrices.

1. Gaussian elimination: We successively transform the SLAE $A\vec{x} = \vec{b}$ into systems $A^{(l)}\vec{x} = \vec{b}^{(l)}$, where $A^{(l)}$ has only zero entries in the rth column for $r \leq l$ from the $(r + 1)$th row on, by means of linear combinations of the current lth equation with the $n - l$ last equations. Eventually, an equation permutation is performed between the lth equation and one of the following $n - l$ equations at the l-th step, where $1 \leq l \leq n - 1$. More precisely, let $A^{(1)} = A$ and $\vec{b}^{(1)} = \vec{b}$ if $a_{1,1} \neq 0$. Otherwise, $A^{(1)}$ and $\vec{b}^{(1)}$ result from a **partial pivoting**, which is an operation consisting of interchanging the first equation with the r-th equation, where $r > 1$ is such that $|a_{r,1}| = \max_{2 \leq i \leq n} |a_{i,1}|$. Then, we transform the SLAE into the equivalent system $A^{(2)}\vec{x} = \vec{b}^{(2)}$, where $a_{i,1}^{(2)} = 0$ for $i = 2, \ldots, n$, $a_{i,j}^{(2)} = a_{i,j}^{(1)} - a_{1,j}^{(1)} a_{i,1}^{(1)} / a_{1,1}^{(1)}$ and $b_i^{(2)} = b_i^{(1)} - b_1^{(1)} a_{i,1}^{(1)} / a_{1,1}^{(1)}$.

More generally, assuming that l steps have already been achieved, resulting in system $A^{(l)}\vec{x} = \vec{b}^{(l)}$, at the $(l + 1)$-th step for $1 \leq l \leq n - 1$ we endeavour to transform the system into an equivalent SLAE $A^{(l+1)}\vec{x} = \vec{b}^{(l+1)}$. In this aim, in case $a_{l,l}^{(l)} = 0$, we first do partial pivoting by searching for the largest nonzero coefficient $a_{r,l}^{(l)}$ in absolute value for $l + 1 \leq r \leq n$. Still denoting the resulting system by $A^{(l)}\vec{x} = \vec{b}^{(l)}$, we compute for $1 \leq l \leq n - 1$,

$$a_{i,l}^{(l+1)} = 0 \ \text{ for } \ i = l + 1, \ldots, n;$$

$$a_{i,j}^{(l+1)} = a_{i,j}^{(l)} - a_{l,j}^{(l)} a_{i,l}^{(l)}/a_{l,l}^{(l)} \ \text{ for } \ i = l + 1, \ldots, n \quad \text{and} \quad j = l + 1, \ldots, n;$$

$$b_i^{(l+1)} = b_i^{(l)} - b_l^{(l)} a_{i,l}^{(l)}/a_{l,l}^{(l)} \ \text{ for } \ i = l + 1, \ldots, n.$$

Once the $(n - 1)$th transformation is achieved, we come up with the SLAE $A^{(n)}\vec{x} = \vec{b}^{(n)}$ whose matrix is upper triangular. Its diagonal can have no zero entries; otherwise, matrix $A^{(n)}$ would be singular. But such a situation is ruled out by the fact that this SLAE must have a unique solution. We may therefore determine $\vec{x}$ by back substitution, that is, by

$$x_n = [a_{n,n}^{(n)}]^{-1} b_n^{(n)}; \ x_{n-i} = [a_{n-i,n-i}^{(n)}]^{-1} \left[b_{n-i}^{(n)} - \sum_{j=n-i+1}^{n} a_{n-i,j}^{(n)} x_j \right] \ \text{ for } \ i = 1, \ldots, n - 1.$$

The systematic use of partial pivoting is recommended, even if the **pivot** at the lth step of Gaussian elimination (i.e., the lth diagonal coefficient of $A^{(l)}$) is not zero. This is because pivoting helps reducing the harmful effects of round-off error propagation. In practice, the lth row of $A^{(l)}$ is interchanged with its $\sigma(l)$th row for a certain $\sigma(l) > l$ such that $|a_{\sigma(l),l}^{(l)}|$ attains a maximum among $|a_{i,l}^{(l)}|$ for $i \geq l$.

2. *Crout's method:* This method results from an interpretation of Gaussian elimination slightly different from what we saw above. Actually one observes that, as a by-product of Gaussian elimination, a lower triangular matrix L with diagonal coefficients equal to one is built up. This matrix fulfils $LU = PA$, where U is the upper triangular matrix $A^{(n)}$ and P is the permutation matrix that stores the (eventual) row interchanges in the successive transforms of A carried out in a partial pivoting process. For instance, if only the first and $\sigma(1)$th row of A with $\sigma(1) > 1$ had been interchanged at the first step, P would equal the elementary permutation matrix P_1 having unit diagonal coefficients, except in the first and $\sigma(1)$th positions, and all off-diagonal coefficients equal to zero, except those equal to one in the positions $(\sigma(1); 1)$ and $(1; \sigma(1))$. If besides this at the second step, the second and $\sigma(2)$th rows of $A^{(2)}$ had been interchanged with $\sigma(2) > 2$, then we would have $P = P_2 P_1$, where P_2 is a matrix similar to P_1 except that 2 and $\sigma(2)$ would play the role of 1 and $\sigma(1)$, respectively. Generally speaking, P equals the product $P_{n-1} \ldots P_2 P_1$, where $P_l = I$ if no row permutation occurs at the lth step of the eliminations; otherwise P_l is a permutation matrix, namely, a modification of I in the lth and $\sigma(l)$th row and column of I, with $\sigma(l) > l$, as explained above for P_1 and P_2. As for matrix L, its entry in the ith row and the lth column equals $a_{i,l}^{(l)}/a_{l,l}^{(l)}$, for $i = l, \ldots, n$. In view of this factorisation of PA, it is easy to see that the SLAE to solve can be written as $LU\vec{x} = P\vec{b}$. Then, as long as U and L are duly stored, we have two systems with triangular matrices to solve, namely, $L\vec{z} = P\vec{b}$ followed by $U\vec{x} = \vec{z}$, which can be done by forth substitution for the former and back substitution for the latter. The use of this procedure is recommended, for instance, in case many SLAEs with the same matrix A have to be solved. Indeed, in this way running a Gaussian elimination for every new right-side vector is avoided. A short program to solve several SLAEs with the same matrix by Crout's method is supplied in Chapter 4, applied to the case of banded matrices. We recall that an $n \times n$ matrix A is a banded matrix with a half-band

width $m < n$ if $a_{i,j} = 0$ for $|i - j| \geq m$. The author strongly encourages the reader to examine this program thoroughly, in order to consolidate her or his understanding of Crout's method.

3. Cholesky's method: If matrix A is symmetric and positive-definite, it admits a decomposition of the type $U^T U$, where U is an invertible real upper triangular matrix. This is known as Cholesky's factorisation. U can be determined recursively as follows:

$$u_{1,1} = \sqrt{a_{1,1}};\ u_{1,j} = [u_{1,1}]^{-1} a_{1,j}\ \text{ for } j = 2, \ldots, n;$$

$$\text{For } i = 2, \ldots, n, u_{i,i} = \sqrt{a_{i,i} - \sum_{k=1}^{i-1} u_{k,i}^2}$$

$$\text{and if } i < n, u_{i,j} = [u_{i,i}]^{-1} \left[a_{i,j} - \sum_{k=1}^{i-1} u_{k,j} u_{k,i} \right] \quad \text{for } j = i + 1, \ldots, n.$$

Once U is determined, all that is left to do is solve two SLAEs with a lower and an upper triangular matrix, namely, $U^T \vec{z} = \vec{b}$, and then $U\vec{x} = \vec{z}$, by forth and back substitution, respectively.

The brief presentation of techniques for solving SLAEs given here will be completed in the chapters with considerations inherent to typical linear systems resulting from the application of numerical methods for boundary value differential equations. This will be done more particularly in the final sections of Chapters 1 and 4, which are specifically devoted to this topic.

1

Getting Started in One Space Variable

A journey of a thousand miles
begins with a single step.

Lao Tzu

In this chapter, we give an initial presentation of the three numerical methods to be studied throughout the book, as the most popular to solve PDEs: the finite difference method (**FDM**), the finite element method (**FEM**) and the finite volume method (**FVM**). The first type of method acts directly on the differential equation in its standard form, also called the equation's **strong form**. In contrast, both the FEM and the FVM are applied to integral forms of the equation called a **weak form** or a **variational form** for the former, and a **conservative form** for the latter. From this point of view, the FVM is viewed by some authors as a variant of the FEM, since it corresponds to a particular way of solving the problem, once it is rewritten in a suitable integral form.

From the material provided in this chapter, the reader will certainly figure out that, as long as the points of the **discretisation lattice** at which solution approximations are determined coincide for the three approaches, and moreover the problem to solve is sufficiently simple in terms of geometry, data, nonlinearities and so on, the underlying SLAEs are very similar. As far as the FEM and the FVM are concerned, for example, this has been observed by several authors for years (see e.g. Idelsohn and Oñate [102]). Nevertheless, the three methods are conceptually different, and for this reason it is worth distinguishing them from each other whatever the case.

In the subsequent chapters, we shall elaborate a little more about the concept of discretisation lattices and the relationship of a numerical method to them in more general frameworks. For the moment, we will do this in the case where the solution depends on a single variable x. More specifically, we will set ourselves in the simplest possible framework of a model two-point boundary value ODE with zero (i.e. homogeneous) boundary conditions, which will serve as

Numerical Methods for Partial Differential Equations:
An Introduction Finite Differences, Finite Elements and Finite Volumes, First Edition. Vitoriano Ruas.
© 2016 John Wiley & Sons, Ltd. Published 2016 by John Wiley & Sons, Ltd.
Companion Website: www.wiley.com/go/ruas/numericalmethodsforpartial

a model for introducing the principles that the FDM, FEM and FVM are based upon. In doing so, a first attempt will be made to provide some insight on the right choice of techniques for the computation of the numerical solution, in more challenging situations encountered in practice.

Chapter outline: In Section 1.1, the model ODE is introduced, together with features of its three different formulations to be exploited in the sequel. In Sections 1.2, 1.3 and 1.4, we describe the basic FDM, FEM and FVM, respectively, to solve this model problem. Section 1.4 is subdivided into three subsections. In Subsection 1.4.1, we describe a version of the FVM strongly connected to the two other methods; in Subsection 1.4.2, another version of the FVM is presented in the usual way for this method; and, in Subsection 1.4.3, some connections between the FVM and the FEM are highlighted. In Section 1.5, we extend the methods' application considered in the previous sections to the case of inhomogeneous (i.e. nonzero) boundary conditions. In Section 1.6, the solution of SLAEs resulting from the application of the three methods to the model problem is addressed, and a numerical example is given.

1.1   A Model Two-Point Boundary Value Problem

Let us consider the elongational deformations of an elastic bar of length L, having one of its ends fixed, say the left end, and the other end free. The bar is subject to a distribution of forces acting along its length, represented by a function $g(x), 0 \leq x \leq L$. Referring to Figure 1.1, we denote by $u(x)$ the resulting length variation of the elastic bar at a point x, or, equivalently, the displacement undergone in the longitudinal direction $\vec{e}_x$ at this point by $u(x)\vec{e}_x$.

Provided the deformations of the elastic bar take place in the small strain regime[1], the physical problem just described can be set in the form of the following two-point boundary value ODE for the unknown function $u(x)$ (see e.g. [181]):

$$\begin{cases} -(pu')' = g & \text{in } (0, L) \\ u(0) = 0 & \text{(fixed left end)} \\ u'(L) = 0 & \text{(free right end)} \end{cases} \tag{P$'_1$}$$

where p is a strictly positive function representing the local stiffness of the material the elastic bar is made of, and u' denotes $\frac{du}{dx}$.

In some situations, spring-like effects on the bar have to be taken into account. This means that a term of the form qu must be added to the left side of (P$'_1$) in order to properly model its deformation, q being a given non-negative function representing the spring rate. This in turn may be associated with another given force F. Representing by f the conjugate action of g and F, this leads to the following equation:

The model two-point boundary value problem

$$\boxed{\begin{aligned} -(pu')' + qu &= f \quad \text{in } (0, L) \\ u(0) &= 0 \qquad \text{(fixed left end)} \\ u'(L) &= 0 \qquad \text{(free right end)} \end{aligned}} \tag{P$_1$}$$

---

[1] In the present case, this means that the absolute value of the first-order derivative of u is small, say, nowhere greater than 1%.

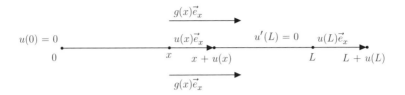

Figure 1.1 Elongation $u(x)$ for a longitudinal loading $g(x)$.

We shall overlook a more elaborate presentation of this elastic bar model, for our goal here is only a comprehensive numerical study of equation (P_1). We refer to [186] for a fine description of the problem's physical modelling in more detail[2].

Actually equation (P_1), eventually with different boundary conditions, is a simplified mathematical model of several other phenomena or processes. In Chapter 7, we consider one of them (cf. Example 7.4).

Whatever real-life problem it models, the differential equation (P_1) is often employed by authors as a basic model to introduce numerical methods for PDEs (see. e.g. [119] and [186], among many others). As already pointed out, this is also what we do in the remainder of this chapter. (P_1) is called the equation's **strong form**. This means that the differential equation is to be regarded in the usual pointwise sense, which applies at least under the assumption that the functions p and q, representing the medium's physical characteristics, satisfy for certain constants α, β, A and B:

$$A \geq p(x) \geq \alpha > 0 \quad \text{and } B \geq q(x) \geq \beta \geq 0, \forall x \in [0, L] \tag{1.1}$$

Now without caring about regularity assumptions, neither on the data p, q and f, nor on the solution u of (P_1), we shall rewrite this problem in two different integral forms. We only assume that all the operations performed for this purpose are feasible and well defined, postponing or leaving to texts on analysis of differential equations the discussion about the conditions under which it is really so.

Let us first consider the **conservative form** of (P_1) relying upon the set $\mathcal{V}$ consisting of all the open subdomains of Ω, generically denoted by ω, that are connected and have a non-zero measure. More concretely, any $\omega \in \mathcal{V}$ is necessarily an open interval contained in $(0, L)$, including of course this interval itself. Clearly enough, the integral in any $\omega \in \mathcal{V}$ of both sides of the differential equation in (P_1) coincides, that is, this problem implies that the following relations are satisfied:

$$\begin{cases} \int_{a_\omega}^{b_\omega} [-(pu')' + qu] dx = \int_{a_\omega}^{b_\omega} f dx \ \forall \omega \in \mathcal{V} \\ u(0) = 0 & \text{(fixed left end)} \\ u'(L) = 0 & \text{(free right end)} \end{cases} \tag{P'_2}$$

---

[2] The differential equation in (P_1) is the *Sturm equation* $a_0 u'' + a_1 u' + a_2 u = a_3$ for given functions a_0, a_1, a_2 and a_3 satisfying certain properties. The Sturm equation can be recast in the form appearing in (P_1) called the *Liouville form*. For this reason, it is sometimes referred to as a *Sturm–Liouville problem*. However, classically the latter stems from a PDE, that governs the free or the forced vibrations of a string (see e.g. [57]). This PDE known as the *equation of the vibrating string*, will be studied in Chapter 3. If the applied forces are periodic in time, it can be reduced to the ODE (P_1). However, in this case q is negative, which makes the problem fundamentally different from the one studied here.

where a_ω and b_ω denote the left and right end of ω, respectively.

Conversely, intuitively enough one can infer that problem (P_2') implies (P_1). As a matter of fact, this can be rigorously established by means of the theory of Lebesgue integration (cf. [206] or [110]). In short, both problems are equivalent.

We may further develop the integral relation in (P_2'), in order to obtain a new integral-differential equation also equivalent to (P_1), namely:

The conservative form of model problem (P_1)

$$
\begin{array}{ll}
[pu'](a_\omega) - [pu'](b_\omega) + \int_{a_\omega}^{b_\omega} qu\,dx = \int_{a_\omega}^{b_\omega} f\,dx \quad \forall \omega \in \mathcal{V}. & \\
u(0) = 0 & \text{(fixed left end)} \\
u'(L) = 0 & \text{(free right end)}
\end{array}
\qquad (P_2)
$$

Problem (P_2) corresponds to the usual form the FVM is applied to.

Let us next consider the usual **weak form** of equation (P_1), also known as its **standard Galerkin variational form**. First, we multiply both sides of the differential equation with a **test function** v whose required properties will be specified in due course. For the moment, let us just say that the expression 'test-function' means that v is supposed to behave somehow like the solution u, in the sense that u itself could be one of such functions. In simpler terms, we say that v sweeps a set V – actually, a vector space consisting of functions – whose properties are specified hereafter.

Next, integrating the resulting relation over the interval $(0, L)$ and using integration by parts, we obtain

$$
-pu'v\Big|_0^L + \int_0^L pu'v'\,dx + \int_0^L quv\,dx = \int_0^L fv\,dx
$$

for every (test) function v.

First of all, we note that since $u'(L) = 0$, if we require that v vanish at the origin like u itself, we come up with

The (standard Galerkin) variational form of (P_1)

$$
\int_0^L pu'v'\,dx + \int_0^L quv\,dx = \int_0^L fv\,dx, \quad \forall v \in V
\qquad (P_3)
$$

where u also belongs to V. Notice that all the integrals above must carry a meaning. Hence, it is necessary to require that f together with every function in V and its first-order derivative are square integrable in the sense of Lebesgue in the interval $(0, L)$, as we will see in Section 2.1. This implies that we are actually defining V to be

$$
V = \{v / v, v' \in \mathcal{L}^2(0, L), v(0) = 0\}
\qquad (1.2)
$$

where $\mathcal{L}^2(0, L)$ is the set of those functions f such that f^2 has a finite (Lebesgue) integral in $(0, L)$. We can assert that vector space V defined by equation (1.2) is the ideal choice in the framework of the standard Galerkin variational formulation (P_3), even though in many cases the solution u satisfies additional boundary conditions and has finer differentiability properties. This is because with such a choice we can deal with the widest possible class of problems carrying a physical meaning, modelled by equation (P_1), as seen in this chapter.

A justification of the fact that this choice is also the right one from the mathematical point of view lies beyond the scope of this book. For a comprehensive discussion on this point, the author refers to [182].

Remark 1.1 *As proven in classical books on Lebesgue integration (see e.g. [206]), every v such that $v' \in \mathcal{L}^2(0, L)$ is continuous[3]. Therefore, there is no contradiction in prescribing $v(0) = 0$. For discussions on the conditions under which the above variational formulation (P₃) carries a precise mathematical sense, we refer for instance to reference [136]. For mathematical tools supporting this theory, we refer to reference [179].* ∎

Now what have we gained by writing problem (P₁) in the conservative form (P₂) or in the variational form (P₃)?

First of all, we note that the highest derivative order that appears in both (P₂) and (P₃) is one, while it is two in (P₁). Thus if, for instance, we choose to approximate u by a function belonging to the class of piecewise polynomials like the FEM does, it is readily seen that ordinary continuous functions may be used for (P₃), whereas continuously differentiable ones would be required for (P₁). This is because it is not possible to differentiate twice a function that has discontinuous first-order derivatives at certain points in $(0, L)$.

Another advantage of formulation (P₃) over (P₁) is the fact that the boundary condition $u'(L) = 0$ may be disregarded in the former, for it is implicitly satisfied. Indeed, if we still assume that all the operations performed below are legitimate, we have from (P₃)

$$pu'v\Big|_0^L - \int_0^L (pu')'vdx + \int_0^L quvdx = \int_0^L fvdx \tag{1.3}$$

for every $v \in V$. Let us choose $v \in V$ such that $v(L) = 0$, but otherwise arbitrary. In doing so, we obtain

$$\int_0^L \left[-(pu')' + qu - f\right]vdx, \quad \forall v \in V \text{ such that } v(L) = 0$$

Following reference [126], here the reader can take it for granted that the class of such functions v is wide enough for the above relation to imply that

$$-(pu')' + qu = f \text{ almost everywhere in } (0, L)$$

From this result equation, (1.3) simply becomes

$$p(L)u'(L)v(L) - p(0)u'(0)v(0) = 0, \quad \forall v \in V$$

Since $v(0) = 0$, choosing now $v \in V$ such that $v(L) = 1$, we immediately infer that $u'(L) = 0$ as required, as p is strictly positive by assumption.

Another clear advantage of (P₃) over (P₁) and (P₂) is related to the regularity of the datum p. Even assuming that the material of the bar is homogeneous, the function p varies with its cross section. Eventually, the latter could change shape abruptly in such a way that p is a discontinuous function. Although, on the one hand, this does not bring about any difficulty as

---

[3] More precisely, v is absolutely continuous.

far as problem (P_3) is concerned, more care is needed in handling the term $(pu')'$ in equation (P_1) or the term $[pu'](a_w) - [pu'](b_w)$ in equation (P_2) in this case.

Finally, a more tricky but fundamental point: formulation (P_3) (and to a lesser extent (P_2)) is more general than equation (P_1) from the physical point of view. In order to clarify this assertion, let us consider a distribution of forces f_ε that have a resultant P, uniformly acting on a very small length ε around the abscissa $x = L/2$. In current applications, such a system of forces is represented by $P\delta_{\frac{L}{2}}$, namely, a single force of modulus equal to P applied at the point given by $x = L/2$.[4] Of course, if we stick to the first definition of f_ε, all the problems (P_1), (P_2) and (P_3) carry a meaning. Actually, denoting by u_ε the corresponding solution, problem (P_3) writes

$$\int_0^L pu_\varepsilon' v' \, dx + \int_0^L qu_\varepsilon v \, dx = \int_{\frac{L}{2}-\frac{\varepsilon}{2}}^{\frac{L}{2}+\frac{\varepsilon}{2}} \frac{P}{\varepsilon} v \, dx, \quad \forall v \in V$$

Now, since v is a continuous function, the mean value theorem for integrals implies that the right side of the above equation tends to $Pv(L/2)$ as ε goes to zero. Otherwise stated, the variational problem (P_3) is perfectly well defined in the case of the single force applied to a particle located at point $x = L/2$, as

$$\int_0^L pu' v' \, dx + \int_0^L quv \, dx = Pv(L/2), \quad \forall v \in V, u \text{ itself in } V$$

As for problem (P_2), if we consider a subset V_ε of V consisting of intervals that either fully contain $(L - \varepsilon/2, L + \varepsilon/2)$ or have an empty intersection with it, we have

$$[pu'](a_w) - [pu'](b_w) + \int_{a_w}^{b_w} qu \, dx = \int_{a_w}^{b_w} f \, dx \quad \forall w \in V_\varepsilon. \tag{1.4}$$

This means that the left side of equation (1.4) equals P if $w \in V_\varepsilon$ is such that $w \cap (L/2 - \varepsilon/2, L/2 + \varepsilon/2) \neq \emptyset$ and equals 0 otherwise.

Since ε is bound to tend to zero, problem (1.4) will be as general as (P_2) for $\varepsilon = 0$. Indeed, it suffices to modify the conservative form in such a way that the integral $\int_w f \, dx$ is replaced by P if w contains $L/2$ and by 0 otherwise.

On the other hand, the equation (P_1) in this case could only be based on an equality of the type $-(pu')' + qu = P\delta_{\frac{L}{2}}$, whose exact meaning is unclear for numerical purposes.

Since forces applied pointwise may act everywhere and may also be combined, we conclude that there are at least infinitely many systems of forces that are admissible for the models (P_3) and (P_2) of the elongational deformation of a bar, although they do not correspond to a differential equation of the form (P_1), at least as far as functions defined in accordance with the classical concept are concerned [179].

Summarising, we have exhibited four advantages of formulation (P_3) over (P_1), among which at least two also apply to (P_2), in spite of the fact that a rather simple linear ODE was used as a model for the purpose of this comparison.

---

[4] In old days, this distribution of forces used to be called the 'Dirac function' $\delta_{\frac{L}{2}}$ multiplied by P, where $\delta_{\frac{L}{2}}$ would be such that $\delta_{\frac{L}{2}}(x) = 0$, $\forall x \in [0, L]$, $x \neq L/2$, $\delta_{\frac{L}{2}}(L/2) = +\infty$, and "$\int_0^L \delta_{\frac{L}{2}} \, dx$" $= 1$.

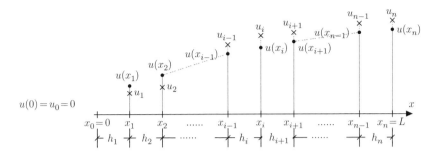

Figure 1.2   An FD grid with exact and approximate values of $u(x)$ at grid points x_i.

1.2   The Basic FDM

In this section, we introduce one of the oldest and simplest numerical methods to solve equation (P_1): the **Three-point FDM**. The basic idea is to search for values of the unknown function u only at the points in $[0, L]$ belonging to a certain (finite) discretisation lattice $\mathcal{G} := \{x_0, x_1, x_2, \ldots, x_{n-1}, x_n\}$, which is called an **FD grid** in this case. An FD grid is illustrated in Figure 1.2.

For the sake of simplicity, we first consider that both p and q are constant. In this case, problem (P_1) becomes

$$\begin{cases} -pu'' + qu = f & \text{in } (0, L) \\ u(0) = 0 & \text{(fixed left end)} \\ u'(L) = 0 & \text{(free right end)} \end{cases} \tag{1.5}$$

Of course, such a problem can be solved by hand provided f is sufficiently simple. However, in order to attain our goal, pretending this is not the case, we head for a numerical solution.

Let the set of $n + 1$ points of $\mathcal{G}$ be such that $x_0 = 0$, $x_n = L$ and $x_i > x_{i-1}$ for every $i \in \{1, 2, \ldots, n - 1, n\}$. For the moment, we assume that these points are equally spaced. Thus, setting $h = L/n$, we have $x_i = ih$, for $i \in \{0, 1, 2, \ldots, n - 1, n\}$, so that $x_i - x_{i-1} = h$ for every $i \in \{1, 2, \ldots, n - 1, n\}$. If the function f is twice continuously differentiable, the solution u of (P_1) will be four times continuously differentiable. So, if $i \neq 0$ and $i \neq n$, using a standard Taylor expansion we have

$$\begin{cases} u(x_{i+1}) = u(x_i) + hu'(x_i) + h^2 u''(x_i)/2 + h^3 u'''(x_i)/6 + h^4 u^{iv}(\xi_i^+)/24 \\ \text{and} \\ u(x_{i-1}) = u(x_i) - hu'(x_i) + h^2 u''(x_i)/2 - h^3 u'''(x_i)/6 + h^4 u^{iv}(\xi_i^-)/24, \end{cases} \tag{1.6}$$

where ξ_i^+ and ξ_i^- are suitable points belonging to the intervals $[x_i, x_{i+1}]$ and $[x_{i-1}, x_i]$, respectively. Adding up the two relations of equation (1.6), after simple manipulations we come up with

$$\begin{cases} -u''(x_i) = \dfrac{2u(x_i) - u(x_{i+1}) - u(x_{i-1})}{h^2} + R_i(u)h^2 \\ \text{where} \\ R_i(u) = [u^{iv}(\xi_i^+) + u^{iv}(\xi_i^-)]/24, \end{cases} \tag{1.7}$$

On the other hand, from equation (1.5) it holds that

$$- pu''(x_i) + qu(x_i) = f(x_i) \text{ for } i = 1, 2, \ldots, n - 1. \tag{1.8}$$

Hence, equation (1.7) means that, up to the term $R_i(u)h^2$, we may replace in equation (1.8) the second derivative of u at x_i by the **FD**, that is, the **finite difference** $[2u(x_i) - u(x_{i+1}) - u(x_{i-1})]/h^2$. Otherwise stated, provided u is four times differentiable in $[0, L]$, the following relations hold among the values of u at the grid points formed by the x_is:

$$\begin{cases} \text{For } i = 1, 2, \ldots, n - 1, \\ p\dfrac{2u(x_i) - u(x_{i+1}) - u(x_{i-1})}{h^2} + qu(x_i) = f_h(x_i) \\ \text{where} \\ f_h(x_i) = f(x_i) - pR_i(u)h^2. \end{cases} \tag{1.9}$$

Our assumptions on u allow us to verify quite easily that $R_i(u)$ can be bounded independently of n. Thus, provided n is large enough, or equivalently h is sufficiently small, the **finite difference equation** $p\frac{2u(x_i)-u(x_{i+1})-u(x_{i-1})}{h^2} + qu(x_i) = f(x_i)$ is satisfied at every point x_i for $i = 1, \ldots, n - 1$ up to a small additive term whose magnitude goes to zero as fast as h^2 when n increases indefinitely. However, though certainly convincing, such an argument is not satisfactory from both the practical and the mathematical points of view. Indeed, we are looking for values of the unknown u at least at the grid points, and relations (1.9) cannot be exploited for this purpose since the function f_h, in spite of being very close to f, is not exactly known. Nevertheless, the above development naturally leads to the conceptual construction of the FDM: if the values $u(x_i)$ do not satisfy an exploitable relation, there should exist some values u_i close to them (i.e., **approximations** of theirs) that do (cf. Figure 1.2). More precisely, the u_is are meant to fulfil

$$\begin{cases} \text{For } i = 1, 2, \ldots, n - 1, \\ p\dfrac{2u_i - u_{i+1} - u_{i-1}}{h^2} + qu_i = f(x_i) \end{cases} \tag{1.10}$$

However, there are other features of problem (P$_1$) which have not been incorporated into our description of the FDM yet, such as the boundary conditions $u(0) = 0$ and $u'(L) = 0$. Whereas the former is naturally taken into account by simply setting $u_0 = 0$, the latter requires somewhat more careful considerations. A possibility that seems quite natural comes from the first (resp. second) relation of equation (1.6) for $i = n - 1$ (resp. $i = n$). Indeed, this tells us that $u'(L) = [u(x_n) - u(x_{n-1})]/h + T_n(u)h$, where $T_n(u)$ accounts for the terms involving derivatives of u of order higher than one. This suggests that the approximate values u_{n-1} and u_n must satisfy $u_n - u_{n-1} = 0$. However, as we will see later on, such a choice is not compatible with the quality of approximation provided by the way we approximate the differential equation itself. The introduction of the concept of a fictitious point $x_{n+1} := (n + 1)h = L + h$ lying outside the domain $(0, L)$ gives rise to a better approximation of the boundary condition at $x = L$. We associate with this point a fictitious exact value of u, $u(x_{n+1})$ satisfying a relation of the form $p\frac{2u(x_n)-u(x_{n+1})-u(x_{n-1})}{h^2} + qu(x_n) = f_h(x_n)$, where $f_h(x_n)$ stands for a natural extension of f_h to $x = L$, which we decline to specify.

Then we mimic as fictitiously the first relation of equation (1.6), taking $i = n$ this time, and subtract the second one from the resulting relation to conclude that a (fictitious) value $u(x_{n+1})$ would satisfy $[u(x_{n+1}) - u(x_{n-1})]/h = S_n(u)h^2$, where $S_n(u)$ accounts for terms involving derivatives of u of order higher than two. This looks like a more reasonable way to deal with the boundary condition $u'(L) = 0$ in our FD analog of (P_1): we introduce an additional unknown u_{n+1} and then eliminate it by simply making $u_{n+1} - u_{n-1} = 0$. Recalling equation (1.10), this leads as naturally to the following extension of the approximation of the differential equation at point $x = x_n = L$:

$$\begin{cases} p\dfrac{2u_n - u_{n+1} - u_{n-1}}{h^2} + qu_n = f(x_n) \\ \text{with } u_{n+1} = u_{n-1}. \end{cases} \tag{1.11}$$

Finally combining equations (1.10) and (1.11), and using the term **uniform grid** to qualify an equally spaced grid, we come up with

The three-point FD analog of (P_1) with a uniform grid

$$\boxed{\begin{array}{l} \text{Find } u_i, i = 1, 2, \ldots, n \text{ satisfying} \\[4pt] p\dfrac{2u_i - u_{i+1} - u_{i-1}}{h^2} + qu_i = f(x_i) \\[4pt] \text{with } u_0 = 0 \text{ and } u_{n+1} = u_{n-1}. \end{array}} \tag{1.12}$$

where u_i is an **approximation** of $u(x_i)$.

A question that immediately arises concerns the quality of such an approximation. In principle we can assert that everything is fine, but the choice $u_{n+1} = u_{n-1}$ might not be the best possible in this respect. For the moment we refrain from specifying what an optimal way to define u_{n+1} should be, among other quality considerations. These are left to Chapter 2, which is specifically devoted to this subject.

Going back to equation (1.12), let us take a closer look at this problem's nature. First of all, we note that equation (1.12) is nothing but a SLAE with n unknowns and n equations, whose right side is the vector $[f(x_1), f(x_2), \ldots, f(x_n)]^T$. Therefore, it possesses a unique solution if and only if $f(x_i) = 0$ for all $i \in \{1, 2, \ldots, n\}$, implies that $u_i = 0$ for $i = 1, 2, \ldots, n$. Actually this implication holds true, according to the following argument.

Let M be the subscript of (one of) the point(s) at which $|u_i|$ attains its maximum value. If $M = 1$, since $f_1 = 0$, taking $i = 1$ in equation (1.12) we have $|u_1| = p|u_2|/(2p + qh^2) \leq |u_2|/2$, since by assumption $p > 0$ and $q \geq 0$. Then, noting that necessarily $|u_2| \leq |u_1|$, we conclude that $|u_1| = 0$ and the result is thus established. If $M > 1$, recalling that $f(x_M) = 0$, we infer from equation (1.12) that $|u_M|$ is at most equal to the mean value of $|u_{M-1}|$ and $|u_{M+1}|$. Since none of these values can be greater than $|u_M|$ by assumption, both are necessarily equal to $|u_M|$. Thus, the maximum value is also attained for $i = M - 1$. If $M - 1 = 1$ we are finished, but if not we may apply the same argument to $|u_{M-1}|$, thereby concluding that the maximum value of $|u_i|$ is also equal to $|u_{M-2}|$. Then step by step we will reach the equation $|u_1| = p|u_2|/(2p + qh^2)$, knowing that both $|u_1|$ and $|u_2|$ must take the maximum absolute value of all the u_is. This implies that $|u_1| = |u_2| = 0$, and hence all the u_is must equal zero.

Another issue to address at this stage is the form of the SLAE (equation (1.12)). By inspection and after division by a factor of two of the nth equation, we easily infer from equation (1.12) that the vector $\vec{u}_h := [u_1, u_2, \ldots, u_n]^T$, consisting of the n approximate values of u, satisfies a SLAE with matrix $A_h = \{a_{i,j}\}$ and right-side vector $\vec{b}_h = \{b_i\}$ of the form

SLAE for the Three-point FD scheme with a uniform grid

$$
\boxed{
\begin{aligned}
&A_h \vec{u}_h = \vec{b}_h \\
&\text{where} \\
&a_{i,i} = 2p/h^2 + q \text{ if } i \neq n; \qquad\qquad a_{n,n} = p/h^2 + q/2; \\
&a_{i,i+1} = a_{i+1,i} = -p/h^2 \ \forall i < n; \quad a_{i,j} = 0 \text{ if } |i-j| > 1; \\
&\text{and} \\
&b_i = f(x_i) \text{ if } i \neq n; \qquad\qquad\qquad b_n = f(x_n)/2.
\end{aligned}
}
\tag{1.13}
$$

The reader might observe that, thanks to the multiplication by $1/2$ of the nth equation resulting from equation (1.12), A_h became a symmetric matrix.

Before pursuing our presentation of the FDM, it seems instructive to give an idea of the aspect of both matrix A_h and vector $\vec{b}_h$, considering the particular case where p and q are constant. Taking $n = 6$, for instance, we display in Array 1.1 the matrix A_h in this case, whereas vector $\vec{b}_h$ is exhibited in equation (1.14).

$$
A_h =
\begin{bmatrix}
\frac{2p}{h^2}+q & -\frac{p}{h^2} & 0 & 0 & 0 & 0 \\
-\frac{p}{h^2} & \frac{2p}{h^2}+q & -\frac{p}{h^2} & 0 & 0 & 0 \\
0 & -\frac{p}{h^2} & \frac{2p}{h^2}+q & -\frac{p}{h^2} & 0 & 0 \\
0 & 0 & -\frac{p}{h^2} & \frac{2p}{h^2}+q & -\frac{p}{h^2} & 0 \\
0 & 0 & 0 & -\frac{p}{h^2} & \frac{2p}{h^2}+q & -\frac{p}{h^2} \\
0 & 0 & 0 & 0 & -\frac{p}{h^2} & \frac{p}{h^2}+\frac{q}{2}
\end{bmatrix}
$$

Array 1.1

$$
\vec{b}_h = \left[f(x_1), \ f(x_2), \ f(x_3), \ f(x_4), \ f(x_5), \ f(x_6)/2 \right]^T.
\tag{1.14}
$$

Notice that most coefficients of A_h are zero (at least for $n > 5$). This kind of matrix is called a **sparse matrix**. Actually matrix A_h, besides being sparse, is also a **tridiagonal matrix** since $a_{i,j} = 0$ whenever $|i-j| > 1$.

In order to extend the FDM to more general cases, it is wise to regard the FD used to approximate u'' at x_i in the light of a different interpretation:

If we introduce intermediate points $x_{i-1/2}$ given by $(i-1/2)h$, for $i = 1, \ldots, n$, we may define approximations of the first-order derivative of u at those points by the FD $D_u^1(x_{i-1/2}) := [u(x_i) - u(x_{i-1})]/h$. Then, naturally enough, we define an approximation of the second-order derivative of u at x_i by another FD using such approximations of the first-order derivatives at $x_{i-1/2}$ and $x_{i+1/2}$, thereby obtaining $D_u^2(x_i) := [D_u^1(x_{i+1/2}) - D_u^1(x_{i-1/2})]/h$, for $i = 1, 2, \ldots, n-1$. Of course, if we consider the context of an FD scheme, we must replace by u_i

the value of $u(x_i)$ in the expression of $D^1_u(x_{i-1/2})$, $i = 1, 2, \ldots, n$, thereby obtaining another approximation $\delta_{h,i-1/2}$ of $u'(x_{i-1/2})$. Then, the approximation of the second-order derivative denoted by $\Delta_{h,i}$ instead of $D^2_u(x_i)$, will be given by $\Delta_{h,i} := [\delta_{h,i+1/2} - \delta_{h,i-1/2}]/h$, $i = 1, 2, \ldots, n - 1$. In this case, using the concept of a fictitious point, we may extend to $i = n$ the definition of $\delta_{h,i+1/2}$, and hence the one of $\Delta_{h,i}$.

Now for several reasons, one might be interested in using a non–equally spaced grid – or **non-uniform grid** – to solve problem (P_1). For instance, this could happen if the datum f was very irregular in a certain region, in which it would be advisable to place more grid points in an attempt to improve the local accuracy of the method. In this case, using a fictitious point which we define to be $x_{n+1} := L + h_n$, we set $h_i := x_i - x_{i-1}$, for $i = 1, 2, \ldots, n + 1$, and approximate the first derivative of u at $x_{i-1/2} := (x_{i-1} + x_i)/2$ by the incremental ratio

$$D^1_u(x_{i-1/2}) := [u(x_i) - u(x_{i-1})]/h_i.$$

Then quite naturally, for $i = 1, 2, \ldots, n$ the second-order derivative of u at x_i is approximated by **the divided difference** using the local **intermediate grid size**, i.e., by

$$D^2_u(x_i) := \frac{D^1_u(x_{i+1/2}) - D^1_u(x_{i-1/2})}{(h_{i+1} + h_i)/2}.$$

Noticing that the values of u at the grid points x_i are not available, and using instead corresponding approximations u_i, we are led to another approximation of $u''(x_i)$, still denoted by $\Delta_{h,i}$, for $i = 1, 2 \ldots, n$, given by

$$\Delta_{h,i} := \frac{\delta_{h,i+1/2} - \delta_{h,i-1/2}}{(h_{i+1} + h_i)/2}$$

with

$$\delta_{h,i-1/2} := [u_i - u_{i-1}]/h_i,$$

where $\delta_{h,i-1/2}$ is an approximation of $u'(x_{i-1/2})$ for $i = 1, 2, \ldots, n$, with the obvious fictitious extension $\delta_{h,n+1/2}$.

All this leads to the following problem to approximate equation (P_1) in the case of constant coefficients p and q:

The three-point FD analog of (P_1) with a non-uniform grid

$$
\boxed{
\begin{array}{l}
\text{Find } u_i, i = 1, 2, \ldots, n \text{ satisfying} \\[4pt]
-p\Delta_{h,i} + qu_i = f(x_i) \\[4pt]
\text{with } u_0 = 0 \text{ and } u_{n+1} = u_{n-1} \\[4pt]
\text{where} \\[4pt]
\Delta_{h,i} := \dfrac{\delta_{h,i+1/2} - \delta_{h,i-1/2}}{(h_{i+1} + h_i)/2} \\[10pt]
\text{with } \delta_{h,i-1/2} := [u_i - u_{i-1}]/h_i.
\end{array}
}
\qquad (1.15)
$$

Like equation (1.12), equation (1.15) is a SLAE with an $n \times n$ matrix $A_h = \{a_{i,j}\}$ and right-side vector $\vec{b}_h = \{b_i\}$, and unknown vector $\vec{u}_h = [u_1, u_2, \ldots, u_n]^T$. We obtain a SLAE with a non symmetric matrix, **symmetrisable** through the multiplication of each equation by the corresponding intermdiate grid size, namely,

The SLAE for the Three-point FD scheme with a non-uniform grid

$$
\begin{aligned}
&A_h \vec{u}_h = \vec{b}_h \\
&\text{where} \\
&a_{i,j} = 0 && \text{if } |i - j| > 1; \\
&a_{i,i} = 2p/(h_i h_{i+1}) + q && \text{for } 1 \le i < n; \; a_{n,n} = p/h_n^2 + q/2; \\
&a_{i,i+1} = -2p/[h_{i+1}(h_i + h_{i+1})] && \text{for } 1 \le i < n; \\
&a_{i,i-1} = -2p/[h_i(h_i + h_{i+1})] && \text{for } 1 < i < n; \; a_{n,n-1} = -p/h_n^2; \\
&b_i = f(x_i) && \text{for } 1 \le i < n; \; b_n = f(x_n)/2.
\end{aligned}
$$

(1.16)

The reader might prove that system (1.16) has a unique solution (cf. Exercise 1.1).

While, on one hand, the extension of either equation (1.12) or (1.15) to the case where q is not constant is straightforward, on the other hand this task requires some care as far as p is concerned. For this reason we postpone such an extension, since it can be carried out more easily in the light of the other two discretisation methods studied in this chapter. This is more especially the case of the FEM, which is next presented.

1.3   The Piecewise Linear FEM ($\mathcal{P}_1$ FEM)

As already pointed out, the FEM applies to the variational form of problem (P$_1$), namely, equation (P$_3$). Here, we no longer assume any differentiability property of p in $(0, L)$. We keep only the assumption that this function is bounded above and below away from zero in $[0, L]$. However, although this is by no means necessary, we make the more than reasonable hypothesis from the physical point of view that p is discontinuous only at a certain finite number of points y_i, $i = 1, 2, \ldots, M$, with $y_1 < y_2 < \ldots < y_M$. Here, the method's discretisation lattice is called a **mesh** denoted by $\mathcal{T}_h$, consisting of n closed intervals $T_i := [x_{i-1}, x_i]$ called **elements**, with $0 = x_0 < x_1 < \ldots < x_{n-1} < x_n = L$. The mesh points S_i whose abscissae are the x_is are called **nodes**. Notice that the nodes are the grid points of a non-uniform FD grid. Recalling the definition of h_i in Section 1.2, we further define a characteristic length of the mesh $\mathcal{T}_h$ called

The mesh step size h

$$
h := \max_{i \in \{1, 2, \ldots, n\}} h_i.
$$

(1.17)

If all the elements of the mesh have the same length h, we say that $\mathcal{T}_h$ is a **uniform mesh**.

Now, for every element $T_j \in \mathcal{T}_h$, we denote by $\mathcal{P}_1(T_j)$ the set of polynomials of degree less than or equal to one defined in T_j. Using the Kronecker symbol δ_{ij}[5], we introduce a set of **shape functions** φ_i for $i = 0, 1, \ldots, n-1, n$ through the

---

[5] $\delta_{ij} = 1$ if $i = j$, and $\delta_{ij} = 0$ if $i \ne j$.

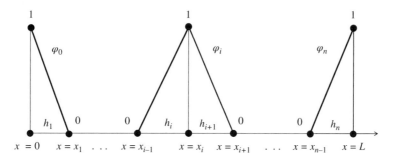

Figure 1.3 Shape functions φ_i for $0 < i < n$, φ_0, and φ_n.

Shape function definition

$$\varphi_i \in \mathcal{P}_1(T_j) \ \forall T_j \in \mathcal{T}_h; \qquad \varphi_i(x_j) = \delta_{ij} \tag{1.18}$$

This definition is illustrated in Figure 1.3, which displays the shape function φ_i for a generic i different from 0 and n, together with φ_0 and φ_n. Actually, equation (1.18) allows us to derive quite easily the following analytic expression of φ_i, for $1 \le i < n$:

$$\varphi_i(x) = \begin{cases} 0 & \text{if } x \notin [x_{i-1}, x_{i+1}] \\ (x - x_{i-1})/h_i & \text{if } x \in [x_{i-1}, x_i] \\ (x_{i+1} - x)/h_{i+1} & \text{if } x \in [x_i, x_{i+1}]. \end{cases} \tag{1.19}$$

The reader should complete the definition of the shape functions by determining the analytic expressions of φ_0 and φ_n. The space spanned by these $n+1$ shape functions is denoted by W_h. Now, instead of u, we shall search for an approximation u_h associated with $\mathcal{T}_h$ of the form:

$$u_h = \sum_{j=1}^{n} u_j \varphi_j. \tag{1.20}$$

The coefficients u_j in equation (1.20) are nothing but the values $u_h(x_j)$, as one can easily check using property (1.18). Actually, they will play the role of approximations of $u(x_j)$ in the same way as in the FDM. Now, instead of sweeping the whole space V, we restrict the variational problem to the space V_h consisting of functions of the form $\sum_{i=1}^{n} v_i \varphi_i$, where the v_is are arbitrary real numbers. In other words, V_h is the space spanned by the φ_is for $i > 0$, and in this sense u_h itself belongs to V_h.

In Figure 1.4, we show the general aspect of a function in V_h, such as u_h. Since by construction $\varphi_i(0) = 0$ for every i, except for $i = 0$, we observe that all the functions v in the space V_h vanish at $x = 0$. For a FEM based on the standard Galerkin formulation, the solution u_h together with test functions $v \in V_h$ should belong to the space V defined in Section 1.1. The fact that all of them vanish at $x = 0$ is half a way to fulfil this condition. Another related question is why the shape functions φ_i, and consequently any $v \in V_h$, must be continuous.

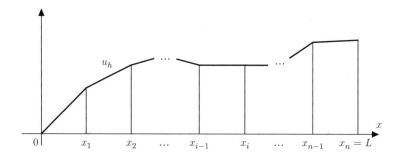

Figure 1.4   An illustration of a typical function belonging to FE space V_h.

The reader is encouraged to think about it to find the right answer. She or he is referred to Subsection 7.1.2 for a hint.

Now we set

The $\mathcal{P}_1$ FE approximation of ($\mathbf{P}_1$)

Find $u_h \in V_h$ such that

$$\int_0^L [pu_h' v' + qu_h v]dx = \int_0^L fv dx \ \forall v \in V_h \qquad (1.21)$$

In particular, equation (1.21) must hold for $v = \varphi_i$ with $i = 1, 2, \ldots, n$. Then, replacing v with φ_i and using equation (1.20) together with well-known properties of integrals, we come up with the **Piecewise linear FE** scheme given by equation (1.22), whose unknowns u_j are approximations of $u(x_j)$, u being the solution of ($\mathbf{P}_3$). This scheme is also called

The $\mathcal{P}_1$ FE scheme to solve ($\mathbf{P}_1$)

Find u_j for $j = 1, 2, \ldots, n$ such that

$$\sum_{j=1}^n u_j \int_0^L [p\varphi_j' \varphi_i' + q\varphi_j \varphi_i]dx = \int_0^L f\varphi_i dx \text{ for } i = 1, 2, \ldots, n. \qquad (1.22)$$

Conversely, multiplying both sides of the above equation by an arbitrary real number v_i and adding up from $i = 1$ up to $i = n$, we conclude quite easily that equation (1.22) implies (1.21). As a consequence, both problems can be viewed as equivalent, if we identify u_h with its coefficients in equation (1.20), even if rigorously they are different entities, since u_h is a function defined at every point of $[0, L]$, and the u_js are only the values of u_h at a finite set of points x_j.

Here again, equation (1.22) is nothing but a SLAE with an $n \times n$ matrix $A_h = \{a_{i,j}\}$ and right-side vector $\vec{b}_h = \{b_i\}$, whose unknown vector $\vec{u}_h$ is $[u_1, u_2, \ldots, u_n]^T$. More concretely, we have to solve

The SLAE for the $\mathcal{P}_1$ FE scheme

$$\begin{aligned} &A_h \vec{u}_h = \vec{b}_h \\ &\text{where} \\ &a_{i,j} = \int_0^L [p\varphi'_j\varphi'_i + q\varphi_j\varphi_i]dx \\ &\text{and} \\ &b_i = \int_0^L f\varphi_i dx. \end{aligned} \qquad (1.23)$$

This means that equation (1.22) can be recast in the same form as equation (1.13). Notice that matrix A_h is necessarily symmetric.

Next, we examine the well-posedness of equation (1.22). As we know, this is ensured provided $\vec{b}_h = \vec{0}$ implies $\vec{u}_h = \vec{0}$. Since systems (1.22) and (1.21) are equivalent, if $\vec{b}_h = \vec{0}$ we have

$$\int_0^L [pu'_h v' + qu_h v]dx = 0 \ \forall v \in V_h.$$

Hence, taking $v = u_h$ we conclude that

$$\int_0^L [p(u'_h)^2 + q(u_h)^2]dx = 0.$$

On the other hand, according to our assumptions, $p(x) \geq \alpha > 0$ and $q(x) \geq 0$ for every $x \in [0, L]$. Therefore, the left side of the above expression is strictly positive, unless u'_h vanishes identically, which is thus the only possibility for the above relation to hold. Hence, u_h must be constant in $[0, L]$, and since by construction $u_h(0) = 0$, we must have $u_h \equiv 0$. This implies that $\vec{u}_h = \vec{0}$.

Similarly to the case of the FDM with an equally spaced grid, let us exhibit the matrix A_h in the particular case where p and q are constant. Referring to Figure 1.3 and to equation (1.19), first of all we note that the product of two functions φ_i and φ_j, or of their derivatives, is identically zero whenever $|i - j| > 1$. On the other hand, from equation (1.19) the product of φ_i and φ_{i-1} (or of its derivatives) vanishes identically in every element but T_i. It immediately follows that, like in the case of the FDM for a uniform grid, A_h besides being symmetric is a tridiagonal matrix.

Let us first compute $a_{i+1,i}$ for $1 \leq i < n$. Using equation (1.19), we trivially obtain

$$a_{i+1,i} = \int_{x_i}^{x_{i+1}} h_{i+1}^{-2}[-p + q(x - x_i)(x_{i+1} - x)]dx$$

Integrating the quadratic function above in (x_i, x_{i+1}), we readily derive

$$a_{i+1,i} = -ph_{i+1}^{-1} + qh_{i+1}/6.$$

Similarly, we can compute $a_{i,i}$ for $1 \leq i < n$ by

$$a_{i,i} = \int_{x_{i-1}}^{x_i} h_i^{-2}[p + q(x - x_{i-1})^2]dx + \int_{x_i}^{x_{i+1}} h_{i+1}^{-2}[p + q(x_{i+1} - x)^2]dx$$

which yields

$$a_{i,i} = p(h_i^{-1} + h_{i+1}^{-1}) + q(h_i + h_{i+1})/3.$$

Without any difficulty, we infer from the above calculation of $a_{i,i}$ that

$$a_{n,n} = ph_n^{-1} + qh_n/3.$$

An illustration of this matrix divided by h is supplied in Array 1.2 for $n = 6$ and a uniform mesh with size h.

$$\frac{A_h}{h} = \begin{bmatrix} \frac{2p}{h^2} + \frac{2q}{3} & -\frac{p}{h^2} + \frac{q}{6} & 0 & 0 & 0 & 0 \\ -\frac{p}{h^2} + \frac{q}{6} & \frac{2p}{h^2} + \frac{2q}{3} & -\frac{p}{h^2} + \frac{q}{6} & 0 & 0 & 0 \\ 0 & -\frac{p}{h^2} + \frac{q}{6} & \frac{2p}{h^2} + \frac{2q}{3} & -\frac{p}{h^2} + \frac{q}{6} & 0 & 0 \\ 0 & 0 & -\frac{p}{h^2} + \frac{q}{6} & \frac{2p}{h^2} + \frac{2q}{3} & -\frac{p}{h^2} + \frac{q}{6} & 0 \\ 0 & 0 & 0 & -\frac{p}{h^2} + \frac{q}{6} & \frac{2p}{h^2} + \frac{2q}{3} & -\frac{p}{h^2} + \frac{q}{6} \\ 0 & 0 & 0 & 0 & -\frac{p}{h^2} + \frac{q}{6} & \frac{p}{h^2} + \frac{q}{3}. \end{bmatrix}$$

Array 1.2

On the other hand, in general the calculation of $\vec{b}_h$ is not as simple as in the case of the FDM, unless we use numerical quadrature. This is possible if f is continuous in $[0, L]$. In this case, since b_i is the sum of the integrals of $f\varphi_i$ over elements T_i and T_{i+1}, without any essential loss in accuracy, we may apply the trapezoidal rule (see e.g. [105]) in each of them, thereby obtaining an approximation $\tilde{b}_i$ of b_i given by

$$\tilde{b}_i = [f(x_{i-1})\varphi_i(x_{i-1}) + f(x_i)\varphi_i(x_i)]h_i/2 + [f(x_{i+1})\varphi_i(x_{i+1}) + f(x_i)\varphi_i(x_i)]h_{i+1}/2.$$

Since $\varphi_i(x_j) = 0$ if $i \neq j$, we derive

$$\tilde{b}_i = f(x_i)(h_i + h_{i+1})/2 \text{ for } i < n$$

and, quite easily,

$$\tilde{b}_n = f(x_n)h_n/2.$$

Notice that this is the same right side as in the case of the FD scheme (1.15) if we multiply the ith equation by $(h_i + h_{i+1})/2$ for $i < n$ and the nth equation by h_n. In particular, if the mesh is uniform, we obtain the same right side as equation (1.14) multiplied by h.

Of course, we can also use the trapezoidal rule to integrate the terms $\int_{x_{i-1}}^{x_i} qg(x)dx$ in the expression of $a_{i,i}$ or $a_{i,i-1}$, where $g(x) = [(x - x_{i-1})/h_i]^2$ or $g(x) = (x - x_{i-1})(x_i - x)/h_i^2$. Then, quite easily, we establish that the resulting approximate values $\tilde{a}_{i,j}$ of the nonzero entries of matrix A_h correspond to the same entries given by equation (1.16) multiplied by $(h_i + h_{i+1})/2$ for $i < n$ and otherwise by $h_n/2$. In particular in the case of a uniform mesh, the matrix A_h/h is of the form displayed in Array 1.1 (for $n = 6$). Otherwise stated, at least if both p and q are constant, the use of the trapezoidal rule to integrate all the functions involved in the calculation of matrix A_h and right-side vector $\vec{b}_h$ results in a SLAE identical to the one of the FDM for a non-uniform grid, whose points coincide with the nodes of $\mathcal{T}_h$. Actually, even in the case where p and q are not constant, but arbitrary continuous functions, the use of the same trapezoidal rule to compute the **FE matrix** A_h induces an identical **FD analog** of (P_1) based on the same non-uniform lattice. The reader could derive such an **FD scheme** as Exercise 1.2, and verify that it yields relations (1.16) whenever p and q are constant.

The reader has certainly observed that if the trapezoidal rule is used to approximate the integral $\int_0^L q\varphi_i\varphi_{i\pm1}dx$, the corresponding terms of the coefficient $a_{i,i\pm1}$ will be zero. The resulting diagonalisation procedure in connection with the P_1 FEM is known as **mass lumping**. In this book this technique will serve several times as a tool to obtain interesting properties of FE-based numerical schemes.

Another similar situation of practical interest is the one where p is only piecewise continuous in $[0, L]$, q being piecewise continuous or not. In this case, it may be practical to construct the FE mesh (resp. FD grid) in such a way that the set of discontinuity points $\{y_1, y_2, \ldots, y_M\}$ forms a subset of $\{x_1, x_2, \ldots, x_{n-1}\}$. In other words, we assume that the discontinuity points of p coincide with mesh nodes (resp. grid points). Then, here again, at least for a continuous q, the FE scheme resulting from the use of the trapezoidal rule as an approximate quadrature method in each element corresponds to an efficient Three-point FD scheme with a non-uniform grid, to solve equation (P_1) for a discontinuous p.

Remark 1.2 *Such an assertion is based on the fact that here pu' is a differentiable function everywhere, even if neither p nor u' is differentiable at the discontinuity points of p. Then, it is reasonable to apply FDs to approximate directly $[pu']'$ and not u'' like we did in Section 1.2. The reader will be able to figure this out when deriving the corresponding FE scheme.* ∎

Finally, if q too is only piecewise continuous, the trapezoidal rule may be applied to integrals in the intersection of the elements with intervals in which q is continuous. An equivalent and natural FD scheme can also be thus derived, in the case where the discontinuities of q coincide with mesh nodes or grid points, as the reader may quite easily check taking the case of a single discontinuity.

1.4 The Basic FVM

The FVM applies to formulation (P_2) of the differential equation for a particular type of subsets ω called **control volumes** (**CVs**). Although there is only one standard formulation of the basic FVM, for any set of disjoint CVs, in the literature two approaches are commonly distinguished: **Vertex-centred FVM** and **Cell-centred FVM**. In the former, the CVs are intervals specified hereafter, containing the nodes of the mesh $\mathcal{T}_h$; it yields discrete problems quite close to those considered in Sections 1.1 and 1.2 for the FDM and the FEM. In the latter, the CVs are the elements of $\mathcal{T}_h$ themselves. Of course, since they are based on the same principles, both methods are very similar. However, even in the case of the one-dimensional problem (P_1), the Cell-centred FVM does not resemble the Three-point FDM or the P_1 FEM.

1.4.1 The Vertex-centred FVM

Starting from the FE mesh $\mathcal{T}_h$ (or the associated non-uniform grid), in this version of the FVM the CVs are the intervals $V_j := (x_{j-1/2}, x_{j+1/2})$ where $x_{j-1/2} := (x_{j-1} + x_j)/2$ for $j = 1, 2, \ldots, n-1$, which we complete with $V_n := (x_{n-1/2}, x_n)$ and $V_0 := (x_0, x_{1/2})$. The

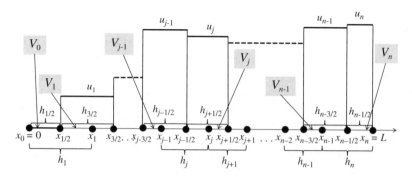

Figure 1.5 CVs V_j for the Vertex-centred FVM and approximations u_j of u.

measure of each CV for $1 \le j \le n-1$ denoted by $h_{j-1/2}$ is given by $(h_{j-1}+h_j)/2$ for $j < n$ and by $h_1/2$ and $h_n/2$ for V_0 and V_n, respectively. Now, we assume that in each V_j, u is approximated by a constant function whose value is denoted here again by u_j. In order to take into account the essential boundary condition $u(0) = 0$, we set $u_0 = 0$. In Figure 1.5, illustrations of these definitions and notations are supplied.

Now we apply the conservation principle in (P$_2$) to each V_j. For the sake of simplicity, we first assume that p is constant and approximate u' at the left end of V_j by $(u_j - u_{j-1})/h_j$ for $j = 1, 2, \ldots, n$ and at the right end of V_j by $(u_{j+1} - u_j)/h_{j+1}$ if $1 \le j < n$. The latter expression is replaced by zero if $j = n$. This gives rise to the following:

Discrete balance equation in V_j for a constant p

> For $j < n$:
>
> $$p\left[\frac{(u_j - u_{j-1})}{h_j} - \frac{u_{j+1} - u_j}{h_{j+1}}\right] + \int_{x_{j-1/2}}^{x_{j+1/2}} [qu_j - f]dx = 0;$$
>
> For $j = n$:
>
> $$p\frac{u_j - u_{j-1}}{h_j} + \int_{x_{j-1/2}}^{x_j} [qu_j - f]dx = 0;$$
>
> with $u_0 = 0$.

(1.24)

This is again a SLAE with n equations and unknown vector $\vec{u}_h = [u_1, u_2, \ldots, u_n]^T$. Assuming that both f and q are continuous at the mesh nodes, let us apply a modified Gaussian mid-point rule (see e.g. [29] or [208]) to approximate their respective integrals on the left and right side of the first equation of (1.24). More precisely, the mid-point of the interval $(x_{j-1/2}, x_{j+1/2})$ for $j < n$ (resp. $(x_{n-1/2}, x_n)$ for $j = n$) is shifted to x_j (resp. x_n) to obtain $q(x_j)h_{j-1/2}$ and $f(x_j)h_{j-1/2}$ for $j < n$ (resp. $q(x_n)h_n/2$ and $f(x_n)h_n/2$ for $j = n$). In doing so, we come up with a scheme derived from equation (1.24) that can be written in the form of a SLAE with an $n \times n$ matrix $A_h = \{a_{j,i}\}$ and right-side vector $\vec{b}_h = \{b_j\}$ derived from equation (1.24). A_h is a symmetric matrix whose entries vanish except $a_{j,j}$ and $a_{j,j-1} = a_{j-1,j}$, for $j = 1, 2, \ldots, n$. These nonzero entries together with the components b_j of $\vec{b}_h$ are the same as for the FEM in the same case, provided the trapezoidal rule is employed to approximate the integrals in each element of the mesh, assuming of course that both q and f are continuous (cf. Section 1.2). This assertion can be verified as Exercise 1.3.

On the other hand, if p is an arbitrary continuous function, in the FV analog of (P_2) we approximate $[pu'](x_{j-1/2})$ by $p(x_{j-1/2})(u_j - u_{j-1})/h_j$. Then, instead of equation (1.24), we have the

Discrete balance equation in V_j for a continuous p

For $1 \leq j < n$:

$$\frac{p(x_{j-1/2})(u_j - u_{j-1})}{h_j} - \frac{p(x_{j+1/2})(u_{j+1} - u_j)}{h_{j+1}} + \int_{x_{j-1/2}}^{x_{j+1/2}} [qu_j - f]dx = 0;$$

For $j = n$: $\qquad\qquad\qquad\qquad\qquad\qquad\qquad\qquad\qquad$ (1.25)

$$\frac{p(x_{j-1/2})(u_j - u_{j-1})}{h_n} + \int_{x_{j-1/2}}^{x_j} [qu_j - f]dx = 0$$

with $u_0 = 0$.

Referring to Figure 1.5, and as seen in Section 4.4, in the FVM approach it is generally considered that unknown values of u are approximated at one point per CV, which is called the CV's **representative point**. The corresponding approximations u_j are extended to the whole CV, thereby giving rise to the piecewise constant approximating function u_h. In the case of the Vertex-centred method, the representative point of CV V_j is x_j, for $j = 1, 2, \ldots, n$.

Now, if the same quadrature rules as in the case of a constant p are employed to compute the integrals involving q and f, that is, the trapezoidal rule for the FEM and the appropriate one-point rules for the FVM, then equation (1.25) corresponds to the $\mathcal{P}_1$ FE scheme, provided the mid-point rule is used to approximate the integrals $\int_{x_{j-1}}^{x_j} p[\varphi_j']^2 dx$ or $\int_{x_{j-1}}^{x_j} p[\varphi_j' \varphi_{j-1}']dx$ for $1 \leq j \leq n$. The reader may easily check that this assertion holds true as Exercise 1.4. Moreover, the resulting modification of equation (1.25) suggests a possible way to discretise equation (P_1) by the FDM with a non-uniform grid for a non-constant p. Summarising, in this case, all three schemes reduce to the following:

Unified FD–FE–FV scheme to solve (P_1) for continuous p and q

For $1 \leq j < n$:

$$\frac{p(x_{j-1/2})(u_j - u_{j-1})}{h_j h_{j+1/2}} - \frac{p(x_{j+1/2})(u_{j+1} - u_j)}{h_{j+1} h_{j+1/2}} + u_j q(x_j) = f(x_j);$$

For $j = n$: $\qquad\qquad\qquad\qquad\qquad\qquad\qquad\qquad\qquad$ (1.26)

$$\frac{p(x_{j-1/2})(u_j - u_{j-1})}{h_j^2} + u_j \frac{q(x_j)}{2} = \frac{f(x_j)}{2};$$

with $u_0 = 0$.

As a conclusion, at least in the one-dimensional case, the Vertex-centred FVM is practically equivalent to the $\mathcal{P}_1$ FE scheme with the same mesh, or to the Three-point FD scheme with a non-uniform grid, whose points are the mesh nodes.

Incidentally, such an equivalence also applies to the case of a discontinuous p, as long as $\mathcal{T}_h$ is constructed in such a way that the set $\{y_1, y_2, \ldots, y_M\}$ of discontinuity points of p is a subset of the set of mesh nodes (i.e. of grid points). In doing so, p is necessarily uniquely defined at the mid-points of each element, and we may apply the balance equation (1.25). Then, here again, if we approximate the integrals of $f\varphi_i$ and $q\varphi_j\varphi_i$ by the trapezoidal rule;

approximate $\int_{x_{j-1/2}}^{x_{j+1/2}} q dx$, $\int_{x_{j-1/2}}^{x_{j+1/2}} f dx$, $\int_{x_{n-1/2}}^{x_n} q dx$ and $\int_{x_{n-1/2}}^{x_n} f dx$ by the 'shifted' mid-point quadrature rule specified above; and use the mid-point quadrature rule to approximate the integral of $p\varphi'_j\varphi'_i$ in each element, we realise that in this case, too, the resulting FE scheme and the Vertex-centred FV scheme derived from equation (1.25) coincide. It is interesting to note again that this scheme suggests an appropriate way to discretise (P$_1$) by the FDM in the case where p has discontinuities at grid points.

1.4.2　The Cell-centred FVM

We recall that, in this case, the CVs are the intervals T_j of the mesh $\mathcal{T}_h$ (also called **cells**). Then, in each cell, the function u is approximated by a constant function equal to $u_{j-1/2}$. In Figure 1.6, these definitions and notations are illustrated.

By definition, the representative point of CV T_j is $x_{j-1/2}$ for $j = 1, \ldots, n$. We also set $x_{-1/2} = x_0$ and $x_{n+1/2} = x_n$ because, for convenience, we will consider an extended set of representative points by adding $x_{-1/2}$ and $x_{n+1/2}$, respectively, for fictitious CVs $T_0 := \{0\}$ and $T_{n+1} : [L, L + h_n]$ lying outside $(0, L)$. The value of u_h equals zero in the former and coincides with $u_{n-1/2}$ in the latter, as will be discussed here. Notice that, in the scheme, $\frac{u_{j+1/2}-u_{j-1/2}}{h_{j+1/2}}$ stands for the first-order derivatives of u at the cell end-point x_j, for $j = 1, 2, \ldots, n-1$; at $x = x_0$, this derivative is approximated by $\frac{u_{1/2}-u_0}{h_{1/2}}$, with $h_{1/2} := h_1/2$ and $u_0 = 0$, which is a way to take into account the boundary condition $u(x_0) = 0$. Finally, we have to set to zero a suitable approximation of $u'(L)$. Like in the FDM, we could add an approximation $u_{n+1/2}$ of u at the fictitious representative point $L + h_n/2$ of T_{n+1} and then enforce $\frac{u_{n+1/2}-u_{n-1/2}}{h_n} = 0$. However, here this is needless for it suffices to take a zero flux on the right end of CV T_n, and this boundary condition will be automatically incorporated into the discrete analog of (P$_2$) (cf. (1.27)). Thus, assuming that p is continuous, this gives rise to the following:

Discrete balance equation in CV T_j for a non-constant p

For $j = 1$:
$$p(x_{j-1})\frac{u_{j-1/2}-u_{j-1}}{h_{j-1/2}} - p(x_j)\frac{u_{j+1/2}-u_{j-1/2}}{h_{j+1/2}} + \int_{x_{j-1}}^{x_j} [qu_{j-1/2} - f] dx = 0;$$

For $1 < j < n$:
$$p(x_{j-1})\frac{u_{j-1/2}-u_{j-3/2}}{h_{j-1/2}} - p(x_j)\frac{u_{j+1/2}-u_{j-1/2}}{h_{j+1/2}} + \int_{x_{j-1}}^{x_j} [qu_{j-1/2} - f] dx = 0; \qquad (1.27)$$

For $j = n$:
$$p(x_{j-1})\frac{u_{j-1/2}-u_{j-3/2}}{h_{j-1/2}} + \int_{x_{j-1}}^{x_j} [qu_{j-1/2} - f] dx = 0;$$

with $u_0 = 0$.

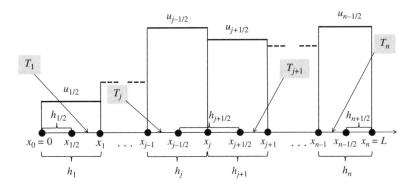

Figure 1.6 CVs T_j for the Cell-centred FVM and approximations $u_{j-1/2}$ of u.

It is a common practice to use the mid-point quadrature rule in order to approximate the integrals of both q and f, if these functions are continuous in each cell. On the other hand, in situations of practical interest, p is often discontinuous at M points of a set $\{y_1, y_2, \ldots, y_M\}$. However, in general, physics of the phenomenon being modelled requires that the **flux** pu' across those discontinuity points of p be continuous. One of the features of the FVM is to mimic conservation properties satisfied by the exact solution of the model equation. In the case under study, this can be achieved if we proceed as follows. Assume that the mesh is constructed in such a way that the intersection of the set of discontinuity points of p with the set $\{x_1, x_2, \ldots, x_{n-1}\}$ is not empty. Let us denote by p_j^- and p_j^+ the values of p at mesh node x_j from the left and from the right, respectively, for $j = 1, 2, \ldots, n-1$. We set $\tilde{p}_j := (p_j^- + p_j^+)/2$ completed with $\tilde{p}_0 = p(x_0)$ and $\tilde{p}_n = p(x_n)$. Of course, if p is continuous at x_j, then $\tilde{p}_j = p(x_j)$.

Then, we may set up the following Cell-centred FV approximation of (P$_1$), which obviously extends to functions p everywhere continuous in $[0, L]$, namely:

Cell-centred FV scheme to solve (P$_1$) accommodating discontinuous p

For $j = 1$:
$$\tilde{p}_{j-1}\frac{u_{j-1/2}-u_{j-1}}{h_j h_{j-1/2}} - \tilde{p}_j\frac{u_{j+1/2}-u_{j-1/2}}{h_j h_{j+1/2}} + q(x_{j-1/2})u_{j-1/2} = f(x_{j-1/2});$$

For $1 < j < n$:
$$\tilde{p}_{j-1}\frac{u_{j-1/2}-u_{j-3/2}}{h_j h_{j-1/2}} - \tilde{p}_j\frac{u_{j+1/2}-u_{j-1/2}}{h_j h_{j+1/2}} + q(x_{j-1/2})u_{j-1/2} = f(x_{j-1/2}); \qquad (1.28)$$

For $j = n$:
$$\tilde{p}_{j-1}\frac{u_{j-1/2}-u_{j-3/2}}{h_j h_{j-1/2}} + q(x_{j-1/2})u_{j-1/2} = f(x_{j-1/2});$$

with $u_0 = 0$.

Remark 1.3 *In reference [68], a more elaborated treatment of discontinuous p is proposed, which supposedly has better approximation properties.* ∎

The reader may exhibit the entries of the $n \times n$ matrix A_h and of the n-component right-side vector $\vec{b}_h$ of the SLAE with n equations corresponding to equation (1.28), whose vector of n unknowns is $[u_{1/2}, u_{3/2}, u_{5/2}, \ldots, u_{n-1/2}]^T$ (Exercise 1.5). The fact that this system has a unique solution can be established by means of a simple variant of the argument employed in the case of the FDM for an equally spaced grid, as long as the CVs have a fixed measure h and p is constant. Indeed, it suffices to let $u_{j-1/2}$ play the role of u_i in that case. If the cells have a variable measure and p is not constant and is eventually discontinuous, at the price of some additional though straightforward calculations, the existence and uniqueness of a solution can also be shown to hold (cf. Exercise 1.6).

Remark 1.4 *The conservation of the **numerical fluxes** from both sides of the cell end-points enforced in scheme (1.28) is a key property to establish its reliability. This means that it is capable of yielding meaningful approximations of the exact solution. In Chapters 5 and 7, we will see this in more detail for related FV schemes.* ∎

1.4.3   Connections to the Other Methods

To conclude this section, we observe that the FVMs, we have studied so far can be viewed as an FEM of a particular type, in the sense that both are based on a variational formulation. The only real difference is that such a variational formulation applies only to the approximate problem (i.e. the discrete problem) and has no real counterpart in the framework of the exact problem (i.e. the continuous problem). Let us see how this works.

Starting from the Cell-centred FVM, we consider a space U_h associated with the mesh $\mathcal{T}_h$, consisting of functions whose value is a constant w_j in each element T_j. Then, we introduce the concept of **discrete derivative** of a function $w \in U_h$ in each interval $(x_{j-3/2}, x_{j-1/2})$. This quantity is defined by $w'^h := (w_j - w_{j-1})/h_{j-1/2}$, for $j = 2, \ldots, n$ completed with $w'^h(x) := w_1/h_{1/2}$ for $x \in (0, x_{1/2})$ and $w'^h = 0$ for $x \in (x_{n-1/2}, L)$. In doing so, we endeavour to search for a function $u_h \in U_h$ such that

$$\int_0^L [pu_h'^h w'^h + qu_h w]dx = \int_0^L fw\,dx \ \forall w \in U_h. \tag{1.29}$$

Let now ψ_i be a *shape function* of U_h, namely, the function whose value is one in T_i and zero elsewhere. Clearly enough, every function in $w \in U_h$ can be expanded as $w(x) = \sum_{i=1}^n w_i\psi_i(x) \ \forall x \in (0, L)$. Hence, using the description of the FEM as a guide, we take in equation (1.29) w successively equal to ψ_i. Clearly, for $i = 2, \ldots, n-1$, $\psi_i'^h(x) = 1/h_{i-1/2}$ if $x \in (x_{i-3/2}, x_{i-1/2})$, $\psi_i'^h(x) = -1/h_{i+1/2}$ if $x \in (x_{i-1/2}, x_{i+1/2})$ and $\psi_i'^h(x) = 0$ elsewhere. Moreover, $\psi_1'^h(x) = 1/h_{1/2}$ if $x \in (0, x_{1/2})$, $\psi_1'^h(x) = -1/h_{3/2}$ if $x \in (x_{1/2}, x_{3/2})$ and $\psi_1'^h(x) = 0$ otherwise. Finally, $\psi_n'^h(x) = 1/h_{n-1/2}$ if $x \in (x_{n-3/2}, x_{n-1/2})$ and $\psi_n'^h(x) = 0$ otherwise.

Therefore, from equation (1.29), we derive

$$
\begin{cases}
\displaystyle\int_0^{x_{1/2}} p\, u_h'^h/h_{1/2}dx - \int_{x_{1/2}}^{x_{3/2}} p\, u_h'^h/h_{3/2}dx + \int_0^{x_1} q\, u_h dx = \int_0^{x_1} f dx; \\[2ex]
\text{For } i = 2, \ldots, n-1: \\[1ex]
\displaystyle\int_{x_{i-3/2}}^{x_{i-1/2}} p\, u_h'^h/h_{i-1/2}dx - \int_{x_{i-1/2}}^{x_{i+1/2}} p\, u_h'^h/h_{i+1/2}dx + \int_{x_{i-1}}^{x_i} q\, u_h dx = \int_{x_{i-1}}^{x_i} f dx \\[2ex]
\displaystyle\int_{x_{n-3/2}}^{x_{n-1/2}} p\, u_h'^h/h_{n-1/2}dx + \int_{x_{n-1}}^{x_n} q\, u_h dx = \int_{x_{n-1}}^{x_n} f dx.
\end{cases}
\tag{1.30}
$$

Now, noting that $u_h = \sum_{j=1}^n u_j \psi_j$ and recalling the definition of $u_h'^h$, from equation (1.30), we obtain

$$
\begin{cases}
\displaystyle\int_0^{x_{1/2}} \frac{pu_1}{h_{1/2}^2}dx - \int_{x_{1/2}}^{x_{3/2}} \frac{p(u_2 - u_1)}{h_{3/2}^2}dx + \int_0^{x_1} q\, u_h dx = \int_0^{x_1} f dx; \\[2ex]
\text{For } 1 < i < n; \\[1ex]
\displaystyle\int_{x_{i-3/2}}^{x_{i-1/2}} \frac{p(u_i - u_{i-1})}{h_{i-1/2}^2}dx - \int_{x_{i-1/2}}^{x_{i+1/2}} \frac{p(u_{i+1} - u_i)}{h_{i+1/2}^2}dx + \int_{x_{i-1}}^{x_i} q\, u_h dx = \int_{x_{i-1}}^{x_i} f dx \\[2ex]
\displaystyle\int_{x_{n-3/2}}^{x_{n-1/2}} \frac{p(u_n - u_{n-1})}{h_{n-1/2}^2}dx + \int_{x_{n-1}}^{x_n} q\, u_h dx = \int_{x_{n-1}}^{x_n} f dx.
\end{cases}
\tag{1.31}
$$

Further assuming that p is continuous in $[0, x_{1/2})$ and in $(x_{i-3/2}, x_{i-1/2})$ for $i = 2, \ldots, n$, if we use a one-point quadrature formula to integrate p in these intervals, and recalling that $u_h = u_i$ in T_i, we obtain

$$
\begin{cases}
\displaystyle\frac{p(0)u_1}{h_{1/2}} - \frac{p(x_1)(u_2 - u_1)}{h_{3/2}} + u_1 \int_0^{x_1} q\, dx = \int_0^{x_1} f dx; \\[2ex]
\text{For } 1 < i < n; \\[1ex]
\displaystyle\frac{p(x_{i-1})(u_i - u_{i-1})}{h_{i-1/2}} - \frac{p(x_i)(u_{i+1} - u_i)}{h_{i+1/2}} + u_i \int_{x_{i-1}}^{x_i} q dx = \int_{x_{i-1}}^{x_i} f dx \\[2ex]
\displaystyle\frac{p(x_{n-1})(u_n - u_{n-1})}{h_{n-1/2}} + u_n \int_{x_{n-1}}^{x_n} q dx = \int_{x_{n-1}}^{x_n} f dx.
\end{cases}
\tag{1.32}
$$

Now, if we renumber the u_is in such a manner that u_i becomes $u_{i-1/2}$ and further apply one-point numerical quadrature to compute the integrals of q and f, we find out that equation (1.32) is nothing but the Cell-centred FV scheme (1.28). As a conclusion, the Cell-centred FVM is derived from a suitable weak formulation of equation (P_1) at a discrete level, using the simplest possible type of representation of the solution, namely, piecewise constant functions.

As for the Vertex-centred FVM, a similar conclusion applies, but the argument is a little more tricky. In this case, we may consider a piecewise linear approximate solution $u_h := \sum_{j=1}^n u_j \varphi_j$, to be determined by solving a discrete variational problem similar to that of equation (1.22), in which φ_i is replaced by $\bar\varphi_i$, where $\bar\varphi_i$ is a piecewise constant function whose value is one in the CV $(x_{i-1/2}, x_{i+1/2})$ and zero elsewhere for $i = 1, 2, \ldots, n-1$; and $\bar{\varphi}_n$ equals one in $(x_{n-1/2}, L)$ and zero elsewhere. For such functions, we also define a discrete derivative analogous to the one of functions in U_h by $\bar{\varphi}_i'^h(x) = 1/h_i$ if $x \in (x_{i-1}, x_i)$ and $\bar{\varphi}_i'^h = -1/h_{i+1}$ if $x \in (x_i, x_{i+1})$; $\bar{\varphi}_i'^h(x) = 0$ elsewhere in $(0, L)$ for $i = 1, 2, \ldots, n-1$; and $\bar{\varphi}_n'^h(x) = 1/h_n$ for $x \in (x_{n-1/2}, L)$ and $\bar{\varphi}_n'^h(x) = 0$ elsewhere. After some rather fastidious calculations, which we voluntarily skip for the sake of conciseness, we conclude that the discrete variational problem analogous to equation (1.22) is

$$
\begin{cases}
\text{Find } u_j \text{ for } j = 1, 2, \ldots, n \text{ such that} \\
\displaystyle\sum_{j=1}^{n} u_j \int_0^L [p\varphi_j' \bar{\varphi}_i'^h + q\varphi_j \bar{\varphi}_i] dx = \int_0^L f\bar{\varphi}_i dx \text{ for } i = 1, 2, \ldots, n.
\end{cases}
\tag{1.33}
$$

Equation (1.33) is nothing but the Vertex-centred FV scheme (1.25), provided suitable one-point quadrature rules are employed to evaluate the integrals in equation (1.33). For this reason, the resulting scheme is also called the finite volume–finite element scheme (see e.g. [68]).

1.5   Handling Nonzero Boundary Conditions

The model boundary value problem we have studied in this chapter so far has only **homogeneous boundary conditions** (i.e. zero boundary conditions). Although such a situation effectively occurs in the case of the elongational deformations of a bar, in many practical applications the boundary conditions are **inhomogeneous**. In the case of equation (P_1), this corresponds to

$$
\begin{cases}
-[pu']' + qu = f \text{ in } (0, L); \\
u(0) = a; \\
p(L)u'(L) = b,
\end{cases}
\tag{1.34}
$$

where a and b are given real numbers. In physical terms, this means that a displacement equal to a is prescribed at the bar's left end, and that a force of intensity equal to b acts on its right end.

Let us briefly examine the modifications that have to be introduced in the numerical schemes studied in this chapter, in order to take into account such data. We expect that, as a by-product, this will give the necessary insight on how to deal with other types of boundary conditions, homogeneous or not.

To begin with, we consider the case where only the **Dirichlet boundary condition** is inhomogeneous, namely, $u(0) = a$. As far as the FDM and both types of FVM are concerned, this is simply a matter of setting $u_0 = a$ instead of $u_0 = 0$ in the corresponding schemes. Whatever the case, this implies that only the right side of the first equation of the corresponding scheme will have to be adjusted by the addition of a coefficient multiplied by a. The reader will face no difficulty to identify such a coefficient in each case.

As for the FEM, the necessary modification is a little more subtle, although in the end the same kind of conclusion applies: setting again $u_0 = a$ and recalling the function φ_0, we first rewrite the approximate solution as

$$
u_h = \sum_{j=0}^{n} u_j \varphi_j.
$$

Now plugging this expression of u_h into equation (1.21) and taking $v = \varphi_i$, we note that for $i = 1$ only, a new term $b_0 := u_0 \int_0^{x_1} [p\varphi_0'\varphi_1' + q\varphi_0\varphi_1] dx$ will appear on the left side of the resulting equation. Hence, such an inhomogeneous boundary condition is taken into account by simply adding to the original right-side b_1 the value $-b_0$. Using quadrature formulae as prescribed in Section 1.3, we conclude that the correction $-b_0$ plays the same role as in the FDM and the FVM.

Next, we switch to the case of the inhomogeneous **Neumann boundary condition** $[pu'](L) = b$. In the case of the FDM, this is only a matter of redefining the fictitious approximate value u_{n+1} by means of the natural relation:

$$p(L)\frac{u_{n+1} - u_{n-1}}{2h_n} = b.$$

Then, the value $u_{n+1} = 2bh_n/p(L) + u_{n-1}$ is plugged into the nth equation of scheme (1.12) or (1.15). As we should point out, in the case of uniform grids, at least for academic purposes, it is advisable to improve the accuracy of this approximation. For instance, one might adapt to an inhomogeneous Neumann boundary condition at $x = L$ the trick to be used in Chapter 2, in the homogeneous case.

In the case of the FVM, we proceed similarly by modifying only the last equation of the Vertex-centred or the Cell-centred scheme (1.25) or (1.28) (cf. Exercise 1.8). The natural thing to do here is to subtract (resp. add) to the left (resp. right) side of this equation the quantity b, in order to take into account the term stemming from the integration of $-[pu']$ in the rightmost CV. The reader might easily determine in Exercise 1.7 the final form of the thus-modified right-side vector $\vec{b}_h$ in the case of each method (FDM or FVM).

Finally, we consider the FEM. First of all, if we want the inhomogeneous boundary condition $p(L)u'(L) = b$ to be implicitly satisfied in the variational formulation (P$_3$), we have to add to the right side the term $bv(L)$. Indeed, when we integrate by parts the first term of the integral on the left side of this equation in order to recover (P$_1$), we first obtain $-[pu']' + qu = f$ by choosing v such that $v(L) = 0$, like in the homogeneous case. Then, using this equation and taking an arbitrary v in V such that $v(L) \neq 0$, we obtain $[pu'v](L) = bv(L)$. This obviously implies that $[pu'](L) = b$. Such a modification is carried to the discrete analog (1.21) of (P$_3$), which leads to a correction in the last equation of (1.22) only. This is because on the one hand $b\varphi_i(L) = 0$ for every i less than n, and on the other hand $b\varphi_n(L) = b$. In short, in order to take into account the inhomogeneous Neumann boundary condition $[pu'](L) = b$, here again it is necessary to correct only the last equation of (1.22) by adding b to former b_n.

Several variants of (P$_1$) aimed at completing the above presentation are considered in Section 6.1. More precisely, boundary conditions other than Dirichlet at $x = 0$ and Neumann at $x = L$ are prescribed, among which lie Dirichlet or Neumann boundary conditions at both ends. Robin boundary conditions, that is, conditions of the type $u'(0) + cu(0) = a$ and/or $u'(L) + du(L) = b$ for given real numbers a, b, c, d, are also treated.

1.6 Effective Resolution

So far, we have described numerical solution schemes without caring about their practical implementation. We know that all of them lead to a SLAE, whose solution for particular field data generates the numerical values the schemes are designed to provide. To close this chapter, we consider some relevant aspects of this process in a twofold manner. First of all, we address some generalities on the form of the SLAEs to be solved, thereby drawing some conclusions on right choices of solution methods. Next, we give some numerical examples by solving problems of the form (P_1) with the Cell-centred FVM.

1.6.1   Solving SLAEs for one-dimensional problems

As we saw in this chapter, all the three discretisation methods under study reduce to the solution of a SLAE $A_h \vec{u}_h = \vec{b}_h$ with n unknowns and n equations. Since we could assert beforehand that all these systems have a unique solution, in principle solving them is no problem. However, in practical terms, it is mandatory to take into account the particular form of the matrix system; otherwise, the resolution cost could become excessive if n is very large. Indeed, A_h is a **sparse matrix**, akin to the case of most numerical methods for solving boundary value differential equations. Actually, here matrix A_h is a symmetric tridiagonal matrix since $a_{i,j} = 0$ whenever $|i - j| > 1$. In any event, a SLAE whose matrix is sparse should preferably be solved by methods that preserve its sparsity structure. This means that, in the underlying matrix manipulations, one should create the least possible, amount of new nonzero entries. Possibly, one should restrict the positions of nonzero entries to the original ones all the way. In this manner, storage requirements are reduced to a minimum, and computational costs do not increase unnecessarily.

In the class of iterative methods for solving SLAEs, in most cases it is possible to keep the matrix as it was at the beginning of the solution process. However, these methods are subject to the convergence (or not) of the successive approximations to the solution vector $\vec{u}_h$. In the framework of the FD, FE and FV schemes for the problem we considered, classical iterative methods for solving SLAEs such as **Gauss–Seidel**, **successive overrelaxation** and **conjugate gradient** do converge, since matrix A_h besides being symmetric is **positive-definite** and **strictly diagonally dominant** if $q > 0$. As we know, the former property means that all the eigenvalues of A_h are strictly positive, and the latter that $a_{i,i} > \sum\limits_{j=1, j \neq i}^{n} |a_{i,j}|$, for every i in $\{1, 2, \ldots, n\}$. In both cases, the spectral radius of the matrix B_h corresponding to those iterative methods is strictly less than one, and hence the iterations converge. Referring to the preliminary section on linear algebra, we recall that B_h is the matrix such that $\vec{u}_h^k = B_h \vec{u}_h^{k-1} + \vec{c}_h$, where $\vec{u}_h^k$ are successive approximations of $\vec{u}_h$ for $k = 1, 2, \ldots$, and $\vec{c}_h$ is a suitable vector of $\Re^n$ derived from $\vec{b}_h$.

On the other hand, one might prefer the use of direct methods, such as **Crout's method** or **Cholesky's method**. As we know, they always lead to the exact solution vector, except for round-off errors. This is because direct methods are based on the decomposition of the original matrix A_h into the product of two matrices S_h and R_h, S_h being lower triangular and R_h upper triangular. In the case of **banded matrices**, such as tridiagonal matrices, the nonzero entries of both factor matrices also lie within a band. In the absence of **pivoting** in **Gaussian elimination**, the latter has the same width as the band of the original matrix. This means that, for a tridiagonal matrix A_h, both S_h and R_h are bi-diagonal matrices[6]. The reader can easily check this assertion by applying the formulae supplied in the preliminary section on linear

---

[6] As long as no partial pivoting is necessary in Gaussian eliminations, which in principle is the case of the matrices considered in this chapter.

algebra. In short, the system is rewritten as $S_h R_h \vec{u}_h = \vec{b}_h$, or into the form of two systems $R_h \vec{u}_h = \vec{z}$ and $S_h \vec{z} = \vec{b}_h$, to be solved one after the other in a straightforward manner, since both S_h and R_h are triangular matrices.

Unfortunately, there is little room here for being more specific about all the relevant aspects of methods for solving the SLAE resulting from the application of FD, FE or FV discretisations to solve a boundary value ODE like (P$_1$). That is why we refer to the vast literature on the subject, including classical books such as references [199], [63] and [121] (volumes 1 and 2), among many others. Nevertheless, we shall go back to large SLAE solving in connection with numerical methods for PDEs in Section 4.4.

1.6.2   Example 1.1: Numerical Experiments with the Cell-centred FVM

In Chapter 2, we shall carry out a numerical study of the Three-point FDM, the $\mathcal{P}_1$ FEM and the Vertex-centred FVM, in the light of the reliability results that we will formally establish for the three methods. Here, we check the numerical behaviour of the Cell-centred FVM (equation (1.28)) in the solution of two test problems for the same ODE as (P$_1$) taking $L = 1$, eventually assorted with inhomogeneous boundary conditions (equation (1.34)). We consider only integrable f, because otherwise the integral of f in some CV may not be finite, and thus the method could not be applied as such.

The main point in the numerical results presented here is the observation of the maximum absolute error decay as the meshes vary, in such a way that the maximum of the mesh steps h_j decreases as n increases. The absolute errors are computed at the method's representative points (i.e. at the cell centres). We use two families of non-uniform meshes with $n = 2^m \times 10$ CVs, for $m = 0, 1, 2, 3, 4$. In the first family, there are only two different values of h_j, namely, $h_j = 0.8/n$ for odd values of j and $h_j = 1.2/n$ for even values of j. In the second family of meshes, we took h_j randomly equal to one out of three possible values proportional to 1, 2 and 3, for at least one CV, and thus for $(n - 2)$ CVs at the most. We denote by F_1 the first family of meshes and by F_2 the second family. Gaussian elimination was used to solve the underlying SLAEs. In Tables 1.1 and 1.2, the displayed values were rounded to the fifth significant figure.

Table 1.1   Maximum absolute errors at CV representative points for test problem 1

n $\rightarrow$	10	20	40	80	160
F_1 $\rightarrow$	$0.14979 \ 10^{-2}$	$0.41053 \ 10^{-3}$	$0.10746 \ 10^{-3}$	$0.27488 \ 10^{-4}$	$0.69511 \ 10^{-5}$
F_2 $\rightarrow$	$0.22546 \ 10^{-2}$	$0.62997 \ 10^{-3}$	$0.16685 \ 10^{-3}$	$0.42821 \ 10^{-4}$	$0.10845 \ 10^{-4}$

Table 1.2   Maximum absolute errors at CV representative points for test problem 2

n $\rightarrow$	10	20	40	80	160
F_1 $\rightarrow$	$0.15786 \ 10^{-1}$	$0.47961 \ 10^{-2}$	$0.24109 \ 10^{-2}$	$0.12088 \ 10^{-2}$	$0.60528 \ 10^{-3}$
F_2 $\rightarrow$	$0.47917 \ 10^{-2}$	$0.13459 \ 10^{-2}$	$0.34463 \ 10^{-3}$	$0.87015 \ 10^{-4}$	$0.21863 \ 10^{-4}$

Test-problem 1: Continuous p and q are considered, namely, $p(x) = q(x) = e^x$. Taking $f \equiv 1$, $a = 1$ and $b = -1$, the exact solution is given by $u(x) = e^{-x}$.

As one can infer from Table 1.1 for both families of meshes, the maximum absolute errors are roughly divided by four as n is multiplied by two. According to the concepts to be introduced in Chapter 2, this behaviour suggests that even for non-uniform meshes the Cell-centred FVM is *second-order convergent* in the pointwise sense, although such a scheme's property will not be formally established in this book.

Test-problem 2: We consider discontinuous p and q, namely, $p(x) = 1/2$ if $0 \le x \le 1/2$ and $p(x) = 1$ if $1/2 < x \le 1$, and $q(x) = x/u(x)$ where u is the manufactured solution given by $u(x) = x(2-x)$ if $0 \le x \le 1/2$ and $u(x) = -x^2/2 + x - 3/8$ if $1/2 < x \le 1$. Notice that in the case of F_1, the discontinuity point of p is always an end-point of two neighbouring cells. Incidentally, according to equation (1.28), whenever p is discontinuous at x_j we take $\tilde{p}_j = [p_j^- + p_j^+]/2$ instead of $p(x_j)$ in the scheme. Watching Table 1.2, we realise that the errors in the case of non-uniform meshes with CVs having systematically $x = 1/2$ as an end-point decrease by a factor of two from a level to the other. On the other hand, surprisingly enough, this factor doubles in the case of the randomly generated meshes, for which the discontinuity point of p is never a CV end-point x_j. This suggests that the latter is a better approach to handle a discontinuous p. However, results of this nature must be the object of more careful and rigorous analyses, such a those conducted in Chapter 2.

1.7 Exercises

1.1 Prove that system (1.16) has a unique solution.

1.2 Derive the FD scheme to solve equation (P_1) with continuously varying p, q and f based on a non-uniform grid, equivalent to a P_1 FE scheme derived as follows: the trapezoidal rule is employed to compute all the underlying integrals in the intervals between two neighbouring grid points.

1.3 Assume that in the P_1 FE formulation, the trapezoidal quadrature rule is employed to compute the integrals involving q and f. Find the one-point quadrature formula to compute the integrals of q and f in the formulation (1.24) of the Vertex-centred FVM, for which both methods correspond to the same SLAE.

1.4 In the case of a continuously varying p, assume that the integrals $\int_{x_{j-1}}^{x_j} p[\varphi_j']^2 dx$, $\int_{x_j}^{x_{j+1}} p[\varphi_j']^2 dx$ and $\int_{x_{j+1}}^{x_j} p[\varphi_j' \varphi_{j+1}'] dx$ are approximated by the mid-point quadrature rule for all appropriate j. Check that the equivalence under the other conditions specified in the previous exercise, between the resulting P_1 FEM and thus-modified FVM (equation (1.25)), which is equation (1.26), still holds true.

1.5 Exhibit the entries of the $n \times n$ matrix A_h and of the n-component right-side vector $\vec{b}_h$ of the SLAE with n equations corresponding to equation (1.28).

1.6 Prove the existence and uniqueness of a solution to equation (1.28).

1.7 Give the final form of the right-side vector $\vec{b}_h$ for the Three-point FDM (equation (1.15)) and the Vertex-centred FVM in case the boundary condition $pu'(L) = b$ holds.

1.8 Modify the FV scheme (equation (1.28)) in order to accommodate the inhomogeneous boundary conditions in equation (1.34).

1.9 For $L = 1$, consider a right side of the differential equation (P_1) equal to 1 plus the Dirac distribution δ_c, where $c \in (0, 1]$. Take $p \equiv 1$, $u(0) = a > 0$, $u'(1) = 0$ and $q = 1/s(x)$, where s is the strictly positive function in $[0, 1]$ defined by $s(x) = a + x$ if $0 \leq x < c$ and $s(x) = a + c$ if $1 \geq x \geq c$. Replace u by s in the variational formulation of problem (1.34), and show that s is the solution to this problem for $c < 1$. What happens if $c = 1$? Give an interpretation of this case in terms of inhomogeneous Neumann boundary conditions. Assume that the problem is being solved by the $\mathcal{P}_1$ FEM with a mesh constructed in such a way that c is the abscissa of a mesh node. Check that whatever the mesh satisfying such a condition the corresponding FE solution u_h equals s.[7]

---

[7] As seen in Chapter 2, this property extended to any piecewise linear function establishes the method's consistency: if the solution to the differential equation is a continuous piecewise linear function and the mesh matches the transition points of its different linear pieces, then the underlying FEM yields the exact solution.

2

Qualitative Reliability Analysis

> nothing at all takes place in
> the universe in which some
> rule of maximum or
> minimum does not appear.
>
> Leonhard Euler

Since we are dealing with techniques providing approximations of the exact solution, the following fundamental question arises: How reliable are they in terms of accuracy? In this chapter, we endeavour to give an appropriate answer to this question in the framework of the model problem studied in the previous chapter. As a by product, the basic steps to be followed in order to handle more complex situations in the subsequent chapters will be disclosed.

Generally speaking, the user of numerical methods for differential equations should possibly know in advance what can be expected from the values obtained through the solution of the underlying discrete problems, as compared to the corresponding values of the exact solution. If we stick to a given discretisation lattice, it is not always easy to give a satisfactory answer to the above question. That is why, as a rule, the best way to ensure that a numerical method is capable of generating accurate approximations is to establish that the finer the discretisation lattice, the closer we come to the exact values we are searching for. Basically, this means that the number of calculation points attached to the grids or meshes increases indefinitely. However, it is advisable (and in some cases mandatory) that this process be accompanied by a certain uniformity criterion, at least for enhanced efficiency. Here we consider that, unless there is a specific reason to do otherwise, the points at which approximations are calculated, though increasing in number, do not become too dense in a certain region of the problem definition domain, to the detriment of any other. Mathematically, this is translated by a certain **uniformity** of a discretisation lattice. For instance, in case the one-dimensional problem (p1) is solved by the FDM with non-uniform grids, we will say we are working with a **quasi-uniform family of grids** if all the grids $\mathcal{G}_n = \{x_0, x_1, \cdots, x_{n-1}, x_n\}$ under consideration are such that the ratio between the minimum ρ and the maximum h over i of $h_i = |x_i - x_{i-1}|$, commonly known as the **grid size**, is bounded below by a constant $c > 0$ independent of the grid. c is called the

Numerical Methods for Partial Differential Equations:
An Introduction Finite Differences, Finite Elements and Finite Volumes, First Edition. Vitoriano Ruas.
© 2016 John Wiley & Sons, Ltd. Published 2016 by John Wiley & Sons, Ltd.
Companion Website: www.wiley.com/go/ruas/numericalmethodsforpartial

uniformity constant of the family. In particular, this bound must hold as h goes to zero, that is, as n goes to infinity. In case the FVM or the FEM is employed with meshes $\mathcal{T}_h$, taking the same definitions of h and ρ, we come up with the similar concept of a **quasi-uniform family of meshes** if $\rho/h \geq c > 0$ for every mesh in the family. Here again, the parameter h (called the **mesh size** for such a family) may become as small as we wish.

A very important quantity in any reliability study of a discretisation method is the **error of the numerical solution**. Basically, the error is a positive measure of the difference between the exact solution and the approximate solution. However, there are infinitely many ways to define such a measure. In order to be objective, it is advisable to choose a type of measure which is practical while being sufficiently representative of the approximation process as a whole. For instance, if only specific grid points are considered to measure the error, one might be disregarding relevant behaviors taking place elsewhere. That is why it is usually preferable to use global error measures, in the sense that somehow the contribution of the errors in every region of the problem definition domain is taken into account. This leads to the concept of **norm**, which can be viewed as a measure of distance between two elements in a given set. In the case of FD errors, the norm will be a measure of the distance between the vector $\vec{u}_h$ of approximate values of the solution at the grid points and the vector of exact values at the same points. In the case of the FEM, the norm will measure the distance between two functions, namely, the exact solution u and its approximation u_h. In the case of the FVM, one can use either the former or the latter, by defining the underlying piecewise constant approximating functions u_h.

Once we are equipped with an appropriate norm, the main purpose of the reliability analysis of a numerical method to solve differential equations of a certain type is to study the conditions under which the error of the approximate solution measured in this norm goes to zero, as the grid or mesh sizes tend to zero. This property is commonly called the **convergence of a numerical method**.

Chapter outline: In Section 2.1, we extend to more general vector spaces the concept of norm recalled in the preliminary section on linear algebra. The same is done therein for inner products. According to the celebrated **Lax–Richtmyer equivalence theorem** [123], the convergence in a certain norm of a numerical method to solve a linear differential equations occurs if the associated scheme is **stable** in the sense of the same norm and **consistent** with a **strictly positive order** in terms of the discretisation parameter(s). Section 2.2 is devoted to the presentation of the concept of **stability in norm**. Consistency and corresponding order are concepts exemplified by relation (1.9). They are formalised in Section 2.3. In Section 2.4, we explain why the combination of stability and consistency with an strictly positive order results in convergence. Numerical examples given in Subection 2.4.4 using the FDM, the FEM and the FVM introduced in Chapter 1 closes this chapter.

2.1 Norms and Inner Products

Before going into reliability analyses of discretisation methods of problem (p1), it is useful to formalise the concepts of norm and inner product in vector spaces, like the one of m-component real vectors, or of functions of a certain class, for example the set V defined in Section 1.3.

In abstract terms, we consider here that E is a real vector space with null element 0_E.

2.1.1   *Normed Vector Spaces*

E is said to be a **normed vector space**, or equivalently a space equipped with a **norm**, if E is a real vector space; and with every $e \in E$ we can associate a real number $\| e \|_E$ having the following properties:

(i) $\|e\|_E \geq 0, \forall e \in E$ and $\|e\|_E = 0$ if and only if $e = 0_E$;
(ii) $\|\alpha e\|_E = |\alpha| \|e\|_E, \forall \alpha \in \mathfrak{R}, \forall e \in E$;
(iii) $\|e_1 + e_2\|_E \leq \|e_1\|_E + \|e_2\|_E, \forall e_1, e_2 \in E$,

where property (i) means positive-definiteness of the norm, (ii) is the homogeneity property and (iii) is known as the **triangle inequality**. Let us give some examples.

Three norms over the space $\mathfrak{R}^n$ of n-component real vectors $\vec{c} = [c_1, c_2, \cdots, c_n]^T$ were defined in the preliminary section on linear algebra, namely, the norms

$$\|\vec{c}\|_\infty \stackrel{def}{=} \max_{i \in \{1,2,\cdots,n\}} |c_i| \qquad (\text{maximum norm or } l^\infty - \text{norm})$$

$$\|\vec{c}\|_1 \stackrel{def}{=} \sum_{i=1}^n |c_i| \qquad (l^1 - \text{norm})$$

$$\|\vec{c}\|_2 \stackrel{def}{=} \left[\sum_{i=1}^n |c_i|^2 \right]^{\frac{1}{2}} \qquad (\text{Euclidean norm or } l^2\text{-norm}).$$

The Euclidean norm $\|\vec{c}\|_2$ is also known as the **modulus** of $\vec{c}$ and corresponds to the length of this vector in $\mathfrak{R}^2$ or $\mathfrak{R}^3$. Actually, this fact explains the denomination of property (iii), for in the case of $\mathfrak{R}^2$ the length of $\vec{c} + \vec{d}$ is necessarily bounded by the sum of the lengths of $\vec{c}$ and $\vec{d}$, as illustrated in Figure 2.1.

For coherence, we employ in this subsection the notation $\|\vec{c}\|_2$ for the modulus of a vector $\vec{c} \in \mathfrak{R}^n$. However, in the remainder of this text, we shall rather use the more popular and shorter notation $|\vec{c}|$ to represent such a quantity.

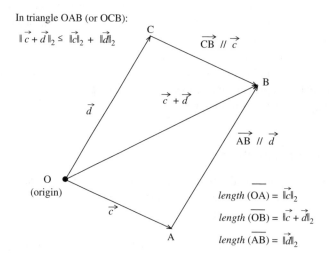

In triangle OAB (or OCB):
$$\|\vec{c} + \vec{d}\|_2 \leq \|\vec{c}\|_2 + \|\vec{d}\|_2$$

$\overrightarrow{CB} \,//\, \vec{c}$

$\overrightarrow{AB} \,//\, \vec{d}$

$\text{length } (\overline{OA}) = \|\vec{c}\|_2$

$\text{length } (\overline{OB}) = \|\vec{c} + \vec{d}\|_2$

$\text{length } (\overline{AB}) = \|\vec{d}\|_2$

Figure 2.1   The triangle inequality for the Euclidean norm in $\mathfrak{R}^2$.

The following norms can be defined over the space of continuous functions f in the closed interval $[0, L]$, denoted by $C[0, L]$:

$$\|f\|_{0,\infty} \stackrel{def}{=} \max_{x \in [0,L]} |f(x)| \qquad \text{(maximum norm or } \mathcal{L}^\infty\text{-norm)}$$

$$\|f\|_{0,1} \stackrel{def}{=} \int_0^L |f(x)| dx \qquad (\mathcal{L}^1\text{-norm)}$$

$$\|f\|_{0,2} \stackrel{def}{=} \left[\int_0^L |f(x)|^2 dx \right]^{\frac{1}{2}} \qquad (\mathcal{L}^2\text{-norm)}$$

Remark 2.1 *It is well-known that all six expressions above fulfil the conditions (i)–(iii). Actually, only property (i) for the norms $\|\cdot\|_{0,1}$ and $\|\cdot\|_{0,2}$ cannot be established without fully exploiting the continuity of functions in $C[0, L]$. This is because such expressions do not really define norms for spaces containing discontinuous functions such as $\mathcal{L}^2(0, L)$. Indeed, if $f \in \mathcal{L}^2(0, L)$ and $\|f\|_{0,2} = 0$, then f is a function that vanishes almost everywhere in $(0, L)$ i.e., everywhere in this interval, except at most at points that form a set whose total measure is equal to zero–in short, a null set (e.g., a countable set of points). But since the norm $\|\cdot\|_{0,2}$ is quite natural and handy for square integrable functions, in order to overcome this difficulty, instead of $\mathcal{L}^2(0, L)$ itself, one usually works with a suitable associated quotient space $L^2(0, L)$ (see e.g. [206]). Nevertheless, as most authors do, one may identify $\mathcal{L}^2(0, L)$ and $L^2(0, L)$ by abusively regarding the latter as a space of functions not to be distinguished from each other, as long as they are identical almost everywhere in $(0, L)$.* ∎

In complement to this brief introduction to the concept of norm, we observe that a real valued expression $|e|_E$ defined for elements e in a vector space E that fulfils (ii) and (iii) and such that $|e|_E \geq 0, \forall e \in E$ is called a **semi-norm** of E.

2.1.2 Inner Product Spaces

We shall be particularly concerned about a subclass of normed spaces whose norm is defined by means of an **inner product**. A real vector space E is said to be an **inner product space**, or, equivalently, to be equipped with an inner product, if with every pair $(d; e)$ of elements belonging to E we associate a real number $(d|e)$ called the inner product of d and e, satisfying the following properties:

(I) $(e|e) \geq 0 \; \forall e \in E$ and $(e|e) = 0$ if and only if $e = 0_E$ (positive-definiteness);
(II) $(d|e) = (e|d) \; \forall d, e \in E$ (symmetry);
(III) $(\alpha_1 d_1 + \alpha_2 d_2 | e) = \alpha_1 (d_1|e) + \alpha_2 (d_2|e) \; \forall \alpha_1, \alpha_2 \in \mathfrak{R}, \; \forall d_1, d_2, e \in E$ (linearity).

For every inner product, the so-called **Cauchy–Schwarz inequality** holds, namely:

$$|(d|e)| \leq \sqrt{(d|d)}\sqrt{(e|e)} \; \forall d, e \in E.$$

Indeed, owing to properties (I)–(III), we have $\forall d, e \in E$ and $\forall \alpha \in \mathfrak{R}$:

$$(d + \alpha e | d + \alpha e) = (d|d) + 2\alpha(d|e) + \alpha^2(e|e) \geq 0$$

Since the quadratic function of α on the left-hand side of the above inequality can only be non-negative for every α if

$$|2(d|e)|^2 - 4(d|d)(e|e) \leq 0,$$

the result follows.

To any given inner product $(\cdot|\cdot)$ corresponds a norm defined by

$$\|e\| \overset{def}{=} \sqrt{(e|e)} \; \forall e \in E$$

On the one hand, properties (i) and (ii) trivially hold for such a norm as a consequence of (I) and (II). On the other hand, it can be checked rather easily (cf. Exercise 2.1) that (iii) is also satisfied by virtue of the Cauchy–Schwarz inequality. For example, the norm $\|\vec{v}\|_2$ of $\vec{v} = [v_1, v_2, \cdots, v_n]^T \in \mathfrak{R}^n$ is derived from the inner product $(\vec{u}|\vec{v}) := \sum_{i=1}^{n} u_i v_i$. Moreover, we know from elementary geometry that for $n = 2$ or $n = 3$, if we denote by θ the angle between $\vec{u}$ and $\vec{v}$, it holds

$$(\vec{u}|\vec{v}) = \|\vec{u}\|_2 \, \|\vec{v}\|_2 \cos \, \theta. \tag{2.1}$$

The norm $\|\cdot\|_{0,2}$ of $C[0, L]$, in turn, is associated with the following inner product:

$$(f|g)_0 \overset{def}{=} \int_0^L f(x)g(x)dx$$

Notice that the expression $(\cdot|\cdot)_0$ is not an inner product for $\mathcal{L}^2(0, L)$, since rigorously property (I) does not hold in this case. However, like in the case of the norm $\|\cdot\|_{0,2}$ (and also of $\|\cdot\|_{0,1}$), we may replace $\mathcal{L}^2(0, L)$ with $L^2(0, L)$, over which $(\cdot|\cdot)_0$ does become an inner product, provided we regard this space as a function space in the sense considered in the previous subsection.

As a by-product of the Cauchy–Schwarz inequality applied to $L^2(0, L)$, we can assert that whenever two functions f and g belong to $L^2(0, L)$, their product belongs to the space $L^1(0, L)$ of (Lebesgue) integrable functions in $(0, L)$. Indeed,

$$\pm \int_0^L f(x)g(x)dx \leq \sqrt{\int_0^L f^2(x)dx} \sqrt{\int_0^L g^2(x)dx}, \forall f, g \in L^2(0, L).$$

Every norm $\| \cdot \|$ associated with an inner product $(\cdot|\cdot)$ obeys the so-called *Parallelogram Law* (cf. [132]), namely:

$$\|d + e\|^2 + \|d - e\|^2 = 2(\|d\|^2 + \|e\|^2) \; \forall d, e \in E.$$

Indeed, the above identity can be verified in a straightforward manner by developing the two terms on its left side and making use of properties (II) and (III).

However, there are norms that correspond to no inner product. For instance, this is the case of the maximum norm. In order to justify this assertion, it suffices to consider the following counterexample:

Let $f(x) = x/L$ and $g(x) = (L - x)/L$ both belonging to $C[0, L]$.
We have $\|f\|_{0,\infty} = \|g\|_{0,\infty} = 1$. Moreover, $\|f + g\|_{0,\infty} = 1$ and $\|f - g\|_{0,\infty} = 1$. Therefore, the Parallelogram Law is violated.

Remark 2.2 *Actually, the converse of the Parallelogram Law is also true: If a norm $\| \cdot \|$ defined on a vector space E satisfies the Parallelogram Law for every $e, d \in E$, then it is necessarily associated with an inner product $(\cdot|\cdot)$ on the same space. More specifically, the latter is given by $(d|e) = \frac{1}{4}(\|d + e\|^2 - \|d - e\|^2)$ (see e.g. [28]).* ■

Remark 2.3 *Some norms or inner products may not be well suited to a certain space, as the corresponding expression could be unbounded or undefined for subclasses of elements within this space. For instance, the expression $(f|g)_0$ is not defined for all the functions f and g in $\mathcal{L}^1(0, L)$ (take e.g. $f(x) = g(x) = x^{-\frac{1}{2}}$), and the maximum norm cannot be a norm over $\mathcal{L}^2(0, L)$ (take e.g. $f(x) = x^{-\frac{1}{4}}$).* ∎

2.2 Stability of a Numerical Method

We say that a numerical method is stable in terms of a certain norm, if the approximate solution measured in this norm is bounded independently of the corresponding discretisation parameter(s) by a constant multiplied by a suitable measure of the problem data. In mathematical terms, let $\|U_d\|_A$ be a certain norm of the solution U_d obtained by the discretisation of a linear problem, and $\|D\|_B$ be another norm of the problem data D. If we can assert that there exists a constant $C_{AB} > 0$, independent of any particular discretisation level under consideration, such that

$$\|U_d\|_A \leq C_{AB} \|D\|_B,$$

then we say that the discretisation method is stable in the sense of the norm $\|\cdot\|_A$. Notice that in all the cases studied in this book (and also in the case of most numerical methods for solving linear PDEs), the solution U_d is completely defined by a linear combination with a finite number n_D of real coefficients, to be determined by solving a SLAE with a square $n_D \times n_D$ matrix. This kind of system has to be solved just once if the problem is independent of time, like the model problem of Chapter 1. If one is dealing with a time-dependent problem like those to be considered in Chapter 3 onwards, in some cases a SLAE has to be solved at every time step.

Regardless, **stability** in the above sense implies **existence and uniqueness** of the solution U_d (eventually at every time step). This is because a SLAE has a unique solution if and only if the only possible solution for a null right-side vector is the null vector. Since this implies that $U_d \equiv 0$ as a linear combination of zero coefficients, there can be no other solution U_d' corresponding to the same data set D, for in this case by linearity we would have $\|U_d - U_d'\|_A \leq C_{AB} \|D - D\|_B = 0$, which implies that $U_d = U_d'$. To complete the argument, it suffices to observe that uniqueness implies existence in the case of a SLAE.

It is important to note that, strictly speaking, there is no need for the constant C_{AB} to be independent of discretisation parameters. However, for the purpose of a convergence study based on the underlying stability result, it is usually required that this constant be independent of the discretisation level, characterised by grid or mesh sizes in most cases. Two examples of stability in the above sense are given right away.

2.2.1 Stability in the Maximum Norm

As a first study of stability in norm, let us consider the counterpart of scheme (1.12) in the case of an inhomogeneous Dirichlet boundary condition, namely,

$$\begin{cases} \text{Find } u_i, i = 1, 2, \cdots, n \text{ satisfying} \\ p\frac{2u_i - u_{i+1} - u_{i-1}}{h^2} + qu_i = f(x_i) \\ \text{with } u_0 = a \text{ and } u_{n+1} = u_{n-1}. \end{cases} \quad (2.2)$$

This scheme is also equivalent to a SLAE with the same matrix as in the case of equation (1.12). Its right-side vector in turn is the same as in the homogeneous case, except for the first component, whose value is now $f(x_1) + pa/h^2$.

For example, if we use the maximum norm of $\Re^n$ for both the numerical solution and $\vec{f}_h := [f(x_1), f(x_2), \cdots, f(x_n)]^T$ assuming that f is bounded in $[0, L]$, we shall prove that the scheme is stable in terms of this norm. We use $\vec{f}_h$ here instead of $\vec{b}_h$ defined in Section 1.2 for convenience. Actually, the components of both vectors coincide, except $b_n (= f(x_n)/2)$. More precisely, we will establish the validity of the

Stability inequality for the Three-point FD scheme

$$\boxed{\begin{array}{l} \| \vec{u}_h \|_\infty \le |a| + CF \\ \text{where } F = \| \vec{f}_h \|_\infty \text{ and } C = L^2/(2p). \end{array}} \tag{2.3}$$

We recall that $\vec{u}_h$ is the vector of approximate values of u at the grid points x_i for $1 \le i \le n$. Hence, this means in particular that the numerical solution cannot grow without control as we compute with finer and finer grids, since the norm of $\vec{f}_h$ depends only on f, and hence not on h.

The stability of scheme (2.2) in the above sense is a consequence of the following result:

The discrete maximum principle (DMP): If $f(x_i) \le 0$ for all i, then $\max_{1 \le i \le n} u_i \le a$ if $q = 0$ and $\max_{1 \le i \le n} u_i \le \max[0, a]$ if $q > 0$.

We can verify that the DMP holds by an argument quite similar to the one employed in Chapter 1 in order to establish that equation (1.12) has a unique solution: Let us first treat the case $q > 0$. Setting $\nu = (qh^2 + 2p)/(2p)$, by assumption we have $\nu > 1$. Let M be such that $u_M = \max_{1 \le i \le n} u_i$. If $u_M \le 0$, then the maximum is non-positive and half a way is done. Let us then assume that $u_M > 0$. If $M = n$, then from the nth equation of (2.2) we infer that $\nu u_n \le u_{n-1}$. Thus, the only possibility is $u_M = u_n = 0$, which is now discarded. Hence, $M \ne n$ and if $M > 0$, equation (2.2) implies that $\nu u_M \le (u_{M+1} + u_{M-1})/2$. In this case, at least one out of u_{M-1} and u_{M+1} must be strictly greater than u_M, which is impossible by assumption. Hence, if $u_M > 0$, M must be zero, and thus $u_M = a$. As a result, the remaining half way is also done.

The case where $q = 0$ can be treated in a similar manner and is left as Exercise 2.2.

Observing that $\min_{1 \le i \le n} u_i = - \max_{1 \le i \le n}[-u_i]$, the validity of the DMP implies that the following **discrete minimum principle** is also true: If $f(x_i) \ge 0$ for all i, then $\min_{1 \le i \le n} u_i \ge a$ if $q = 0$ and $\min_{1 \le i \le n} u_i \ge \min[0, a]$ if $q > 0$.

Notice that if the distribution of forces is positive, the bar will continuously stretch away from its left end. However, its minimum (negative) elongation will take place precisely there if $a < 0$, as both physics and mathematics indicate. On the other hand, if only a displacement $a > 0$ is applied at the section $x = 0$, physics also tells us that the elongation process will be alleviated along the bar. However, since it is subject to a tension $f \ge 0$, the elongation can nowhere be negative. The case $a = 0$ is even more eloquent, for whenever f is positive, the physical solution certainly corresponds to increasing positive section displacement as one approaches the bar's right end. This means again that the minimum value of the displacement is 0, precisely at $x = 0$. The discrete minimum principle tells us that the numerical scheme reproduces these properties at the discrete level. Otherwise stated, in practical terms, the

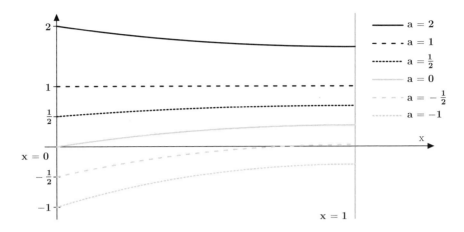

Figure 2.2   An illustration of the solution u for $f \geq 0$ and different values of $a = u(0)$.

DMP (and consequently the stability), means that somehow the numerical solution is able to mimic a physically admissible solution. In mathematical terms, this property is illustrated in Figure 2.2, where the aspect of the displacement function u in the case of a positive f is shown for different values of a both positive and negative and for $a = 0$. The graph of u takes the approximate shape of a parabola, with a horizontal tangent at $x = L$ and a minimum value a guaranteed at $x = 0$, if $a \leq 0$. On the other hand, a strictly positive a is not necessarily the minimum value of u. However, according to the minimum principle, the curve representing u cannot cut the x axis.

Now, in order to apply the DMP to the stability study, let us assume that $\phi(x) = [(x - L)/L]^2/(2p)$ is the solution of equation (p1). Notice that this corresponds to a right-hand side f of the differential equation equal to $-[p\phi']' + q\phi$ supplemented with the inhomogeneous Dirichlet boundary condition $\phi(0) = 1/(2p)$, which happens to be the maximum value of ϕ in $[0, L]$. Applying the operations on the left side of scheme (2.2) to the exact values of ϕ and taking into account that $\phi[(n + 1)h] = \phi[(n - 1)h]$, we obtain after straightforward calculations:

$$\frac{p[2\phi(ih) - \phi(ih - h) - \phi(ih + h)]}{h^2} + q\phi(ih) = -\frac{1}{L^2} + q\phi(ih) \qquad (2.4)$$

Let us define the n-component vectors $\vec{w}^+$ and $\vec{w}^-$ by $w_i^+ = +u_i + FL^2[\phi(ih) - 1/(2p)]$ and $w_i^- = -u_i + FL^2[\phi(ih) - 1/(2p)]$, $i = 1, 2,, \cdots, n$. Then, owing to equations (2.4) and (2.2), we have

$$\begin{cases} \dfrac{p[2w_i^+ - w_{i-1}^+ - w_{i+1}^+]}{h^2} + qw_i^+ = f(ih) - F + qFL^2\left[\phi(ih) - \dfrac{1}{2p}\right] \leq 0, \\[4mm] \dfrac{p[2w_i^- - w_{i-1}^- - w_{i+1}^-]}{h^2} + qw_i^- = -f(ih) - F + qFL^2\left[\phi(ih) - \dfrac{1}{2p}\right] \leq 0. \end{cases} \qquad (2.5)$$

Hence, by virtue of the DMP, $w_i^+ \leq \max\{0, a + FL^2[\phi(0) - 1/(2p)]\} = \max[0, a] \leq |a|$ and $w_i^- \leq \max\{0, -a + FL^2[\phi(0) - 1/(2p)]\} = \max[0, -a] \leq |a|$ for every i. However,

$\pm u_i = w_i^{\pm} + FL^2[1/(2p) - \phi(ih)] \le w_i^{\pm} + FL^2/(2p) \le |a| + FL^2/(2p)$, since $\phi(x) \ge 0$ for every x and $w_i^{\pm} \le |a|$. It immediately follows that we have stability in the sense of equation (2.3) with $C = L^2/(2p)$.

The DMP can be used in an analogous manner to study the stability in the sense of equation (2.3) of the variant equation (1.15) for a non-uniform grid, and also to several other schemes for constant p and q or not. This comment also extends to all those schemes duly adapted to the case of inhomogeneous boundary conditions, such as the equation (1.25) and (1.28) FV schemes and the linear FE scheme. However, for at least one reason, the bound (2.3) may be unsatisfactory. This happens, for instance, if p is very small as compared to other data and more especially to q. That is why we conclude this subsection by giving another stability result of a wider scope, as long as $q > 0$, since it applies to all those schemes for different types of boundary conditions and nonzero functions q and p. In particular, it will be very useful in Chapter 3. Denoting the vector $[u_0, u_1, u_2, \cdots, u_{n-1}, u_n]^T \in \mathfrak{R}^{n+1}$ by $\vec{u}$, let q_i, p_i, p_i^+ and p_i^- be given non-negative coefficients for $i = 0, \cdots, n$, satisfying $q_i \ge \beta > 0$, for every i, together with $2p_i = p_i^+ + p_i^-$ for $0 \le i \le n$ and $p_0^- = p_n^+ = 0$. Assume that for every $\vec{g} = [g_0, g_1, \cdots, g_n]^T$ and $\vec{f} = [f_0, f_1, \cdots, f_n]^T$ in $\mathfrak{R}^{n+1}$, we have

$$(q_i + 2p_i)u_i - p_i^+ u_{i+1} - p_i^- u_{i-1} = q_i g_i + f_i \text{ for } i = 0, 1, 2, \cdots, n, \tag{2.6}$$

where u_{-1} and u_{n+1} are meaningless. Then, $\vec{u}$ is bounded as follows:

$$\|\vec{u}\|_{\infty} \le \|\vec{g}\|_{\infty} + \|\vec{f}\|_{\infty}/\beta. \tag{2.7}$$

Indeed, since $q_i > 0$, we can rewrite equation (2.6) as

$$(1 + 2p_i/q_i)u_i = (p_i^+ u_{i+1} + p_i^- u_{i-1})/q_i + g_i + f_i/q_i \text{ for } i = 0, 1, 2, \cdots, n.$$

Using three times the triangle inequality for absolute values, we obtain

$$(1 + 2p_i/q_i)|u_i| \le (p_i^+ |u_{i+1}| + p_i^- |u_{i-1}|)/q_i + |g_i| + |f_i|/q_i \text{ for } i = 0, 1, 2, \cdots, n.$$

Therefore, we trivially have

$$(1 + 2p_i/q_i)|u_i| \le \|\vec{u}\|_{\infty}(p_{i+1}^+ + p_{i-1}^-)/q_i + \|\vec{g}\|_{\infty} + \|\vec{f}\|_{\infty}/\beta \text{ for } i = 0, 1, 2, \cdots, n.$$

Hence, using the properties of the coefficients p_i^- and p_i^+, we derive

$$(1 + 2p_i/q_i)|u_i| \le 2p_i/q_i \|\vec{u}\|_{\infty} + \|\vec{g}\|_{\infty} + \|\vec{f}\|_{\infty}/\beta \text{ for } i = 0, 1, 2, \cdots, n.$$

Letting u_M be a component of $\vec{u}$ such that $|u_M| = \|\vec{u}\|_{\infty}$, we have

$$(1 + 2p_M/q_M)\|\vec{u}\|_{\infty} \le 2p_M/q_M \|\vec{u}\|_{\infty} + \|\vec{g}\|_{\infty} + \|\vec{f}\|_{\infty}/\beta \text{ for } i = 0, 1, 2, \cdots, n.$$

which yields equation (2.7).

Notice that this result applies in particular to equation (2.2) if $q > 0$. Indeed, in this case, $p_i^+ = p_i^- = p/h^2$, $p_i = p/h^2$ for $i = 1, 2, \cdots, n-1$, $p_0 = p_0^+ = 0$, $p_n^- = 2p_n = 2p/h^2$, $q_0 = 1$, $g_0 = a$, $f_0 = 0$ and $g_i = 0$, $q_i = q$, $f_i = f(x_i)$ for $i = 1, 2, \cdots, n$. Hence, we have established the following:

Stability inequality for the Three-point FD scheme with $q > 0$

$$\max_{1 \leq i \leq n} |u_i| \leq |a| + \max_{1 \leq i \leq n} |f(x_i)|/\min[q, 1], \qquad (2.8)$$

Equation (2.8) is fairly equivalent to equation (2.3), except if $p << \min[q, 1]$ or $\min[q, 1] << p$.

2.2.2   Stability in the Mean-square Sense

Now, we switch to an example of stability in terms of norms of the mean square type, such as L^2 norms. Since this is more natural in the context of variational forms, we will only carry out this kind of study for the FE scheme (equation (1.22)). However, for the sake of generality, we will consider its modification in order to incorporate inhomogeneous boundary conditions at both ends of $(0, L)$, with arbitrary p and q satisfying our initial assumptions. This requires a working space consisting of functions which are sums of functions in V with constants. This space is known in the literature as the **Sobolev space** $H^1(0, L)$ [1].

First, we write down the problem to solve followed by the corresponding FE approximate problem, namely,

$$\begin{cases} \text{Find } u \in H^1(0, L) \text{ such that } u(0) = a \text{ and} \\ \int_0^L (pu'v' + quv)dx = \int_0^L fvdx + bv(L) \ \forall v \in V \end{cases} \qquad (2.9)$$

$$\begin{cases} \text{Find } u_h \in W_h \text{ such that } u_h(0) = a \text{ and} \\ \int_0^L (pu_h'v' + qu_hv)dx = \int_0^L fvdx + bv(L) \ \forall v \in V_h. \end{cases} \qquad (2.10)$$

Let us take $v = u_h - a$. Since $v \in V_h$ because $v(0) = 0$ by construction, from equation (2.10) we obtain

$$\int_0^L [(pu_h')^2 + qu_h^2]dx = \int_0^L [aqu_h + f(u_h - a)]dx + b[u_h(L) - a]. \qquad (2.11)$$

Recalling the definition of the L^2-norm and the bounds of p and q, we derive

$$\alpha \int_0^L [u_h']^2 dx \leq \int_0^L [B|a||u_h| + |f(u_h - a)|]dx + |b|(|u_h(L)| + |a|). \qquad (2.12)$$

On the other hand, for every function in $v \in V$ and $\forall y \in (0, L]$, we have

$$v^2(y) = \int_0^y [v^2]'(x)dx.$$

Therefore, $\forall y \in (0, L]$

$$v^2(y) = 2 \int_0^y [vv'](x)dx.$$

Then, by virtue of the Cauchy–Schwarz inequality, this implies that

$$v^2(y) \leq 2 \left\{ \int_0^y v(x)^2 dx \right\}^{1/2} \left\{ \int_0^y [v'(x)]^2 dx \right\}^{1/2} \forall y \in [0, L]. \qquad (2.13)$$

or

$$v^2(y) \leq 2 \| v \|_{0,2} \| v' \|_{0,2} \ \forall y \in [0, L].$$ (2.14)

Two useful results follow from equation (2.14), namely,

$$|v(L)| \leq [2 \| v \|_{0,2} \| v' \|_{0,2}]^{1/2} \ \forall v \in V,$$ (2.15)

$$\text{and} \int_0^L v^2(y) dy \leq \int_0^L [2 \| v \|_{0,2} \| v' \|_{0,2}] dy \ \forall v \in V,$$

or, equivalently,

$$\| v \|_{0,2}^2 \leq 2L \| v \|_{0,2} \| v' \|_{0,2} \ \forall v \in V,$$

which finally yields

The Friedrichs–Poincaré inequality

$$\boxed{\| v \|_{0,2} \leq 2L \| v' \|_{0,2} \ \forall v \in V.}$$ (2.16)

Now, we plug into equation (2.12) the inequality $\int_0^L |u_h| dx \leq L^{1/2} \| u_h \|_{0,2}$, which holds thanks to the Cauchy–Schwarz inequality. It follows that

$$\alpha \| u_h' \|_{0,2}^2 \leq 2L^{3/2} B|a| \| u_h' \|_{0,2} + \int_0^L |f u_h| dx + |a| \int_0^L |f| dx + |b|(|u_h(L)| + |a|).$$

Next, we manipulate the right side above by using both equations (2.15) and (2.16). More precisely, the Cauchy–Schwarz inequality, combined with the obvious relation $u_h(L) = \int_0^L u_h'(x) dx + a$, leads successively to

$$\alpha \| u_h' \|_{0,2}^2 \leq 2L(B|a|L^{1/2} + \| f \|_{0,2}) \| u_h' \|_{0,2} + L^{1/2}|a| \| f \|_{0,2} + |b|(|u_h(L)| + |a|),$$

$$\alpha \| u_h' \|_{0,2}^2 \leq 2L(B|a|L^{1/2} + \| f \|_{0,2}) \| u_h' \|_{0,2} + L^{1/2}|a| \| f \|_{0,2} + |b|$$
$$(2L^{1/2} \| u_h' \|_{0,2} + |a|),$$

or, in more compact form:

$$\begin{cases} \alpha \| u_h' \|_{0,2}^2 \leq D \| u_h' \|_{0,2} + |a|(L^{1/2} \| f \|_{0,2} + |b|) \\ \text{where } D = 2L(B|a|L^{1/2} + \| f \|_{0,2}) + 2|b|L^{1/2}. \end{cases}$$ (2.17)

Now we use **Young's inequality**,[1]

$$|sr| \leq (s^2\delta + r^2/\delta)/2 \ \forall s, r \in \mathfrak{R} \text{ and } \forall \delta > 0.$$

Then, taking $s = \| u_h' \|_{0,2}$, $r = D$ and $\delta = \alpha$, we readily derive from equation (2.17):

$$\alpha \| u_h' \|_{0,2}^2 \leq D^2/\alpha + 2|a|(L^{1/2} \| f \|_{0,2} + |b|).$$ (2.18)

After straightforward calculations, equation (2.18) yields

---

[1] A trivial consequence of the inequality: $(s\sqrt{\delta} + r/\sqrt{\delta})^2 \geq 0$ for every pair of real numbers $(s; r)$ and $\forall \delta > 0$.

A first mean-square stability inequality for the $\mathcal{P}_1$ FEM

$$\| u_h' \|_{0,2} \leq C[|a| + \| f \|_{0,2} + |b|],$$
(2.19)

for a suitable constant C depending only on α, B and L (determination of a fine expression for C from equations (2.17) and (2.18) is proposed to the reader as Exercise 2.3). Now, it is clear that a stability result similar to equation (2.19) also holds for the L^2-norm of u_h, owing to the Friedrichs–Poincaré inequality, that is,

A second mean-square stability inequality for the $\mathcal{P}_1$ FEM

$$\| u_h \|_{0,2} \leq C'[|a| + \| f \|_{0,2} + |b|].$$
(2.20)

Indeed, equation (2.16) implies that $\| u_h - a \|_{0,2} \leq 2L \| (u_h - a)' \|_{0,2} = 2L \| u_h' \|_{0,2}$. On the other hand, using property (iii) of a norm, we easily obtain $\| u_h - a \|_{0,2} \geq \| u_h \|_{0,2} - |a| L^{1/2}$. Therefore, $\| u_h \|_{0,2} \leq 2L \| u_h' \|_{0,2} + L^{1/2}|a|$. Then, combining this with equation (2.19), we obtain equation (2.20) with $C' = 2LC + L^{1/2}$. Under an additional condition on q specified below, combining equations (2.19) and (2.20), one can easily establish that a similar result also holds for

The energy norm of the bar

$$\| v \|_e := \left\{ \frac{1}{2} \int_0^L [p(v')^2 + qv^2](x)dx \right\}^{1/2} \quad \forall v \in V.$$
(2.21)

The fact that $\| \cdot \|_e$ is effectively a norm on space V is an immediate consequence of the properties of p and q. Notice that the energy norm is only a semi-norm on $H^1(0, L)$, and in the case of an inhomogeneous boundary condition $u_h(0) = a$. Thus, $u_h \in H^1(0, L)$, but $u_h \notin V$. Nevertheless, if there exists $\beta > 0$ such that

$$q(x) \geq \beta \ \forall x \in [0, L]$$

the energy norm is a norm on $H^1(0, L)$, and also on V even in case not. This is due to the fact that this norm is associated with the **energy inner product** given by

$$(u|v)_e := \frac{1}{2} \int_0^L [pu'v' + quv](x)dx.$$
(2.22)

Actually, the energy norm carries this name because it expresses the **total internal energy** of the bar given by $I_e(v) := (v|v)_e$ for a given state of longitudinal displacement v. The term $\frac{1}{2} \int_0^L [(pv')(x)]^2 dx/2$ of I_e accounts for the elongational stiffness of the bar, and the term $\frac{1}{2} \int_0^L [(qv)(x)]^2 dx/2$ stands for the stored energy due to a spring effect on the bar subject to a longitudinal displacement v.

Summarising what we saw in this subsection, for a suitable mesh independent constant C_e, we have

A stability inequality for the $\mathcal{P}_1$ FEM in the energy norm

$$\| u_h \|_e \leq C_e[|a| + \| f \|_{0,2} + |b|].$$
(2.23)

It is interesting to note that all the steps leading to equation (2.23) apply as well to the exact solution u, and thus we also have

$$\| u \|_e \leq C_e[|a| + \| f \|_{0,2} + |b|].$$ (2.24)

The stability result (equation (2.23)) states that, similarly to the longitudinal displacement u, the total internal energy of the bar for its FE counterpart u_h remains controlled by the applied loads represented by a, b and f, whatever the mesh. In short, here again the stability of the discrete model means that somehow it mimics the physical behavior of the exact model.

In complement to the above study, the reader may derive as Exercise 2.4 the precise expression of C_e resulting from a suitable combination of equations (2.19) and (2.20).

Remark 2.4 *The square of the energy norm minus the* **work of external forces** $\int_0^L fv \, dx$ *is the* **total potential energy** *of the bar subject to a density of forces f, for a given* **admissible longitudinal displacement** *v. The term 'admissible' means that, besides satisfying the kinematic condition $v(0) = 0$, the total potential energy in terms of v is finite. The latter condition requires that $v \in H^1(0, L)$, which justifies the choice of space V in the variational formulation (P_3), from the point of view of classical continuum mechanics. Actually, the solution u is the displacement function that minimises the total potential energy (see e.g. [131]). Incidentally, we observe that the boundary condition $u'(L)$ is satisfied by the minimising displacement u, but not necessarily by an arbitrary admissible displacement v. In this sense, the different nature of boundary conditions $u(0) = 0$ and $u'(L) = 0$ is not merely mathematical. In solid mechanics, they are called* **essential boundary conditions** *and* **natural boundary conditions**. *These expressions characterise that the former are prescribed to any admissible displacement, while the latter is a balance of force condition to be satisfied by the minimising displacement.* ∎

To conclude, we observe that the stability of both the Vertex-centred and the Cell-centred FVM can be studied following the same principles as those applying to the FDM. It suffices to start from the corresponding schemes, and apply the maximum principle that holds for equations (1.24) or (1.27) in both cases. Checking this assertion is left to the reader as Exercise 2.5.

2.3   Scheme Consistency

In this section, we will be concerned about determining the order of the so-called **local truncation error** inherent to a given numerical method in terms of relevant discretisation parameters, such as grid or mesh sizes. By definition, this error is the difference between the left and the right side of the numerical scheme if the approximate values are replaced by the corresponding exact values. This difference is also commonly called the **scheme's residual**. In case the residuals tend to zero as the discretisation parameters go to zero, we say that the method is **consistent**. The **order of consistency** of the method will result from a global evaluation of the local truncation errors as seen below. Let us consider two examples.

2.3.1   Consistency of the Three-point FD Scheme

For simplicity, we study only the case of a constant p. Moreover, for convenience, we use $n \times n$ matrix B_h instead of A_h defined in Section 1.2, both matrices being identical except for the last row of B_h, which equals the last row of A_h multiplied by two. Recalling the presentation of the FDM in Section 1.2, let us replace u_i by $u(x_i)$ in the FD scheme (1.12) for $i = 1, 2, \cdots, n$. Denoting by $\vec{u}_h$ the vector $[u(x_1), u(x_2), \cdots, u(x_n)]^T$ of $\mathfrak{R}^n$, we define the **residual vector** $\vec{r}_h(u) = [r_{h,1}(u), r_{h,2}(u), \cdots, r_{h,n}(u)]^T$ by

$$\vec{r}_h(u) = B_h \vec{u}_h - \vec{f}_h.$$

Recalling equations (1.7) and (1.9), we easily infer that for $i < n$,

$$r_{h,i}(u) = p \frac{2u(x_i) - u(x_{i-1}) - u(x_{i+1})}{h^2} + qu(x_i) - f(x_i) = -pR_i(u). \qquad (2.25)$$

On the other hand, since $u'(L) = 0$, using a standard Taylor expansion we have

$$2u(x_{n-1}) = 2u(x_n) + h^2 u''(x_n) - h^3 u'''(x_n)/3 + h^4 u^{(iv)}(\xi_n)/12$$

where ξ_n is a suitable abscissa in the interval $[x_{n-1}, x_n]$. Therefore, we have

$$r_{h,n}(u) = 2p \frac{u(x_n) - u(x_{n-1})}{h^2} + qu(x_n) - f(x_n) = -pR_n(u), \qquad (2.26)$$

where $R_n(u) = h u'''(x_n)/3 - h^2 u^{(iv)}(\xi_n)/12$. Hence, provided the derivatives of u up to the fourth order are continuous in $[0, L]$, the **local truncation errors** for the FDM with a uniform grid can be bounded as follows:

$$|r_{h,i}(u)| \leq p \| u^{(iv)} \|_{0,\infty} h^2/12 \ \forall i < n \qquad (2.27)$$

$$|r_{h,n}(u)| \leq p(\| u''' \|_{0,\infty} h/3 + \| u^{(iv)} \|_{0,\infty} h^2/12).$$

Then, by definition, we say that the local truncation error for the FDM with an equally spaced grid is of order two at the grid points x_i for $i < n$, and of order one at the grid point $x_n = L$. As we will see in Section 2.3.2, the truncation errors are directly linked to the accuracy of the discretisation method. Therefore, the above conclusion is not satisfactory, since for a small h, the error at grid point x_n might pollute the accuracy elsewhere. In other words, in spite of the fact that all the local truncation errors but one are of order two, the order of consistency of the scheme will be one, since the order of the local truncation error at x_n is one, that is, the dominant order. That is why, for explanatory purposes, it is advisable to modify the definition of the fictitious unknown u_{n+1} in the following fashion.

First of all, we observe that f must be continuously differentiable in $[0, L]$, otherwise, there is no way for u to be more than two times differentiable. Thus, we have $-pu'''(L) + qu'(L) = f'(L)$, and taking into account the boundary condition at $x = L$, we have: $u'''(L) = -f'(L)/p$. Therefore, if we replace u_{n+1} with $u_{n-1} - h^3 f'(L)/3p$, the local truncation error at x_n turns to $\underline{r}_{h,n}(u)$ given by

$$\begin{cases} \underline{r}_{h,n}(u) = 2p \dfrac{u(x_n) - u(x_{n-1})}{h^2} + qu(x_n) - f(x_n) + hf'(x_n)/3 = -p\underline{R}_n(u) \\ \text{where } \underline{R}_n(u) := -h^2 u^{(iv)}(\xi_n)/12, \end{cases} \qquad (2.28)$$

as one can easily check. The above correction in the last equation gives rise to another Three-point FD scheme, for which a better consistency result holds, namely,

A modified Three-point FD scheme

$$p\frac{2u_i - u_{i-1} - u_{i+1}}{h^2} + q\underline{u}_i = \underline{f}_i \text{ for } i = 1, 2, \cdots, n, \text{ with } \underline{u}_0 = a \text{ and } \underline{u}_{n+1} = \underline{u}_{n-1},$$

$$\underline{f}_i = f(x_i) \text{ for } i = 1, 2, \cdots, n - 1 \text{ and } \underline{f}_n = f(x_n) - hf'(x_n)/3.$$

(2.29)

Let us denote by $\underline{\vec{u}}_h$ the corresponding solution vector, $\underline{\vec{u}}_h = [\underline{u}_1, \underline{u}_2, \cdots, \underline{u}_n]^T$, which is the solution of the SLAE

$$B_h \underline{\vec{u}}_h = \underline{\vec{f}}_h \qquad (2.30)$$

where $\underline{\vec{f}}_h = [\underline{f}_1, \underline{f}_2, \cdots, \underline{f}_n]^T$. Accordingly, we denote the underlying residual vector by $\vec{r}_h(u) = [\underline{r}_{h,1}(u), \underline{r}_{h,2}(u), \cdots, \underline{r}_{h,n}(u)]^T$, satisfying $\vec{r}_h(u) := B_h \underline{\vec{u}}_h - \underline{\vec{f}}_h$. Noticing that $\underline{r}_{h,i}(u) := r_{h,i}(u)$ for $i = 1, 2, \cdots, n - 1$, combining equations (2.27) and (2.28), we establish the

Consistency of the modified Three-point FD scheme (2.29)

$$|\underline{r}_{h,i}(u)| \leq p \, \| \, u^{(iv)} \|_{0,\infty} h^2/12 \text{ for } 1 \leq i \leq n. \qquad (2.31)$$

The other way around, we can assert that, at least for constant p and q and $b = 0$, the Three-point FD scheme (1.12) with an equally spaced grid is **second-order consistent**, provided we set $u_{n+1} = u_{n-1} - h^3 f'(L)/3p$. Moreover, taking this modification into account, the reader can easily verify that a stability inequality of the same type as equation (2.3) holds for scheme (2.29), provided F is replaced by $\underline{F} := \max_{1 \leq i \leq n} |\underline{f}_i|$.

2.3.2   Consistency of the $\mathcal{P}_1$ FE Scheme

As a model, we consider the case of homogeneous Neumann boundary conditions at $x = L$ and inhomogeneous Dirichlet boundary conditions at $x = 0$. In the study of the FEM, it is more natural to treat consistency and truncation errors in a variational framework. This means that now we plug the exact values of u at the nodal points on the left side of the approximate variational problem (1.22), and then subtract from the result its right side. Let us denote by V_h^+ the subset of W_h consisting of functions whose value at $x = 0$ equals a. We are actually replacing u_h with a function $\tilde{u}_h \in V_h^+$, whose nodal value at x_i for $i > 0$ is $u(x_i)$. Using the inner product notation for the integrals in equation (1.22), it is convenient to define two **residual functions** $r_h(u)$ and $s_h(u)$ such that

$$(r_h(u)|v)_0 + (s_h(u)|v')_0 = (p\tilde{u}_h'|v')_0 + (q\tilde{u}_h|v)_0 - (f|v)_0 \text{ for any } v \in V_h.$$

Recalling that the above right side vanishes $\forall v \in V_h$, if we replace $\tilde{u}_h$ by the exact solution u, we obtain

$$(r_h(u)|v)_0 + (s_h(u)|v')_0 = (p[\tilde{u}_h - u]'|v')_0 + (q[\tilde{u}_h - u]|v)_0 \text{ for any } v \in V_h. \qquad (2.32)$$

In order to appropriately express the functions $[\tilde{u}_h - u]'$ and $\tilde{u}_h - u$, we resort to the elementary interpolation theory. Referring to Figure 2.3, we first notice that in each element

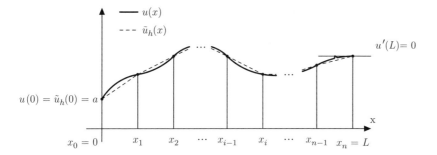

Figure 2.3 The function u and its interpolate $\tilde{u}_h$ in the FE space V_h^+.

$T_i = [x_{i-1}, x_i]$, $\tilde{u}_h$ is the linear interpolating function of u at x_{i-1} and x_i, whose expression is

$$\tilde{u}_h(x) = \frac{(x - x_{i-1})u(x_i) + (x_i - x)u(x_{i-1})}{x_i - x_{i-1}}.$$

Let us assume that u'' belongs to $L^2(0, L)$, or equivalently that $u \in H^2(0, L)$, using a **Sobolev space** notation (cf. [1]). Since this implies that both u'' and $\varphi u''$ are integrable functions in $(0, L)$ for every linear function φ, the following identities hold (please check!):

$$u(x) = u(x_i) - (x_i - x)u'(x) - \int_x^{x_i} (x_i - s)u''(s)ds \text{ and}$$

$$u(x) = u(x_{i-1}) + (x - x_{i-1})u'(x) - \int_{x_{i-1}}^x (s - x_{i-1})u''(s)ds.$$

Recalling that $h_i = x_i - x_{i-1}$, suitable combinations of both identities allow us to derive quite easily

$$[\tilde{u}_h - u](x) = \frac{(x - x_{i-1}) \int_x^{x_i}(x_i - s)u''(s)ds + (x_i - x) \int_{x_{i-1}}^x(s - x_{i-1})u''(s)ds}{h_i},$$

$$(2.33)$$

$$\text{and } [\tilde{u}_h - u]'(x) = \frac{\int_x^{x_i}(x_i - s)u''(s)ds - \int_{x_{i-1}}^x(s - x_{i-1})u''(s)ds}{h_i}. \qquad (2.34)$$

Plugging equations (2.33) and (2.34) into (2.32), we readily obtain for $x \in (x_{i-1}, x_i)$:

$$\begin{cases} [r_h(u)](x) = \dfrac{q(x)}{h_i} \left[(x - x_{i-1}) \displaystyle\int_x^{x_i} (x_i - s)u''(s)ds \right. \\ \qquad\qquad\qquad \left. + (x_i - x) \displaystyle\int_{x_{i-1}}^x (s - x_{i-1})u''(s)ds \right] \\ \text{and} \\ [s_h(u)](x) = \dfrac{p(x)}{h_i} \left[\displaystyle\int_x^{x_i} (x_i - s)u''(s)ds - \displaystyle\int_{x_{i-1}}^x (s - x_{i-1})u''(s)ds \right]. \end{cases} \qquad (2.35)$$

Rather fastidious though straightforward calculations reported below lead to the following estimates for the L^2-norm of both residual functions, thereby characterising the

Consistency of the $\mathcal{P}_1$ FE scheme

$$
\boxed{
\begin{aligned}
\| \, s_h(u) \|_{0,2} &\le C_s h \, \| \, u'' \|_{0,2} \\
\| \, r_h(u) \|_{0,2} &\le C_r h^2 \, \| \, u'' \|_{0,2}.
\end{aligned}
}
\tag{2.36}
$$

C_s is a mesh-independent constant proportional to A. C_r in turn is estimated below as proportional to B, following basically the same arguments that apply to C_s.

Let us prove equation (2.36) from (2.35). First, we consider the case of $s_h(u)$. We have

$$
\| \, s_h(u) \, \|_{0,2}^2 \le \left[\frac{A}{ch} \right]^2 \sum_{i=1}^{n} \int_{x_{i-1}}^{x_i} \left[\int_{x}^{x_i} (x_i - s) u''(s) ds - \int_{x_{i-1}}^{x} (s - x_{i-1}) u''(s) ds \right]^2 dx.
$$

c being the **uniformity constant** of the family of meshes in use (notice that $\min\limits_{1 \le i \le n} h_i / \max\limits_{1 \le i \le n} h_i \ge c$ for every n).

Since $(s \pm r)^2 \le 2(s^2 + r^2)$ and both (x, x_i) and (x_{i-1}, x) are subsets of (x_{i-1}, x_i), this yields

$$
\| \, s_h(u) \, \|_{0,2}^2 \le \frac{2A^2}{(ch)^2} \times
$$

$$
\sum_{i=1}^{n} \int_{x_{i-1}}^{x_i} \left\{ \left[\int_{x_{i-1}}^{x_i} (x_i - s) |u''(s)| ds \right]^2 + \left[\int_{x_{i-1}}^{x_i} (s - x_{i-1}) |u''(s)| ds \right]^2 \right\} dx.
$$

or, noting that $\forall s \in (x_{i-1}, x_i)$, $|s - x_{i-1}| \le h$ and $|x_i - s| \le h$,

$$
\| \, s_h(u) \, \|_{0,2}^2 \le \frac{4A^2}{c^2} \sum_{i=1}^{n} \int_{x_{i-1}}^{x_i} \left[\int_{x_{i-1}}^{x_i} |u''(s)| ds \right]^2 dx,
$$

Thus,

$$
\| \, s_h(u) \, \|_{0,2}^2 \le \frac{4A^2 h}{c^2} \sum_{i=1}^{n} \left[\int_{x_{i-1}}^{x_i} |u''(x)| dx \right]^2.
$$

On the other hand, by virtue of the Cauchy–Schwarz inequality applied to the L^2-inner product in (x_{i-1}, x_i), for any square integrable function g in this interval, we have $| \int_{x_{i-1}}^{x_i} g(x) dx |^2 \le \int_{x_{i-1}}^{x_i} dx \int_{x_{i-1}}^{x_i} |g(x)|^2 dx = h_i \int_{x_{i-1}}^{x_i} |g(x)|^2 dx$. Hence, we further obtain

$$
\| \, s_h(u) \, \|_{0,2}^2 \le \frac{4A^2 h}{c^2} \sum_{i=1}^{n} h_i \int_{x_{i-1}}^{x_i} |u''(x)|^2 dx.
$$

This finally leads to

$$
\| \, s_h(u) \|_{0,2} \le \frac{2Ah}{c} \left[\int_{0}^{L} |u''(x)|^2 dx \right]^{1/2} = C_s h \, \| \, u'' \|_{0,2}
$$

with $C_s = 2A/c$.

The case of $r_h(u)$ can be dealt with in a similar manner. That is why we skip some details which call on the same arguments as above. We have

$$\| r_h(u) \|_{0,2}^2 \leq \left[\frac{B}{ch} \right]^2 \times$$

$$\sum_{i=1}^{n} \int_{x_{i-1}}^{x_i} \left[(x - x_{i-1}) \int_x^{x_i} (x_i - s)u''(s)ds + (x_i - x) \int_{x_{i-1}}^x (s - x_{i-1})u''(s)ds \right]^2 dx,$$

or

$$\| r_h(u) \|_{0,2}^2 \leq \frac{2B^2}{c^2} \sum_{i=1}^{n} \int_{x_{i-1}}^{x_i} \left\{ \left[\int_x^{x_i} (x_i - s)u''(s)ds \right]^2 \right.$$

$$\left. + \left[\int_{x_{i-1}}^x (s - x_{i-1})u''(s)ds \right]^2 \right\} dx.$$

This yields

$$\| r_h(u) \|_{0,2}^2 \leq 2 \left[\frac{Bh}{c} \right]^2 \sum_{i=1}^{n} \int_{x_{i-1}}^{x_i} \left\{ \left[\int_x^{x_i} |u''(s)|ds \right]^2 + \left[\int_{x_{i-1}}^x |u''(s)|ds \right]^2 \right\} dx,$$

and, further,

$$\| r_h(u) \|_{0,2}^2 \leq h \left[\frac{2Bh}{c} \right]^2 \sum_{i=1}^{n} \left[\int_{x_{i-1}}^{x_i} |u''(s)|ds \right]^2.$$

Then, using the Cauchy–Schwarz inequality applied to integrals in (x_{i-1}, x_i), we obtain

$$\| r_h(u) \|_{0,2}^2 \leq \left[\frac{2Bh^2}{c} \right]^2 \sum_{i=1}^{n} \int_{x_{i-1}}^{x_i} |u''(x)|^2 dx,$$

or, equivalently,

$$\| r_h(u) \|_{0,2} \leq C_r h^2 \| u'' \|_{0,2},$$

with $C_r = 2B/c$.

Summarising, the estimates in equation (2.36) of the global truncation error suggest that the FEM is a consistent method with order two in terms of the L^2-norm of the error, and of order one in terms of the same norm of the error first-order derivative. However, in a variational framework, these conclusions are not as clear as in the local sense applying to the FDM. That is why we prefer to clarify the concept of order of an FEM in the following section, which deals precisely with error estimates for discretisation methods, besides their convergence. Incidentally, we note that an elementary variant of equation (2.36) promptly yields estimates for the interpolation error in the norms of $L^2(0, L)$ and the semi-norm of $H^1(0, L)$, namely,

Error estimates for the $\mathcal{P}_1$ FE interpolate $\tilde{v}_h$ of $v \in \mathbf{H}^2(0, L)$

$$\boxed{\begin{array}{l} \| v - \tilde{v}_h \|_{0,2} \leq \tilde{C}_0 h^2 \| v'' \|_{0,2} \\ \| [v - \tilde{v}_h]' \|_{0,2} \leq \tilde{C}_1 h \| v'' \|_{0,2}. \end{array}}$$

$$(2.37)$$

where $\tilde{C}_0$ and $\tilde{C}_1$ are mesh-independent constants.

Incidentally, we point out that finer estimates for C_s and C_r independent of c can be obtained. These estimates are left to the reader as Exercise 2.6. Notice that resulting values of $\tilde{C}_1$ will not account for whether or not a quasi-uniform family of meshes is in use. We strongly suggest the reader to determine fine expressions for both constants in Exercise 2.7.

To conclude this section, we note that the order of consistency of the Vertex-centred FVM can be determined in the same way as in the case of the FDM. Indeed, the corresponding scheme can be viewed as an FD scheme with a non-uniform grid according to equation (1.26). However, the order of consistency of this (FD or FV) scheme is one instead of two, since for a non-uniform grid it is not possible to get rid of the terms involving u''' in the Taylor expansions.

On the other hand, the Cell-centred FVM works differently, for the order of its local truncation errors is zero. Actually, this scheme cannot be labelled as a consistent scheme by means of Taylor expansions, or more vulgarly in the 'FD sense'. However, the fluxes it is based upon are consistent in another sense, and from this fact it is possible to guarantee that the Cell-centred FVM is completely reliable. We refer to Example 1.1 for an illustration of this assertion. We omit a detailed study of this property here for the sake of brevity, preferring to postpone it to Chapters 5 and 7, where it will be carried out in a similar but wider context. However, we encourage the reader to practice the techniques employed in this section, by determining the local truncation errors for both FVMs. Incidentally, this is the purpose of Exercise 2.8.

2.4   Convergence of the Discretisation Methods

A numerical method for solving a certain type of differential equation cannot be viewed as a reliable one, unless it possesses a property called **convergence**. In simple terms, this means that the numerical values provided by the method will tend in some sense to the corresponding values of the exact solution, as the method's discretisation parameters approach zero. Assume that we can measure in a certain norm $\| \cdot \|$ both the relevant values of the approximate solution generically denoted by $U_{\mathcal{H}}$ produced by a discretisation method, and the corresponding values of the exact solution denoted by U, where $\mathcal{H}$ represents the set of discretisation parameters. The method is said to converge if $\| U_{\mathcal{H}} - U \|$ tends to zero as all the parameters in $\mathcal{H}$ go to zero. Moreover, we say that the **order of convergence** of the discretisation method is $\mu_l > 0$ in terms of the lth discretisation parameter, say H_l, if the above value goes to zero as fast as $H_l^{\mu_l}$ does, making the utopian assumption that all the other parameters have already tended to zero. Of course, in case there is only one discretisation parameter, or in case all the discretisation parameters are linked to each other by certain relations, the concept of order of convergence can be mastered more realistically, since they are supposed to tend to zero all together in a prescribed manner. An expression bounding $\| U - U_{\mathcal{H}} \|$, possibly implying the method's convergence, is called an **error estimate**.

Equivalently, we say that the approximate solution $U_{\mathcal{H}}$ converges to U as all the parameters in $\mathcal{H}$ go to zero if the above condition is fulfilled. For example, in the FDM, $\mathcal{H}$ is reduced to the grid size h, and $U_{\mathcal{H}}$ is the vector of grid values $\vec{u}_h$. Then, naturally enough, U is the vector $\vec{u}_h = [u(x_1), u(x_2), \cdots, u(x_n)]^T$, and we may check the method's convergence using any norm of $\Re^n$, in particular the maximum norm. Actually, this is the way we will study the convergence of the FDM hereafter.

Now pushing further the maxim quoted in the introduction of this chapter [123] regarding convergence of a numerical method, in this section we give examples that significantly illustrate it. More precisely, we will establish convergence in certain norms by combining the method's stability in the sense of these norms with a strictly positive order of consistency.

2.4.1   Convergence of the Three-point FDM

Here, we endeavour to show that the FDM with an equally spaced grid to solve the modification of (P_1) accommodating the boundary condition $u(0) = a$ converges in the maximum norm of $\Re^n$ with order two, as long as the solution u has some suitable regularity properties. More specifically, we want to establish that $\| \, \vec{\tilde{u}}_h - \underline{\vec{\tilde{u}}}_h \|_\infty$ goes to zero like a term of the form Ch^2 as h goes to zero, for a suitable constant C depending on u. This means in particular that the approximate values at the grid points will be closer and closer to the corresponding values of the exact solution as h goes to zero, or as n goes to infinity. This result applies to the modification (2.29) of equation (1.12), for the order of consistency is only one in the case of the latter scheme. Let us define the vector $\vec{\bar{u}}_h := \vec{\tilde{u}}_h - \vec{\tilde{u}}_h$. Noticing that $\underline{u}_0 = \tilde{u}_0 = a$, we have $\bar{u}_0 = 0$. Moreover, since $B_h \underline{\vec{\tilde{u}}}_h - \underline{\vec{f}}_h = \vec{0}$, and from the consistency analysis in the previous section $B_h \vec{\tilde{u}}_h - \underline{\vec{f}}_h = \vec{r}_h(u)$, we have $B_h \vec{\bar{u}}_h = \vec{r}_h(u)$. This means that the vector $\vec{\bar{u}}_h$ is the solution provided by FD scheme (1.12) with $a = 0$, and the right side is the vector $\vec{r}_h(u)$ instead of $\vec{f}_h$. But, according to our stability result in equation (2.3), we must have

$$\| \, \vec{\bar{u}}_h \|_\infty \leq C \, \| \, \vec{r}_h(u) \|_\infty.$$

Thus recalling equation (2.31), provided the fourth derivative of u is bounded in $[0, L]$, we immediately establish the

Error estimate for the modified Three-point FD scheme (equation (2.29))

$$\boxed{\| \, \vec{\tilde{u}}_h - \underline{\vec{\tilde{u}}}_h \|_\infty \leq Cph^2 \, \| \, u^{(iv)} \|_{0,\infty}/12.} \tag{2.38}$$

This means that the modified FD scheme (2.29) is **second-order convergent** in the maximum norm of the grid point values. Logically enough, by the same arguments we would conclude that scheme (equation (1.12)) is first-order convergent in the same norm. However, while on the one hand **the order of consistency may be sufficient for the same order of convergence to hold, on the other hand it is by no means necessary**. Actually (1.12) turns out to be **second-order convergent**, but the proof of this result is more subtle (cf. Exercise 7.1). Similarly, the reader can figure out quite easily that in the case of non-uniform grids, both schemes are first-order consistent, and hence to the least first order convergent in this norm. The wonder is that numerical experiments show second order convergence even in this case, provided u is sufficiently smooth.

For a more concrete interpretation of equation (2.38), let us assume that we are using nested grids in such a way that h is divided by two from a given discretisation level to the next. Confining ourselves to values of n greater than two, equation (2.38) implies that $u_{n/2}$ will tend to $u(L/2)$ like an $O(1/n^2)$ term, or an $O(h^2)$ term, as n increases or h decreases indefinitely. Of course, if we consider the process to be infinite, this conclusion applies to every grid point at any discretisation level. However, convergence in the above sense also has favourable consequences on the quality of the numerical solution as related to any point in the equation's definition domain, as we will see in Chapter 3 (cf. Remark 3.2).

2.4.2   Convergence of the $\mathcal{P}_1$ FEM

In order to establish convergence results for the $\mathcal{P}_1$ FEM applied to (P_3), we will follow the very same recipe already exploited for the FDM. However, in order to do so, it is convenient to consider the following variant of (P_3):

$$\left\{\begin{array}{l} \text{Given } f \text{ and } g \in L^2(0, L) \text{ find } w \in V \text{ such that} \\[2mm] \displaystyle\int_0^L pw'v'dx + \int_0^L qwv \; dx = \int_0^L fv \, dx + \int_0^L gv'dx, \; \forall v \in V \end{array}\right. \qquad (P_3')$$

The idea is to add to the right side a term that mimics the one appearing on the left side of (P_3), but not on its right side. Problem (P_3') has a unique solution, and we may consider its linear FE approximation denoted by $w_h \in V_h$, that is, the solution of

$$\left\{\begin{array}{l} \text{Find } w_h \in V_h \text{ such that} \\[2mm] \displaystyle\int_0^L pw_h'v'dx + \int_0^L qw_h v \; dx = \int_0^L fv \, dx + \int_0^L gv'dx, \; \forall v \in V_h \end{array}\right. \qquad (2.39)$$

Since we are considering the case where $a = b = 0$, the following stability inequalities hold for w_h:

$$\left\{\begin{array}{l} \| \, w_h' \, \|_{0,2} \leq C_1[\| \, f \, \|_{0,2} + \| \, g \, \|_{0,2}], \\[2mm] \| \, w_h \, \|_{0,2} \leq C_1'[\| \, f \, \|_{0,2} + \| \, g \, \|_{0,2}]. \end{array}\right. \qquad (2.40)$$

A simple adaptation of the stability analysis for equation (2.10) carried out in Section 2.1, leading to equations (2.19) and (2.20), allows us to establish stability inequality (equation (2.40)) without any significant difficulty (cf. Exercise 2.9).

Now we recall that $\tilde{u}_h$ is the interpolate of u in V_h^+ defined in Subsection 2.3, together with the consistency results (equation (2.36)). Notice that in the particular case we are considering, $V_h^+ = V_h$. If we replace w_h on the left side of equation (2.39) by $\tilde{u}_h - u_h$, this corresponds to taking $f = r_h(u)$ and $g = s_h(u)$. Therefore applying the stability result (equation (2.40)) we readily obtain

$$\left\{\begin{array}{l} \| \, [\tilde{u}_h - u_h]' \, \|_{0,2} \leq C_1(\| \, r_h(u) \, \|_{0,2} + \| \, s_h(u) \, \|_{0,2}), \\[2mm] \| \, [\tilde{u}_h - u_h] \, \|_{0,2} \leq C_1'(\| \, r_h(u) \, \|_{0,2} + \| \, s_h(u) \, \|_{0,2}), \end{array}\right. \qquad (2.41)$$

Recalling equation (2.36), this leads to the following estimate:

$$\| \, \tilde{u}_h - u_h \, \|_{0,2} + \| \, [\tilde{u}_h - u_h]' \, \|_{0,2} \leq \max[C_1, C_1'][C_r h^2 + C_s h] \| \, u'' \, \|_{0,2}. \qquad (2.42)$$

The error estimate in equation (2.42) should be sufficient to persuade the user that the $\mathcal{P}_1$ FEM converges with order one in the L^2-norm of the derivative of u, by requiring only that the second-order derivative of u is square integrable in $(0, L)$. However, in this case, we can go further than with the FDM, since we can exhibit the order with respect to norms of $u - u_h$. Indeed, since $u - u_h = (u - \tilde{u}_h) + (\tilde{u}_h - u_h)$, using the triangle inequality we derive

$$\| \, u - u_h \, \|_{0,2} \leq \| \, \tilde{u}_h - u \, \|_{0,2} + \| \, \tilde{u}_h - u_h \, \|_{0,2}$$

together with a similar relation for the derivatives. Then, noticing that estimates of the same type as equation (2.42) hold for the interpolation error $u - \tilde{u}_h$ according to equation (2.37), and since $h^2 \leq Lh$, we finally obtain for a suitable mesh-independent constant $\tilde{C}$ the

Error estimate for the $\mathcal{P}_1$ FEM

$$\| u - u_h\|_{1,2} \leq \tilde{C}h \parallel u'' \parallel_{0,2},$$
$$\text{where } \| u - u_h\|_{1,2} := [\parallel u - u_h \parallel_{0,2}^2 + \parallel [u - u_h]' \parallel_{0,2}^2]^{1/2} \tag{2.43}$$

Otherwise stated, we have just established that the $\mathcal{P}_1$ FEM is first-order convergent in the L^2-norms of both the function and its first-order derivative. The reader is encouraged to check the path leading to the value of $\tilde{C}$ in terms of L, α, A and B, in order to consolidate the understanding of such convergence result (cf. Exercise 2.10).

The part of equation (2.43) stating that the $\mathcal{P}_1$ FEM is first-order convergent in the L^2-norm is not optimal. Actually, provided we can assert that u'' belongs to $L^2(0, L)$ *for all* $f \in L^2(0, L)$, it is possible to prove that the order of convergence of $\| u - u_h \|_{0,2}$ as h goes to zero is two. Such a regularity of u'' holds under some suitable assumptions on both p' and q. Here, in order to simplify the argument, besides the boundedness assumptions on p and q of Section 1.1, we assume that p' is bounded and discontinuous to the most at a countable set of points in $(0, L)$. Moreover, for the sake of brevity, we consider only the case where $a = b = 0$, leaving the extension to the general case as Exercise 2.11. Let us verify the assertion that $u'' \in L^2(0, L)$ under such assumptions.

First, we note that p must be continuous and satisfy the corresponding condition in equation (1.1). Then, from (P_1), we have

$$pu'' = -f + qu - p'u'.$$

Therefore, letting C be an upper bound of $|p'(x)|$ in $(0, L)$, we trivially have

$$|u''(x)| \leq (|f(x)| + B|u(x)| + C|u'(x)|)/\alpha \ \forall x \in (0, L).$$

By a simple application of the Cauchy–Schwarz inequality in $\Re^3$, we easily obtain

$$|u''(x)|^2 \leq 3(|f(x)|^2 + B^2|u(x)|^2 + C^2|u'(x)|^2)/\alpha^2 \ \forall x \in (0, L). \tag{2.44}$$

Now, the stability inequality (equation (2.24)) in energy norm that holds for (P_1) implies that

$$\| u' \|_{0,2} \leq \frac{\sqrt{2}C_e}{\sqrt{\alpha}} \parallel f \parallel_{0,2}. \tag{2.45}$$

Moreover, from the Friedrichs–Poincaré inequality, we have

$$\| u \|_{0,2} \leq \frac{2\sqrt{2}C_e L}{\sqrt{\alpha}} \parallel f \parallel_{0,2}. \tag{2.46}$$

Then, after straightforward manipulations, it follows from equations (2.44), (2.45) and (2.46) that there exists a constant C'' depending only on p, q and L (the reader should determine this constant as Exercise 2.12), such that

$$\| u'' \|_{0,2} \leq C'' \parallel f \parallel_{0,2}, \tag{2.47}$$

and we are done.

Next, we proceed to the finer estimate of the error $\| u - u_h \|_{0,2}$ using a so-called **duality argument**. Provided $u \neq u_h$ we can write:

$$\| u - u_h \|_{0,2} = \frac{(u - u_h|u - u_h)_0}{\| u - u_h \|_{0,2}} \tag{2.48}$$

Now, let v be the solution of the **adjoint problem** [2], namely, equation (P$_1$) taking $f = u - u_h$. Since $u - u_h \in L^2(0, L)$, we have $v'' \in L^2(0, L)$ and, from equation (2.47), $\| v'' \|_{0,2} \leq C''$ $\| u - u_h \|_{0,2}$. Then, from the definition of v and equation (2.48), we have

$$\| u - u_h \|_{0,2} \leq C'' \frac{(u - u_h| - [pv']' + qv)_0}{\| v'' \|_{0,2}}$$

or yet, using integration by parts and recalling equation (2.22),

$$\| u - u_h \|_{0,2} \leq 2C'' \frac{(u - u_h|v)_e}{\| v'' \|_{0,2}} \tag{2.49}$$

Let $\tilde{v}_h \in V_h$ be the interpolate of v at the mesh nodes. Still assuming that the integral of $f v_h$ in $(0, L)$ is computed exactly, we obviously have $(u|\tilde{v}_h)_e = (u_h|\tilde{v}_h)_e$. Plugging this relation into equation (2.49), it follows that

$$\| u - u_h \|_{0,2} \leq 2C'' \frac{(u - u_h|v - \tilde{v}_h)_e}{\| v'' \|_{0,2}}$$

From Subsection 2.2.2, we know that, as long as $q > 0$, $(\cdot|\cdot)_e$ (resp. $\| \cdot \|_e$) is an inner product (resp. a norm) on the space $H^1(0, L)$ of those functions in $L^2(0, L)$, whose first-order derivatives also belong to $L^2(0, L)$. Moreover, it can be easily shown that $\| w \|_e \leq \sqrt{\max[A, B]/2} \, \| w \|_{1,2} \, \forall w \in H^1(0, L)$. Then, using this inequality together with the Cauchy–Schwarz inequality applied to $(\cdot|\cdot)_e$, we further obtain

$$\| u - u_h \|_{0,2} \leq C'' \max[A, B]/2 \, \| u - u_h \|_{1,2} \frac{\| v - \tilde{v}_h \|_{1,2}}{\| v'' \|_{0,2}} \tag{2.50}$$

Finally resorting to equation (2.37), we can assert that there exists a constant C^* depending neither on the mesh nor on u, such that $\| v - \tilde{v}_h \|_{1,2} \leq C^* h \, \| v'' \|_{0,2}$. Recalling equation (2.43), this readily yields the

Error estimate for the $\mathcal{P}_1$ FEM in the L^2-norm

$$\boxed{\| u - u_h \|_{0,2} \leq C_0 h^2 \, \| u'' \|_{0,2} ,} \tag{2.51}$$

where $C_0 = \tilde{C} C^* C'' \max[A, B]$.

2.4.3   Remarks on the Convergence of the FVM

In this subsection, we highlight the convergence properties of the FVM. Basically, the convergence of the Vertex-centred FVM can be studied in the same manner as the FDM, that is, in the sense of the maximum norm. This is because, similarly to the FD scheme with a non-uniform grid, this FV scheme also satisfies a DMP. However, from the formal point of view, there is a

---

[2] Here, the adjoint problem has the same form as the differential equation (P$_1$) it is related to, but it is not always so. By definition, the adjoint problem of a boundary value problem $\mathcal{L}u = f \in L^2(0, L)$, $\mathcal{L}$ being a differential operator, is $\mathcal{L}^* v = g$ for some given function $g \in L^2(0, L)$, where $\mathcal{L}^*$ is the differential operator such that $(u|\mathcal{L}^* v)_0 = (\mathcal{L}u|v)_0$ for all $u, v \in H^1(0, L)$ satisfying suitable additional integrability and boundary conditions, not necessarily identical.

fundamental difference between both methods. Indeed, in the case of the FVM, we are dealing with piecewise constant functions and not only with values at a finite set of grid points.
However, such a subtlety plays no important role in the convergence analysis.

More concretely, according to equation (1.26), the Vertex-centred FV scheme is nothing
but the Three-point FD scheme with a non-uniform grid. Therefore, even if both methods are
conceptually different, in practical terms the corresponding convergence results can only be
identical. For example, in the case of constant p and q, and a uniform mesh, provided $u^{(iv)}$
is bounded in $[0, L]$, the Vertex-centred FVM is first-order convergent in the maximum norm.
Here, the expected result is the

Error estimate for the Vertex-centred FVM with a uniform mesh

$$\| u - u_h\|_{0,\infty} \leq C_v[h \, \| u'\|_{0,\infty} + h^2 \, \| u^{(iv)}\|_{0,\infty}] \tag{2.52}$$

where C_v is a mesh-independent constant; and u_h is the piecewise constant function whose
value in the jth FV is u_j. Estimate (2.52) can be derived using the triangle inequality:
$$\| u - u_h\|_{0,\infty} \leq \| \underline{u}_h - u_h\|_{0,\infty} + \| u - \underline{u}_h\|_{0,\infty},$$

where $\underline{u}_h$ is the piecewise constant function whose value in the jth CV V_j is $u(x_j)$. Indeed,
from Subsection 2.4.1, we know that $\| \underline{u}_h - u_h\|_{0,\infty} \leq C_d h^2 \, \| u^{(iv)}\|_{0,\infty}$ for a suitable
mesh-independent constant C_d. Furthermore, from the theory of polynomial interpolation
(cf. [158]), for a suitable mesh-independent constant C_I it holds that
$$\max_{x\in V_j} |u(x) - \underline{u}_h(x)| \leq C_I h \max_{x\in[0,L]} |u'(x)| \; \forall j.$$

This readily implies equation (2.52).

If the mesh is not uniform, the same qualitative results can be derived on the basis of the
corresponding ones applying to the FDM. Still taking constant p and q for simplicity and
using a DMP, the reader may check that, in this case, the method's convergence relies on the
following:

Error estimate for the Vertex-centred FVM with a non-uniform mesh

$$\| u - u_h\|_{0,\infty} \leq C_v'h[\| u'\|_{0,\infty} + \| u'''\|_{0,\infty}] \tag{2.53}$$

for a suitable mesh-independent constant C_v'.

As far as the Cell-centred FVM is concerned, except for particular cases, the situation is
quite different. We can also derive a DMP for this scheme, and the stability in the maximum
norm easily follows. However now, even for a uniform mesh, in contrast to the inner cells,
the local truncation error in the cell $(0, h)$ is not even an $O(h)$. In order to check this, let us
write two Taylor expansions about the point $x = h/2$, assuming that the third derivative of u
is continuous in $(0, L)$. For suitable $\xi^- \in (0, h/2)$ and $\xi^+ \in (h/2, 3h/2)$, we have

$$0 = u(0) = u(h/2) - (h/2)u'(h/2) + (h/2)^2 u''(h/2)/2 - (h/2)^3 u'''(\xi^-)/6$$

and

$$u(3h/2) = u(h/2) + hu'(h/2) + h^2 u''(h/2)/2 + h^3 u'''(\xi^+)/6.$$

Hence, taking a constant p, we have

$$p\{u(h/2)/(h/2) + [u(h/2) - u(3h/2)]/h\}/2 = -3phu''(h/2)/4 + O(h^2),$$

where the term $O(h^2)$ stands for $ph^2[u'''(\xi^-)/24 - u'''(\xi^+)/6]$.On the other hand, the scheme's residual at $x = h/2$ in the pointwise sense is given by

$$[r_h(u)]_{1/2} := p\{u(h/2)/(h/2) + [u(h/2) - u(3h/2)]/h\}/h + qu(h/2) - f(h/2)$$

Thus, $[r_h(u)]_{1/2}$ is seen to equal $pu''(h/2)/4 + O(h)$. Such an $O(1)$ term is not sufficient to establish the convergence in the maximum norm of $\vec{u}_h$ using the same logic as in Subsection 2.4.1. Actually, it has been observed for a long time that the local truncation errors in the FD sense can be much larger than the true absolute errors at CVs' representative points (see e.g. [62]). Fortunately, convergence results in the maximum norm can be proven to hold for Cell-centred FV schemes using more tricky arguments (cf. [68] and references therein), rather than a mere evaluation of local truncation errors. It is even possible to prove second-order convergence, under suitable regularity assumptions on u, as long as the CV representative points are located at their centres (cf. [74]). However, the reader should be aware of the fact that equivalent results cannot be proven for the higher dimensional counterparts of the Cell-centred FVM.

We refer to reference [68] for further details on all those issues related to the FVM. In this book, we confine ourselves to establishing the following result applying to equation (1.28):

Error estimate for the Cell-centred FVM with a non-uniform mesh

$$\boxed{\| u - u_h \|_{0,2} \le C_c(u)h,} \tag{2.54}$$

where $C_c(u)$ is a constant depending only on L, p, q and the derivatives of u, u_h being the piecewise constant function whose value in the jth cell is $u_{j-1/2}$ (cf. Subsection 1.4.2). Incidentally, we prefer postponing the justification of equation (2.54) to Chapter 5, as a sort of by-product of results proven to hold for two-dimensional FV schemes analogous to the one-dimensional Cell-centred FV scheme. After examining this material, the reader will be able to check the validity of this error estimate rather easily, and obtain a precise expression for $C_c(u)$, in the case of a constant p (cf. Exercise 5.17). Finally, it is worth commenting that, in the literature, the quality of the FVM is frequently evaluated by using (discrete) mean-square norms, instead of the maximum norm. For a more comprehensive study of error estimates for the FVM, the author refers to reference [68].

Remark 2.5 *In view of the numerical results of Example 1.1, one could effectively conjecture that error estimate (2.54) is suboptimal. Notice, however, that in this case the errors were evaluated only at the representative points, where **superconvergence** takes place. We refer to test problems 1 and 3 of Example 2.1 hereafter for further explanations on this effect.* ∎

2.4.4   Example 2.1: Sensitivity Study of Three Equivalent Methods

The aim of this example is to illustrate the convergence rate of the FDM, the P_1 FEM and the Vertex-centred FVM for solving problem (P_1) in the interval $(0, 1)$ by means of numerical values. In particular, we shall deal with the case where the three methods are equivalent, which

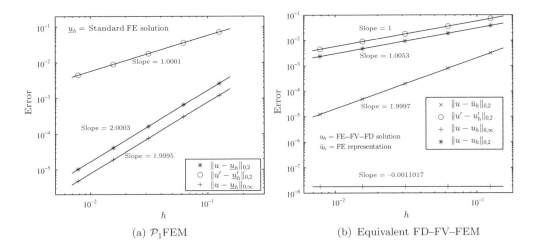

Figure 2.4   Errors for test problem 1.

means that they give rise to the same SLAE. Besides experiments performed under the conditions estimates (2.38), (2.43), (2.51) and (2.52) are valid, we consider situations not addressed in the theory developed in this chapter. More specifically, we take the cases of a discontinuous p and of an unbounded q, for which the exact solution u is known. Moreover, in order to figure out what can be expected in terms of convergence in case f is not in $L^2(0, 1)$ (even though not too 'wild'), we perform a **sensitivity analysis** pretending that the exact solution to one of the test problems is unknown. This means that we observe at which rate the numerical solutions obtained with a sequence of nested grids or meshes tend in different norms to the one computed with a mesh much finer than all of them. Three test problems are solved whose exact solution is known, with both smooth and irregular data. In the fourth test problem, the theoretical solution is unknown and a sensitivity analysis was carried out. In all cases, uniform meshes with n intervals were used for $n = 2^m$, with $m = 3, 4, 5, 6, 7$. For each test problem, we display in the subfigure on the right the errors of the solution computed by means of the three equivalent FD–FV–FE methods. We recall that such an equivalence occurs when the trapezoidal rule is employed to calculate the integrals of $qu_h v$ and fv in each element in the FE solution, and the CV representative points are selected for the application of the one-point quadrature rule to approximate the integrals of q and f in each cell in the Vertex-centred FV solution. The logarithm of the absolute errors measured in different norms are plotted against the logarithm of $h := 1/n$. These log–log plots are shown in Figures 2.4–2.6 and 2.7 for test problems 1, 2, 3 and 4, respectively. The straight lines whose slopes roughly correspond to the actual convergence rate are the best fitted to the error data in the least-squares sense. More precisely, we computed the L^2-norm of the solution and its first-order derivative errors (i.e. the errors in the L^2- and in the H^1-semi-norm), and also in the discrete L^∞-norm, that is, the error maximum absolute value restricted to the mesh nodes. The errors $u - u_h$ for the standard FE solution u_h are plotted on the left (part a), while the errors $u - u_h$ for the FD–FV–FE solution u_h are displayed on the right (part b). In the latter case, we also show the evolution of the L^2-norm of $u - \bar{u}_h$ where $\bar{u}_h$ is the piecewise linear function that takes the value of u_h at the mesh nodes.

In notational terms, we replaced the discrete first-order derivative $u_h^{'h}$ by the exact derivative $\bar{u}_h'$, but for uniform meshes the result is the same, as the reader can easily check. Both in the error computations and in the evaluation of integrals inherent to the standard FEM, numerical integration by Simpson's rule in each interval was employed.

Test problem 1: In this test, we took $p = 0.5$, $f(x) = 1 + x$ and a bounded continuous q, namely, $q = x/u(x)$, in such a way that our convergence results apply. The manufactured exact solution is $u(x) = x(2 - x)$. As one can infer from Figure 2.4, the numerical solution at the grid points or mesh nodes are exact practically up to machine precision. This occurs because the exact solution is a polynomial of degree two, and an FD solution reproduces exactly such a function at the grid points, since its third-order derivative vanishes identically (cf. Subsection 2.4.1). The order increase owing to this property of the FDM is carried over to the FE or the FV approximation of a sufficiently smooth function at the mesh nodes. This effect is called **superconvergence** or **supraconvergence**. For more details on this issue, we refer, for instance, to [204], [71] and [21].

Test problem 2: We took the same exact solution u as in test problem 2 of Example 1.1 for a discontinuous p, $q = x/u(x)$ and $f(x) = 1 + x$. We recall that $-[pu']' = 1$ and that the exact solution is given by $u(x) = x(2 - x)$ if $0 \leq x \leq 1/2$ and $u(x) = -x^2/2 + x - 3/8$ if $1/2 < x \leq 1$. Notice that u does not belong to $H^2(0, 1)$. Figure 2.5 points to deteriorated convergence rates for both implemented methods. More precisely, the convergence in both the L^2 and the L^∞ senses are roughly of the first order only, while a drop in value apparently from one to the inverse $\Phi - 1$ of the **golden ratio** (or **golden number**) $\Phi = (\sqrt{5} + 1)/2$ can be noticed in the case of the H^1-semi-norm. The one-point decrease in the L^2 convergence rate could be explained by the fact that the convergence rate in H^1 appears twice in the argument leading to the corresponding estimate. But this point requires a more careful analysis, since logically the drop would be rather from two to a value about twice $\Phi - 1$ in this case.

Test problem 3: We were given the same smooth manufactured solution $u(x) = 2x - x^2$ as in test problem 1, but this time we took $p \equiv 0.5$ and $q(x) = 1/xu(x)$. Although $u \in H^2(0, 1)$

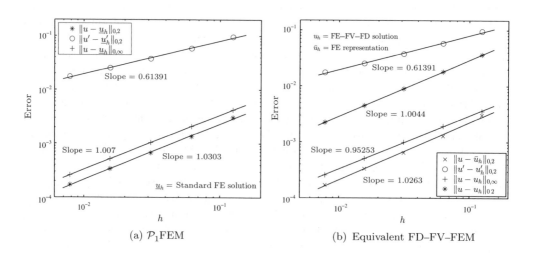

(a) $\mathcal{P}_1$FEM   (b) Equivalent FD–FV–FEM

Figure 2.5 Errors for test problem 2.

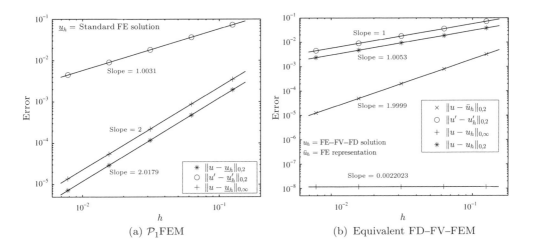

Figure 2.6   Errors for test problem 3.

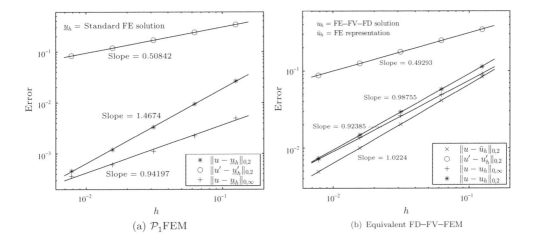

Figure 2.7   Errors for test problem 4.

this case does not really fit into the convergence theory we worked out in this chapter, because neither q is bounded, nor $f \in L^2(0,1)$. Nevertheless, the convergence behaviour of the numerical solution is exactly the optimal one observed in test problem 1, as one can infer from Figure 2.6. Notice that in this test problem, and also in test problem 4, $f = 1 + 1/x$ and q behaves like an $O(1/x^2)$ near the origin. However, both singularities can be removed when computing the left- and right-side terms $\int_0^1 q u_h v \, dx$ and $\int_0^1 f v \, dx$, since both u_h and v are of the form cx for a certain constant c in the leftmost element. In this respect, we recall that in order to let the three methods coincide, we must use a quadrature formula to approximate these integrals. In this way, the singularities of q and f at $x = 0$ are also avoided in practice, if we enforce $[fv](0) = c$ and $[qu_h v](0) = c'$ for suitable constants c and c' in the numerical quadrature formula in case $x = 0$ is one of the quadrature points. Notice that none of these singularities is a problem for the FD or the Vertex-centred FV schemes, since values of q or f at $x = 0$ are nowhere necessary. We refer to Remark 2.6 hereafter for further comments on this test problem.

Test problem 4: We took $q \equiv 1$, but p is the same discontinuous function as in test problem 2 and f is as in test problem 3. It follows that, in this case, the solution u does not belong to $H^2(0,1)$ either. As already pointed out, the convergence rates are observed as if the exact solution was the one obtained by the standard FEM with $n = 2^{12} = 4096$. It is interesting to underline that, in spite of data similarity, the observed convergence rates are quite different from those of test problem 2. Indeed, here the one of the numerical solutions by the standard FEM in $H^1(0,1)$ lowers from one to $1/2$ instead of $\Phi - 1$. Moreover, the convergence rate of the same solution in the L^2-norm is one point greater than in the $H^1(0,1)$-norm, which is less logical than in the case of test problem 2. This could suggest that a kind of superconvergence effect is occurring here, since q is more regular than in the latter case. As for the solution obtained by the three equivalent methods, similar rates are observed, except the aforementioned superconvergence in $L^2(0,1)$. But this is rather coherent with the poorer approximation properties of piecewise constant functions as compared to piecewise linear ones.

Remark 2.6 *Classical principles of continuum mechanics require that the total energy $I_e(v)$ of a bar undergoing longitudinal deformations be finite for every admissible displacement $v \in V = \{v|\, v \in H^1(0, L), v(0) = 0\}$. We recall that $I_e(v) = (v|v)_e - (f|v)_0$, where $(u|v)_e = \frac{1}{2} \int_0^L [pu' v' + quv](x) dx$ (cf. (2.21) and related expressions that follow). In the case of test problem 4, this holds true thanks to the facts that $v(0) = 0$ and q is bounded. Indeed, $\int_0^1 f(x)v(x)\, dx = \int_0^1 (1 + 1/x) \left[\int_0^x v'(s) ds \right] dx$, and by the Cauchy–Schwarz inequality,*

$$\int_0^1 f(x)v(x)dx \le \int_0^1 \left(\sqrt{x} + \frac{1}{\sqrt{x}} \right) \sqrt{\int_0^x |v'(s)|^2 ds}\ dx$$

This implies that

$$\int_0^1 f(x)v(x)dx \le \| v' \|_{0,2} \int_0^1 \left(\sqrt{x} + \frac{1}{\sqrt{x}} \right) dx = \frac{8}{3} \| v' \|_{0,2}.$$

In practical terms, this result means that the work of the load density $f(x) = 1 + 1/x$, increasing without a limit as one approaches the end where the bar is kept fixed, is always bounded. Notice that this result would be false if the same end had undergone a given nonzero displacement, say, equal to a. Indeed, in this case, any admissible displacement v would have to satisfy $v(0) = a$, and the work $(f|v)_0$ would be unbounded (please check!). In order to draw similar conclusions for test problem 3, more elaborated arguments are necessary, since in this case q is also unbounded. But this problem is rather academic, as we are dealing with a smooth manufactured solution. For this reason, we refrain from further commenting on it. ∎

   The observed convergence rates for test problems 2, 3 and 4 require further investigation, but they can certainly be explained using the mathematical tools supplied in this chapter.

2.5 Exercises

2.1 Check that, whatever the inner product $(\cdot|\cdot)$ defined on a linear vector space V, the expression $(v|v)^{1/2}$ defines a norm of all $v \in V$.

2.2 Show that the DMP holds for the Three-point FD scheme with a non uniform grid applied to problem (1.34), taking $b = 0$, $q \equiv 0$ and a constant p.

2.3 Give fine upper bounds for the constants C in equation (2.19) and C' in equation (2.20) in terms of α, B and L.

2.4 Give a fine upper bound for the constant C_e in equation (2.23), by combining equations (2.19) and (2.20).

2.5 Check that a DMP, analogous to the one that holds for the Three-point FDM in case $q(x) > 0$ for every $x \in [0, L]$, applies to the modification of the FV schemes (1.26) and (1.28) in order to incorporate the inhomogeneous boundary condition $u(0) = a$. Conclude the pointwise stability of both schemes in the same sense as for the FDM.

2.6 Give new expressions for the constants C_s and C_r bounding the L^2-norms of the residual functions $s_h(u)$ and $r_h(u)$ proportional to A and B, but not involving the uniformity constant c of a quasi-uniform family of meshes eventually in use. Conclude that the assumption that this family be quasi-uniform is needless for estimating the error $\| u_h - u \|_{1,2}$ by a term of the form $Ch \| u'' \|_{0,2}$ (hint: use the fact that both $|x - x_i|$ and $|x - x_{i-1}|$ are bounded above by $h_i \, \forall x \in T_i$).

2.7 Give expressions for the constants $\tilde{C}_0$ and $\tilde{C}_1$ in (2.37), independent of the uniformity constant c of a quasi-uniform family of meshes.

2.8 Determine the local truncation errors for the FVMs (1.26) and (1.28) for sufficiently differentiable p and u. What can be concluded about the consistency of each one of these methods? Check in particular the case of $x_{1/2}$. What conclusions can be drawn for scheme (1.28)?

2.9 Prove the validity of equation (2.40) by exhibiting the constants C_1 and C_1'.

2.10 Check the path leading to an expression for the constant $\tilde{C}$ in equation (2.42), in terms of L, α, A and B.

2.11 u being a solution to equation (1.34), verify the validity of the regularity result $u'' \in L^2(0, L)$ under the boundedness assumptions on p and q of Section 1.1, further assuming that $|p'|$ is bounded by a constant C_p' and discontinuous to the most at a finite set of points in $(0, L)$. Determine a constant C_2 in terms of L, B, α and C_p' such that $\| u'' \|_{0,2} \le C_2[|a| + |b| + \| f \|_{0,2}]$. Do not assume that q is bounded below in $[0, L]$ by $\beta > 0$.

2.12 Determine an expression for the constant C'' in equation (2.47) in terms of L, B, α and C_p'.

2.13   The purpose of this exercise is to set up a second-order Three-point FD scheme in the maximum norm to solve equation (1.34). The scheme should be a modification of the FD scheme proposed in Exercise 1.7, analogous to equation (2.29). Develop all the steps leading to a second-order convergence result.

2.14   Derive a first-order error estimate for the $\mathcal{P}_1$ FEM (2.10) to solve equation (2.9) in the norm $\| \cdot \|_{1,2}$, assuming that $u \in H^2(0, L)$.

2.15   Based on the result of Exercise 2.14, derive a second-order error estimate in the L^2-norm for equation (2.10), using a duality argument and assuming a suitable regularity of p, q and p'.

3

Time-dependent Boundary Value Problems

> Euler calculated without apparent effort, as men breathe, or as eagles sustain themselves in the wind.
>
> François Arago

In this chapter, we apply the techniques and related concepts introduced in Chapters 1 and 2, in order to study PDEs that model transient phenomena or processes. For obvious reasons, it is not our intention to be exhaustive about such a vast subject. Like in the two previous chapters, the idea is to address the essential aspects of the most popular numerical techniques to treat this class of problems. In order to achieve this at an introductory level, it is sufficient to consider some classical PDEs, whose domain of definition is a finite one-dimensional spatial domain $(0, L)$ and a time interval $(0, \Theta]$, Θ being finite or not.

The three types of discretisation methods studied in Chapter 1 are well adapted to treat the solution dependency on the single space variable x. As we saw, in this simpler case, for a given set of points at which the solution is approximated, the basic FD and FE schemes, together with the Vertex-centred FVM, yield the same SLAE if some tricks are used to derive them. We recall that equivalence of these schemes was achieved by means of numerical quadrature. Therefore, in order to focus on aspects inherent to the complementary time discretisation, we will address in this chapter mainly the FDM for the space discretisation, keeping in mind that in doing so we are implicitly considering the FEM and the Vertex-centred FVM as well. Nevertheless, in Subsection 3.3.4, we illustrate the most widespread way to use the latter methods in the framework of time-dependent problems, that is, a combination of the integral formulation in space with an FD time discretisation.

Also, in the aim of avoiding non-essential difficulties, we will consider only equations with constant physical coefficients. For the same reason, we will deal basically with uniform grids, in both space and time.

Numerical Methods for Partial Differential Equations:
An Introduction Finite Differences, Finite Elements and Finite Volumes, First Edition. Vitoriano Ruas.
© 2016 John Wiley & Sons, Ltd. Published 2016 by John Wiley & Sons, Ltd.
Companion Website: www.wiley.com/go/ruas/numericalmethodsforpartial

The two relevant situations correspond to what is known in the literature as PDEs of the **parabolic** and **hyperbolic** types, thoroughly studied from the mathematical point of view in classical text books such as reference [77] for the former and reference [109] for the latter. The first type of equation is significantly represented by the so-called **heat equation**, whose unknown is the temperature u of a bar with length equal to L, made of a medium with heat conductivity p, undergoing a heat conduction process. Denoting by ∂_t the partial derivative of a function v with respect to t, and by ∂_{xx} its second-order partial derivative with respect to x, we write as follows:

The heat equation

$$\partial_t u - p\partial_{xx} u = f \text{ in } (0, L) \times (0, \Theta], \tag{3.1}$$

where f is a given function well defined for all x and for all t, and assumed to be bounded. Equation (3.1) is a well-posed problem, provided it is supplemented with an **initial condition** at $t = 0$ and boundary conditions at $x = 0$ and $x = L$ for every t. Here, for the sake of simplicity, we shall deal only with Dirichlet boundary conditions. Therefore, equation (3.1) is supplemented with

Initial and boundary conditions for the heat equation

$$
\begin{aligned}
u(x, 0) &= u_0(x) \ \forall x \in [0, L] \\
u(0, t) &= a(t) \text{ and } u(L, t) = b(t) \ \forall t \in (0, \Theta].
\end{aligned}
\tag{3.2}
$$

where a, b and u_0 are given functions assumed to be bounded and everywhere defined in the respective definition intervals.

From the physical point of view, these conditions mean

- The temperature of the bar at a distance x from its left end equals $u_0(x)$ when the process starts;
- At every time t, the temperature of the bar is controlled at its ends, in such a way that it equals $a(t)$ at the left end and $b(t)$ at the right end.

Function f in turn represents a known heat source applied along the bar evolving in time.

The PDEs of the hyperbolic type are significantly represented by the **equation of the vibrating string**, whose unknown is the transversal deflection u of a string with length equal to L, made of a medium with lateral stiffness r and mass density per unit of length equal to ρ, subject to initial deflection conditions, and eventually a forcing function g such as gravity. The equation governing the vibrations of the string under these conditions is

The equation of the vibrating string

$$\partial_{tt} u - p\partial_{xx} u = f \text{ in } (0, L) \times (0, \Theta] \tag{3.3}$$

where $p = r/\rho$ and $f = g/\rho$, the meaning of $\partial_{tt} u$ being obvious. Here again, equation (3.3) is a well-posed problem, provided it is supplemented with (initial) conditions at $t = 0$ and (boundary) conditions at $x = 0$ and $x = L$ for every t. We will consider conditions frequently encountered in practical applications, namely,

Initial and boundary conditions for the vibrating string equation

$$
\begin{array}{|l|}
\hline
u(x,0) = u_0(x) \text{ and } \partial_t u(x,0) = u_1(x) \ \forall x \in [0,L] \\
u(0,t) = 0 \text{ and } u(L,t) = 0 \ \forall t \in (0,\Theta]. \\
\hline
\end{array}
\qquad (3.4)
$$

where u_0 and u_1 are given functions assumed to be bounded in $[0,L]$.

From the physical point of view, these conditions mean:

- The position of the string at the section located at a distance x from its left end is known to be $u_0(x)$ when the motion starts;
- The initial (lateral) velocity of the string at a distance x from its left end is known to be $u_1(x)$;
- Both ends of the string are kept fixed all along the motion.

It is worthwhile noting that equation (3.3) can be rewritten in the form of a system of first-order PDEs. For instance, setting $o = \pm\sqrt{p}$, $v = -\partial_x u$ and $w = o^{-1}\partial_t u$, the meaning of $\partial_x u$ being obvious, v and w are seen to satisfy a

System of first-order equations equivalent to equation (3.3)

$$
\begin{array}{|l|}
\hline
\partial_t v + o\partial_x w = 0 \\
\partial_t w + o\partial_x v = o^{-1} f \\
\hline
\end{array}
\qquad (3.5)
$$

supplemented with the initial conditions $v(x,0) = -u_0'(x)$ and $w(x,0) = o^{-1}u_1(x)$ $\forall x \in [0,L]$. The boundary conditions for w can be directly obtained from u, that is, $w(0,t) = w(L,t) = 0 \ \forall t \in (0,\Theta]$. Strictly speaking, boundary conditions for v would be super-abundant, for after all system (3.5) is equivalent to a second-order PDE. Nevertheless, since this could be helpful in a numerical solution scheme, we may apply Neumann boundary conditions to v stemming from equation (3.5), according to the following considerations. Assume that u is sufficiently smooth in $(0,L) \times (0,\Theta]$, and require that f be at least continuous. Since $o\partial_x v(0,t) = -\partial_t w(0,t) + o^{-1} f(0,t)$ and $\partial_t w(0,t) = 0$ for all t, we have $\partial_x v(0,t) = p^{-1}f(0,t)$, and similarly $\partial_x v(L,t) = p^{-1}f(L,t)$ for all t.

External forces can often be neglected in practice, and for this reason in all the sequel we will take $f \equiv 0$. In this case, system (3.5) becomes a two-component vector counterpart of

A paradigm of first-order hyperbolic PDEs: the transport equation

$$
\begin{array}{|l|}
\hline
\partial_t u + o\partial_x u = 0 \ \forall t \in (0,\Theta] \text{ and } \forall x \in (0,L). \\
\hline
\end{array}
\qquad (3.6)
$$

Equation (3.6) is commonly known as the **linear transport equation** for the reasons explained hereafter. Its numerical solution gave rise to fundamental papers on discretisation methods. Among them, the work of Lax and Richtmyer in the 1950s [123] is a remarkable example. We refer to the vast literature on numerical methods for **conservation laws** to know more about classical hyperbolic systems and their numerical solution. Indeed, the physical conservation laws such as mass or energy conservation are often expressed by systems of first-order hyperbolic equations. Hence, although equation (3.3) is a second-order hyperbolic PDE, we will deal with the above first-order equation (equation (3.6)) in Section 3.2, since it gives insight on the proper way to handle equation (3.3) and the equivalent first-order system (3.5).

Incidentally, we note that equation (3.6) has a unique solution when suitable initial and boundary conditions are prescribed. In Section 3.2, we will consider the following ones ensuring problem well-posedness:

Initial and boundary conditions for the transport equation

$$
\begin{aligned}
&u(x,0) = u_0(x) \ \forall x \in [0,L] \text{ and} \\
&u(0,t) = 0 \ \forall t \in (0,\Theta], \text{if } o > 0, \text{ or} \\
&u(L,t) = 0 \ \forall t \in (0,\Theta], \text{if } o < 0.
\end{aligned}
\tag{3.7}
$$

Before pursuing, we should emphasise that the vibration of a string is far from being the only application of equation (3.3). For instance, it also governs the phenomenon of wave propagation. For this reason, it is also known as the **wave equation**. For more information on hyperbolic PDEs we refer to reference [109].

Chapter outline: In Section 3.1, we deal with the space–time discretisation of equations (3.1)–(3.2), while in Section 3.2 we do the same for equations (3.6)–(3.7). In Sections 3.3 and 3.4, we study the stability and the convergence of the corresponding numerical schemes, respectively. Some complements on the equation of the vibrating string are addressed in Section 3.5, where a numerical example is also given.

3.1 Numerical Solution of the Heat Equation

Let us first consider the following semi-discretisation in space of equations (3.1)–(3.2). Given the equally spaced FD grid $\mathcal{G} = [0, h, 2h, \dots, (n-1)h, nh = L]$, we denote by $\hat{u}_i(t)$ an FD approximation of the solution u at the grid point $x = ih$, $1 \le i \le n-1$, at time t. These approximations are supplemented with the exact values $\hat{u}_0(t) := a(t)$ and $\hat{u}_n(t) := b(t)$. Then, like in Section 1.2, we approximate $\partial_{xx}u$ at the grid point $x = ih$ for every t by the FD $[2\hat{u}_i(t) - \hat{u}_{i-1}(t) - \hat{u}_{i+1}(t)]/h^2$. Naturally enough, equation (3.1) at point (ih, t) is approximated by

$$
\hat{u}_i'(t) + p[2\hat{u}_i(t) - \hat{u}_{i-1}(t) - \hat{u}_{i+1}(t)]/h^2 = f(ih, t) \ \forall t \in (0, \Theta], \text{for } i = 1, 2, \dots, n-1,
\tag{3.8}
$$

where g' represents the first-order derivative of a function $g(t)$. Notice that equation (3.8) is nothing but a system of linear ODEs in terms of the independent variable t, for the $(n-1)$-component unknown vector $\vec{\hat{u}}(t) := [\hat{u}_1(t), \hat{u}_2(t), \dots, \hat{u}_{n-1}(t)]^T$, having $\vec{\hat{u}}(0) = [u_0(h), u_0(2h), \dots, u_0(nh - h)]^T$ as the initial condition. Recalling the notations used in Chapter 1 for the equally spaced FDM, with a constant p and $q = 0$, in concise form this system writes

$$
\begin{cases}
[\vec{\hat{u}}(t)]' + A_h^+ \vec{\hat{u}}(t) = \vec{b}_h(t) \ \forall t \in (0, \Theta] \\
\text{with } \vec{\hat{u}}(0) = \vec{u}^0,
\end{cases}
\tag{3.9}
$$

where $\hat{u}_0(t) = a(t)$, $\hat{u}_n(t) = b(t)$, $\vec{b}_h(t) := [f(h, t), f(2h, t), \dots, f(nh - h, t)]^T$ and $\vec{u}^0 := [u_0(h), u_0(2h), \dots, u_0(nh - h)]^T$. Notice that matrix A_h^+ is similar but not identical to the

$n \times n$ matrix introduced in Section 1.2. In order to incorporate the Dirichlet boundary conditions at $x = 0$ and $x = L$, in equation (3.9) A_h^+ is an $(n - 1) \times (n + 1)$ matrix, whose columns are numbered from 0 through n. The column numbered as the jth column, for j between one and $n - 1$, is the same as the one of the matrix corresponding to the FD discretisation of (P_1). The 0th and the nth columns in turn are the vectors of $\Re^{n-1}$ $[-p/h^2, 0, \ldots, 0]^T$ and $[0, \ldots, 0, -p/h^2]^T$, respectively.

In general, an analytic solution to system (3.9) cannot be found. Therefore, we endeavour to solve it numerically as well. As the reader certainly knows, there are countless efficient methods for the numerical solution of initial value first-order systems of ODEs, such as the Euler methods, Runge–Kutta methods, and multistep methods. The literature on the subject is consequently profuse, and if we are to give just one reference we could cite reference [84]. However, here the best choices are those that, while being simple to implement, provide an accuracy compatible with the method employed for the space discretisation. For this reason, many users choose **Euler methods of the first order**, enabling either an **explicit** or an **implicit** solution. Another technique widely in use is the **Crank–Nicolson method**. It is of the second-order in time and gives rise to an implicit resolution. We will exploit these three possibilities hereafter.

3.1.1   Implicit Time Discretisation

Let Θ be finite and l be an integer strictly greater than one. We define a **time step** $\tau = \Theta/l$, with which we associate a (time) FD grid $\mathcal{H} = [0, \tau, 2\tau, \ldots, l\tau = \Theta]^T$. Now, for every $i \in \{1, 2, \ldots, n-1\}$ and for $k \in \{1, 2, \ldots, l\}$, we define an approximation of the time derivative of $\hat{u}_i$ at a certain time $t \in [(k-1)\tau, k\tau]$ to be specified for each method, by the FD:

$$D_i^k(u) := \frac{\hat{u}_i(k\tau) - \hat{u}_i([k-1]\tau)}{\tau}. \tag{3.10}$$

However, since the vector $\vec{\hat{u}}(t)$ is not known, here again the above definition is useless for practical purposes. Instead, we consider that we wish to determine approximations u_i^k of $\hat{u}_i(k\tau)$, and hence of $u(ih, k\tau)$, which satisfy a time-discrete analog of equation (3.9). We consider two possibilities in this subsection.

In the first method, we assign to the FD (equation (3.10)) the role of approximating the time derivative at $t = k\tau$. This results in a fully discretised **implicit** FD scheme to approximate equations (3.1)–(3.2), known as

The Backward Euler scheme (BES)

$$\boxed{\begin{array}{l} \frac{u_i^k - u_i^{k-1}}{\tau} + p \frac{2u_i^k - u_{i-1}^k - u_{i+1}^k}{h^2} = f(ih, k\tau) \\ \text{for } k = 1, 2, \ldots, l \text{ and } i = 1, 2, \ldots, n-1, \\ \text{with } u_i^0 = u_0(ih) \text{ and } u_0^k = a(k\tau), u_n^k = b(k\tau). \end{array}} \tag{3.11}$$

Notice that the underlying problem to solve for each value of k is a SLAE whose vector of unknowns is $\vec{u}^k := [u_1^k, u_2^k, \ldots, u_{n-1}^k]^T$. Assuming at first that $a(t) = b(t) = 0 \ \forall t$, this vector is determined, taking as data both $\vec{b}_h^k := [f(h, k\tau), f(2h, k\tau), \ldots, f(nh - h, k\tau)]^T$ and $\vec{u}^{k-1}$. In compact form, using the $(n-1) \times (n-1)$ identity matrix I, $\vec{u}^k$ is seen to be the solution of the following system:

$$(I/\tau + A_h)\vec{u}^k = \vec{u}^{k-1}/\tau + \vec{b}_h^k. \tag{3.12}$$

where in this case $A_h = \{a_{i,j}\}$ is the $(n-1) \times (n-1)$ symmetric tridiagonal matrix, whose nonzero entries are given by

$$a_{i,i} = 2p/h^2 \text{ for } 1 \le i \le n-1 \quad a_{i-1,i} = a_{i,i-1} = -p/h^2 \text{ for } 2 \le i \le n-1. \tag{3.13}$$

Another possibility is to consider that the FD (equation (3.10)) is an approximation of the time derivative at time $t = (k - 1/2)\tau$. Since we are neither defining nor computing approximations of $u(ih, [k-1/2]\tau)$, the natural thing to do is to let the mean value $(u_i^{k-1} + u_i^k)/2$ play this role. Setting $f_i^k := f(ih, k\tau)$ for $k = 0, 1, \ldots, l$ and $i = 1, \ldots, n-1$, this leads to the following implicit fully discretised FD scheme to approximate equations (3.1)–(3.2), known as

The Crank–Nicolson scheme (CNS)

$$
\boxed{
\begin{aligned}
&\frac{u_i^k - u_i^{k-1}}{\tau} + p\frac{2u_i^{k-1/2} - u_{i-1}^{k-1/2} - u_{i+1}^{k-1/2}}{h^2} = f_i^{k-1/2} \\
&\text{where } u_i^{k-1/2} := (u_i^{k-1} + u_i^k)/2 \text{ and } f_i^{k-1/2} := (f_i^{k-1} + f_i^k)/2, \\
&\text{for } k = 1, 2, \ldots, l \text{ and } i = 1, 2, \ldots, n-1, \\
&\text{with } u_i^0 = u_0(ih) \text{ and } u_0^k = a(k\tau), u_n^k = b(k\tau).
\end{aligned}
}
\tag{3.14}
$$

The problem to solve for each value of k is again a SLAE, whose vector of unknowns is $\vec{u}^k := [u_1^k, u_2^k, \ldots, u_{n-1}^k]^T$. If we assume again that $a(t) = b(t) = 0 \; \forall t$, $\vec{u}^k$ is the solution of the following system:

$$[I/\tau + A_h/2]\vec{u}^k = [I/\tau - A_h/2]\vec{u}^{k-1} + \vec{b}_h^{k-1/2}. \tag{3.15}$$

where $\vec{b}_h^{k-1/2} := [f_1^{k-1/2}, f_2^{k-1/2}, \ldots, f_{n-1}^{k-1/2}]^T$. Equations (3.12) and (3.15) tell us that for implicit methods, such as those studied in this subsection, at every time step a SLAE has to be solved, but the corresponding matrices remain fixed. Therefore, they can be factorised once for all before the time marching starts. In addition to this advantage, these methods will be shown to be **unconditionally stable** in specific senses, which means that the corresponding schemes are stable whatever discretisation parameters h and τ we choose. This property broadly makes up for the intrinsically necessary computational effort.

Notice that the matrices in both equations (3.12) and (3.15) are tridiagonal, and therefore these systems can be easily solved by iterative methods and advantageously by direct methods, since their matrices can be factorised into the product of bidiagonal triangular matrices once for all at the first step.

The reader may examine as Exercise 3.1 the necessary modifications in both schemes written in matrix form (equations (3.12) and (3.15)), in order to accommodate inhomogeneous boundary conditions. As a guide she or he could use the $(n-1) \times (n+1)$ matrix A_h^+ defined at the beginning of this section.

3.1.2   Explicit Time Discretisation

A tempting alternative to the implicit methods considered in the previous subsection to solve the differential system in equation (3.8) or equivalently equation (3.9) is an **explicit resolution**. By such an expression we mean that, in contrast to the Backward Euler method and the Crank–Nicolson method, at every time step each component of the solution vector $\vec{u}^k$ is determined independently of the others. This can be achieved by considering that the FD (equation (3.10)) is an approximation of the time derivative at time $(k-1)\tau$. This leads to the following explicit fully discretised FD scheme to approximate equations (3.1)–(3.2), namely,

The Forward Euler scheme (FES)

$$
\boxed{
\begin{aligned}
&\frac{u_i^k - u_i^{k-1}}{\tau} + p\frac{2u_i^{k-1} - u_{i-1}^{k-1} - u_{i+1}^{k-1}}{h^2} = f(ih, k\tau - \tau) \\
&\text{for } k = 1, 2, \ldots, l \text{ and } i = 1, 2, \ldots, n-1, \\
&\text{with } u_i^0 = u_0(ih) \text{ and } u_0^k = a(k\tau), u_n^k = b(k\tau).
\end{aligned}
}
\tag{3.16}
$$

In the case of homogeneous boundary conditions, at every time step, the above relations in matrix form writes as

$$
\vec{u}^k = [I - \tau A_h]\vec{u}^{k-1} + \tau \vec{b}_h^{k-1}, \tag{3.17}
$$

where A_h is the same $(n-1) \times (n-1)$ matrix given by equation (3.13). Since there is no matrix multiplying the unknown vector $\vec{u}^k$ on the left side of equation (3.17), it can be determined component by component by sweeping the spatial grid points one after the other. Nevertheless, it would be surprising that such a method works so much better than its implicit counterparts, equations (3.11) and (3.14), without any disadvantage. As we show by means of the example below, there are indeed limits beyond which explicit methods are unreliable.

Let $a = 0$, $b = 0$, $f = 0$ and $u_0(x) = \sin(m\pi x/L)$, m being a non-negative integer. For such data, the solution of problems (3.1)–(3.2) is given by $u(x, t) = e^{-\lambda t}\sin(m\pi x/L)$, with $\lambda = p(m\pi/L)^2$ as one can easily check. Let us apply scheme (3.16) to approximate u. Noticing that the abscissae of the grid points are iL/n for $i = 0, 1, \ldots, n$, at the first time step we obtain

$$
u_i^1 = \sin(m\pi i/n) + \tau p\{\sin[m\pi(i+1)/n] - 2\sin(m\pi i/n) + \sin[m\pi(i-1)/n]\}/h^2.
$$

On the other hand, from well-known trigonometric identities,

$$
\sin[m\pi(i+1)/n] + \sin[m\pi(i-1)/n] = 2\sin(m\pi i/n)\cos(m\pi/n).
$$

Thus, $u_i^1 = \{1 - 2p\tau[1 - \cos(m\pi/n)]/h^2\}\sin(m\pi i/n)$, that is, $u_i^1 = \eta(h, \tau, m, L)u_i^0$ for $i = 1, 2, \ldots, n-1$, where $\eta(h, \tau, m, L) = 1 - 2p\tau[1 - \cos(mh\pi/L)]/h^2$.

Since η depends neither on i nor on k, by mathematical induction it follows that necessarily

$$
u_i^k = [\eta(h, \tau, m, L)]^k u_i^0 \quad \forall i \in \{1, 2 \ldots, n-1\} \text{ and } \forall k \geq 1. \tag{3.18}
$$

Now assuming that $m = n$, $\eta = 1 - 4p\tau/h^2$. Therefore, if $2p\tau/h^2 > 1$, as k varies, the absolute value of the numerical solution will increase indefinitely. This is an inadequate response, since if $f = 0$, the solution of the heat equation is supposed to decrease in absolute value as time increases, as pointed out hereafter. To conclude these preliminary stability considerations, we should add that such a choice of u_0 is not as restrictive as it may seem. This is because every u_0 such that $u_0(0) = u_0(L) = 0$ can be expanded into an absolutely convergent **Fourier series** of the form:

$$
u_0(x) = \sum_{m=1}^{\infty} c_m \sin(m\pi x/L)
$$

for suitable real coefficients c_m (see e.g. [57]). By linearity, the numerical solution will be an absolutely convergent series whose terms are of the form given in equation (3.18), pre-multiplied by the c_ms. Therefore, if $|\eta| > 1$ for some values of m, with $n > 1$, the corresponding terms will gradually pollute the numerical solution as a whole–so much so that, as k increases, approximate solution absolute values will increase here and there, until the numerical solution becomes completely spoilt. On the other hand, if $|\eta| \leq 1$, these values will gradually decrease. For a detailed explanation about this process, we refer to reference [67].

Summarizing, like most explicit schemes for time integration of time-dependent PDEs, the Forward Euler scheme is only reliable if the time step is bounded above by a constant multiplied by the square of the spatial grid size. Actually, there is an explanation for this, in connection with the expected behaviour of a solution to the heat equation. We will see all this in more detail in the next subsection.

3.1.3   Example 3.1: Numerical Behaviour of the Forward Euler Scheme

In this example, we solve the heat equation by the FES (equation (3.17)), for the data $L = 1$, $p = 1$, $f = 0$, $u_0 = 1$, $a = 0$ and $b = 0$. In Figures 3.1 and 3.2, we illustrate what can be expected from an explicit solution scheme depending on whether a stability condition is satisfied or not. They reflect the numerical solution of the heat equation with a very coarse grid. More precisely, we let k vary from 1 up to 4 taking $h = 1/4$. In Figure 3.1a and 3.1b, we illustrate the numerical solution for τ equal to $1/8$ and $1/16$, respectively, while in Figure 3.2a and 3.2b, we do the same for τ equal to $1/32$ and $1/64$, respectively. As one can see, Figure 3.1 exhibits unstable numerical results, whereas in Figure 3.2 stable responses are depicted. In order to better understand the ongoing processes for these four different values of the time step τ, the curious reader could figure out by her or himself the following behaviours, by pushing the calculations two or three steps further:

- $\tau = 1/8$ $\rightarrow$ widely oscillating solution;
- $\tau = 1/16$ $\rightarrow$ moderately oscillating solution;
- $\tau = 1/32$ $\rightarrow$ fast decreasing solution;
- $\tau = 1/64$ $\rightarrow$ slowly decreasing solution.

As we will see in Section 3.3, the solution of the heat equation satisfies a **maximum principle** (cf. [156]). In the case where $f = 0$ and $u_0 > 0$, like in the above example, this implies that the temperature decreases until it attains a zero limiting value everywhere, in principle at infinite time. In other words, this means that the final temperature will be the one at which both ends of the bar are maintained[1]. Only smaller values of τ reproduce such a mathematically and physically acceptable temperature evolution. Notice, however, that the smaller the τ, the slower the temperature will decrease, which is logical since larger values of τ correspond to time discretisations not so close to reality.

---

[1] As a reference these temperatures are zero, but this is just a matter of scale as they could as well be equal to some value $\theta \neq 0$. In this case, the initial value of u would have to be adjusted to $u_0 + \theta$, thereby yielding a solution equal to $u + \theta$.

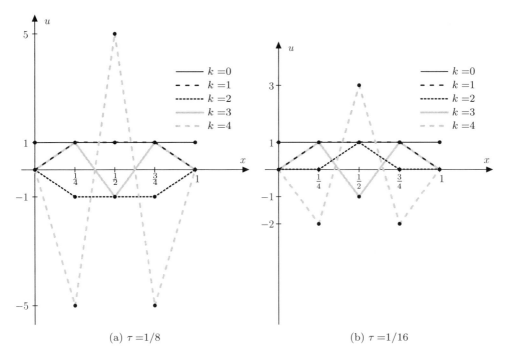

(a) $\tau = 1/8$ (b) $\tau = 1/16$

Figure 3.1   Unstable solution of the heat equation by the Forward Euler scheme for $h = 1/4$

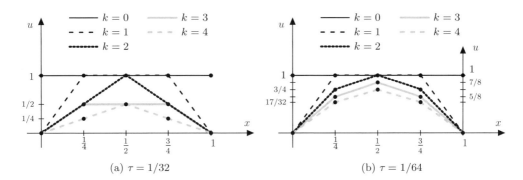

(a) $\tau = 1/32$ (b) $\tau = 1/64$

Figure 3.2   Stable solution of the heat equation by the Forward Euler scheme for $h = 1/4$

Summarizing, we have showed in this example that scheme (3.17) is unreliable if $\tau >$ $h^2/(2p)$. The other way around, according to the study to be conducted in Section 3.3, it is always reliable as long as τ is bounded by $h^2/(2p)$. However, this implies that the time step must be very small if one uses fine spatial grids. As a consequence, in this case the numerical resolution will advance very slowly in time. As previously announced, this is the price to pay for using an explicit scheme.

3.2   Numerical Solution of the Transport Equation

From the numerical point of view, considering equations (3.6)–(3.7) will definitely lead to conclusions typical of hyperbolic equations, including second-order ones, such as equations (3.3)–(3.4). That is why we confine ourselves to studying the (first-order) transport equation in its simplest form. This equation carries this name because it governs the transport with velocity o of a quantity, whose initial distribution along the x-axis is u_0 (possibly satisfying $u_0(0) = 0$ if $o > 0$ or $u_0(L) = 0$ if $o < 0$). In case the space domain is extended to the whole real axis, and at least as long as u_0 is differentiable in $\Re$, the unique solution of equation (3.6) is given by $u(x, t) = u_0(x - ot)$, as one can easily check. This means that at time $t > 0$, the value of u_0 at a position x will have been carried to the position $x + ot$, which is perfectly normal for a transport phenomenon with constant velocity o.

We will be mostly concerned about explicit time integration methods. This is because, in contrast to the case of parabolic equations, the solution of hyperbolic equations by explicit methods is not so costly, as we will see in Section 3.3. Moreover, for the same reason as in Section 3.1, we will study in detail only FD total discretisations. Nevertheless, in some situations it will be instructive to consider FE and FV counterparts of the space discretisation. We keep assuming in this section that we are given an equally spaced grid $\mathcal{G} = [0, h, 2h, \ldots, (n - 1)h, nh = L]$, together with a time step $\tau = \Theta/l$. Then, starting from $u_j^0 = u_0(jh)$, we wish to determine approximations u_j^k of $u(jh, k\tau)$ for $k = 1, 2, \ldots, l$, and $j = 1, 2, \ldots, n$ setting $u_0^k = 0$ if $o > 0$, or $j = 0, 1, \ldots, n - 1$ setting $u_n^k = 0$ if $o < 0$.

3.2.1   Natural Schemes

Like in the case of the Forward Euler scheme, it seems quite natural to approximate the time derivative at $(jh, [k - 1]\tau)$ by $(u_j^k - u_j^{k-1})/\tau$. As for the space derivative at the same point, a simple Taylor expansion about $(jh, [k - 1]\tau)$ indicates that the centred FD $(u_{j+1}^{k-1} - u_{j-1}^{k-1})/(2h)$ for $1 \leq j \leq n - 1$ is very tempting, since it yields a second-order local truncation error. Notice that a modification of this approximation of the x-derivative would be necessary to treat the case $j = n$ if $o > 0$ and $j = 0$ if $o < 0$, but for the moment we disregard this issue by considering the case where only initial conditions are prescribed and the transport equation holds in the whole real axis Ox. In doing so, we can focus on the scheme's behavior for any integer j, namely,

$$\frac{u_j^k - u_j^{k-1}}{\tau} + o\frac{u_{j+1}^{k-1} - u_{j-1}^{k-1}}{2h} = 0. \tag{3.19}$$

Here, the pure trigonometric initial condition considered in Subsection 3.1.2 would not be conclusive regarding the stability of the above scheme. On the other hand, if we enlarge the range of the transport equation to the complex domain, we will be able to draw pertinent conclusions on this issue. As a matter of fact, we will apply the so-called **Von Neumann stability criterion**, by taking an initial datum $u_0 = e^{imx}$ with $i = \sqrt{-1}$. Let $o > 0$ and $a := o\tau/(2h)$. First, we observe that equation (3.19) leads to the following expression of u_j^k for every j and $k > 1$:

$$u_j^k = u_j^{k-1} + a(u_{j+1}^{k-1} - u_{j-1}^{k-1})$$

Applying this formula to the initial values $u_j^0 = e^{imjh}$, after straightforward calculations we obtain at the first time step,

$$u_j^1 = C(a, m, h)u_j^0 \text{ for every } j, \text{ where } C(a, m, h) := 1 + 2ai \sin(mh).$$

It immediately follows that

$$u_j^k = [C(a, m, h)]^k u_j^0 \text{ for every } j.$$

We know that the exact solution for such an initial condition is given by $u(x, t) = e^{im(x - ot)}$, whose modulus equals one for all x and t. Therefore, the modulus of the numerical solution should grow indefinitely at no grid point. Unfortunately, undesirable growth is likely to happen to scheme (3.19), since $|C(a, m, h)| = \sqrt{1 + 4a^2 \sin^2(mh)}$. In other words, whatever we do with h and τ, inevitably we will be far from a physically acceptable solution in one way or another.

The situation will be completely different if, instead of the explicit approach in equation (3.19), we use its implicit version, namely,

$$\frac{u_j^k - u_j^{k-1}}{\tau} + o\frac{u_{j+1}^k - u_{j-1}^k}{2h} = 0. \tag{3.20}$$

for $j = 1, 2, \ldots, n - 1$ assorted with suitable adaptations in order to approximate equation (3.6) at $x = L$ if $o > 0$ and at $x = 0$ if $o < 0$. In this aim, we set as naturally,

$$\begin{cases} \frac{u_j^k - u_j^{k-1}}{\tau} + o\frac{u_j^k - u_{j-1}^k}{h} = 0 \text{ for } j = n \text{ if } o > 0, \\ \frac{u_j^k - u_j^{k-1}}{\tau} + o\frac{u_{j+1}^k - u_j^k}{h} = 0 \text{ for } j = 0 \text{ if } o < 0. \end{cases} \tag{3.21}$$

The combination of equations (3.20) and (3.21) corresponds to the following implicit scheme to determine the u_j^ks for $k = 1, 2, \ldots, l$:

$$\begin{cases} u_j^k + o\tau u_{j+1}^k/(2h) - o\tau u_{j-1}^k/(2h) = u_j^{k-1} \text{ for } 1 \le j \le n - 1, \\ (1 + o\tau/h)u_j^k - o\tau u_{j-1}^k/h = u_j^{k-1} \text{ for } j = n \text{ if } o > 0, \\ (1 - o\tau/h)u_j^k + o\tau u_{j+1}^k/h = u_j^{k-1} \text{ for } j = 0 \text{ if } o < 0. \end{cases} \tag{3.22}$$

It is interesting to check that this scheme generates no change of sign for a non-negative (resp. non-positive) u_0, taking $\tau = 1/4$, $L = 1$, $n = 4$, $o = 1$ and $u_0 = 1$, for instance. Notice that in this case, it is necessary to solve a 4×4 linear system in order to determine the solution at every time step, whose matrix is

$$A_{1/4} = \begin{bmatrix} 1 & 1/2 & 0 & 0 \\ -1/2 & 1 & 1/2 & 0 \\ 0 & -1/2 & 1 & 1/2 \\ 0 & 0 & -1 & 2 \end{bmatrix}$$

and whose right side at the first step is $[1, 1, 1, 1]^T$. The solution at this step is $\dfrac{[17, 24, 27, 28]^T}{29}$. In order to push further the calculations, we suggest that the reader compute the inverse of $A_{1/4}$. In doing so, she or he would find out that the components of the solution always remain strictly positive (except for $u_0^k = 0$). Moreover, the solution absolute values never increase without control. Of course, those computations prove nothing, for this is only a particular case. Nevertheless, we can say that such a favourable behaviour does not happen by chance; it is due to the fact that equation (3.22) is unconditionally stable (see e.g. [67]). However, we refrain from going into details on this point, because there are many explicit schemes for solving hyperbolic equations that work well under very reasonable conditions, as seen further in this chapter.

3.2.2   The Lax Scheme

This scheme is obtained by modifying the way the time derivative is approximated in equation (3.19). More specifically, we replace u_j^{k-1} by $(u_{j+1}^{k-1} + u_{j-1}^{k-1})/2$, though only for $j = 1, 2, \ldots, n-1$. This gives rise to the following relations for $k = 1, 2, \ldots, l$:

The Lax scheme

$$
\begin{aligned}
&\frac{u_j^k - (u_{j+1}^{k-1} + u_{j-1}^{k-1})/2}{\tau} + o\frac{u_{j+1}^{k-1} - u_{j-1}^{k-1}}{2h} = 0 \text{ for } 1 < j < n-1, \\
&\frac{u_j^k - u_j^{k-1}}{\tau} + o\frac{u_j^{k-1} - u_{j-1}^{k-1}}{h} = 0 \text{ for } j = n \text{ if } o > 0, \\
&\frac{u_j^k - u_j^{k-1}}{\tau} + o\frac{u_{j+1}^{k-1} - u_j^{k-1}}{h} = 0 \text{ for } j = 0 \text{ if } o < 0.
\end{aligned}
\tag{3.23}
$$

Equation (3.23) allows one to determine in an explicit manner approximations u_j^k of u at $x = jh$ and at the kth time step by

$$
\begin{cases}
u_j^k = [(1 - o\tau/h)u_{j+1}^{k-1} + (1 + o\tau/h)u_{j-1}^{k-1}]/2 \text{ for } 1 \le j \le n-1 \\
u_j^k = [(1 - o\tau/h)u_j^{k-1} + o\tau u_{j-1}^{k-1}/h \text{ for } j = n \text{ if } o > 0, \\
u_j^k = [(1 + o\tau/h)u_j^{k-1} - o\tau u_{j+1}^{k-1}/h \text{ for } j = 0 \text{ if } o < 0.
\end{cases}
\tag{3.24}
$$

It follows that, for the same test case, no value u_j^k can be negative if $|o|\tau \le h$. Of course, this does not allow us to conclude that this scheme is reliable whatever the datum u_0. Nevertheless, we will see in Section 3.3 that it is always stable if the above condition holds, and in Section 3.4 that it is convergent under the same condition. Scheme (3.23), or yet (3.24), is known as the **Lax scheme**. In principle, this scheme prevents the numerical solution from growing in absolute value as k increases if $|o|\tau \le h$. This assertion is corroborated by an argument exploited in Subsection 3.2.3, where a scheme with the very same property will be studied. For the moment, the reader might check this no-growth property by applying the Lax scheme to the data u_0, L, τ and h tested for the explicit scheme (3.19).

Equation (3.23) is only a first-order scheme in time as seen in Section 3.4. It can be improved in several manners, among which lies a modification known as the **Lax–Wendroff scheme**. For further details about this second-order scheme in both space and time, among several other possibilities to solve efficiently first-order hyperbolic PDEs, we refer to the vast literature on the subject such as reference [67].

3.2.3   Upwind Schemes

We next present another first-order scheme, this time in space too at every grid point. Akin to equation (3.23), the scheme is designed in order to avoid solution growth, among other shortcomings. Now the space derivative at point $(jh, k\tau)$ is approximated by $(u_j^k - u_{j-1}^k)/h$ if $o > 0$ and by $(u_{j+1}^k - u_j^k)/h$ if $o < 0$, for $k = 1, 2, \ldots, l$. In more compact form, this approximation can be rewritten as

The Upwind scheme

Setting $D_o(k\tau, jh) := o^+(u_j^k - u_{j-1}^k)/h + o^-(u_{j+1}^k - u_j^k)/h$ with
$o^+ := \max[0, o]$ and $o^- := \min[0, o]$,

determine u_j^k for $k = 1, 2, \ldots, l$ by $\dfrac{u_j^k - u_j^{k-1}}{\tau} + D_o([k-1]\tau, jh) = 0$

for $j = 1, \ldots, n$ with $u_0^{k-1} = 0$ if $o > 0$, and $j = 0, \ldots, n-1$ with $u_n^{k-1} = 0$ if $o < 0$.

$$(3.25)$$

This means that, at the kth time step, u_j^k is given by

$$u_j^k = [1 - (o^+ - o^-)\tau/h]u_j^{k-1} + [o^+ u_{j-1}^{k-1} - o^- u_{j+1}^{k-1}]\tau/h \qquad (3.26)$$

for $1 \le j \le n$ with $u_0^k = 0$ if $o > 0$, and $0 \le j \le n-1$ with $u_n^k = 0$ if $o < 0$, for all k. Notice that the values u_{n+1}^{k-1} and u_{-1}^{k-1} are never used, but we keep them in equation (3.26) in order to unify the treatment of all grid points, setting them to zero for all k, for the sake of clarity. The no-growth property of scheme (3.26) holds if $|o|\tau/h \le 1$. Indeed, in this case the three coefficients of the approximations of u at the $(k-1)$th step are non-negative and form a **partition of unity**, since their sum equals one. This property implies that

$$u_j^k \le \max[u_{j-1}^{k-1}, u_j^{k-1}, u_{j+1}^{k-1}] \text{ and } u_j^k \ge \min[u_{j-1}^{k-1}, u_j^{k-1}, u_{j+1}^{k-1}].$$

for every j and k. Both inequalities can be established by arguments similar to those exploited in Chapter 2.

The denomination *upwind* for scheme (3.25) is due to the fact that the second point used to approximate the space derivative at a given point $x_j = jh$ is always located in the upwind direction with respect to the velocity o. In the next two sections, we will see that both equations (3.23) and (3.25) are stable and convergent provided $|o|\tau/h \le 1$.

3.2.4 Extensions to the FVM and the FEM

To complete this presentation, we briefly address possible FV and FE counterparts of the explicit schemes (3.23) and (3.25). An Upwind scheme can be easily designed and implemented in the context of both types of FVM. It suffices to consider the **upwind CV** in the approximation of the space derivative. More specifically, recalling the definitions of Section 1.4, for $k = 1, 2, \ldots, l$, and sticking to the academic distinction between the two types of CV we considered there, the following schemes can be proposed.

Referring to Figure 1.5, for the Vertex-centred FVM the unknowns are constant approximations u_j^k of u in the CVs V_j at time $k\tau$, for $j = 0, 1, \ldots, n$, where $u_j^0 = u_0(x_j)$. We set $u_0^k = 0$ if $o > 0$ and $u_n^k = 0$ if $o < 0$ for every k. In this case, the upwind CV with respect to $V_j = (x_{j-1/2}, x_{j+1/2})$ is V_{j-1} if $o > 0$ and V_{j+1} if $o < 0$. Recalling that the length of V_j is $h_{j+1/2} = (h_j + h_{j+1})/2$ for $j = 1, 2, \ldots, n-1$ and those of V_0 and V_n are $h_{1/2} = h_1/2$ and $h_{n+1/2} = h_n/2$, respectively, the term $o\partial_x u$ at time $k\tau$ is approximated in V_j by $[o^+(u_j^k - u_{j-1}^k) + o^-(u_{j+1}^k - u_j^k)]/h_{j+1/2}$ for $j = 1, 2, \ldots, n$ if $o > 0$ and $j = 0, 1, \ldots, n-1$ if $o < 0$. Then, integrating in CV V_j, we obtain for $k = 1, 2, \ldots$:

The Upwind Vertex-centred FV scheme

Determine u_j^k satisfying,

$$h_{j+1/2}\frac{u_j^k - u_j^{k-1}}{\tau} + o^+[u_j^{k-1} - u_{j-1}^{k-1}] + o^-[u_{j+1}^{k-1} - u_j^k] = 0,$$
for $j = 1, 2, \ldots, n$ with $u_0^k = u_0^{k-1} = u_{n+1}^{k-1} = 0$ if $o > 0$,
and $j = 0, 1, \ldots, n-1$ with $u_{-1}^{k-1} = u_n^{k-1} = u_n^k = 0$ if $o < 0$.

(3.27)

At the kth time step, the unknowns are explicitly determined by

$$\begin{cases} u_j^k = [1 - \tau(o^+/h_{j+1/2} - o^-/h_{j+1/2})]u_j^{k-1} + \tau[o^+ u_{j-1}^{k-1}/h_{j+1/2} - o^- u_{j+1}^{k-1}/h_{j+1/2}], \\ \text{for } j = 1, 2, \ldots, n \text{ with } u_0^k = u_0^{k-1} = u_{n+1}^{k-1} = 0 \text{ if } o > 0, \\ \text{and } j = 0, 1, \ldots, n-1 \text{ with } u_{-1}^{k-1} = u_n^{k-1} = u_n^k = 0 \text{ if } o < 0. \end{cases}$$

(3.28)

Like before in equations (3.27) and (3.28), $u_{-1}^{k-1}, u_{n+1}^{k-1}$, though never used, were set to zero for all k.

We refer to Figure 1.6 for the notation we adopted to formulate the Cell-centred FVM with a slight difference in the notation of the approximations of u, namely, subscript $-1/2$ is used instead of 0 and $n + 1/2$ replaces n. The unknowns are constant approximations $u_{j-1/2}^k$ of u in the CVs $T_j = (x_{j-1}, x_j)$ at time $k\tau$, for every k, and in principle for $j = 0, 1, \ldots, n$, except for $j = 0$ if $o > 0$ and for $j = n$ if $o < 0$. We also set $u_{j-1/2}^0 = u_0(x_{j-1/2})$, for $j = 0, 1, \ldots, n$, with $x_{j-1/2} = (x_j + x_{j-1})/2$ for $j = 1, 2, \ldots, n$, $x_{-1/2} = x_0 = 0$ and $x_{n+1/2} = x_n = L$. In this case, the upwind CV with respect to T_j is T_{j-1} if $o > 0$ and T_{j+1} if $o < 0$. Recalling that the measure of T_j is $h_j = x_j - x_{j-1}$ for $j = 1, 2, \ldots, n$, the term $o\partial_x u$ at time $k\tau$ is approximated in T_j by $[o^+(u_{j-1/2}^k - u_{j-3/2}^k) + o^-(u_{j+1/2}^k - u_{j-1/2}^k)]/h_j$ for $j = 1, 2, \ldots, n$ if $o > 0$ and $j = 0, 1, \ldots, n-1$ if $o < 0$. Then, integrating in CV T_j, we obtain for $k = 1, 2, \ldots$:

The Upwind -centred FV scheme

Determine $u_{j-1/2}^k$ satisfying,

$$h_j\frac{u_{j-1/2}^k - u_{j-1/2}^{k-1}}{\tau} + o^+[u_{j-1/2}^{k-1} - u_{j-3/2}^{k-1}] + o^-[u_{j+1/2}^{k-1} - u_{j-1/2}^{k-1}] = 0,$$
for $j = 1, 2, \ldots, n$ with $u_{-1/2}^k = u_{-1/2}^{k-1} = u_{n+1/2}^{k-1} = 0$ if $o > 0$,
and $j = 0, 1, \ldots, n-1$ with $u_{-3/2}^{k-1} = u_{n-1/2}^{k-1} = u_{n-1/2}^k = 0$ if $o < 0$.

(3.29)

At the kth time step, the unknowns are explicitly determined by

$$\begin{cases} u_{j-1/2}^k = [1 - \tau(o^+ - o^-)/h_j]u_{j-1/2}^{k-1} + \tau[o^+ u_{j-3/2}^{k-1}/h_j - o^- u_{j+1/2}^{k-1}/h_j], \\ \text{for } j = 1, 2, \ldots, n \text{ with } u_{-1/2}^k = u_{-1/2}^{k-1} = u_{n+1/2}^{k-1} = 0 \text{ if } o > 0, \\ \text{and } j = 0, 1, \ldots, n-1 \text{ with } u_{-3/2}^{k-1} = u_{n-1/2}^{k-1} = u_{n-1/2}^k = 0 \text{ if } o < 0. \end{cases}$$

(3.30)

Akin to u_{-1}^{k-1} and u_{n+1}^{k-1} in equation (3.28), the values $u_{-3/2}^{k-1}$ and $u_{n+1/2}^{k-1}$ are never used, but were set to zero for all k in equations (3.29) and (3.30).

Like equation (3.26), both schemes above avoid solution unphysical growth or oscillations provided $\tau|o| \leq ch$, where c is the constant of the quasi-uniform family of meshes in use, and h is the maximum CV length. This result can be established as Exercise 3.2. On the other hand, in contrast to equation (3.26), except for particular situations such as uniform meshes, none of the above schemes attain a first-order of consistency in both space and time in the pointwise sense. This fact rules out any simple proof of convergence in the maximum norm, even though it does not prevent the schemes from being convergent. In this respect, we refer to reference [68].

An interesting FE analog of the Lax scheme, strictly avoiding uncontrolled growth of solution absolute values, results from a combination of the **mass lumping** technique and the exact integration of coefficients attached to the time derivative. This scheme has been mainly exploited by Kawahara and co-workers (cf. [114]) in intensive fluid flow simulations. In order to see how this works, let us assume again that the mesh is uniform with n elements and mesh size $h = L/n$. Assuming that $o > 0$, for instance, and recalling the notations of Sections 1.3 and 2.2, a natural semi-discretisation in space by the linear FEM of equations (3.6)–(3.7) is

$$\text{Find } u_h \in V_h \text{ such that } (\partial_t u_h | v)_0 + o(\partial_x u_h | v)_0 = 0 \ \forall v \in V_h.$$

Like in the case of the FDM, if we want to solve the above problem, in principle we have to further proceed to a time discretisation. In this aim, we first expand $u_h(x, t)$ as the sum $\sum_{j=1}^{n} u_j(t)\varphi_j(x)$. Then, by a natural and standard procedure for an explicit solution, we consider that at time $(k-1)\tau$ the time derivative of u_h is approximated by $(u_h^k - u_h^{k-1})/\tau$, where $u_h^k(x) = \sum_{j=1}^{n} u_j^k \varphi_j(x)$ for every $x \in [0, L]$, the u_j^ks being the nodal unknowns at the kth step.

After setting $v = \varphi_i$ for $i = 1, 2, \ldots, n$, the u_j^ks are easily found to be given by

$$\sum_{j=1}^{n} u_j^k (\varphi_j | \varphi_i)_0 - \sum_{j=1}^{n} u_j^{k-1} [(\varphi_j | \varphi_i)_0 + o\tau(\varphi_j' | \varphi_i)_0] = 0. \tag{3.31}$$

Now, if the mass lumping technique is used to compute the integrals $(\varphi_j | \varphi_i)_0$ on both sides of the above relation, it is not difficult to conclude that the resulting explicit scheme for computing u_j^k will be the same as equation (3.19) (cf. Exercise 3.3). Therefore, we have to discard this possibility. However, if the resolution is to remain explicit, the simplest choice is to use **mass lumping** on the left side. On the other hand, we note that exact integration of $(\varphi_j | \varphi_i)_0$ on the right side does not prevent the resolution from being explicit. That is exactly what we will do, in order to recover our FE analog of the Lax scheme. However, here a word of caution is in order: in principle, this recipe introduces an inconsistency in the space–time discretisation, due to the unbalance between both sides of the scheme. Thus, it has to be duly validated by means of a careful consistency analysis. Anyway, recalling the coefficients determined in Section 1.3 for the integrals of the product $\varphi_j \varphi_i$, and setting $u_0^k = 0$ if $o > 0$ and $u_n^k = 0$ if $o < 0$ for all $k > 0$, after straightforward calculations, which the reader is encouraged to carry out as Exercise 3.4, we come up with

An FE analog of the Lax scheme for uniform meshes

For $j = 1, 2, \ldots, n-1$,
$u_j^k - 2u_j^{k-1}/3 - [1/6 + o\tau/(2h)]u_{j-1}^{k-1} - [1/6 - o\tau/(2h)]u_{j+1}^{k-1} = 0,$
and   (3.32)
$u_j^k - [2/3 - o\tau/h]u_j^{k-1} - [1/3 + o\tau/h]u_{j-1}^{k-1} = 0 \text{ for } j = n \text{ if } o > 0;$
$u_j^k - [2/3 + o\tau/h]u_j^{k-1} - [1/3 - o\tau/h]u_{j+1}^{k-1} = 0 \text{ for } j = 0 \text{ if } o < 0.$

After expressing the u_j^ks in terms of the u_j^{k-1}s in equation (3.32), by inspection we can state that, provided $|o|\tau/h \leq 1/3$, all the coefficients on the right side of the resulting relations are non-negative. Moreover, they form a partition of unity for all j. Hence, using the same argument as before, we conclude that the FE scheme (equation (3.32)), though being different from the Lax scheme, behaves very much like it. In particular, neither changes of sign can be observed for a positive (resp. negative) u_0, nor can the numerical solution grow in absolute value as k increases. Although both properties still hold whatever the case, unfortunately this FE scheme is no longer consistent for non-uniform meshes. Therefore, some modifications are necessary in order to recover consistency and hence convergence in this case. We refer to reference [172] for further details, and also to Example 7.3.

3.3   Stability of the Numerical Models

As we saw in Chapter 2, the stability of a numerical method is not a purely mathematical property. In some sense, it must lead to the conclusion that the numerical solution is able to conform to a physical behaviour, close to the one the mathematical model represents.

In the case of the heat equation, the basic property to be searched for states that, as it evolves, the absolute value of the temperature always remains controlled by the maximum in absolute value of the heat source f, its value at the bar ends and its value u_0 at the initial time. In particular, this means that in the absence of heat sources, if the temperature is kept fixed at both ends, say $u(0) = 0$ and $u(L) = 0$, the maximum of $|u|$ at every time is bounded by the maximum of $|u_0|$. Finally, if under the same conditions the initial temperature is non-negative (resp. non-positive), then the temperature will gradually decrease (resp. increase) as time goes by.

All these properties are a consequence of the following **maximum principle** mathematically described in several texts (cf. [101], [112] or [156]):

If $f \leq 0$ (corresponding to *heat sinks*), then for any $\Theta > 0$,

$$\max_{t\in[0,\Theta],x\in[0,L]} u(x,t) \leq \max\{\max_{t\in[0,\Theta]} \max[a(t),b(t)], \max_{x\in[0,L]} u_0(x)\}. \tag{3.33}$$

If $f \geq 0$ (corresponding to *heat soaring*), then for any $\Theta > 0$,

$$\min_{t\in[0,\Theta],x\in[0,L]} u(x,t) \geq \min\{\min_{t\in[0,\Theta]} \min[a(t),b(t)], \min_{x\in[0,L]} u_0(x)\} \tag{3.34}$$

For the proof of these results, we refer to references [101], [77] and [156], for example.

Combining equation (3.33) and (3.34), we readily conclude that if $f = 0$,

$$\max_{t\in[0,\Theta],x\in[0,L]} |u(x,t)| \leq \max\{\max_{t\in[0,\Theta]} \max[|a(t)|,|b(t)|], \max_{x\in[0,L]} |u_0(x)|\} \tag{3.35}$$

Summing up, the physical interpretation of these principles in simple terms is that, in a pure heat conduction process undergone by a certain medium, a local 'hot-spot' cannot develop spontaneously in its interior when no heat sources are present. That is the property we expect to recover from numerical schemes too.

As far as the transport equation is concerned, at least solution boundedness has to be mimicked if the initial conditions are bounded. In addition to this, one should possibly target the capture of the initial profile u_0 being transported with velocity o, as the process goes on.

3.3.1   Schemes for the Heat Equation

For the sake of brevity, we will study only the Euler schemes, since rather simple stability analyses lead to realistic results. Nevertheless, this choice will allow us to highlight the main differences between explicit and implicit schemes, as far as stability is concerned.

Let us first consider the FES. We saw in Section 3.1 that whenever $\tau > h^2/(2p)$, the numerical solution obtained with this scheme may be unphysical. Now we will establish that whenever τ is bounded by $h^2/(2p)$, scheme (3.17) is stable in terms of the maximum norm. Setting $\gamma = p\tau/h^2$, first we recall that at the kth time step, this scheme is

$$u_i^k = (1 - 2\gamma)u_i^{k-1} + \gamma u_{i+1}^{k-1} + \gamma u_{i-1}^{k-1} + \tau f(ih, [k-1]\tau) \text{ for } i = 1, 2, \ldots, n-1$$

with $u_0^k = a(k\tau)$, $u_n^k = b(k\tau)$ and $u_i^0 = u_0(ih)$. Hence, if $\gamma \leq 1/2$, the coefficients on the right side of the above relation are all non-negative and form a partition of unity. This implies that, provided $\gamma \leq 1/2$, for all i between 1 and $n-1$,

$$u_i^k \leq \max[u_{i-1}^{k-1}, u_i^{k-1}, u_{i+1}^{k-1}] + \tau \max_{1 \leq j \leq n-1} f(jh, [k-1]\tau),$$

$$u_i^k \geq \min[u_{i-1}^{k-1}, u_i^{k-1}, u_{i+1}^{k-1}] + \tau \min_{1 \leq j \leq n-1} f(jh, [k-1]\tau) \tag{3.36}$$

It follows that

$$\max_{1 \leq i \leq n-1} |u_i^k| \leq \max_{0 \leq i \leq n} |u_i^{k-1}| + \tau \max_{1 \leq i \leq n-1} |f(ih, [k-1]\tau)|. \tag{3.37}$$

Setting $k = 1$ in equation (3.37), we easily conclude that

$$\max_{0 \leq i \leq n} |u_i^1| \leq \max[\max(|a(\tau)|, |b(\tau)|), \max_{0 \leq i \leq n} |u_i^0|] + \tau \max_{1 \leq i \leq n-1} |f(ih, 0)|.$$

Now, assume that for a certain $k > 2$,

$$\max_{0 \leq i \leq n} |u_i^{k-1}| \leq \max[\max_{1 \leq m \leq k-1} \max(|a(m\tau)|, |b(m\tau)|), \max_{0 \leq i \leq n} |u_i^0|]$$

$$+ (k-1)\tau \max_{1 \leq i \leq n-1, 1 \leq m \leq k-1} |f(ih, [m-1]\tau)|.$$

Then, using equation (3.37), we obtain

$$\max_{0 \leq i \leq n} |u_i^k| \leq \max[\max_{1 \leq m \leq k} \max(|a(m\tau)|, |b(m\tau)|), \max_{0 \leq i \leq n} |u_i^0|]$$

$$+ k\tau \max_{1 \leq i \leq n-1;\ 1 \leq m \leq k} |f(ih, [m-1]\tau)|. \tag{3.38}$$

Thus, by mathematical induction, equation (3.38) is valid for an arbitrary $k \geq 1$.

Now, setting by $\vec{u}^k = [u_0^k, u_1^k, \ldots, u_n^k]^T$, and since necessarily $k\tau \leq \Theta$, according to equation (3.38) the following result holds for all $k \in [1, 2, \ldots, l]$:

Stability inequality for the FES in the maximum norm

> if $\tau = \Theta/l$ and $h = L/n$ satisfy $\tau \leq h^2/(2p)$ then
> $$\max_{1 \leq k \leq l} \| \vec{u}^k \|_\infty \leq \max[\max_{1 \leq k \leq l} \max(|a(k\tau)|, |b(k\tau)|), \| \vec{u}^0 \|_\infty]$$
> $$+ \Theta \max_{1 \leq i \leq n-1;\ 0 \leq k \leq l-1} |f(ih, k\tau)|. \tag{3.39}$$

Otherwise stated, if $\tau \leq h^2/(2p)$, the solution provided by the Forward Euler scheme is stable in the l^∞-norm. This stability result is a discrete counterpart of the one satisfied by the exact solution of the heat equation. This appears very clearly if we take $f = 0$ and recall the properties of the heat equation given at the beginning of this section.

To conclude, the reader could use equation (3.36) to prove discrete maximum and minimum principles analogous to those that hold for the heat equation, as in Exercise 3.5.

Next we establish that, in contrast to the Forward Euler scheme, the Backward Euler scheme is unconditionally stable in the l^∞-norm.

First, we recall that in order to determine the u_i^ks at the kth time step, the SLAE corresponding to equation (3.11) must be solved. If we compare the system matrix to the matrix associated with the FDM described in Chapter 2 for a uniform grid and inhomogeneous boundary conditions, we realise that, except for the number of unknowns, the nonzero entries are the same, provided we set $q = 1/\tau$. Thus, at every time step, we can apply the DMP. In order to do so, we first multiply every equation by τ and note that the ith component of the right-side vector $\vec{b}_h$ at the kth time step is expressed by $u_i^{k-1} + \tau f(ih, k\tau)$, while the left side writes as, $(1 + 2p\tau/h^2)u_i^k - p\tau(u_{i+1}^k + u_{i-1}^k)/h^2$. Hence, the Backward Euler scheme at the kth step can be recast in the form of equation (2.6) with $q_i = 1$ for all i, $\vec{g} = [a(k\tau), u_1^{k-1}, u_2^{k-1}, \dots, u_{n-1}^{k-1}, b(k\tau)]^T$, $\vec{f} = [0, f(h, k\tau), f(2h, k\tau), \dots, f([n-1]h, k\tau), 0]^T$, $p_i = p\tau/h^2$, $p_i^+ = p_i^- = p\tau/h^2$ for $i = 1, 2, \dots, n-1$ and $p_0 = p_n = p_0^+ = p_n^- = 0$. Therefore, using equation (2.7), we immediately obtain

$$\max_{0\leq i\leq n} |u_i^k| \leq \max[|a(k\tau)|, |b(k\tau)|, \max_{0\leq i\leq n} |u_i^{k-1}|] + \tau \max_{1\leq i\leq n-1} |f(ih, k\tau)|. \tag{3.40}$$

Then, for $k = 1$, we have

$$\max_{0\leq i\leq n} |u_i^1| \leq \max[|a(\tau)|, |b(\tau)|, \max_{0\leq i\leq n} |u_i^0|] + \tau \max_{1\leq i\leq n-1} |f(ih, \tau)|. \tag{3.41}$$

Now, let us assume that for a certain $k > 2$,

$$\left\{ \begin{array}{l} \max\limits_{0\leq i\leq n} |u_i^{k-1}| \leq \max\{ \max\limits_{1\leq m\leq k-1} \max[|a(m\tau)|, |b(m\tau)|], \max\limits_{0\leq i\leq n} |u_i^0|\} \\ \qquad + (k-1)\tau \max\limits_{1\leq i\leq n-1, 1\leq m\leq k-1} |f(ih, m\tau)|. \end{array} \right. \tag{3.42}$$

Then, using equation (3.40), we readily obtain

$$\left\{ \begin{array}{l} \max\limits_{0\leq i\leq n} |u_i^k| \leq \max\{ \max\limits_{1\leq m\leq k} \max[|a(m\tau)|, |b(m\tau)|], \max\limits_{0\leq i\leq n} |u_i^0|\} \\ \qquad + k\tau \max\limits_{1\leq i\leq n-1, 1\leq m\leq k} |f(ih, m\tau)|. \end{array} \right. \tag{3.43}$$

It follows by mathematical induction that equation (3.43) holds for any $k \geq 1$. Taking into account that $k\tau \leq \Theta$, we finally conclude that it holds the following:

Stability inequality for the BES in the maximum norm

$$\boxed{\begin{array}{l} \text{For every } \tau = \Theta/l \text{ and } h = L/n, \\ \max\limits_{1\leq k\leq l} \|\vec{u}^k\|_\infty \leq \max[\max\limits_{1\leq k\leq l} \max(|a(k\tau)|, |b(k\tau)|), \|\vec{u}^0\|_\infty] \\ + \Theta \max\limits_{1\leq i\leq n-1;\ 1\leq k\leq l} |f(ih, k\tau)|. \end{array}} \tag{3.44}$$

In short, a stability result similar to equation (3.39) for the FES also holds for the BES, but in the latter case without any restriction on h and τ.

Remark 3.1 *A stability inequality qualitatively equivalent to equation (3.44) applies to the Crank–Nicolson scheme, though under the assumption that $\tau \leq 3h^2/(2p)$ (see e.g. [190]). However, this scheme is known to be stable, irrespective of the values of the two discretisation parameters. In particular, this is true in the mean-square sense, but for the moment we omit any discussion on this point, and refer to reference [80] or to Section 5.2 hereafter for more details.* ■

3.3.2   The Lax Scheme for the Transport Equation

To begin with, we note that conditions like $|o|\tau/h \leq c$, identified in Section 3.2 as critical for a reliable numerical solution of equations (3.6)–(3.7), have a meaningful physical interpretation in terms of a transport phenomenon. Indeed, assuming for example that $o > 0$, the exact solution at $t = k\tau$ and $x = jh$ depends only on $u_0(jh - ok\tau)$. This means that in terms of the space–time grid, the values u_i^0 for $x = ih$ in the neighbourhood of $jh - ok\tau$ must somehow influence the calculation of u_j^k; otherwise, there is no way for them to be transported to the kth time step. Otherwise stated, if it is not so, the physical meaning of the transport equation will be lost at the numerical level. The possible lack of such an influence is illustrated in Figure 3.3 in the case where $\tan \beta = \tau/h > \tan \alpha = o^{-1}$: assuming for instance that $jh - ok\tau > ih$ for $i = 0$, the value u_0^0 will not influence at all the calculation of u_j^k as it should. Instead, the wrong value $u_{j-k}^0 = u_0([j - k]h)$ with $j > k$ will be 'transported' to u_j^k. In Figure 3.4, we locate the grid points influencing the calculation of u_j^k at the time level $k - 1$ inside the triangle with upper vertex at $(jh, k\tau)$, taking the Lax scheme as an example. Then, it is clear that if the slope h/τ is less than o, there will be no way for the value of u_0 at the appropriate grid point ih at time level 0 to reach the kth level. This means that the Lax scheme violates physics, unless $h/\tau \geq o$. A similar argument applies to all the other schemes we have considered in this chapter to solve equations (3.6)–(3.7).

On the one hand, the stability analysis given below is restricted to the Lax scheme, but on the other hand it encompasses the case of inhomogeneous boundary conditions of the type $u(0, t) = a$ if $o > 0$ or $u(L, t) = b$ if $o < 0$ $\forall t$.

We know that at every time step, the approximate values at the grid points $(jh, k\tau)$ are given by

$$u_j^k = [1/2 + o\tau/(2h)]u_{j+1}^{k-1} + [1/2 - o\tau/(2h)]u_{j-1}^{k-1} \text{ for } 1 \leq j \leq n - 1. \qquad (3.45)$$

with $u_0^k = a$ if $o > 0$ or $u_n^k = b$ if $o < 0$, and $u_j^0 = u_0(jh)$. Then, provided $|o|\tau \leq h$, both coefficients in brackets in equation (3.45) are strictly positive. Recalling that $\vec{u}^k = [u_0^k, u_1^k, \ldots, u_{n-1}^k, u_n^k]^T$, it follows that if this condition is fulfilled, $|1/2 + o\tau/(2h)| + |1/2 - o\tau/(2h)| = 1$, and hence,

$$|u_j^k| \leq \max[|u_{j+1}^{k-1}|, |u_{j-1}^{k-1}|] \text{ for } 1 \leq j \leq n - 1. \qquad (3.46)$$

Recalling equation (3.23) and the definition $u_{-1}^k = u_{n+1}^k = 0$ for all k, since $u_0^k = u_0^{k-1} = a$ if $o > 0$ and $u_n^k = u_n^{k-1} = b$ if $o < 0$, it is as easy to establish that under the condition $|o|\tau \leq h$,

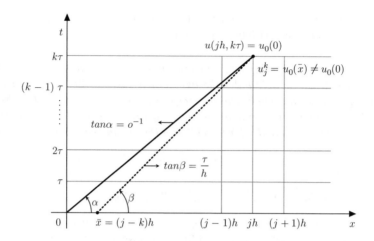

Figure 3.3   Transport of a wrong initial value at $x = 0$ to time level k

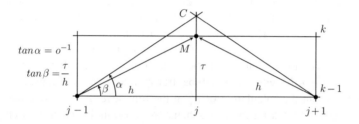

Figure 3.4   Grid points at time level $k - 1$ influencing the calculation of u_j^k

for $j = 0$ and $j = n$, we have

$$|u_j^k| \leq \max[|u_{j+1}^{k-1}|, |u_j^{k-1}|, |u_{j-1}^{k-1}|] \text{ for } j = 0 \text{ and } j = n. \qquad (3.47)$$

Taking the maximum over the grid points on the right side of equations (3.46) and (3.47), and then on the left side of the resulting relations, we come up with

$$\| \vec{u}^k \|_\infty \leq \| \vec{u}^{k-1} \|_\infty \ \forall k > 0. \qquad (3.48)$$

Then, applying repeatedly equation (3.48) from $k = 1$ up to a generic value of k, we establish the

Stability inequality for the Lax scheme in the maximum norm

$$\boxed{\begin{array}{l} \text{if } \tau \leq h/|o| \text{ then,} \\ \| \vec{u}^k \|_\infty \leq \| \vec{u}^0 \|_\infty \ \forall k, 1 \leq k \leq l. \end{array}} \qquad (3.49)$$

The requirement that $\tau \leq h/|o|$, limiting the choice of the time step with respect to the mesh size, is known as the **Courant–Friedrichs–Levy condition** or more commonly the **CFL condition**. This is because it is generally acknowledged that these three authors first identified it as a crucial reliability condition in a work dating back to 1928 [55].

As pointed out above, the CFL stability condition allows us to mimic the property that the solution does not grow as time increases. It is also sufficient to establish that any non-negative (resp. non-positive) initial datum u_0 will remain non-negative (resp. non-positive) and bounded away from zero for all k. Addressing this issue is the object of Exercise 3.6.

Another interesting exercise (cf. Exercise 3.7) is to prove that the stability result (equation (3.49)) also holds for the Upwind scheme.

In order to complete the study conducted in this subsection, it is wise to handle an equation more general than equation (3.6), namely,

$$\partial_t u + o\partial_x u = f \tag{3.50}$$

with the same kind of boundary and initial conditions as in equation (3.6), where f is a given bounded function in $[0, L] \times [0, \Theta]$. Like in the case of the Forward Euler scheme, it is natural to modify the Lax scheme (equation (3.23)) or the Upwind scheme (equation (3.25)) through the addition to the right side of the jth relation at the kth time step of the term $f(jh, [k - 1]\tau)$.

Inspired by the arguments employed in the previous subsection for the Euler schemes, the reader could quite easily verify that the stability result (equation (3.49)) for both schemes above extends to the case of equation (3.50) to yield (cf. Exercise 3.8)

$$\begin{cases} \text{if } \tau \leq h/|o| \text{ then,} \\ \| \vec{u}^k \|_\infty \leq \| \vec{u}^0 \|_\infty + \Theta \max_{0 \leq j \leq n;\ 0 \leq i \leq k-1} |f(jh, i\tau)| \; \forall k, 1 \leq k \leq l. \end{cases} \tag{3.51}$$

3.4   Consistency and Convergence Results

This introductory study of time-dependent problems is completed in this section with convergence results for some of the schemes presented in Sections 3.1 and 3.2.

3.4.1   Euler Schemes for the Heat Equation

Similarly to what we did in Section 2.3 for problem (P_1), we first determine the local truncation errors. In the case under study, we must quantify the residual of the Euler schemes at a certain grid point, when we replace the approximate values u_j^k by the exact values $u(jh, k\tau)$, assuming that the exact solution is sufficiently differentiable in $[0, L] \times [0, \Theta]$.

As far as the Forward Euler scheme is concerned, the right choice is to determine the residual $E_j^{k-1}(u)$ at point $(jh, [k - 1]\tau)$ at the kth time step.

Assuming that $\partial_{tt} u$ is continuous in $[0, L] \times [0, \Theta]$, a classical Taylor expansion yields

$$[u(ih, k\tau) - u(ih, [k - 1]\tau)]/\tau = \partial_t u(ih, [k - 1]\tau) + \tau \partial_{tt} u(ih, \xi_i^k)/2, \tag{3.52}$$

for a certain $\xi_i^k \in ([k - 1]\tau, k\tau)$, $i = 1, 2, ..., n-1$.

On the other hand, assuming that $\partial_{xxxx} u$ is continuous in $[0, L] \times [0, \Theta]$, similarly to Section 2.3, we obtain

$$\begin{cases} p\{2u(ih,[k-1]\tau) - u([i+1]h,[k-1]\tau) - u([i-1]h,[k-1]\tau)\}/h^2 \\ = -p\partial_{xx}u(ih,[k-1]\tau) - ph^2[\partial_{xxxx}u(\underline{\chi}_{i-1}^k,[k-1]\tau) + \partial_{xxxx}u(\underline{\chi}_{i+1}^k,[k-1]\tau)]/24 \end{cases} \tag{3.53}$$

for certain $\underline{\chi}_{i-1}^k \in ([i-1]h, ih)$ and $\underline{\chi}_{i+1}^k \in (ih, [i+1]h)$.

Combining equations (3.52) and (3.53), we conclude that

$$\begin{cases} E_i^{k-1}(u) := -f(ih,[k-1]\tau) + [u(ih,k\tau) - u(ih,[k-1]\tau)]/\tau \\ +p\{2u(ih,[k-1]\tau) - u([i+1]h,[k-1]\tau) - u([i-1]h,[k-1]\tau)\}/h^2 \\ = \{\partial_t u(ih,[k-1]\tau) - p\partial_{xx}u(ih,[k-1]\tau) - f(ih,[k-1]\tau)\} \\ +\tau\partial_{tt}u(ih,\underline{\xi}_i^k)/2 - ph^2[\partial_{xxxx}u(\underline{\chi}_{i-1}^k,[k-1]\tau) + \partial_{xxxx}u(\underline{\chi}_{i+1}^k,[k-1]\tau)]/24. \end{cases} \tag{3.54}$$

The term in curly brackets on the right side of equation (3.54) vanishes because the heat equation must hold at every grid point. This leads to the following:

Bound for the Forward Euler scheme local truncation error

$$\boxed{\begin{aligned} &|E_i^{k-1}(u)| \leq \tau \max_{0\leq x\leq L; 0\leq t\leq\Theta} |\partial_{tt}u(x,t)| + ph^2 \max_{0\leq x\leq L;\ 0\leq t\leq\Theta} |\partial_{xxxx}u(x,t)|/12 \\ &\text{for } i=1,\dots,n-1 \text{ and } k=1,\dots,l. \end{aligned}} \tag{3.55}$$

In an entirely analogous manner, for the Backward Euler scheme, we obtain the following expression for the residual $I_i^k(u)$, this time at point $(ih, k\tau)$:

$$\begin{cases} I_i^k(u) := -f(ih,k\tau) + u(ih,k\tau) - u(ih,[k-1]\tau) \\ +p\{2u(ih,k\tau) - u([i-1]h,k\tau) - u([i-1]h,k\tau)\}/h^2 \\ = \{\partial_t u(ih,k\tau) - p\partial_{xx}u(ih,k\tau) - f(ih,k\tau)\} \\ -\tau\partial_{tt}u(ih,\overline{\xi}_i^k)/2 - ph^2[\partial_{xxxx}u(\overline{\chi}_{i-1}^k,k\tau) + \partial_{xxxx}u(\overline{\chi}_{i+1}^k,k\tau)]/24, \\ \text{for } \overline{\xi}_i^k \in ([k-1]\tau, k\tau), \overline{\chi}_{i-1}^k \in ([i-1]h, ih) \text{ and } \overline{\chi}_{i+1}^k \in (ih, [i+1]h). \end{cases} \tag{3.56}$$

Thus, $I_i^k(u)$ can be bounded in the same manner as $E_i^{k-1}(u)$, that is,

Bound for the Backward Euler scheme local truncation error

$$\boxed{\begin{aligned} &|I_i^k(u)| \leq \tau \max_{0\leq x\leq L; 0\leq t\leq\Theta} |\partial_{tt}u(x,t)| \\ &+ph^2 \max_{0\leq x\leq L;\ 0\leq t\leq\Theta} |\partial_{xxxx}u(x,t)|/12 \text{ for } i=1,\dots,n-1 \text{ and } k=1,\dots,l. \end{aligned}} \tag{3.57}$$

Summarizing equations (3.55) and (3.57) means that the local truncation error of both Euler methods with uniform grids is of the first-order in terms of τ and of the second-order in terms of h. In view of this, they are consistent methods.

Now, we combine the stability result (equation (3.39)) and the consistency result (equation (3.55)) to establish that the Forward Euler scheme provides convergent approximations in the maximum norm, under the stability condition $\tau \leq h^2/(2p)$.

First, we denote the defect $u(ih, k\tau) - u_i^k$ at grid point $(ih, k\tau)$ by $\bar{u}_i^k$, for $i = 0, 1, \dots, n$ and $k = 0, 1, \dots, l$. If we replace the u_i^ks by the $\bar{u}_i^k$s in scheme (3.16), similarly to Section 2.3, taking into account equation (3.54) we have

$$\begin{cases} \bar{u}_i^k - \bar{u}_i^{k-1} + p\tau(2\bar{u}_i^{k-1} - \bar{u}_{i+1}^{k-1} - \bar{u}_{i+1}^{k-1})/h^2 = \tau E_i^{k-1}(u) \\ \text{for } i=1,\dots,n-1 \text{ and } k=1,\dots,l. \end{cases} \tag{3.58}$$

On the other hand, since $\tau \leq h^2/(2p)$ setting $\vec{\bar{u}}^k := [\bar{u}_0^k, \bar{u}_1^k, \ldots, \bar{u}_n^k]^T$, we can apply equation (3.39), thereby obtaining

$$\begin{cases} \max_{1 \leq k \leq l} \| \vec{\bar{u}}^k \|_\infty \leq \max[\| \vec{\bar{u}}^0 \|_\infty, \max_{0 \leq k \leq l-1} \max \left(|\bar{u}_0^k|, |\bar{u}_n^k| \right) \\ + \Theta \max_{1 \leq i \leq n-1;\ 0 \leq k \leq l-1} |E_i^k(u)|. \end{cases} \tag{3.59}$$

However, by definition $\vec{\bar{u}}^0 = \vec{0}$ and $\bar{u}_0^k = \bar{u}_n^k = 0$ for every k, since both the initial conditions and the boundary conditions are exactly satisfied. Therefore, we have

$$\max_{1 \leq k \leq l} \| \vec{\bar{u}}^k \|_\infty \leq \Theta \max_{1 \leq i \leq n-1;\ 0 \leq k \leq l-1} |E_i^k(u)|. \tag{3.60}$$

Finally, using equation (3.55), and assuming that $\partial_{tt} u$ and $\partial_{xxxx} u$ are uniformly bounded by $2C(u)$ and $12C'(u)/p$, respectively, in $[0, L] \times [0, \Theta]$ at every grid point $(ih, k\tau)$, we derive the

Pointwise error estimate for the Forward Euler scheme

Provided $\tau \leq h^2/(2p)$,
$|u(ih, k\tau) - u_i^k| \leq \Theta[C(u)\tau + C'(u)h^2]$, for $i = 1, \ldots, n-1, k = 1, \ldots, l$. \hfill (3.61)

Noting that τ must vary like a multiple of h^2 as we let both parameters decrease, we can say that the Forward Euler scheme with uniform grids is of the second-order in terms of the spatial grid size.

Now in view of equations (3.44) and (3.57), an estimate of the same type applies *verbatim* to the Backward Euler scheme, though this time without any restrictive condition relating τ and h, that is, the

Pointwise error estimate for the Backward Euler scheme

for every τ and h,
$|u(ih, k\tau) - u_i^k| \leq \Theta[C(u)\tau + C'(u)h^2]$, for $i = 1, \ldots, n-1, k = 1, \ldots, l$. \hfill (3.62)

In short, the Backward Euler scheme with uniform grids is of the second-order in space and of the first-order in time.

Remark 3.2 *A useful interpretation of the estimates (equations (3.61) and (3.62)) in terms of convergence is as follows: If we assign to every $x^* \in (0, L)$ and $t^* \in (0, \Theta]$ a set of grid points of successively refined grids tending to (x^*, t^*) with spatial grid size h and grid time step τ as small as we wish, then the corresponding approximations will tend to $u(x^*, t^*)$. This is because $|u(x^*, t^*) - u_i^k| \leq |u(x^*, t^*) - u(ih, k\tau)| + |u(ih, k\tau) - u_i^k|$, and in this case the first term is tending to zero by construction and the second one too, owing to equation (3.61) or (3.62).* ∎

3.4.2   Schemes for the Transport Equation

In the previous subsection, we presented the road map to establish convergence results for a FD scheme, once its stability in a certain norm is demonstrated. Now we apply the same kind of argument to schemes for the transport equation (3.6) or (3.50), but rather briefly. In particular, we will unify as much as possible the treatment of the Lax scheme and the Upwind scheme.

First assuming that the second-order partial derivatives of u are defined and continuous in $[0, L] \times [0, \Theta]$, using a classical Taylor expansion about point $(jh, [k-1]\tau)$ for $1 \leq j \leq n-1$, we note that

for some $\sigma_j^k \in ([j-1]h, [j+1]h)$

$$\{u([j-1]h, [k-1]\tau) + u([j+1]h, [k-1]\tau)\}/2 = u(jh, [k-1]\tau) + h^2 \partial_{xx} u(\sigma_j^k, [k-1]\tau)$$

and for suitable $\overline{\zeta}_j^{k-1}, \underline{\zeta}_j^k \in ([j-1]h, [j+1]h)$

$$\{u([j+1]h, [k-1]\tau) - u([j-1]h, [k-1]\tau)\}/(2h)$$

$$= \partial_x u(jh, [k-1]\tau) + h[\partial_{xx} u(\overline{\zeta}_j^k, [k-1]\tau) - \partial_{xx} u(\underline{\zeta}_j^k, [k-1]\tau)]/4.$$

Hence, recalling equation (3.52) and using the fact that $h < L$, after straightforward calculations, if $1 \leq j \leq n-1$, the residual $H_j^{k-1}(u)$ at grid point $(jh, [k-1]\tau)$ for the Lax scheme applied to equation (3.50) is found to be

$$\begin{cases} H_j^{k-1}(u) = \tau \partial_{tt} u(jh, \xi_j^k)/2 + hL \partial_{xx} u(\sigma_j^k, [k-1]\tau) \\ +oh[\partial_{xx} u(\overline{\zeta}_j^k, [k-1]\tau) - \partial_{xx} u(\underline{\zeta}_j^k, [k-1]\tau)]/4. \end{cases} \tag{3.63}$$

Similarly, the residual $H_n^{k-1}(u)$ if $o > 0$ or $H_0^{k-1}(u)$ if $o < 0$ is of the same form as equation (3.63), except for the fact that one of the terms with a ζ as an argument disappears (checking which one of them according to the sign of o is a good exercise!), and the denominator 4 becomes 2. As a consequence, the residual for the Upwind scheme at point $(jh, [k-1]h)$ also has the latter form, provided adjustments of coefficient o into o^+ or o^- are performed. Whatever the case, it is clear that the following bound holds for the

Local truncation error for the Lax and the Upwind schemes

$$\boxed{|H_j^k(u)| \leq Q\tau \max_{0 \leq x \leq L;\, 0 \leq t \leq \Theta} |\partial_{tt} u(x,t)| + (|o| + L)h \max_{0 \leq x \leq L;\, 0 \leq t \leq \Theta} |\partial_{xx} u(x,t)|} \tag{3.64}$$

for every pair $(j; k)$ corresponding to an unknown value of u, Q being a suitable rational number. It follows from equation (3.64) that both the Lax scheme and the Upwind scheme are consistent with order one in terms of τ and h.

Since these schemes are stable as long as $\tau \leq h/|o|$ according to equation (3.51), under this condition they must be convergent with order one in terms of either h or τ. To make sure that the arguments of the Lax–Richtmyer theorem are duly mastered, the reader is advised to complete this convergence study, by establishing that the following estimate applies to both the Lax scheme and the Upwind scheme:

Pointwise error estimate for the Lax and the Upwind schemes

$$\boxed{\begin{aligned} &\text{Provided } \tau \le h/|o|, \\ &|u(jh, k\tau) - u_j^k| \le \Theta C^*(u)(h + \tau) \text{ for } j = 0, \dots, n, k = 1, \dots, l. \end{aligned}} \qquad (3.65)$$

where $C^*(u) = Q\max[\max\limits_{0 \le x \le L;\, 0 \le t \le \Theta} |\partial_{tt} u(x, t)|, (|o| + L) \max\limits_{0 \le x \le L;\, 0 \le t \le \Theta} |\partial_{xx} u(x, t)|].$

Notice that in practice, τ should be chosen equal to a fixed constant times h. Thus, at least for uniform grids, both schemes can be considered to be of the first-order in terms of the spatial grid size only.

3.5 Complements on the Equation of the Vibrating String (VSE)

We conclude this chapter with some material related to the vibrating string equation (3.3) that comes in complement to what we saw in Section 3.2. More particularly, we study basic explicit schemes both for the equation's standard second-order form and for the equivalent first-order system (equation (3.5)).

3.5.1 The Lax Scheme to Solve the VS First-order System

We already know that all the schemes studied in Section 3.2 can be used to solve equation (3.3) rewritten in the form of the first-order two-unknown system (cf. equation (3.5)). For example, denoting by $(v_j^k; w_j^k)$ the approximation of the pair $(v; w)$ at grid point jh at the kth time step, starting from $(v_j^0; w_j^0) = (v(jh, 0); w(jh, 0))$ with $v(jh, 0) = -[u_0]'(jh)$ and $w(jh, 0) = u_1(jh)/o$, and setting $w_0^k = w_n^k = 0$, for $j = 1, 2, \dots, n - 1$, we approximate the system (equation (3.5)) by

The Lax scheme for the VS first-order system at inner grid points

$$\boxed{\begin{aligned} &\text{For } j = 1, \dots, n - 1: \\ &v_j^k = (v_{j+1}^{k-1} + v_{j-1}^{k-1})/2 - o\tau[(w_{j+1}^{k-1} - w_{j-1}^{k-1})/(2h)]; \\ &w_j^k = (w_{j+1}^{k-1} + w_{j-1}^{k-1})/2 - o\tau[(v_{j+1}^{k-1} - v_{j-1}^{k-1})/(2h)] + \tau f(jh, [k-1])/o \end{aligned}} \qquad (3.66)$$

Scheme (3.66) can be labeled as incomplete because expressions to determine v_j^k for $j = 0$ and $j = n$ are missing. As already pointed out in the introduction of this chapter, Neumann boundary conditions for v at both ends of the string can be applied. Recalling the approximation of these boundary conditions using fictitious grid points, considered in Section 1.2, we are led to the definitions of $v_{-1}^{k-1} := v_1^{k-1} - 2hf(0, [k-1]\tau)/p$ and $v_{n+1}^{k-1} := v_{n-1}^{k-1} + 2hf(L, [k-1]\tau)/p$. On the other hand, we note that it is not possible to define in a similar manner w_{-1}^{k-1} and w_{n+1}^{k-1}. Therefore, we cannot use a centred FD to approximate $\partial_x w$ at $x = 0$ and $x = L$. Taking these remarks into account, we add the following expressions to equation (3.66):

End point equations to complete Lax scheme (3.66)

$$
\begin{aligned}
&v_0^k = (v_1^{k-1} + v_{-1}^{k-1})/2 - o\tau[(w_1^{k-1} - w_0^{k-1})/h]; \\
&v_n^k = (v_{n+1}^{k-1} + v_{n-1}^{k-1})/2 - o\tau[(w_n^{k-1} - w_{n-1}^{k-1})/h], \\
&\text{where} \\
&v_{-1}^{k-1} := v_1^{k-1} - 2hpf(0, [k-1]\tau); \\
&v_{n+1}^{k-1} := v_{n-1}^{k-1} + 2hpf(L, [k-1]\tau).
\end{aligned}
\tag{3.67}
$$

Similarly to the case of equation (3.6), it can be shown that the complete scheme (equations (3.66)–(3.67)) is stable in the maximum norm, provided $|o|\tau \le h$ (cf. [67])

3.5.2 Example 3.2: Numerical Study of Schemes for the VS First-order System

We consider the application of numerical schemes to solve the following system of the form of (equation (3.5)) in $(0,1) \times (0,1)$:

$$
\begin{cases}
\partial_t v + \partial_x w = 0 \ \forall(x,t); \\
\partial_t w + \partial_x v = 0 \ \forall(x,t); \\
w(x,0) = u_1(x) \ \forall x; \\
v(x,0) = -u_0'(x) \ \forall x; \\
w(0,t) = w(1,t) = 0 \ \forall t.
\end{cases}
$$

with $u_0(x) = 2\sin(\pi x)$ and $u_1(x) \equiv 0$. The exact solution is given by $v = -\partial u/\partial x$ and $w = \partial u/\partial t$, where $u(x,t) = \sin[\pi(x+t)] + \sin[\pi(x-t)]$.

For solving the above problem, we applied the Lax scheme (equations (3.66)–(3.67)) with $h = 1/n$ and τ such that $1/\tau$ is an integer m. We also tested the following counterpart of the unstable scheme (equation (3.19)):

Starting from $(v_j^0; w_j^0) = (v[jh,0]; w[jh,0])$ with $v[jh,0] = -u_0'(jh) = -2\pi\cos(\pi x)$ and $w[jh,0] = u_1(jh) = 0$ for $0 \le j \le n$ and setting $w_0^k = w_n^k = 0$ for all k, determine $(v_j^k; w_j^k)$ for $k = 1, \ldots, m$ by

$$
\begin{cases}
\text{For } j = 1, \ldots, n-1: \\
v_j^k = v_j^{k-1} - \tau[(w_{j+1}^{k-1} - w_{j-1}^{k-1})/(2h)]; \\
w_j^k = w_j^{k-1} - \tau[(v_{j+1}^{k-1} - v_{j-1}^{k-1})/(2h)]; \\[4pt]
v_0^k = v_0^{k-1} - \tau[(w_1^{k-1} - w_0^{k-1})/h]; \\
v_n^k = v_n^{k-1} - \tau[(w_n^{k-1} - w_{n-1}^{k-1})/h].
\end{cases}
$$

We took $n = 100$ and $m = 250$, which implies that h and τ satisfy the CFL condition $\tau \le h$ for the above problem. However, we operated significantly apart from the limiting case $\tau = h$. As a result, the Lax scheme, though stable, tends to generate numerical values a little away from the exact solution here and there, as shown in Figure 3.5 (resp. 3.6). The exact values of w (resp. v) at times $t = 0.5$ (label a) and $t = 0.6$ (label b) are displayed in Figures 3.5 and 3.6, together with the corresponding approximate values obtained with the Lax scheme and the unstable scheme. Figures 3.5a and 3.6a indicate that the latter is already showing up some

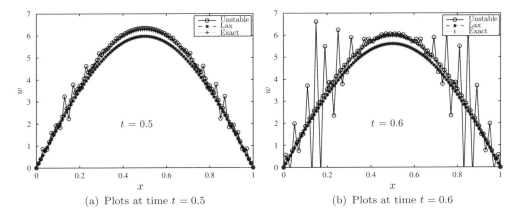

(a) Plots at time $t = 0.5$ (b) Plots at time $t = 0.6$

Figure 3.5 Function $w = \partial u / \partial t$ approximated by the Lax scheme and an unstable scheme

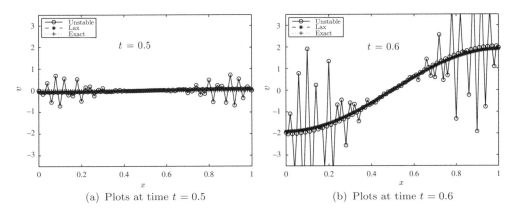

(a) Plots at time $t = 0.5$ (b) Plots at time $t = 0.6$

Figure 3.6 Function $v = -\partial u / \partial x$ approximated by the Lax scheme and an unstable scheme

instabilities at time $t = 0.5$, and from Figures 3.5b and 3.6b one can infer that after just a little more time, the numerical solution obtained with this scheme becomes completely chaotic.

3.5.3 A Natural Explicit Scheme for the VSE

The solution u of equation (3.3) can be successively approximated at grid points $(jh, k\tau)$ for $j = 1, \ldots, n-1$, $k = 1, \ldots, m$, through a post-processing of the values w_j^k given by the Lax scheme (equations (3.66)–(3.67)) to solve the equivalent first-order system (3.5). In this aim, one of the following time numerical integration formulae can be employed, starting from $u_j^0 = u_0(jh)$:

○ Forward formula: $u_j^k = u_j^{k-1} + \tau w_j^{k-1}$;
○ Centred formula: $u_j^k = u_j^{k-1} + \tau (w_j^{k-1} + w_j^k)/2$;

○ Backward formula: $u_j^k = u_j^{k-1} + \tau w_j^k$.

Unfortunately, in all cases the accuracy will be damaged by the accumulation of errors from at least two sources: the numerical time integration itself and the approximation of $\partial u/\partial t$ at the grid points by the w_j^ks. When better approximations of u are required, a good alternative is to use a higher order scheme to solve equation (3.5) such as the **Lax–Wendroff scheme** (see e.g. [67]) and a more accurate time integration formula to determine approximations of u. However, we refrain from further elaborating on the numerical treatment of system (3.5), and for more details on the subject we refer the reader to reference [67].

In view of the above introduction, among other reasons it might be interesting to treat directly the second-order equation (3.3), rather than the first-order system (3.5). That is why we conclude this chapter with a brief presentation of a popular space–time discretisation of equation (3.3) by the FDM. In doing so, we note beforehand that the extension of the resulting scheme can be easily adapted to the FVM or the FEM space discretisations, keeping unchanged the way the time discretisation is performed.

Denoting by u_j^k the approximation of $u(jh, k\tau)$, the main difference from the heat equation lies on the treatment of the time dependence. Since the time derivative is $\partial_{tt}u$, quite naturally we can apply the centred FD operator

$$D_{j,\tau}^{k,2} := (-2u_j^k + u_j^{k+1} + u_j^{k-1})/\tau^2$$

to approximate $\partial_{tt}u$ at point $(jh, k\tau)$ for $k \geq 1$ and $1 \leq j \leq n-1$.

If, like in the case of the first-order equation, we adopt an explicit approach, we can centre the scheme about point $(jh, [k-1]\tau)$, by adjusting the FD operator $D_{j,\tau}^{k,2}$ to the $(k-1)$th step. Then, we proceed to

A two-level time marching scheme for the vibrating string equation

> Starting from u_j^0 and u_j^1, for $1 \leq j \leq n-1$, determine for $k = 2, 3, \ldots, l$,
> $$u_j^k = -p\tau^2(2u_j^{k-1} - u_{j+1}^{k-1} - u_{j-1}^{k-1})/h^2 + 2u_j^{k-1} - u_j^{k-2} + \tau^2 f(jh, [k-1]\tau), \quad (3.68)$$

Notice that a new difficulty arises, since we cannot launch the calculations without knowing the values u_j^1 in advance. However, somehow we still have to satisfy the second initial condition $\partial_t u(x, 0) = u_1(x)$ for $x \in [0, L]$. Although this is not the best thing to do in terms of accuracy, for the sake of simplicity we consider that this condition is applied to determine the u_j^1s in the most straightforward manner, that is, by

$$u_j^1 = u_j^0 + \tau u_1(jh) \text{ for } j = 1, 2, \ldots, n-1, \quad (3.69)$$

Indeed, by a standard Taylor expansion up to the first-order derivative, $u(jh, \tau) = u(jh, 0) + \tau\partial_t u(jh, \tau_j)$ for a certain $\tau_j \in [0, \tau]$. This means that we approximate $\partial_t u(jh, \tau_j)$ by the closest value at our disposal, namely, $u_1(jh) = \partial_t u(jh, 0)$ for $0 \leq j \leq n-1$.

According to celebrated text books (see e.g. [139]), the stability of schemes (3.68)–(3.69) is ensured under the same condition relating the time step τ to the grid size h, that is, the CFL condition. We do not intend to treat this case in detail. Instead, we next carry out a **Von Neumann stability analysis** [41], following arguments similar to those of Subsection 3.2.1. Our goal is to derive at least a **necessary condition** for stability of this scheme. Before starting, it is instructive to point out that every solution to equation (3.3) with $f = 0$ is necessarily of the form $v(x + ot)$ or $v(x - ot)$, where o is the positive square root of p and v

is a twice-differentiable function. For example, if $v(y) = \sin(m\pi y/L)$ for a certain nonzero integer m, the function $u(x,t) = v(x + ot) + v(x - ot)$ also satisfies the Dirichlet boundary conditions $u(0,t) = u(L,t) = 0 \; \forall t \in (0, \Theta]$. Without any loss of essential aspects, we could take $L = 1$ and examine what happens if we approximate a solution of the form $u(x,t) = \sin(m\pi[x + ot]) + \sin(m\pi[x - ot])$ by means of schemes (3.68)–(3.69). Notice that such a solution corresponds to the case where the string is maintained at a position defined by $u_0(x) = 2\sin(m\pi x)$ for a certain nonzero integer m before the motion starts. Then, by freely releasing it, in the absence of forces f and for an initial velocity $u_1 = 0$, the string will vibrate to successively occupy the position $u(x,t)$ given above.

Although we are dealing with real solutions, the Von Neumann stability analysis consists of handling initial data u_0, which are linear combinations with complex coefficients of functions of the form $e^{\pm im\pi x}$ (notice that in the above test problem, $u_0(x) = -i[e^{im\pi x} - e^{-im\pi x}]$).

If, under these conditions, we apply scheme (3.68) with a suitable modification to determine the u_j^1s that we decline to specify, we obtain for $j = 1, 2, \ldots, n - 1$,

$$u_j^2 = -p\tau^2(2u_j^1 - u_{j+1}^1 - u_{j-1}^1)/h^2 + 2u_j^1 - u_j^0$$

Since $u_j^0 = e^{\pm im\pi j/n}$ for m positive or negative, say with $|m| \le n$, we assume that the starting step yields $u_j^1 = \gamma u_j^0$ where γ is a certain function of m, τ, h and p but not of j. Plugging these values into the above expression, by means of some elementary calculations, similarly to Section 3.3, we obtain for every j, $u_j^2 = (2\gamma\beta - 1)e^{\pm im\pi j/n}$, where $\beta = 1 - p\tau^2[1 - \cos(m\pi/n)]/h^2$. Hence, provided $\gamma^2 = 2\beta\gamma - 1$, we have $u_j^2 = \gamma^2 e^{\pm im\pi j/n}$. Now, assume that for a certain $k > 2$, we have $u_j^l = \gamma^l e^{\pm im\pi j/n}$ for every j, with $l = k$ and $l = k - 1$. Then, plugging these values into equation (3.68) and replacing k with $k + 1$, by the same calculations, we easily derive

$$u_j^{k+1} = -2p\tau^2\gamma^k e^{\pm im\pi j/n}[1 - \cos(m\pi/n)]/h^2 + (2\gamma^k - \gamma^{k-1})e^{\pm im\pi j/n}.$$

This means that

$$u_j^k = (2\gamma\beta - 1)\gamma^{k-1}e^{\pm im\pi j/n}.$$

Since $\gamma^2 = 2\gamma\beta - 1$, we have $u_j^{k+1} = \gamma^{k+1}e^{\pm im\pi j/n}$, and by mathematical induction we can state that the numerical solution is given by $u_j^k = \gamma^k e^{\pm im\pi j/n}$ for all j and k as long as γ is a root of the quadratic function $x^2 - 2\beta x + 1$. These roots are given by $\gamma_1 = \beta + \sqrt{\beta^2 - 1}$ and $\gamma_2 = \beta - \sqrt{\beta^2 - 1}$. It is clear that if $\beta < -1$ (resp. $\beta > 1$), then both γ_1 and γ_2 are real numbers satisfying $\gamma_2 < 0$ and $|\gamma_2| > 1$ (resp. $\gamma_1 > 1$), whereas $|\gamma_1| < 1$ (resp. $0 < \gamma_2 < 1$). Now, by linearity, initial data corresponding to $u_j^0 = e^{im\pi j/n} - e^{-im\pi j/n}$ and $u_j^1 = \gamma_1 e^{im\pi j/n} - \gamma_2 e^{-im\pi j/n}$ will yield at the kth step $u_j^k = \gamma_1^k e^{im\pi j/n} - \gamma_2^k e^{-im\pi j/n}$. On the one hand, the first (resp. second) term will tend to zero as k increases, the absolute value of the second (resp. first) term will indefinitely grow and we will gradually diverge from the expected values of the solution to our test problem. If we take, for instance, β just a little less than -1, we observe that at fixed grid points such as $(1/2; \Theta)$, the numerical solution will tend to infinity as k goes to infinity, whatever m (in order to systematically attain $x = 1/2$ as h goes to zero, it suffices to take an increasing even n). Take, for example, $\beta = -1 - \tau$ and $\tau = \Theta/k$ arbitrarily small (i.e. k arbitrarily large). We have $|\gamma_2|^k = (-\gamma_2)^{\Theta/\tau} = [1 + \tau + \sqrt{(1+\tau)^2 - 1}]^{\Theta/\tau} = \{[1 + \sigma]^{1/\sigma}\}^{\Theta(1+\sqrt{2/\tau+1})}$ with $\sigma = \tau + \sqrt{2\tau + \tau^2}$. Notice that σ tends to zero like $\sqrt{\tau}$ as τ goes to zero. Thus, the limit of $|\gamma_2|^k$ as k goes to infinity is $+\infty$, since it equals the limit of

$(1 + \sigma)^{1/\sigma}$ to a power $c\Theta/\sigma$, for a suitable constant $c > 0$ as σ goes to zero (by a similar argument, one can verify that the limit of $|\gamma_1|^k$ as k goes to infinity is zero).

Incidentally, $\beta = 1 - 2p\tau^2 sin^2[m\pi/(2n)]/h^2$, and taking the most unfavorable case, we readily see that $\beta < -1$ implies that $p\tau^2 > h^2$ and the CFL condition $\sqrt{p}\tau \leq h$ is violated.

On the other hand, the largest moduli values of the solution occur for $m = n/2$ if n is even. But, for such a value of m, we have $1 \geq \beta \geq 0$ if $(o^2\tau/h)^2 \leq 1$, which is nothing but the CFL condition. Actually the scheme is stable in the maximum norm in this case. In contrast the natural implicit counterpart of equations (3.68)–(3.69) is stable whatever τ and h. The reader may check this assertion as Exercise 3.9.

Finally, the case $-1 \leq \beta < 0$ is more tricky, because the roots are complex conjugate with moduli equal to one. This certainly rules out any uncontrolled solution growth, even though stability in norm cannot be ensured.

Summarizing, the CFL condition is necessary for the Von Neumann stability of schemes (3.68)–(3.69). It is also sufficient for stability in norm. We refer to reference [139] for a more comprehensive analysis of FD schemes for equation (3.3).

Remark 3.3 *To date, several efficient FD- or FV-based schemes for the numerical treatment of hyperbolic equations not studied in this chapter can be found in the literature. The main principle of such techniques is to simulate as accurately as possible conservation laws of physics modelled by a system of PDEs, which is usually of the hyperbolic type. Most equations in these systems are non-linear and can even have discontinuous solutions. This is the main reason for not addressing them in this book. We refer readers to references [125] or [195] for a comprehensive study of these methods. As an outstanding example of FDM to handle discontinuous solutions to hyperbolic systems, we quote Godunov's celebrated scheme [91], which can be exported to a FV framework.*∎

3.6   Exercises

3.1   Introduce the necessary modifications in the matrix forms (3.12)–(3.13) and (3.15)–(3.13) of the Backward Euler and the Crank–Nicolson FD schemes for the heat equation, in order to accommodate inhomogeneous Dirichlet boundary conditions. More specifically, use the $(n - 1) \times (n + 1)$ extension A_h^+ of the $(n - 1) \times (n - 1)$ matrix A_h whose non-zero entries are defined in equation (3.13).

3.2   Show that both equations (3.27) and (3.29) enjoy the property of avoiding unphysical oscillations and solution uncontrolled growth, as long as $\tau|o| \leq ch$, where c is the constant of the quasi-uniform family of meshes in use (c is such that for all the meshes in the family and for all CVs, V in the mesh $length(V) \geq ch$, where h is the maximum CV length in the mesh).

3.3   Taking a non-uniform mesh, in which the length of element $T_j = [x_{j-1}, x_j]$ is h_j for $j = 1, \ldots, n$, write down the FE scheme (3.31) in case the mass lumping technique is used to compute the integrals $(\varphi_j|\varphi_i)_0$ on both sides of the equations. Show that if the mesh is uniform, the resulting explicit scheme will be the same as equation (3.19), for $1 \leq j \leq n - 1$. What can be concluded about the reliability of such a scheme?

3.4 Compute the integrals in $(0, L)$ leading to an analog of equation (3.32) for a non-uniform mesh.

3.5 Use equation (3.36) to prove discrete maximum and minimum principles for the Forward Euler scheme analogous to those that hold for the heat equation, assuming that $\gamma \leq 1/2$.

3.6 Check that if the CFL condition is fulfilled, the approximations u_j^k obtained by the Lax scheme will remain non-negative (resp. non-positive) for all k, whatever the non-negative (resp. non-positive) initial datum u_0 in $[0, L]$.

3.7 Prove that the stability result (equation (3.49)) also holds for the Upwind scheme.

3.8 Prove the validity of equation (3.51) for the Lax and the Upwind schemes applied to equation (3.50) with a homogeneous boundary condition at either $x = 0$ or $x = L$, under the CFL condition.

3.9 Consider equations (3.3)–(3.4) of the vibrating string with $f \equiv 0$. Check that if the solution u to this equation is sufficiently smooth, the following identity holds:

$$\| \partial_t u(\cdot, t) \|_{0,2}^2 + p \, \| \partial_x u(\cdot, t) \|_{0,2}^2 = \| u_1 \|_{0,2}^2 + p \, \| u_0' \|_{0,2}^2 \ \forall t \in (0, \Theta],$$

where $\| \cdot \|_{0,2}$ is the standard norm of $L^2(0, L)$ and $g(\cdot, t)$ represents the function ${}^t g$ defined in $[0, L]$, given by ${}^t g(x) = g(x, t)$ for a fixed t in $[0, \Theta]$. This property means that, in the absence of external forces f string's total energy remains constant as it evolves in time. Show that whatever τ and h an analogous stability property holds for the numerical solution of equations (3.3) and (3.4) obtained by the following implicit counterpart of scheme (3.68)–(3.69) for $f \equiv 0$:

For $k = 2, 3, \ldots, l$ determine u_j^k for $j = 1, 2, \ldots, n - 1$ such that

$$\frac{u_j^k - 2u_j^{k-1} + u_j^{k-2}}{\tau^2} + p \frac{2u_j^k - u_{j+1}^k - u_{j-1}^k}{h^2} = 0,$$

starting from $u_j^0 = u_0(jh)$ and $u_j^1 = u_j^0 + \tau u_1(jh)$ for every $j = 1, \ldots, n$, where $u_0^k = u_n^k = 0$ for $k = 0, 1, \ldots, l$.

4

Methods for Two-dimensional Problems

Les différences sont finies;
vive la différence!
(The differences are finite.
Long live the difference!)

In this chapter, we introduce basic and popular FDM, FEM and FVM to solve elliptic PDEs posed in a bounded two-dimensional domain. The FDM is a very good option when an equation's domain is rectangular shaped. If this is not so, this method loses a lot of its attractive simplicity. In this case, the FEM or the FVM for plane domains provides efficient options from the computational point of view, since both can be based on arbitrary triangular meshes.

We take as a model the **Poisson equation**, which is by far the most used second-order linear PDE for numerical studies of stationary events that can be completely described in a plane geometry. Regarding the **Laplace equation** as a particular type of Poisson equation, one can say that the latter directly models a wide spectrum of physical phenomena, as diverse as membrane deformations, irrotational flow of incompressible fluids, electric field generation, heat conduction and diffusion processes in general in homogeneous and isotropic media. Furthermore, it lies on the basis of PDE models in countless other scientific and technological applications, either as the dominant part of the governing equation or as the linearised form of a non-linear second-order PDE. In addition to this, several techniques to solve the Poisson equation can be easily extended to more complex variants of it, namely, second-order differential models for heterogeneous and/or anisotropic media, such as porous media. This is particular relevant in applications related to oil recovery, underground dissemination of contaminants and many others.

The Poisson equation also represents the stationary counterpart of innumerable PDE models for time-dependent problems, such as membrane vibrations, wave propagation, as well as the

Numerical Methods for Partial Differential Equations:
An Introduction Finite Differences, Finite Elements and Finite Volumes, First Edition. Vitoriano Ruas.
© 2016 John Wiley & Sons, Ltd. Published 2016 by John Wiley & Sons, Ltd.
Companion Website: www.wiley.com/go/ruas/numericalmethodsforpartial

already mentioned phenomena of processes in transient regime. Finally, this equation can be useful in the study of methodologies to determine vector fields solving systems of second-order PDEs in various domains, such as electromagnetism, elasticity and fluid mechanics, for which it can be viewed as the scalar analog. In short, the study of numerical methods for the Poisson equation is generally viewed as an ideal pattern to provide the necessary insight into their extension to other second-order boundary value problems posed in two-dimensional or higher dimensional domains.

Chapter outline: In Section 4.1, we introduce the uniquely solvable Poisson equations to be dealt with, together with pertaining notations and fundamental facts about these problems. In Section 4.2, the FDM is presented only in the framework of problems posed in rectangular domains. We also show that by means of straightforward extensions, the same FDM applies to more general domains consisting of the union of rectangles, whose edges are parallel to two fixed orthogonal axes. Section 4.3 is devoted to the two-dimensional version of the **Piece-wise linear FEM**, commonly known as the $\mathcal{P}_1$ **FEM**, defined in connection with **triangular meshes**. Among many useful aspects related to this method, we append to the Section 4.3 method's important application to the linear elasticity system. In Section 4.4, we consider two FVMs often called the **Vertex-centred FVM** and the **Cell-centred FVM**, associated with triangular meshes or not. We emphasise that, for a wide class of meshes, the former FVM yields schemes very close to the $\mathcal{P}_1$ FEM. Moreover, we show how the latter FVM can be derived from a **mixed FEM** known as the **lowest order Raviart–Thomas method**. For this reason, the presentation of the Cell-centred FVM will be assorted by a description of the Poisson equation in standard **mixed formulation**. Finally, in Section 4.5 we focus on SLAEs of the particular form originated from the application of the aforementioned discretisation methods. A FORTRAN subroutine for solving such systems by Crout's method is supplied, and a numerical example is given.

4.1   The Poisson Equation

The Poisson equation to be studied is defined in a bounded domain Ω of $\mathfrak{R}^2$ with boundary Γ, as illustrated in Figure 4.1. Γ will always be assumed to have regularity properties compatible with the validity of the operations to be carried out in the sequel. We refrain from being more specific about this point in order to avoid clumsy definitions, which are useless in most practical situations. Let us just say that polygons or smooth curved domains satisfy the required assumptions. Actually, most results to be established hereafter for the two-dimensional case hold for **Lipschitz domains** Ω (see e.g. [1]), such as polygons and many other types of domains encountered in real-life applications.

Now, given a function f with suitable properties to be specified later on in this chapter, we wish to find a function u of the space variables x and y satisfying

The Poisson equation

$$\boxed{-\Delta u = f \text{ in } \Omega,} \qquad (4.1)$$

where Δ is the **Laplacian operator** defined by $\Delta u := \partial_{xx} u + \partial_{yy} u$, the meaning of ∂_{xx} and ∂_{yy} being obvious.

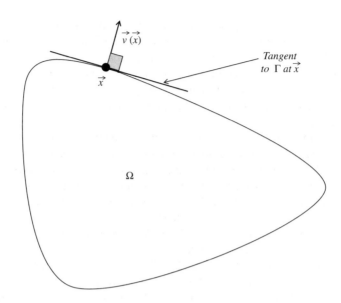

Figure 4.1   A smooth domain Ω with boundary Γ and outer normal $\vec{\nu}(\vec{x})$, $\vec{x} \in \Gamma$

From the mathematical point of view, the Poisson equation plays the role of paradigm for the analysis of **elliptic PDEs**, by means of which the most important properties of this class of differential equations can be derived (see e.g. [156]). As already stressed, we assume that the reader is sufficiently informed about such aspects.

We assume to be working in a Cartesian frame $(O; x, y)$ of $\mathfrak{R}^2$; the components of the position vector $\vec{x} = [x_1, x_2]^T$ of every point of the plane are identified with its coordinates $(x; y)$ in this frame by the relations $x_1 = x$ and $x_2 = y$.

Before pursuing, we explain the meaning of notations or expressions to be used in the remainder of this book, which some readers might not be familiar with:

- $\overline{\Omega}$ is the union of Ω and Γ, that is, the **closure** of Ω;
- A vector $\vec{v} = [v_1, v_2]^T \in \mathfrak{R}^2$ is said to be a **unit vector** if $|\vec{v}| = \sqrt{v_1^2 + v_2^2} = 1$; in this case, we write as well $\vec{v} = (v_1; v_2)$;
- A unit vector $\vec{\nu}(\vec{x}) = (\nu_1; \nu_2)$ attached to a point of Γ with position vector $\vec{x} = [x_1, x_2]^T$ which is not a corner vertex, is called the **outer normal** at $\vec{x}$ if it is orthogonal to the tangent to Γ at this point and is directed outwards Ω. The outer normal is illustrated in Figure 4.1.

Clearly enough, equation (4.1) has infinitely many solutions, unless proper boundary conditions are prescribed for u. For the sake of simplicity, in this chapter we consider mostly zero boundary conditions, applying to two disjoint portions Γ_0 and Γ_1, whose union is Γ, namely,

$$\begin{cases} u = 0 \text{ on } \Gamma_0 \\ \partial_\nu u = 0 \text{ on } \Gamma_1, \end{cases} \tag{4.2}$$

where $\partial_\nu u$ denotes the **outer normal derivative** of u on Γ, which can be expressed by $\partial_\nu u = (\mathbf{grad}\ u \mid \vec{\nu})$. We recall that $(\mathbf{grad}\ u \mid \vec{\nu}) = \partial_x u\ \nu_1 + \partial_y u\ \nu_2$ (cf. Chapter 2).

Γ_0 eventually coincides with the whole Γ, and in any case is assumed to be the union of closed arcs or segments having a strictly positive total length. The end points of the closed portions whose union is Γ_0 are called the **transition points** between Γ_0 and Γ_1.

Whenever Γ_0 is the whole Γ, the problem (equations (4.1)–(4.2)) becomes

The homogeneous Dirichlet problem for the Laplacian

$$\boxed{\begin{aligned} -\Delta u &= f \text{ in } \Omega \\ u &= 0 \text{ on } \Gamma \end{aligned}} \tag{4.3}$$

Otherwise, the problem is

The homogeneous mixed Dirichlet–Neumann problem for the Laplacian

$$\boxed{\begin{aligned} -\Delta u &= f \text{ in } \Omega \\ u &= 0 \text{ on } \Gamma_0 \\ \partial_\nu u &= 0 \text{ on } \Gamma_1, \end{aligned}} \tag{4.4}$$

In both cases, the word **homogeneous** refers to the fact that zero boundary data are assigned to u, in contrast to **inhomogeneous** boundary conditions, where two given functions g_0 and g_1 would replace zero on the right side of the relations satisfied on Γ_0 and Γ_1, respectively, both of them not being identically zero. In this case, there exists a unique non-trivial solution to the **Laplace equation**, which is nothing but equation (4.1) for $f \equiv 0$.

Incidentally, it is important to stress that for sufficiently smooth Ω and f, such as those to be considered hereafter, both problems (4.3) and (4.4) have a unique solution (cf. [109]).

Remark 4.1 *Existence and uniqueness of a solution would not apply to the homogeneous Neumann problem, for instance, which corresponds to equation (4.4) in case Γ_1 is the whole Γ. Indeed, in this case existence of a solution only occurs if the integral of f in Ω vanishes. However, even if this condition is fulfilled, uniqueness does not hold since any solution plus an arbitrary constant is also a solution to the homogeneous Neumann problem.* ■

To close this introductory section, we note that in some calculations it is convenient to view the Laplacian as the composition div **grad**, where div stands for the **divergence operator** acting on a vector field $\vec{v} = [v_x(x,y), v_y(x,y)]^T$ and **grad** is the **gradient operator** acting on a scalar function $u(x,y)$. We recall that the divergence of $\vec{v}$ is the scalar function $div\,\vec{v} := \partial_x v_x + \partial_y v_y$, and that the gradient of u is the vector field **grad** $u := [\partial_x u, \partial_y u]^T$.

4.2   The Five-point FDM

Owing to its simplicity, the FDM treated in this section is certainly the most popular among the discretisation methods to solve second-order elliptic equations posed in rectangular domains.

4.2.1   *Framework and Method Description*

Here we consider only the case of equation (4.3).

Let Ω be the rectangle defined by $0 \le x \le L_x$, $0 \le y \le L_y$ where $L_x > 0$ and $L_y > 0$. We are given a non-uniform rectangular grid $\mathcal{G} = \{(x_i; y_j)\}$, consisting of points in $\overline{\Omega}$, where

$0 \leq i \leq n_x$ and $0 \leq j \leq n_y$ for positive integers $n_x > 1$ and $n_y > 1$ with $x_0 = 0$, $x_{n_x} = L_x$, $y_0 = 0$, $y_{n_y} = L_y$ and $x_{i-1} < x_i$ for $i = 1, \cdots, n_x$, and $y_{j-1} < y_j$ for $j = 1, \cdots, n_y$. We set $h_i = x_i - x_{i-1}$ for $i = 1, \cdots, n_x$ and $k_j = y_j - y_{j-1}$ for $j = 1, \cdots, n_y$. In Figure 4.2, we illustrate a uniform grid consisting of points $\mathcal{G} = \{M_{i,j} = (ih_x; jh_y), i = 0, 1, \cdots, n_x, j = 0, 1, \cdots, n_y\}$, that is for $h_i = h_x$ $\forall i$ and $k_j = h_y$ $\forall j$, with $h_x = L_x/n_x$ and $h_y = L_y/n_y$.

Letting $u_{i,j}$ be an approximation of $u(x_i, y_j)$ or a given boundary value, according to the values of i and j, similarly to the case of the model problem of Chapter 1 discretised by means of a non-uniform grid, the simplest FDM for equation (4.1) is based on the approximation of the second-order derivatives ∂_{xx} and ∂_{yy} at point $(x_i; y_j)$ for $1 \leq i \leq n_x - 1$ and $1 \leq j \leq n_y - 1$ by means of

Divided difference operators for a non-uniform rectangular grid

Setting $h_{i+1/2} = (h_i + h_{i+1})/2$ and $k_{j+1/2} = (k_j + k_{j+1})/2$,

for $i = 1, \cdots, n_x - 1$ and $j = 1, \cdots, n_y - 1$:

$$\delta_x^2(u_{i,j}) := \frac{\delta_x(u_{i+1/2,j}) - \delta_x(u_{i-1/2,j})}{h_{i+1/2}}$$

$$\delta_y^2(u_{i,j}) := \frac{\delta_y(u_{i,j+1/2}) - \delta_y(u_{i,j-1/2})}{k_{j+1/2}}$$

(4.5)

where

$$\delta_x(u_{i-1/2,j}) := \frac{u_{i,j} - u_{i-1,j}}{h_i} \quad \text{for } i = 1, \cdots, n_x \text{ and } j = 1, \cdots, n_y - 1$$

and

$$\delta_y(u_{i,j-1/2}) := \frac{u_{i,j} - u_{i,j-1}}{k_j} \quad \text{for } i = 1, \cdots, n_x - 1 \text{ and } j = 1, \cdots, n_y.$$

Then, assuming that $f(x, y)$ is uniquely defined for all $(x; y) \in \Omega$, a natural FD analog of equation (4.3) is

The Five-point FD scheme with homogeneous Dirichlet boundary conditions

Find $u_{i,j}$ for $1 \leq i \leq n_x - 1$ and $1 \leq j \leq n_y - 1$
satisfying for i and j in the same range:

$$-\Delta_h(u_{i,j}) = f(x_i, y_j)$$

(4.6)

where $\Delta_h(u_{i,j}) := \delta_x^2(u_{i,j}) + \delta_y^2(u_{i,j})$,
with $u_{i,0} = u_{i,n_y} = u_{0,j} = u_{n_x,j} = 0$ for $0 \leq i \leq n_x$ and $0 \leq j \leq n_y$.

As illustrated in Figure 4.3, for expressing an approximation of the Poisson equation at point $M_{i,j} := (x_i; y_j)$ by the Five-point FD scheme (equation (4.6)), the points involved are this point itself plus its four immediate neighbours on the lines given by $x = x_i$ and $y = y_j$. That is why this scheme carries this name.

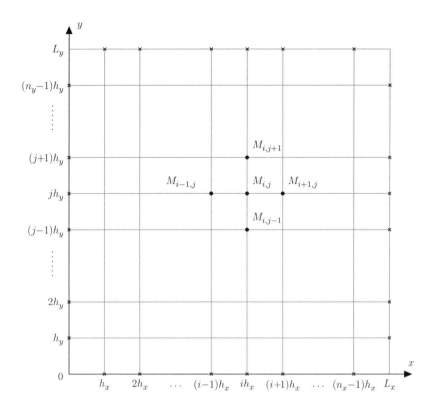

Figure 4.2 A uniform FD grid for a rectangular domain

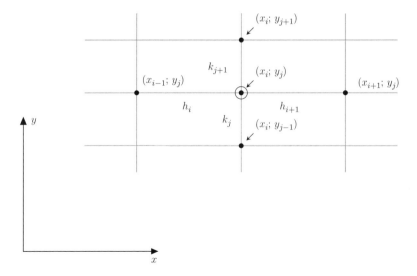

Figure 4.3 An illustration of the Five-point FD scheme

$$
A_h = \begin{bmatrix}
z & z & 0 & 0 & | & z & 0 & 0 & 0 & | & 0 & 0 & 0 & 0 \\
z & z & z & 0 & | & 0 & z & 0 & 0 & | & 0 & 0 & 0 & 0 \\
0 & z & z & z & | & 0 & 0 & z & 0 & | & 0 & 0 & 0 & 0 \\
0 & 0 & z & z & | & 0 & 0 & 0 & z & | & 0 & 0 & 0 & 0 \\
- & - & - & - & & - & - & - & - & & - & - & - & - \\
z & 0 & 0 & 0 & | & z & z & 0 & 0 & | & z & 0 & 0 & 0 \\
0 & z & 0 & 0 & | & z & z & z & 0 & | & 0 & z & 0 & 0 \\
0 & 0 & z & 0 & | & 0 & z & z & z & | & 0 & 0 & z & 0 \\
0 & 0 & 0 & z & | & 0 & 0 & z & z & | & 0 & 0 & 0 & z \\
- & - & - & - & & - & - & - & - & & - & - & - & - \\
0 & 0 & 0 & 0 & | & z & 0 & 0 & 0 & | & z & z & 0 & 0 \\
0 & 0 & 0 & 0 & | & 0 & z & 0 & 0 & | & z & z & z & 0 \\
0 & 0 & 0 & 0 & | & 0 & 0 & z & 0 & | & 0 & z & z & z \\
0 & 0 & 0 & 0 & | & 0 & 0 & 0 & z & | & 0 & 0 & z & z
\end{bmatrix}
$$

Array 4.1

Equation (4.6) is a SLAE with $(n_x - 1) \times (n_y - 1)$ unknowns and the same number of equations. Its matrix is sparse, and if we number the unknowns from $i = 1$ to $i = n_x - 1$, letting j vary from 1 to $n_y - 1$ for each i, we come up with a tridiagonal $(n_x - 1) \times (n_x - 1)$ **block matrix**. The form of such a matrix denoted by A_h is illustrated in Array 4.1 for $n_x = 4$ and $n_y = 5$, where a z represents a generic nonzero entry. As one can see, each diagonal block is an $(n_y - 1) \times (n_y - 1)$ tridiagonal matrix, and each off-diagonal block is an $(n_y - 1) \times (n_y - 1)$ diagonal matrix. In this way, every row has at most five nonzero entries as a Five-point scheme suggests. Besides being sparse and a tridiagonal block matrix. A_h is a **banded matrix** with a half band width l strictly less than its dimension minus one. This means that all entries are zero if they lie outside a band consisting of $l - 1$ contiguous sub-diagonals on each side of the main diagonal. In the case of the Five-point FD scheme with unknowns numbered as indicated above, we have $l = n_y - 1$. This means that the matrix entry in the rth row and the sth column is necessarily zero if $|r - s| > l = n_y - 1$ (cf. Array 4.1).

In Section 4.5, we will say a little more about grid-point numbering and the effective resolution of the corresponding SLAE.

4.2.2   A Few Words on Possible Extensions

In the case where Neumann boundary conditions hold on a portion Γ_1 of Γ, the situation becomes a little more delicate. This is because the grid points located on Γ_1 are also unknowns. So we have to adapt equation (4.6) in order to define discrete analogs of the Poisson equation at those points generically denoted by P, that somehow take into account the condition $[\partial_\nu u](P) = 0$, while avoiding to spoil the order of local truncation errors. Just to make ideas clear, let us assume that Γ_1 is the edge given by $x = L_x$ excluding the points $(L_x, 0)$ and (L_x, L_y). A typical grid point (L_x, y_j) on Γ_1 is illustrated in Figure 4.4. The (Neumann) boundary condition on the normal derivative of u at this point is $\partial_x u(L_x, y_j) = 0$. If we approximate this condition by $(u_{n_x+1,j} - u_{n_x-1,j})/(2h_{n_x}) = 0$, like in the case of the

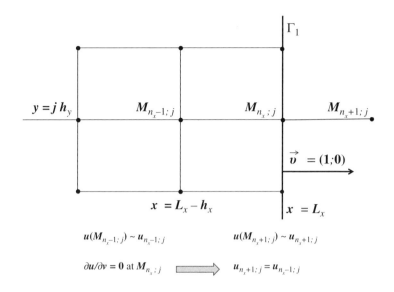

Figure 4.4   FD approximation of homogeneous Neumann boundary conditions on Γ_1

condition $u'(L) = 0$ for the model problem in Chapter 1, by introducing fictitious values $u_{n_x+1,j}$ for $j = 1, 2, \cdots, n_y - 1$, we come up with the scheme

$$\begin{cases} -\Delta_h(u_{i,j}) = f(x_i, y_j) \text{ for } 1 \leq i \leq n_x \text{ and } 1 \leq j \leq n_y - 1, \\ \text{with } u_{n_x+1,j} = u_{n_x-1,j} \text{ and } u_{0,j} = u_{i,0} = u_{i,n_y} = 0, 1 \leq i \leq n_x - 1. \end{cases} \tag{4.7}$$

Since the local truncation error for the inner points is of order one, this approximation of the normal derivative does not cause any erosion, and globally the local truncation error remains of the first order. However, this would not be the case if we were using a uniform grid. In order to overcome the resulting reduction of the scheme's order of consistency from two to one, we have to resort to the Poisson equation by assuming that f is differentiable in Ω, like in the case of the model problem of Chapter 1. However, we skip details on this modification for the sake of conciseness, and propose its study as Exercises 5.1 and 5.2 in a similar case. We refer to reference [67] for further explanations on this technique. We also leave it to the reader the adaptation of scheme (4.7) to the case of Neumann boundary conditions on other edges excluding their end points. We just note that applying only Dirichlet boundary conditions at transition points between both types of boundary conditions is mathematically consistent, but we will spend no time to justify such an issue. Notice also that if a transition point is not a vertex of Ω, the grid should be adjusted in such a manner that it coincides with a grid point.

To conclude, it is worth pointing out that through minor and straightforward adaptations, scheme (4.6) can be applied to domains Ω consisting of the union of rectangles. In Figure 4.5, we give an example of a grid adjusted to a domain of this type. This is also the case of scheme (4.7), provided the Neumann boundary conditions apply to edges of such a domain, excluding transition points between both types of boundary conditions. The FDM also turns out to be a simple and attractive numerical solution technique in the important case of inhomogeneous

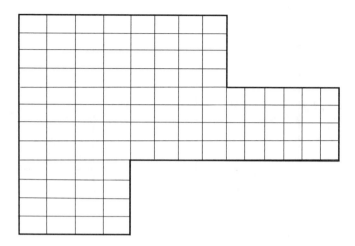

Figure 4.5   FD grid of a domain consisting of assembled rectangular subdomains

(Dirichlet or Neumann) boundary conditions. We refer to Section 1.5 for the modifications to be enforced in order to take such conditions into account. Actually, the extension of equation (4.6) to this case is quite obvious for rectangular domains, so we rather encourage the reader to carry out the pertaining modifications (see also [67] and [190]). Notice that error estimates qualitatively equivalent to those to be given in Subsection 5.1.1 (cf. equation (5.11)) can be expected from thus-adapted schemes.

4.3   The $\mathcal{P}_1$ FEM

The $\mathcal{P}_1$ FEM was introduced by Courant in his celebrated paper [56] published in 1943. It has since been intensively employed in numerical simulations of phenomena governed by elliptic or parabolic equations in two- or three-dimensional space. Later on, it was found to be a good alternative to solve hyperbolic equations as well. By studying the FE solution of the Poisson equation, we will be largely addressing the essential steps of its application to more general elliptic PDEs.

In order to apply the FEM to solve equation (4.3), we first have to recast it in the form of an equivalent variational problem, most commonly its **standard Galerkin formulation**. In this aim, we need some auxiliary material, in particular classical integral relations involving differential operators with respect to the variables x and y. It is also useful to extend the norm $\| \cdot \|_{0,2}$ and associated inner product $(\cdot | \cdot)_0$ introduced in Chapter 2 to the two-dimensional case.

4.3.1   Green's Identities

Fundamental tools to set up variational forms equivalent to a boundary value PDE in a bounded domain Ω of N-dimensional space $\Re^N$ with $N > 1$ are **Green's identities** or formulae, which play the role of classical integration by parts for $N = 1$. These formulae apply to functions having some regularity properties, and minimally to functions that are square integrable in Ω.

Although the concepts that follow trivially extend to any value of N, in this chapter we confine ourselves to the case $N = 2$. The space of square integrable functions in a bounded domain Ω of $\Re^2$ is denoted by $L^2(\Omega)^1$, which is equipped with the inner product $(\cdot|\cdot)_0$ defined as follows: f, g being two functions in $L^2(\Omega)$,

$$(f|g)_0 = \int_\Omega f(x, y)g(x, y)dxdy.$$

The fact that $(f|g)_0$ satisfies the three conditions to be an inner product can be established exactly like in the one-dimensional case. Similarly, if $\vec{f} = [f_1, f_2]^T$ and $\vec{g} = [g_1, g_2]^T$ are two vector fields whose components f_i and g_i belong to $L^2(\Omega)$ for $i = 1, 2$, recalling the inner product $(\cdot|\cdot)$ of two vectors of $\Re^2$, we define the inner product of $\vec{f}$ and $\vec{g}$ still denoted by $(\cdot|\cdot)_0$ as follows:

$$(\vec{f}|\vec{g})_0 = \int_\Omega (\vec{f}(x, y)|\vec{g}(x, y))dxdy$$

(i.e. $(\vec{f}|\vec{g})_0 = \int_\Omega [f_1(x, y)g_1(x, y) + f_2(x, y)g_2(x, y)]dxdy$).

For both a scalar function or a vector field $\mathbf{F}$, we use the notation $\| \mathbf{F} \|_{0,2}$ for the norm given by $(\mathbf{F}|\mathbf{F})_0^{1/2}$.

Let $H^1(\Omega)^2$ be the space of functions $v \in L^2(\Omega)$, whose first-order partial derivatives $\partial_x v$ and $\partial_y v$ also belong to $L^2(\Omega)$. The subspace of $H^1(\Omega)$ consisting of functions that vanish on a closed portion Γ_0 of Γ with nonzero total length will be denoted by V in this chapter. If Γ_0 is the whole Ω, we denote the corresponding subspace of $H^1(\Omega)$ by V_0^3. We define the semi-norm of a function $v \in H^1(\Omega)$ by $|v|_{1,2} := \| \mathbf{grad}\ v\|_{0,2}$. Indeed it is a semi-norm, for $|v|_{1,2} = 0$ implies that both first-order partial derivatives of v vanish (almost everywhere) in Ω, and hence v is constant in Ω but not necessarily zero. However, over V (and necessarily over V_0 too), this semi-norm becomes a norm for obvious reasons. Moreover, the following Friedrichs–Poincaré inequality holds for classes of domains Ω usually encountered in practical applications (cf [128]): For a certain constant $C_P > 0$ depending only on Ω and Γ_0, we have

$$\| v\|_{0,2} \leq C_P|v|_{1,2}\ \forall v \in V. \tag{4.8}$$

Remark 4.2 *Like in the one-dimensional case, the proof of equation (4.8) relies on simple properties of integrals. It is not so difficult in the case of V_0, provided we are dealing with sufficiently smooth functions. It is a little more elaborated in the case of V whenever $V \neq V_0$ and/or of non-smooth functions. The case of smooth functions in V_0 is proposed as Exercise 4.1, and hints are also given on how to deal with other cases (cf. Exercise 4.2).* ∎

---

[1] Here again, we don't distinguish two functions belonging to this space, which differ from each other only on subsets of Ω whose area is zero, such as curves, isolated points, and so on.

[2] This space belongs to the family of Sobolev spaces $W^{m,p}(\Omega)$ for an integer $m \geq 1$ and a real number $p \geq 1$; if $p = 2$ it is more commonly denoted by $H^1(\Omega)$ (cf. [1]).

[3] In the literature, V_0 is commonly denoted by $H_0^1(\Omega)$.

The following result lies on the basis of Green's identities:

First, we recall that for every $\vec{x} \in \Gamma$, $\vec{\nu}(\vec{x}) = (\nu_1; \nu_2)$ is the unit vector normal to Γ directed outwards Ω (cf. Figure 4.1). If Γ is sufficiently smooth, for every (Lebesgue) integrable function v whose first-order partial derivatives are also (Lebesgue) integrable [206] in Ω, it holds that

$$\int_\Omega \frac{\partial v}{\partial x_i} \, dx dy = \oint_\Gamma v \, \nu_i \, ds \tag{4.9}$$

with $i = 1$ or $i = 2$ and $x_1 = x$ and $x_2 = y$.

From relation (4.9), we derive a series of identities for sufficiently smooth Γ.

◇ If both components v_x and v_y of the field $\vec{v}$ together with its **divergence** $div \, \vec{v} :=$ $\dfrac{\partial v_x}{\partial x} + \dfrac{\partial v_y}{\partial y}$ are (Lebesgue) integrable in Ω, then we have

The Divergence Theorem

$$\boxed{\int_\Omega div \, \vec{v} \, dx dy = \oint_\Gamma (\vec{v}(\vec{x}) \mid \vec{\nu}(\vec{x})) \, ds.} \tag{4.10}$$

◇If both components w_x and w_y of a field $\vec{w}$ together with its divergence belong to $L^2(\Omega)$, and $q \in H^1(\Omega)$, we have

The two-dimensional analog of integration by parts

$$\boxed{\int_\Omega (\vec{w} \mid \mathbf{grad} \, q) \, dx dy = \oint_\Gamma q(\vec{w} \mid \vec{\nu}) \, ds - \int_\Omega q \, div \, \vec{w} \, dx dy} \tag{4.11}$$

◇ For $v \in H^1(\Omega)$ and $u \in H^2(\Omega)^4$, $H^2(\Omega)$ being the space consisting of functions whose partial derivatives up to the second order belong to $L^2(\Omega)$, recalling the notation $\partial_\nu u$ for the outer normal derivative of u on Γ given by $(\mathbf{grad} \, u \mid \vec{\nu})$, it holds

First Green's identity

$$\boxed{\int_\Omega (\mathbf{grad} \, u \mid \mathbf{grad} \, v) \, dx dy = \oint_\Gamma \partial_\nu u \, v \, ds - \int_\Omega \Delta u \, v \, dx dy.} \tag{4.12}$$

While equation (4.10) is a trivial consequence of (4.9), equation (4.11) follows from (4.10) by taking $\vec{v} = q \, \vec{w}$, which yields $div \, \vec{v} = (\vec{w} \mid \mathbf{grad} \, q) + q \, div \, \vec{w}$, and hence equation (4.11). Finally, equation (4.12) results from (4.11) by taking $\vec{w} = \mathbf{grad} \, u$, $q = v$, and noting that $div \, \mathbf{grad} \, u = \Delta u$.

Remark 4.3 *The fact that all the above integrals make sense under the assumption that the different functions and fields involved belong to the given spaces, is mainly due to the Cauchy–Schwarz inequality. Indeed, this implies that the product of two square integrable*

---

[4] $H^m(\Omega)$ for a integer $m \geq 1$ is the Sobolev space of functions whose partial derivatives up th the mth order belong to $L^2(\Omega)$ (cf. [1]).

*functions in a bounded domain is integrable in this domain. Clearly enough, this property applies in particular to square integrable functions on Γ. As a matter of fact, in the case of formulae (4.9)–(4.12) it is not necessary to assume anything regarding the degree to which the involved functions are integrable on Γ. This is because the required properties turn out to hold for extensions of functions or fields belonging to the pertinent spaces, to boundaries of open sets in $\mathfrak{R}^2$. These extensions are called **traces**, and establishing the validity of such properties of theirs is not so simple. The interested reader might consult specialised text books such as references [1] and [87] for rigorous studies on traces of scalar functions and vector fields. We also refer to Exercises 6.1, 6.2 and 6.7 for properties of traces of functions in Sobolev spaces.* ■

4.3.2 The Standard Galerkin Variational Formulation

Let us now consider the variational formulation equivalent to problem (4.1) with homogeneous Dirichlet boundary conditions $u = 0$ on Γ or the homogeneous mixed Dirichlet–Neumann boundary conditions in equation (4.2), under the assumption that $f \in L^2(\Omega)$. This formulation is also called the equation's **weak form**.

First, we define

Bilinear form a(·,·) and linear form$F(\cdot)$

$$
\begin{aligned}
a(u,v) &:= \int_\Omega (\mathbf{grad}\ u(x,y) \mid \mathbf{grad}\ v(x,y))dxdy \ \forall u,v \in H^1(\Omega), \\
F(v) &:= \int_\Omega f(x,y)v(x,y)dxdy \ \forall v \in L^2(\Omega)
\end{aligned}
\tag{4.13}
$$

and set

The homogeneous Dirichlet problem for the operator $- \Delta$

$$
\begin{aligned}
&\text{Find } u \in V_0 \text{ such that} \\
&a(u,v) = F(v) \ \forall v \in V_0.
\end{aligned}
\tag{4.14}
$$

The homogeneous mixed Dirichlet–Neumann problem for the operator $- \Delta$

$$
\begin{aligned}
&\text{Find } u \in V \text{ such that} \\
&a(u,v) = F(v) \ \forall v \in V.
\end{aligned}
\tag{4.15}
$$

Let us show that problem (4.14) is equivalent to the PDE (equation (4.3)). Although this is by no means essential, we will do this by assuming that the solution u to the variational problem belongs to $H^2(\Omega)$, which is always true if Ω is convex (cf. [92]). In this case, we may apply equation (4.12), thereby obtaining:

$$
a(u,v) = \oint_\Gamma \partial_\nu u\ v\ ds - \int_\Omega \Delta u\ v\ dxdy = F(v) = \int_\Omega f\ v\ dxdy \ \forall v \in V_0.
\tag{4.16}
$$

Since $v = 0$ on Γ, we have

$$
\int_\Omega - \Delta u\ v\ dxdy = \int_\Omega fv\ dxdy \ \forall v \in V_0.
\tag{4.17}
$$

Here we may use the following result: If two functions g and $\bar{g}$ belonging to $L^2(\Omega)$ are such that $(g|v)_0 = (\bar{g}|v)_0$ for every $v \in V_0$, then $g = \bar{g}$[5].

Applying this result to equation (4.17), we conclude that $-\Delta u = f$ in Ω, and since $u = 0$ on Γ by assumption, we are done.

Notice that the converse is also true. Indeed, assuming that the solution of problem (4.14) belongs to $H^2(\Omega)$, we may reconstruct the original variational problem the other way around. This means that we first multiply both sides of the differential equation with an arbitrary $v \in V_0$, and next we integrate the result in Ω. Finally, we apply First Green's identity to derive problem (4.14). The former is thus seen to be equivalent to the latter, at least as long as we can assert that $u \in H^2(\Omega)$.

Regarding problem (4.15), rigorously we cannot proceed like in the case of problem (4.14) if we want to establish its equivalence with equation (4.4). This is because in general $u \notin H^2(\Omega)$, even if Ω is convex (cf. [92]). However, First Green's identity still applies to the solution of equation (4.15) and $v \in V$ in a sense that lies beyond the scope of this book. Therefore, taking arbitrary $v \in V_0$, we still conclude in the same manner as above that $-\Delta u = f$ in Ω. Now using this equality and considering an arbitrary $v \in V$, though not belonging to V_0, as a result of the application of First Green's identity to equation (4.15), we will have

$$\oint_\Gamma \partial_\nu u \, v \, ds = \int_{\Gamma_1} \partial_\nu u \, v \, ds = 0 \ \forall v \in V.$$

There surely exists a function $v \in V$ such that $v = \partial_\nu u$ on Γ_1, and therefore $\int_{\Gamma_1} |\partial_\nu u|^2 ds = 0$. This means that $\partial_\nu u = 0$ on Γ_1 and hence equation (4.15) implies that u solves equation (4.4).

It is rather easy to see that the converse is also true. It suffices to apply First Green's identity the other way around, assuming a suitable regularity of the solution to equation (4.4).

Notice that in the above, we assumed the existence of a solution to both problems (4.14) and (4.15). This is true according to the celebrated **Lax–Milgram Theorem** [122]. As far as uniqueness is concerned, it suffices to take $f \equiv 0$, and verify that the resulting problems have no other solution than $u \equiv 0$. Indeed if u and $\bar{u}$ are two solutions of problem (4.14) (resp. (4.15)) we have $a(u - \bar{u}, v) = 0$ for every $v \in V_0$ (resp. $v \in V$). Since necessarily $u - \bar{u} \in V_0$ (resp. V) we have $a(u - \bar{u}, u - \bar{u}) = 0$. This implies that $|\mathbf{grad}(u - \bar{u})| = 0$, and hence $u - \bar{u}$ is constant in Ω. Since $u - \bar{u}$ vanishes at least on Γ_0, necessarily $u = \bar{u}$.

4.3.3   Method Description

Referring to the main principles described in Section 1.3, the FEM belongs to the class of **Ritz methods** (cf. [182]), which is based on the replacement of the space V or V_0 in problems (4.15) and (4.14) respectively, by a finite dimensional subspace V_n. Therefore, before describing the

---

[5] This result is a consequence of the *density* of V_0 in $L^2(\Omega)$ (cf. [126]), which means that for every $v \in L^2(\Omega)$ and every $\varepsilon > 0$, there exists $v_\varepsilon \in V_0$ such that $\| v - v_\varepsilon \|_{0,2} \leq \varepsilon$. In the above case, we use such a density as follows: For every $v \in L^2(\Omega)$, $(g - \bar{g}|v)_0 = (g - \bar{g}|v_\varepsilon)_0 + (g - \bar{g}|v - v_\varepsilon)_0 = (g - \bar{g}|v - v_\varepsilon)_0 \leq \| g - \bar{g} \|_{0,2} \| v - v_\varepsilon \|_{0,2}$, by the Cauchy–Schwarz inequality. Using the same argument for $-v$ instead of v, we conclude that $|(g - \bar{g}, v)_0|$ can be made as small as we wish, by taking ε arbitrarily small. Therefore, $(g - \bar{g}|v)_0 = 0$ for every $v \in L^2(\Omega)$, and taking $v = g - \bar{g}$, it follows that $\| g - \bar{g} \|_{0,2}^2 = 0$, that is, $g = \bar{g}$.

FEM, we find it instructive to introduce general facts related to this approach, in which instead of u we search for its approximation $u_n \in V_n$ satisfying:

$$a(u_n, v) = F(v) \ \forall v \in V_n. \tag{4.18}$$

Then, letting $\mathcal{B} := \{\phi_j\}_{j=1}^n$ be a basis of V_n we may write $u_n = \sum_{j=1}^n u_j \phi_j$ for suitable real coefficients u_j. Plugging this sum into equation (4.18), choosing v successively equal to ϕ_i, for $i = 1, 2, \cdots, n$, and using well-known summation properties of integrals, we come up with

The Ritz method to approximate u

$$\sum_{j=1}^n u_j a(\phi_j, \phi_i) = F(\phi_i) \ \text{ for } i = 1, 2 \cdots, n. \tag{4.19}$$

Recalling the arguments in Section 1.3, from equation (4.19) we can reconstruct (4.18), and hence both problems are equivalent. Moreover defining A_n to be the $n \times n$ matrix whose entry $a_{i,j}$ is $a(\phi_j, \phi_i)$ and the vector $\vec{b}_n$ whose ith component b_i is $F(\phi_i)$, the vector of unknown coefficients $\vec{u}_n = [u_1, u_2, \cdots, u_n]^T$ in the expression of u_n in terms of the basis $\mathcal{B}$ is the solution of the following SLAE:

$$A_n \vec{u}_n = \vec{b}_n.$$

This system is clearly equivalent to (4.19), and hence to (4.18) too. It has a unique solution if the symmetric matrix A_n is positive-definite for instance. This is precisely the case of problems (4.15) and (4.14), according to the following argument:

First of all, we denote by $\vec{v} = [v_1, v_2, \cdots, v_n]^T$ the vector of components with respect to $\mathcal{B}$ of an arbitrary function $v \in V_n$. It is easy to see that we can write $(A_n \vec{v} \mid \vec{v}) = a(v, v) = \int_\Omega (\mathbf{grad}\ v \mid \mathbf{grad}\ v) dx dy \geq 0$. Moreover, if $(A_n \vec{v} \mid \vec{v}) = 0$, then necessarily $\mathbf{grad}\ v = \vec{0}$, and since $v \in V$ it must vanish everywhere in Ω. It follows that $\vec{v} = \vec{0}$, and thus A_n is **positive-definite**. Summarising, the existence and uniqueness of a solution obtained by any Ritz method to approximate equation (4.15) or (4.14) are guaranteed. This will also be the case of the FEM, which corresponds to particular choices of space V_n, such as the one described further here.

Since, in the general case, it is not very easy to construct a subset of V (or V_0), in order to simplify the presentation, henceforth unless otherwise stated, Ω **is assumed to be a polygon**. The $\mathcal{P}_1$ FEM is defined in connection with a partition $\mathcal{T}_h$ of $\overline{\Omega}$ into triangles, also called a **triangulation** of Ω. The triangles in $\mathcal{T}_h$, named the partition's **elements**, fulfil the following conditions (cf. Figure 4.6):

- Each $T \in \mathcal{T}_h$ is a closed set (i.e. includes its boundary);
- The intersection of two elements of $\mathcal{T}_h$ is empty, a common vertex or a common edge;
- The union of all the elements $T \in \mathcal{T}_h$ is the whole $\overline{\Omega}$.

The above conditions imply that $\forall T \in \mathcal{T}_h, T \cap \Gamma$ is either the empty set or a subset of the set of vertices or edges of T. In this manner, we first define a function space W_h related to $\mathcal{T}_h$ as follows:

$$W_h := \{v : \ v|_T \in \mathcal{P}_1(T) \ \forall T \in \mathcal{T}_h, v \ \text{ is continuous in } \Omega \cup \Gamma\} \tag{4.20}$$

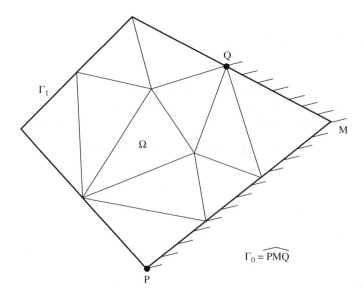

Figure 4.6   Partition of Ω matching two complementary boundary portions Γ_0, Γ_1

whereby $\mathcal{P}_1(T)$ denotes the space of polynomials in two variables defined in T, of degree less than or equal to one. This means that any function in $\mathcal{P}_1(T)$ is of the form $\alpha x + \beta y + \gamma$ for suitable real coefficients α, β and γ, and hence it is a vector space of dimension three. The vertices of the triangles belonging to $\mathcal{T}_h$ are called the **nodes** of the partition $\mathcal{T}_h$.

In the case of the homogeneous Dirichlet problem (equation (4.14)), the space V_n is the subspace V_{h0} of W_h consisting of functions that vanish on Γ. Notice that it suffices to require that such functions vanish at the nodes of the partition lying on Γ. Indeed, by definition, these functions are linear along the segments joining two neighbouring boundary nodes, and every linear function that vanishes at two distinct points of a segment vanishes everywhere on it.

In the case of problem (4.15), we have to require an additional condition on the partition $\mathcal{T}_h$, namely,

• If the total length of Γ_1 is not zero, the set of transition points between Γ_0 and Γ_1 are vertices of triangles in $\mathcal{T}_h$ (see Figure 4.6).

Then, we define the space V_n for the homogeneous mixed Dirichlet–Neumann problem (equation (4.15)) to be the subspace V_h of W_h consisting of functions that vanish everywhere on Γ_0. Using the same argument as above, we can assert that V_h is nothing but the set of functions in W_h whose value equals zero at every vertex of a triangle in the partition $\mathcal{T}_h$ lying on Γ_0. Notice that this includes the transition nodes between Γ_1 and Γ_0, similarly to the boundary grid points in the FDM.

In order to show how to compute with such spaces, we must be able to determine the corresponding basis $\mathcal{B}$. This will lead to the associated matrix A_n and right-hand-side vector $\vec{b}_n$, here denoted by A_h and $\vec{b}_h$, with unknown vector $\vec{u}_h = [u_1, u_2, \cdots, u_{N_h}]^T$ instead of $\vec{u}_n$, where N_h is the number of nodes of $\mathcal{T}_h$ not lying on Γ_0. In the case of the FEM, the basis functions are referred to as **shape functions**. The shape functions for the $\mathcal{P}_1$ FEM belong to a subset of the set of functions φ_j defined as follows. Let the partition have K_h nodes and $\{S_i\}_{i=1}^{K_h}$ be the set of nodes of $\mathcal{T}_h$, numbered in a certain systematic way. Then, we define the

Shape functions for the $\mathcal{P}_1$ FEM

$$
\boxed{\begin{array}{l} \varphi_j \text{ is continuous and linear in every triangle of } \mathcal{T}_h; \\ \varphi_j(S_i) = \delta_{ij} \text{ for } 1 \le j \le K_h \text{ and } 1 \le i \le K_h. \end{array}} \qquad (4.21)
$$

Although this is by no means a standard procedure, here we assume that in the case of the (homogeneous) Dirichlet problem, the node numbering is such that the first N_h nodes are located in the interior of Ω, and the last $K_h - N_h$ nodes are those belonging to Γ. Similarly, in the case of the (homogeneous) mixed Dirichlet–Neumann problem, we assume that the first N_h nodes belong either to the interior of Ω or to Γ_1, while the last $K_h - N_h$ nodes are those located on Γ_0.

In doing so, in the particular case of V_{h0} (resp. V_h) the basis $\mathcal{B}$ of the Ritz method is given by $\mathcal{B} := \{\varphi\}_{j=1}^{N_h}$.

The definition (equation (4.21)) is practical since it immediately leads to an interpretation of the coefficients u_j of u_h: Since $u_h(\vec{x}) = \sum_{j=1}^{N_h} u_j \varphi_j(\vec{x})$ for every $\vec{x} \in \Omega$, setting $\vec{x} = S_i$ for $i = 1, 2, \cdots, K_h$, we obtain $u_h(S_i) = \sum_{j=1}^{N_h} u_j \varphi_j(S_i)$. Then, using equation (4.21), we readily obtain $u_h(S_i) = u_i$ for every i, that is the coefficient u_i is nothing but the value of u_h at node S_i, exactly like in the one-dimensional case (cf. Section 1.3). Notice that, in particular, $u_i = 0$ for $i > N_h$ as required.

An illustration of such a shape function φ_i related to an inner node S_i is provided in Figure 4.7. The surface representing nonzero values of the function is highlighted in grey. Notice that φ_i vanishes identically in all the triangles not having S_i among their vertices, which is an immediate consequence of equation (4.21). For the mixed Dirichlet–Neumann problem, some shape functions φ_j associated with nodes located on Γ_1, such as S_j, have a slightly different shape, as shown in Figure 4.7. One might check without any difficulty that, thanks to the assumption that Ω is a polygon and to the mesh transition rule from Γ_0 to Γ_1, every basis function φ_i vanishes on Γ or on Γ_0 according to the case. This property implies that any function in V_{h0} (resp. V_h) vanishes on Γ (resp. Γ_0) as required, and consequently V_h is a subspace of V and V_{h0} is a subspace of V_0 (cf. Exercise 4.3).

Remark 4.4 *The fact that the set of shape functions φ_j defined by equation (4.21) form a basis of the space W_h of continuous piecewise linear function over a triangular mesh can be established in the following manner. First of all, each φ_i is continuous since its values at the two vertices of an edge common to two neighbouring triangles of the mesh coincide by construction. Therefore, φ_i is continuous at those edge interfaces between two elements of $\mathcal{T}_h$. Moreover, $\{\varphi_i\}_{i=1}^{K_h}$ is a set of linearly independent functions (cf. [132]). Indeed, if $\sum_{j=1}^{K_h} \alpha_j \varphi_j \equiv 0$, we have for every S_i, $\sum_{j=1}^{K_h} \alpha_j \varphi_j(S_i) = 0$, and hence $\alpha_i = 0$ for $i = 1, 2, \cdots, K_h$. This implies that $\dim W_h \ge K_h$. The fact that any function $w \in W_h$ can be written as a linear combination of the $\varphi_i s$ is established by setting $\overline{w} = \sum_{j=1}^{K_h} w(S_j) \varphi_j$. Clearly, $w - \overline{w} \in W_h$ and $(\overline{w} - w)(S_i) = \sum_{j=1}^{K_h} w(S_j) \varphi_j(S_i) - w(S_i) = 0$ for $i = 1, 2, \cdots, K_h$. Hence, the restriction of $w - \overline{w}$ to every $T \in \mathcal{T}_h$ vanishes identically, for it vanishes at its three vertices, and the*

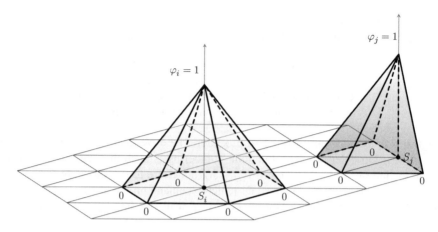

Figure 4.7   Shape functions of the $\mathcal{P}_1$ FEM for an inner node and a boundary node

result follows. The same argument combined with the fact that the shape functions are linearly independent leads to the conclusion that $\mathcal{B}$ is a basis of the subspace V_{h0} (resp. V_h). ∎

To conclude this description of the $\mathcal{P}_1$ FEM, we note that like in Chapter 1, it is necessary to have analytical expressions for the shape functions φ_i available. In this aim, the concept of **barycentric coordinates** of a triangle plays a crucial role. Letting S_1^T, S_2^T, S_3^T be the three vertices of a given triangle T, numbered in the counterclockwise sense, the barycentric coordinate λ_r^T of T related to vertex S_r^T is the linear function that takes the value one at S_r^T and the value zero at the two other vertices. Since φ_i is a linear function that takes the value one at S_i and the value zero at the two other vertices of any triangle T having S_i as a vertex, by definition φ_i in T equals the barycentric coordinate of T related to this vertex. Therefore, if we know the expressions of the barycentric coordinates of the mesh triangles, we can readily exhibit the analytic expression of the shape functions φ_i. Let T be a triangle of $\mathcal{T}_h$ and $(x_r^T; y_r^T)$ be the coordinates of S_r^T, for $r = 1, 2, 3$, in a suitable orthonormal frame corresponding to the variables x and y. Since the local numbering of the vertices of T is performed in the counterclockwise sense, from elementary geometry we know that the area of T is given by

$$area(T) = \frac{1}{2} \begin{vmatrix} 1 & x_1^T & y_1^T \\ 1 & x_2^T & y_2^T \\ 1 & x_3^T & y_3^T \end{vmatrix} \tag{4.22}$$

where $|\mathcal{A}|$ denotes the determinant of a square array $\mathcal{A}$. Using (4.22), we obtain the following expressions for the

Barycentric coordinates of a triangle T with vertices at $(x_r^T; y_r^T)$, $r = 1, 2, 3$

$$\lambda_1^T(x, y) = \frac{1}{2\,area(T)} \begin{vmatrix} 1 & x & y \\ 1 & x_2^T & y \\ 1 & x_3^T & y_3^T \end{vmatrix} \tag{4.23}$$

$$\lambda_2^T(x,y) = \frac{1}{2\,area(T)} \begin{vmatrix} 1 & x_1^T & y_1^T \\ 1 & x & y \\ 1 & x_3^T & y_3^T \end{vmatrix} \qquad\qquad (4.24)$$

$$\lambda_3^T(x,y) = \frac{1}{2\,area(T)} \begin{vmatrix} 1 & x_1^T & y_1^T \\ 1 & x_2^T & y_2^T \\ 1 & x & y \end{vmatrix}. \qquad\qquad (4.25)$$

From the properties of determinants, λ_r^T is a linear function of x and y, and moreover it clearly satisfies $\lambda_r^T(x_s^T, y_s^T) = \delta_{rs}$ for $1 \le r, s \le 3$ as required.

A basic property of the barycentric coordinates is the fact that the set $\{\lambda_r^T\}_{r=1}^3$ forms a basis of the space $\mathcal{P}_1(T)$ (cf. [132]) of linear functions in T. Indeed, since the dimension of $\mathcal{P}_1(T)$ is 3, using the same argument as in Remark 4.2, we can establish that the three barycentric coordinates are linearly independent functions of $\mathcal{P}_1(T)$ (cf. [132]).

Referring to Figure 4.8, and dropping the superscripts T to simplify the notation, we list here several other handy properties of the barycentric coordinates of a triangle T:

1. λ_ν vanishes identically on the edge e_ν opposite to vertex S_ν of T for $\nu = 1, 2, 3$ (because it is linear on e_ν and vanishes at the vertices of T lying on e_ν);
2. For $\nu = 1, 2, 3$, the value of λ_ν at every point of the parallel to e_ν at a distance from S_ν is equal to r times the height h_ν with respect to this vertex, with $0 < r \le 1$, is constant and equals $1 - r$ (this is a simple consequence of the Thales theorem);
3. The sum of the three barycentric coordinates is the constant function equal to one in T (indeed, this sum belongs to $\mathcal{P}_1(T)$ and the only linear function whose value at all the three vertices is one is the constant function $\equiv 1$);

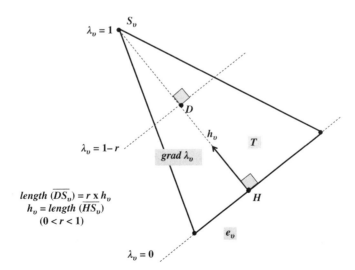

Figure 4.8   Properties 1-, 2- and 6- of a barycentric coordinate λ_ν of triangle T

4. $\sum\limits_{\nu=1}^{3} x_\nu \lambda_\nu(x,y) \equiv x$ (resp. $\sum\limits_{\nu=1}^{3} y_\nu \lambda_\nu(x,y) \equiv y$) (this is because x (resp. y) is a linear function, whose values at the three vertices of T coincide with those given by the corresponding
summation);

5. The sum of the gradients of the three barycentric coordinates is the null vector at every
point (a trivial consequence of 4.);

6. The modulus of **grad** λ_ν is $1/h_\nu$ (indeed, **grad** λ_ν is a vector orthogonal to e_ν and λ_ν is a
linear function that varies linearly from zero to one within a length of h_ν);

7. λ_ν is equal to $1/3$ at the centroid of T for $\nu = 1, 2, 3$ (this property follows from 2. and
elementary geometry, for on the parallel to e_ν through the centroid, we have $r = 2/3$ for
$\nu = 1, 2, 3$).

Remark 4.5 *It is interesting to observe that the concept of barycentric coordinate applies
to the one-dimensional case, too, in which the elements are intervals instead of triangles. In
this case, the two barycentric coordinates for a given interval $T := [x_1^T; x_2^T]$ are illustrated in
Figure 4.9. Notice that these barycentric coordinates are expressed as follows:*

$$\lambda_1^T(x) = \frac{1}{length(T)} \begin{vmatrix} 1 & x \\ 1 & x_2^T \end{vmatrix} ; \quad \lambda_2^T(x) = \frac{1}{length(T)} \begin{vmatrix} 1 & x_1^T \\ 1 & x \end{vmatrix} \qquad (4.26)$$

*and hence they have similar properties such as $\lambda_1^T + \lambda_2^T \equiv 1$, and $\lambda_1^T = \lambda_2^T = 1/2$ at the
mid-point of $[x_1^T, x_2^T]$, and so on. The practical use of one-dimensional barycentric coordinates
will be illustrated in Section 6.1.* ∎

Let us now see how the barycentric coordinates can help to derive analytic expressions for
the shape functions for the FEM in two-dimensional space.

4.3.4 Implementation Aspects

Once we have defined an appropriate basis $\mathcal{B}$ for the space V_{h0} or V_h, in principle nothing
prevents us from solving the FE approximate problems (4.14) or (4.15), that is, from computing
the matrix A_h and the right-side vector $\vec{b}_h$, and consequently from solving the SLAE $A_h \vec{u}_h =$

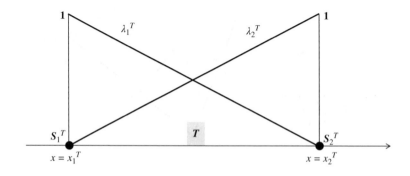

Figure 4.9   The two barycentric coordinates of a segment T

$\vec{b}_h$. Indeed, we are dealing with a particular Ritz method whose implementation was described in the previous subsection in general terms. However, the properties of the shape functions φ_i allow many simplifications in the practical computation of A_h and $\vec{b}_h$. Let us see how this works.

Let M_h be the number of elements in mesh $\mathcal{T}_h$. We recall that the mesh triangles satisfy the compatibility conditions specified in Subsection 4.2.3, and the additional requirement that the transition points between Γ_0 and Γ_1 belong to the set of nodes of the partition, if applicable. We further recall that the total number of vertices of $\mathcal{T}_h$ is K_h, and that in order to apply the FEM to solve equation (4.14) or (4.15), we employ a set $\{\varphi_i\}_{i=1}^{K_h}$ of **continuous piecewise linear** shape functions φ_i, associated with the nodes S_i of $\mathcal{T}_h$ by the relation $\varphi_j(S_i) = \delta_{ij}$. Now the question is: How to compute the matrix coefficients $a_{i,j} = a(\varphi_j, \varphi_i)$ efficiently? A widespread way to carry out these calculations is based on the observation that $a_{i,j}$ is given by

$$a_{i,j} = \sum_{T \in \mathcal{T}_h} \left[\int_T (\mathbf{grad}\ \varphi_j | \mathbf{grad}\ \varphi_i)\ dxdy \right], \tag{4.27}$$

and that, in most triangles of $\mathcal{T}_h$, either φ_i or φ_j vanishes identically. Therefore, it appears fairly reasonable to deal with the problem the other way around, that is, to consider each triangle $T \in \mathcal{T}_h$ separately and compute all the corresponding nonzero values of the term in brackets in equation (4.27). Then, these contributions are added to the appropriate coefficients $a_{i,j}$. In fact, each triangle contains only three mesh nodes, and therefore not more than three shape functions are concerned. This means that the number of possible nonzero values for each triangle is at the most 3^2, but owing to symmetry only *six* different values need to be determined. Actually, this number can be smaller if we are dealing with a triangle intersecting Γ or Γ_0, according to the case, as seen later on in this chapter.

The implementation of this element-by-element procedure for assembling matrix A_h requires a pointer identifying the subscripts i and j each contribution belongs to. In short, one may proceed in the following manner:

1. Set all the initial values of the $a_{i,j}$s to zero.
2. Number the M_h triangles of the mesh, say, $T_1, T_2, \ldots, T_{M_h}$.
3. Number the nodes of the mesh not belonging to Γ (resp. Γ_0), from one through N_h, and next those belonging to Γ (resp. Γ_0) from $N_h + 1$ through K_h.
4. For each triangle T_l, introduce a local numbering of the nodes S_r^l, with r ranging between one and three. The vertices must be numbered in the counterclockwise sense as indicated in Figure 4.10, but otherwise arbitrarily.
5. Construct an **incidence matrix** for the mesh, namely, an $M_h \times 3$ integer array $\mathcal{N} = \{n_{l,r}\}$, such that for $1 \leq l \leq M_h$ and for $1 \leq r \leq 3$, $n_{l,r}$ is the number in the mesh of the node corresponding to the rth (local) node of the lth triangle.
6. For $l = 1, 2, \ldots, M_h$, compute a 3×3 matrix $E^l = \{e_{r,s}^l\}$ called the **element matrix** for triangle T_l, given by

$$e_{r,s}^l = \int_{T_l} (\mathbf{grad}\ \varphi_j | \mathbf{grad}\ \varphi_i)\ dxdy, \quad 1 \leq r, s \leq 3, \tag{4.28}$$

with $i = n_{l,r}$ and $j = n_{l,s}$.

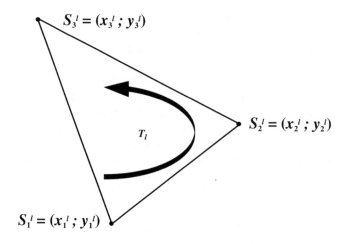

Figure 4.10   Local node numbering

7. For each triangle T_l, add the coefficients $e^l_{r,s}$ to the current value of $a_{i,j}$, for $i = n_{l,r}$ and $j = n_{l,s}$, with $r = 1,2,3$ and $s = 1,2,3$, provided $1 \leq i,j \leq N_h$. If $i > N_h$ or $j > N_h$, discard $e^l_{r,s}$.

In Figure 4.11, we illustrate the construction of the incidence matrix $\mathcal{N}$ for a mesh consisting of 24 triangles with 19 nodes, among which only the seven inner nodes correspond to unknown solution values. For better reading, Arabic figures are used for numbering the nodes, while Roman figures are employed for the elements; the entries of the row corresponding to element XXII are exhibited.

The same element-by-element assembling procedure may be used to compute the right-side vector $\vec{b}_h$. Indeed, we note that

$$b_i = \sum_{l=1}^{M_h} \int_{T_l} f(x,y)\varphi_i(x,y) \; dxdy$$

Since only a few φ_is are concerned by the integrals over T_l for a given l, we may conveniently introduce an **element vector** $\vec{f}^l = [f^l_1, f^l_2, f^l_3]^T$ whose components are defined by

$$f^l_r = \int_{T_l} f(x,y)\varphi_i(x,y) \; dxdy$$

where $i = n_{l,r}$.

Then, starting from zero values of b_i, $1 \leq i \leq N_h$, we sweep the mesh for $l = 1,2, \ldots, M_h$ and compute the element vector $\vec{f}^l$ for each T_l, and add its component f^l_r to the current value of b_i, for $i = n_{l,r}$, provided $1 \leq i \leq N_h$. If $i > N_h$ f^l_r is discarded. At the term of the procedure, the final values of the b_is will be available for $i = 1,2, \ldots, N_h$.

Let us now turn our attention to the details on the computation of the element matrices and element vectors themselves, for the $\mathcal{P}_1$ FEM. Henceforth, we assume that a table of coordinates $(x_i; y_i)$ of all the vertices, S_i of $\mathcal{T}_h$ in a suitable orthonormal frame is available.

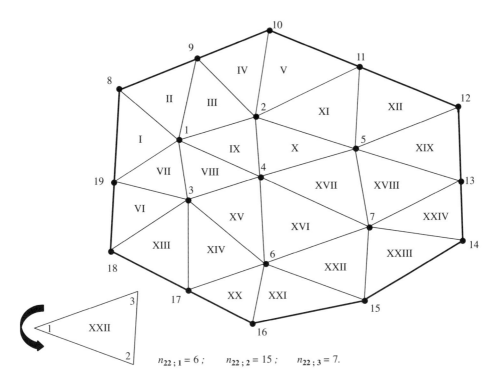

$$n_{22\,;\,1} = 6\,; \qquad n_{22\,;\,2} = 15\,; \qquad n_{22\,;\,3} = 7.$$

Figure 4.11   Mesh numbering and entries of the incidence matrix for the 22nd triangle

Element matrix: This case is quite simple since **grad** φ_i is constant in every triangle T_l. Hence, the area integral in the expression of $e^l_{r,s}$ is just $(\mathbf{grad}\ \varphi_i\,|\,\mathbf{grad}\ \varphi_j)\ area(T_l)$ for $i = n_{l,r}$ and $j = n_{l,s}$.

Let us see how to determine **grad** φ_i in a systematic way in an element T_l having S_i as a vertex. Noticing that φ_i over T_l is just the barycentric coordinate λ^l_r of T_l, with $n_{l,r} = i$, recalling equations (4.23)–(4.25) and letting $(x^l_r; y^l_r) = (x_i; y_i)$ be the local coordinates of the rth vertex S^l_r of T_l, we have

$$\partial_x \lambda^l_r = \frac{\Delta^l_{rx}}{2\ area(T_l)} \quad \text{and} \quad \partial_y \lambda^l_r = \frac{\Delta^l_{ry}}{2 area(T_l)}, \tag{4.29}$$

where for $r = 1, 2, 3$ Δ^l_{rx} equals the determinant of the array obtained by replacing the rth row of the array in the numerator of the expression of λ^l_r by $[0, 1, 0]$. Δ^l_{ry} in turn is given by the determinant of the array obtained by replacing the rth row of the array in the numerator of the expression of λ^l_r by $[0, 0, 1]$. The reader can easily determine the final expression of **grad** λ^l_r for a triangle T_l, in terms of its vertex coordinates.

If a more detailed geometrical description of the triangles is available, a convenient expression of the integrals $\int_{T_m} (\mathbf{grad}\ \varphi_j\,|\,\mathbf{grad}\ \varphi_i)\ dxdy$ can be obtained without resorting to the vertex coordinates. More specifically, letting S_i, S_j, S_k be the three vertices of a mesh triangle T_m for suitable subscripts $1 \le i, j, k \le K_h$, first we recall that **grad** φ_i is oriented from

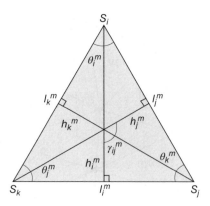

Figure 4.12   Triangle T_m and pertaining angles and lengths

the edge opposite to S_i towards this vertex and $|\textbf{grad } \varphi_i| = 1/h_i^m$, where h_i^m, h_j^m, h_k^m are the heights of T_m corresponding to S_i, S_j and S_k, respectively. Letting l_i^m be the length of the edge of T_m opposite to S_i, we know that $area(T_m) = l_i^m h_i^m/2$. Thus, it is readily seen that $e_{ii}^m = \int_{T_m} |\textbf{grad } \varphi_i|^2 \, dxdy = l_i^m/(2h_i^m)$. Referring to Figure 4.12, by a trivial geometric argument, we finally obtain $e_{ii}^m = (\cot \theta_j^m + \cot \theta_k^m)/2$, where θ_i^m, θ_j^m and θ_k^m are the angles of T_m with vertices at S_i, S_j and S_k, respectively.

On the other hand, recalling a well-known property of the Euclidean inner product of $\Re^2$, for triangle T_m we have $|\textbf{grad } \lambda_i^{T_m}| = 1/h_i^m$, together with $(\textbf{grad } \varphi_i|\textbf{grad } \varphi_j) = 1/(h_i^m h_j^m) \cos \gamma_{ij}^m$ for $i \neq j$, where γ_{ij}^m is the angle indicated in Figure 4.12. Hence,

$$\int_{T_m} (\textbf{grad } \varphi_i|\textbf{grad } \varphi_j)dxdy = \frac{h_i^m l_i^m}{2} \frac{1}{h_i^m} \frac{1}{h_j^m} \cos \gamma_{ij}^m.$$

Noticing that $\gamma_{ij}^m = \pi - \theta_k^m$ and $h_j^m/l_i^m = \sin \theta_k^m$, we obtain $e_{ij}^m = \int_{T_m} (\textbf{grad } \varphi_i|\textbf{grad } \varphi_j)$ $dxdy = -\dfrac{\cot \theta_k^m}{2}$. Interchanging the roles of i, j, k, we obtain the values of all the integrals of the type $\int_{T_m} (\textbf{grad } \varphi_i|\textbf{grad } \varphi_j)dxdy$, and hence all the coefficients of the element matrix E^m in terms of the angles of T_m.

The thus-determined element matrix coefficients allow us to assemble the global matrix A_h, thereby obtaining final expressions for its entry $a_{i,j}$. More precisely, referring to Figure 4.18, for convenience we rename the two triangles that have $S_i S_j$ as a common edge by T_{ij}^1 and T_{ij}^2, and the angle of T_{ij}^l opposite to this edge for $l = 1, 2$ by θ_{ij}^l. Then according to the above calculations we have, for $1 \leq i \leq N_h$ and $1 \leq j \leq K_h$,

$$\begin{cases} a_{i,i} = -\displaystyle\sum_{j=1, j\neq i}^{K_h} a_{i,j} \\ a_{i,j} = -(\cot \theta_{ij}^1 + \cot \theta_{ij}^2)/2 & \text{if } S_j \text{ is a neighbour of } S_i \\ a_{i,j} = 0 & \text{otherwise.} \end{cases} \qquad (4.30)$$

Notice that if $j > N_h$, the value of $a_{i,j}$ should be discarded, for it is not an entry of A_h. However, the knowledge of such values is useful in the above setting since the set of shape functions φ_j for $j = 1, 2, \ldots, K_h$ forms a **partition of unity**, that is, $\sum_{j=1}^{K_h} \varphi_j \equiv 1$. This can be proved as Exercise 4.4 in the same manner as for the barycentric coordinates. Then, $\sum_{j=1}^{K_h} \mathbf{grad}\ \varphi_j \equiv \mathbf{0}$, or, equivalently, $\mathbf{grad}\ \varphi_i = - \sum_{j=1,j\neq i}^{K_h} \mathbf{grad}\ \varphi_j$, and thus we have another manner to determine the diagonal coefficients, namely,

$$
\begin{aligned}
a_{i,i} &= \int_\Omega (\mathbf{grad}\ \varphi_i | \mathbf{grad}\ \varphi_i) dxdy \\
&= \int_\Omega (\mathbf{grad}\ \varphi_i\ | - \sum_{j=1,j\neq i}^{K_h} \mathbf{grad}\ \varphi_j) dxdy \\
&= - \sum_{j=1,j\neq i}^{K_h} \int_\Omega (\mathbf{grad}\ \varphi_i | \mathbf{grad}\ \varphi_j) dxdy = - \sum_{j=1,j\neq i}^{K_h} a_{i,j}.
\end{aligned}
$$

Element right-side vector: Next, we turn our attention to the calculation of the integrals yielding the components f_r^l of the element vector $\vec{f}^l$, namely, $\int_{T_l} f(x,y)\varphi_i(x,y)\ dxdy$ for $i = n_{l,r}$. In most applications, f is a simple function such as a piecewise polynomial of very low degree. In such cases, the integrals above can be calculated exactly. Nevertheless, in order to make this text as self-contained as possible, we assume that f is an arbitrary continuous function in every element of the mesh. Thus, in general, we must resort to numerical quadrature in order to approximate the above integrals in each triangle.

For linear elements, the following one-point Gaussian quadrature formula (see e.g. [184]) should do it, without causing any harm to the order of convergence of the method, at least as long as f is sufficiently smooth (see e.g. [162]). Letting G_l denote the centroid of T_l, the one-point formula is

$$
\int_{T_l} g\ dxdy \simeq g(G_l)area(T_l), \forall g \in C^0(T_l)
$$

In our particular case, this leads to

$$
\int_{T_l} f\ \varphi_i\ dxdy \simeq f(G_l)\ area(T_l)/3\ \forall i, i = n_{l,r}\ \ \text{for a certain } r
$$

· In Section 4.5, we address the crucial issue concerning node and element numbering, in order to achieve the best performances in terms of SLAE resolution.

4.3.5   *The Master Element Technique*

We next present another widespread systematic to compute element matrices and vectors based on a **master element** (cf. [146]). In case the elements are triangles, the standard master element is the unit right triangle $\hat{T}$ in a reference plane equipped with a coordinate system $(\xi; \eta)$ with origin $\hat{O}$. Referring to Figure 4.13, the vertices of $\hat{T}$ are $\hat{S}_1 = (0;0), \hat{S}_2 = (1;0), \hat{S}_3 = (0;1)$.

Just to illustrate how useful the master element can be, we compute below the integral in triangle T of the product of two linear functions p and q. Recalling the properties of the barycentric coordinates λ_r^T associated with the vertices S_r^T of T with Cartesian coordinates $(x_r^T; y_r^T)$

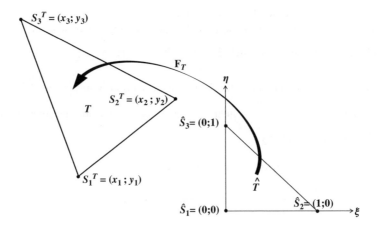

Figure 4.13 The master triangle $\hat{T}$ and a generic triangle T in the mesh

for $r = 1, 2, 3$, and setting $p_r := p(S_r^T)$ and $q_r = q(S_r^T)$, we have $p(x, y) = \sum_{r=1}^{3} p_r \lambda_r^T(x, y)$ and $q(x, y) = \sum_{s=1}^{3} q_s \lambda_s^T(x, y)$. Therefore,

$$\int_T p(x, y) q(x, y) \, dx dy = \sum_{r=1}^{3} \sum_{s=1}^{3} p_r q_s \int_T \lambda_r^T(x, y) \lambda_s^T(x, y) \, dx dy \qquad (4.31)$$

Thus, calculating the integral in T of the product of p and q reduces to the calculation of the integral in T of the product of two barycentric coordinates of T.

Next, we make a change of variables from $(x; y)$ into the variables $(\xi; \eta)$ sweeping the (reference) plane of the master element $\hat{T}$. Let $\mathbf{F}_T$ be the affine invertible mapping from $\hat{T}$ onto T, such that vertex $\hat{S}_r$ is taken to S_r^T, for $r = 1, 2, 3$. More precisely, $\mathbf{F}_T(\xi, \eta) = [(x_2 - x_1)\xi + (x_3 - x_1)\eta + x_1; (y_2 - y_1)\xi + (y_3 - y_1)\eta + y_1]$, as the reader can easily check.

Now, v being an arbitrary integrable function in T, we denote by $\hat{v}$ the function defined in $\hat{T}$ by $\hat{v}(\xi, \eta) = v(x, y)$, where $(x; y) = \mathbf{F}_T(\xi, \eta), \forall (\xi; \eta) \in \hat{T}$. From elementary calculus, we know that

$$\int_T v(x, y) \, dx dy = \int_{\hat{T}} \hat{v}(\xi, \eta) \, J_F^T d\xi d\eta.$$

where J_F^T is the Jacobian of mapping $\mathbf{F}_T$. In the case of an affine mapping, this Jacobian happens to be equal to the quotient between the areas of both domains, that is, $J_F^T = 2area(T)$.

Moreover, we note that $\widehat{\lambda_r^T} = \mu_r$, where the μ_r is the barycentric coordinate of $\hat{T}$ associated with $\hat{S}_r$ for $r = 1, 2, 3$. Actually, by simple inspection, we infer that $\mu_1 = 1 - \xi - \eta$; $\mu_2 = \xi$; $\mu_3 = \eta$. It follows that, for all pairs $(r; s)$,

$$\int_T \lambda_r^T \lambda_s^T \, dx dy = 2area(T) \int_{\hat{T}} \mu_r \mu_s \, d\xi d\eta,$$

Notice that we could as well have chosen, as a master element, the equilateral triangle whose vertices in the reference plane have polar coordinates $(1/2; 0)$, $(1/2; 2\pi/3)$ and $(1/2; 4\pi/3)$.

Then, using the same argument as above, we would have related the integral of $\lambda_r^T \lambda_s^T$ to the integral of the associated barycentric coordinates. By symmetry, these functions are necessarily invariant under rotations of the equilateral triangle of angles multiple of $2\pi/3$, and such operations cannot modify integrals involving them. We conclude that there can be only two possible results for the integral of $\lambda_r^T \lambda_s^T$, namely, a value for $r = s$ and another one for $r \neq s$. Going back to the case of $\hat{T}$, all we have to do is determine (for instance)

$$\int_{\hat{T}} \xi^2 \, d\xi d\eta = \int_0^1 \left[\int_0^{1-\eta} \xi^2 d\xi \right] d\eta = 1/12$$

$$\text{and } \int_{\hat{T}} \xi\eta \, d\xi d\eta = \int_0^1 \eta \left[\int_0^{1-\eta} \xi d\xi \right] d\eta = 1/24.$$

Summarising, after straightforward calculations, we obtain

$$\int_T p(x,y)q(x,y) \, dxdy = \left[\sum_{r=1}^3 p_r q_r + \left(\sum_{r=1}^3 p_r \right) \left(\sum_{s=1}^3 q_s \right) \right] area(T)/12. \qquad (4.32)$$

Remark 4.6 *The above procedure based on the master element can be extended to the calculation of integrals in T of functions of the form $[\lambda_1^T]^{\alpha_1}[\lambda_2^T]^{\alpha_2}[\lambda_3^T]^{\alpha_3}$, where the α_rs are non-negative integers. On the basis of nothing but elementary algebra and sustained application, the following formula is derived (see e.g. [210]):*

$$\int_T [\lambda_1^T]^{\alpha_1}[\lambda_2^T]^{\alpha_2}[\lambda_3^T]^{\alpha_3} dxdy = \frac{2\alpha_1!\alpha_2!\alpha_3!area(T)}{(\alpha_1 + \alpha_2 + \alpha_3 + 2)!}. \quad \blacksquare$$

Remark 4.7 *The use of a master element is mandatory in case the mapping $\mathbf{F}_T$ is not affine. Significant examples of FEM falling into this category, such as **isoparametric FEM**, are addressed in Chapter 6.* $\blacksquare$

4.3.6 *Application to Linear Elasticity*

The FEM was introduced to the engineering community in the mid-1950s as a tool capable of performing fine analyses of complex structures. The main motivation was sharp accuracy data required in projects conducted in the flourishing aircraft industry, entering the big jet era. The work [196] is generally cited as the one in which the FE concept was first introduced in this context. Several authors, most in the 1960s, popularised the method and shaped it to its present form. Outstanding contributions in this direction are those of the authors already cited in the introduction (in alphabetic order, see e.g. [7], [8], [75] and [210]), among many others. In this subsection, we give an overview of the FEM as applied to a basic structural problem: the study of plane deformations of a **homogeneous and isotropic**[6] **elastic solid**. As we should clarify, the main purpose of this presentation is to show what adaptations are needed to solve numerically a problem whose unknown is a vector field–here the **displacement**

---

[6] By definition, an isotropic solid has identical mechanical properties such as stiffness, whatever the direction considered. For instance, pure steel is isotropic, while wood is not.

field–and not a scalar function like in Subsection 4.2.1. However, as a by-product, we will address some important features to be taken into account when solving the **linear elasticity equations**, which model the behavior of elastic solid bodies in the small strain regime. In principle, the case of porous rocks and existing structures of the kind in the nature do not fit the framework considered hereafter, since in general they are neither homogeneous nor isotropic. However, such cases require nothing else than mild modifications in the formulation for homogeneous and isotropic media, in order to take into account the dependence of physical coefficients on space variables and directions.

We describe here the problem to solve in general terms, following celebrated works such as references [93] and [134], among many others. Let a homogeneous plate with constant thickness and mid-plane Ω be fixed on a portion Γ_0 of its boundary. Applying loads $\vec{g} = [g_x, g_y]^T$ on the complementary boundary portion Γ_1 in the plane of Ω, together with a force density $\vec{f} = [f_x, f_y]^T$, distributed over Ω and also parallel to it, a solid particle located at $\vec{x} \equiv (x; y)$ at rest will move to $\vec{x} + \vec{u}(\vec{x})$, $\vec{u} = [u_x, u_y]^T$ being the displacement field (see Figure 4.14).

As long as the material the solid is made of is rigid enough, and provided the loads $\vec{f}$ and $\vec{g}$ remain moderate, its deformation will be very small. This is equivalent to stating that the tensor $\Phi(\vec{u}) := (\mathbf{grad}\ \vec{u} + [\mathbf{grad}\ \vec{u}]^T)/2$ is small, $\mathcal{A}^T$ being the transpose of a tensor $\mathcal{A}^7$. In this case, $\Phi(\vec{u})$–the so-called **symmetric gradient** of $\vec{u}$–represents the **strain tensor** expressed as follows in two dimensions:

$$\Phi(\vec{u}) = \begin{bmatrix} \partial_x u_x & (\partial_x u_y + \partial_y u_x)/2 \\ (\partial_x u_y + \partial_y u_x)/2 & \partial_y u_y \end{bmatrix}. \tag{4.33}$$

The behavior of a homogeneous isotropic elastic solid undergoing small strains is characterised by two strictly positive elasticity coefficients. We may use the pair $(E; \sigma)$, where E is **Young's modulus** representing the body's stiffness and σ is **Poisson's ratio**, a dimensionless coefficient representing the medium's compressibility, whose value lies in the interval $(0, 1/2)$. Here, we shall rather employ the **Lamé coefficients** μ and λ related to E and σ by the expressions $\mu = E/[2(1 + \sigma)]$ and $\lambda = E\sigma/[(1 + \sigma)(1 - 2\sigma)]$.

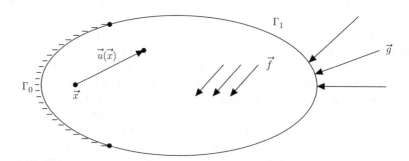

Figure 4.14   Particle displacement of an elastic solid subject to forces $\vec{f}$ and $\vec{g}$

---

[7] We recall that an $N \times N$ tensor is a linear operator from $\Re^N$ onto $\Re^N$ represented by an $N \times N$ matrix A, once a basis of $\Re^N$ is specified.

The stress undergone by the solid for a given displacement field $\vec{u}$ is expressed by **Hooke's law**, namely,

$$\Sigma = 2\mu\Phi(\vec{u}) + \lambda \ div \ \vec{u} \ I, \tag{4.34}$$

Σ being the **Cauchy stress tensor** and $I := \{\delta_{ij}\}$ being the 2×2 identity tensor[8].

Then the relations $\vec{u}$ must satisfy, completed with equation (4.34) are:

$$\begin{cases} -\mathbf{div}\ \Sigma = \vec{f} & \text{in } \Omega \\ \vec{u} = \vec{0} & \text{on } \Gamma_0 \\ \Sigma\vec{\nu} = \vec{g} & \text{on } \Gamma_1. \end{cases} \tag{4.35}$$

where $\mathbf{div}\ \Psi$ is a vector field standing for the **divergence of a tensor** Ψ. In case Ψ is a 2×2 tensor, we have $\mathbf{div}\ \Psi = [\partial_x \Psi_{xx} + \partial_y \Psi_{xy}; \partial_x \Psi_{yx} + \partial_y \Psi_{yy}]^T$. In simple terms, this means that the ith component of this vector is the divergence of the ith row of Ψ, $i = 1, 2$. If we eliminate Σ from equation (4.35) using (4.34), we come up with the following system of PDEs in terms of u_x and u_y:

System of linear elasticity for a homogeneous isotropic solid

$$\begin{array}{ll}
-\mu\Delta u_x - (\mu + \lambda)\partial_x(\partial_x u_x + \partial_y u_y) = f_x & \text{in } \Omega \\
-\mu\Delta u_y - (\mu + \lambda)\partial_y(\partial_x u_x + \partial_y u_y) = f_y & \text{in } \Omega \\
u_x = 0 & \text{on } \Gamma_0 \\
u_y = 0 & \text{on } \Gamma_0 \\
[(2\mu + \lambda)\partial_x u_x + \lambda\partial_y u_y]\nu_1 + \mu(\partial_y u_x + \partial_x u_y)\nu_2 = g_x & \text{on } \Gamma_1 \\
\mu(\partial_y u_x + \partial_x u_y)\nu_1 + [(2\mu + \lambda)\partial_y u_y + \lambda\partial_x u_x]\nu_2 = g_y & \text{on } \Gamma_1.
\end{array} \tag{4.36}$$

In order to be as close as possible to our scalar model (equation (4.1)), we could consider the rather academic case where $\vec{g} = \vec{0}$. However, this is not what we will do, and thus, referring to classical text books such as references [210] and [45], we can write equation (4.36) in the following standard equivalent form, namely, the

Galerkin variational form of the linear elasticity system

$$\begin{array}{l}
\text{Find } \vec{u} \in \mathbf{V} \text{ such that} \\
\mathbf{a}(\vec{u}, \vec{v}) = \mathbf{F}(\vec{v}) \ \forall \vec{v} \in \mathbf{V} \text{ where} \\
\mathbf{a}(\vec{u}, \vec{v}) := \int_\Omega [2\mu(\Phi(\vec{u}) | \Phi(\vec{v})) + \lambda(div \ \vec{u} | div \ \vec{v})] \ dxdy \\
\mathbf{F}(\vec{v}) := \int_\Omega (\vec{f} | \vec{v}) \ dxdy + \int_{\Gamma_1} (\vec{g} | \vec{v}) ds.
\end{array} \tag{4.37}$$

where $\mathbf{V}$ is the space of fields whose components belong to V defined in Subsection 4.2.2, and the inner product $(\Sigma | \Psi)$ of two symmetric tensors Σ and Ψ is to be understood as the sum $\Sigma_{xx} \Psi_{xx} + \Sigma_{yy} \Psi_{yy} + 2\Sigma_{xy} \Psi_{xy}$.

Remark 4.8 *Basically, equation (4.37) is obtained by using vector versions of the divergence theorem (i.e. by applying its scalar version to each component of* $\mathbf{div}\Sigma$*). Checking the equivalence between equations (4.36) and (4.37) is proposed to the reader as Exercise 4.5.* ■

---

[8] Rigorously, Σ is a 3×3 symmetric tensor fulfilling $\Sigma_{i3} = 0$ for $i = 1, 2$ and $\Sigma_{33} = \lambda \ div \ \vec{u}$.

Remark 4.9 *While, on the one hand, the Lamé coefficient μ is always bounded, on the other hand there is no limit for λ, which can thus formally attain the value $+\infty$. This happens when the material the elastic solid is made of is incompressible, which means that every portion of the solid in undeformed state evolves into a portion in its deformed state having the same measure (i.e. the volume or the area in a two-dimensional model). However, in any case the stresses must be finite, and hence if $\lambda = +\infty$ then necessarily $div\ \vec{u} = 0$. But then, the term $\lambda div\ \vec{u}$ in the expression of Σ becomes an additional scalar unknown. This function is commonly represented by $-p$, for p plays the same role as the hydrostatic pressure in incompressible fluid flow. Notice that the unknown p (hydrostatic pressure) comes together with the extra-scalar equation $div\ \vec{u} = 0$. The resulting system for $\vec{u}$ and p consisting of three PDEs is well-posed, like the one governing incompressible flow (see e.g. [188]).* ∎

Now, we apply the Ritz method to solve equation (4.37) based on $\mathcal{P}_1$ FEs with the already defined mesh $\mathcal{T}_h$. This means that the finite-dimensional space V_n is $\mathbf{V}_h$, defined to be the space of displacement fields $\vec{v} = (v_x; v_y)$ whose components belong to the FE space V_h defined in Subsection 4.2.3. Then, the problem to solve is the

$\mathcal{P}_1$ FE analog of the linear elasticity system

$$\boxed{\begin{array}{l} \text{Find } \vec{u}_h = [u_{xh}, u_{yh}]^T \in \mathbf{V}_h \text{such that} \\ \mathbf{a}(\vec{u}_h, \vec{v}) = \mathbf{F}(\vec{v})\ \forall \vec{v} \in \mathbf{V}_h. \end{array}} \tag{4.38}$$

Akin to equation (4.37), the discrete counterpart equation (4.38) also has a unique solution (cf. [45]). Since there are two components per node, the dimension of $\mathbf{V}_h$ is $J_h := 2N_h$, where N_h is the number of mesh nodes belonging to $\Omega \cup \Gamma_1$. The set $\{\vec{\eta}_j\}_{j=1}^{J_h}$, where $\vec{\eta}_{2i-1} = [\varphi_i, 0]^T$ and $\vec{\eta}_{2i} = [0, \varphi_i]^T$ for $i = 1, 2, \dots, N_h$, φ_i being the $\mathcal{P}_1$ FE shape functions, is a natural basis of this space. Indeed, let $\vec{u}_h$ be the vector whose components include the two components of all the unknown nodal values, that is, $[u_{x_i}, u_{y_i}]^T = \vec{u}_h(S_i)$, for $i = 1, 2, \dots, N_h$, called the displacement **degrees of freedom**[9]. These components are assumed to be arranged as $\vec{\underline{u}}_h := [u_{x_1}, u_{y_1}, \dots, u_{x_{N_h}}, u_{y_{N_h}}]^T$. Then, we can expand $\vec{u}_h$ into the sum $\sum_{j=1}^{N_h} [u_{x_j} \vec{\eta}_{2j-1} + u_{y_j} \vec{\eta}_{2j}]$. As a consequence, equation (4.38) is equivalent to the SLAE $\mathbf{A}_h \vec{\underline{u}}_h = \vec{\underline{b}}_h$ with J_h equations and J_h unknowns, where $\vec{\underline{b}}_h = [b_{x_1}, b_{y_1}, \dots, b_{x_{N_h}}, b_{y_{N_h}}]^T$ with $b_{x_i} = \int_\Omega f_x \varphi_i\ dxdy + \oint_{\Gamma_1} g_x \varphi_i\ ds$ and $b_{y_i} = \int_\Omega f_y \varphi_i\ dxdy + \oint_{\Gamma_1} g_y \varphi_i\ ds$, and $\mathbf{A}_h = \{a_{j,l}\}$, for $j, l \in \{1, 2, \dots, J_h\}$, with $a_{j,l} = \mathbf{a}(\vec{\eta}_l, \vec{\eta}_j)$ (please check!).

Assuming that $\vec{f}$ is continuous, the same one-point quadrature formula as in the scalar case can be used to evaluate the terms of the components of $\vec{\underline{b}}_h$ related to $\vec{f}$. As for the integrals $\oint_{\Gamma_1} g_x \varphi_i$ or $\oint_{\Gamma_1} g_y \varphi_i$, in general approximations must be computed. However, the right choice of a numerical quadrature formula and corresponding implementation aspects is more subtle. In Chapter 6, we address this issue more thoroughly.

---

[9] By extension, many authors refer to any set of unknowns in a FE solution as the method's degrees of freedom.

Now recalling equations (4.33) and (4.37), the values of the entries $a_{j,l}$ are given by three types of expressions according to the parity of j and l. More specifically, using symmetry, for $i, k = 1, 2, \ldots, N_h$, we have

$$
\begin{cases}
a_{2i-1,2k-1} & = (2\mu + \lambda)\int_\Omega \partial_x \varphi_k \partial_x \varphi_i \, dxdy + \mu \int_\Omega \partial_y \varphi_k \partial_y \varphi_i \, dxdy \\
\quad a_{2i,2k} & = (2\mu + \lambda)\int_\Omega \partial_y \varphi_k \partial_y \varphi_i \, dxdy + \mu \int_\Omega \partial_x \varphi_k \partial_x \varphi_i \, dxdy \\
a_{2i-1,2k} = a_{2k,2i-1} = & \quad \mu \int_\Omega \partial_x \varphi_k \partial_y \varphi_i \, dxdy + \lambda \int_\Omega \partial_y \varphi_k \partial_x \varphi_i \, dxdy
\end{cases}
$$

The reader is advised to check the above expressions, as a way to become familiar with the treatment of problems whose unknowns are vector fields.

Remark 4.10 *The above systematic for arranging the unknowns is optimal in terms of matrix band width, and is certainly what should be done, whatever method is used to solve the resulting SLAE. In the following section, we will consider a variational problem posed in terms of a vector field and a scalar variable. In that case, we will use another arrangement for the vector of unknowns, as a more convenient way to eliminate the former from the final system, posed only in terms of the latter.* ∎

Finally, a word about implementation in the vector case. We can use the same assembly technique based on the incidence matrix and element matrix and right-side vector as in the scalar case. However, now the element matrix is a 6×6 matrix and the element right side vector has six components, both related to the six degrees of freedom in each element, namely, the two components of the displacement at each vertex. Here again, for the sake of conciseness we do not develop this point, because the procedure is a straightforward extension of the one already described for the scalar case in Subsection 4.2.5. We refer to reference [210] for further details.

4.4   Basic FVM

Throughout this section, we keep assuming that the domain Ω is a polygon.

Conceptually, the FVM is meant to be Cell-centred with disjoint cells covering the whole domain. Like in the one-dimensional case, the idea is to approximate the quantity of interest in each CV $\mathcal{C}$, by a constant function symbolically represented by its value at some point $\mathbf{x}_C$ belonging to $\mathcal{C}$, eventually lying on the boundary of this CV. The main motivation is to represent the flux of relevant physical quantities across cell boundaries in a consistent and visible manner. We refer to reference [68] for both a comprehensive description and a thorough study of this technique for some important classes of problems in both two- and three-dimensional space.

As pointed out in reference [68], the FVM in the two-dimensional case is designed for convex polygonal cells of arbitrary shape, as long as they fulfil some geometric conditions. In simple terms, we can say that the CVs must be designed in such a way that for every pair of CVs $(\mathcal{C}; \mathcal{C}')$ having a common edge e, the points $\mathbf{x}_C$, $\mathbf{x}_{C'}$ can be chosen so as to belong to the same perpendicular r_e to e. We refer to Figure 4.15 for an illustration. Meshes satisfying this condition among others specified in reference [68] are said to be **admissible**. We do not further elaborate on this concept because the FVMs to be studied in this section are naturally associated with meshes satisfying the required condition.

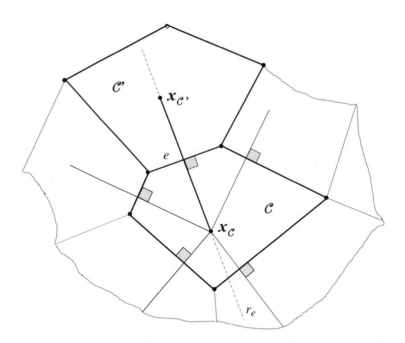

Figure 4.15   Neighbouring polygonal CVs $\mathcal{C}$ and $\mathcal{C}'$ in an admissible partition of Ω

Two basic FVMs to solve the Poisson equation are described in this section. Both methods can be viewed as two-dimensional analogs of the Vertex-centred and the Cell-centred FVMs presented in Chapter 1. In the case of the former, the set of CVs is constructed upon a primal FE triangular mesh. More precisely, to each node S_i of the triangulation corresponds a CV V_i containing S_i.

Only the pure Dirichlet problem (equation (4.3)) will be addressed in detail. The necessary modifications in order to treat the mixed problem (equation (4.4)) will be briefly considered in a few remarks.

4.4.1   The Vertex-centred FVM: Equivalence with the $\mathcal{P}_1$ FEM

Let $\mathcal{T}_h$ be a partition of Ω into triangles, satisfying the same compatibility condition as for the FEM, and moreover forming a Delaunay triangulation. For a simple description of the latter, we refer for instance to reference [213]. Here, we just underline that in the case under consideration the vertices of a polygonal cell are nothing but the circumcentres of the triangles having the corresponding node as a vertex[10]. In the same vein, we circumvent the formal definition of a Delaunay triangulation, by simply confining ourselves to admissible triangulations $\mathcal{T}_h$. These consist of triangles T whose circumcentre belongs to T. Conversely, this condition rules out mesh elements T that have an obtuse angle since its circumcentre lies outside T. We remind the reader that by definition T includes its boundary, and hence triangles T whose circumcentre, lies thereupon–that is, right triangles–are admissible. We assume throughout this

---

[10] We recall that the circumcentre of a triangle is the centre of the circumscribed circle about it.

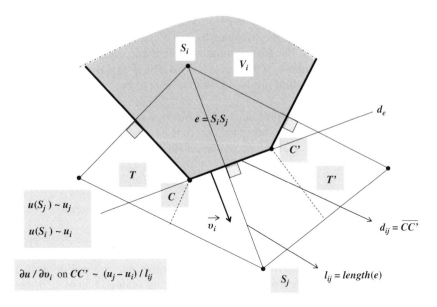

Figure 4.16 Subsets in CV V_i of adjacent triangles T and T' having S_i as a vertex

subsection that $\mathcal{T}_h$ fulfills the non-obtuse angle condition. A mesh having no angle greater than $\pi/2$ is called a **mesh of the acute type**.

Referring to Figure 4.16, we can assert that for an edge e common to two triangles T and T' the edge of a CV V_i having a non-empty intersection with each one of these triangles lies on their common perpendicular bisector d_e. Then, observing Figure 4.17, we see that CV V_i pertaining to vertex S_i is constructed as follows: For every mesh triangle T having S_i as a vertex, $T \cap V_i$ is its portion delimited by the two **perpendicular bisectors** of the edges intersecting at S_i.[11] Since the intersection of the perpendicular bisectors of a triangle is its

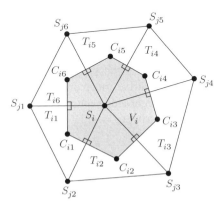

Figure 4.17 Control volume V_i for the Vertex-centred FVM

---

[11] We recall that a perpendicular bisector is a line that forms a right angle with one of the triangle's edges and intersects that edge at its midpoint.

circumcentre C, the CV V_i of the triangle-based Vertex-centred FVM is the polygon whose vertices are the circumcentres of the triangles having S_i as a vertex (cf. Figure 4.17). It is clear that if all the vertices of the triangulation are taken into account, the union of all the non-overlapping control volumes V_i is the whole Ω. This remark explains why the set of control volumes V_i is said to form the **dual mesh** of Ω with respect to the original mesh $\mathcal{T}_h$ called the **primal mesh**.

In the FVM, the solution u is approximated by a function u_h, which is constant in each CV, say equal to u_i in V_i, the value of u_i being set to zero if $i > N_h$. u_h satisfies in a certain sense to be specified later on the integral balance equations that hold for u in the CVs, namely,

$$- \int_{V_i} \Delta u \, dxdy = \int_{V_i} f \, dxdy. \tag{4.39}$$

Using the divergence theorem, the above relation is seen to be equivalent to

$$- \oint_{\partial V_i} \partial_{\nu_i} u \, ds = \int_{V_i} f \, dxdy, \tag{4.40}$$

where ∂V_i denotes the boundary of V_i and ∂_{ν_i} denotes the outer normal derivative on ∂V_i defined everywhere but at the vertices of V_i. Then, referring again to Figure 4.16, we approximate $\partial_{\nu_i} u$ along a side of ∂V_i orthogonal to the edge $\overline{S_i S_j}$, by a constant value equal to $(u_j - u_i)/l_{ij}$, where $l_{ij} = length(\overline{S_i S_j})$. We denote by d_{ij} the Euclidean distance between the circumcentres of the triangles having $\overline{S_i S_j}$ as a common edge. Of course, if both circumcentres do not coincide, the segment joining it is an edge of V_i intersecting the segment $\overline{S_i S_j}$ at the mid-point of the latter, and we have $d_{ij} > 0$. However, if this segment is reduced to the mid-point of $\overline{S_i S_j}$ we have $d_{ij} = 0$. This occurs if both are right triangles and their right angles are opposite to $\overline{S_i S_j}$. In this case, the number of edges of V_i will be smaller than the number of triangles around S_i, but this will do no harm at the computational level.

Let Π_i be the polygon consisting of the union of the mesh triangles having S_i as a vertex. A straight forward application of equation (4.40) leads to N_h equations, to determine the N_h unknowns u_i, that is,

The Vertex-centred FV scheme for the Poisson equation

$$\sum_{j \,|\, S_j \text{ vertex of } \Pi_i} d_{ij}(u_i - u_j)/l_{ij} = \int_{V_i} f \, dxdy \text{ for } i = 1, 2, \ldots, N_h. \tag{4.41}$$

Of course, scheme (4.41) can be set in matrix form, namely, $A_h \vec{u}_h = \vec{b}_h$ where $\vec{u}_h$ is the vector of unknowns $[u_1, u_2, \ldots, u_{N_h}]^T$, the right-side vector $\vec{b}_h = [b_1, b_2, \ldots, b_{N_h}]^T$ is defined by $b_i = \int_{V_i} f \, dxdy$, and the $N_h \times N_h$ matrix $A_h = [a_{i,j}]$ is given by

$$\begin{cases} a_{i,i} = \sum\limits_{j \,|\, S_j \text{ vertex of } \Pi_i} d_{ij}/l_{ij} \\ a_{i,j} = -d_{ij}/l_{ij} \quad \text{if } S_j \text{ is a vertex of } \Pi_i, \\ a_{i,j} = 0 \qquad\qquad \text{otherwise.} \end{cases} \tag{4.42}$$

The interesting thing is that such a matrix A_h happens to be the same as the $\mathcal{P}_1$ FEM matrix given by equation (4.30). Let us see why.

First of all we recall the notation T_{ij}^l for the triangles sharing the vertices S_i and S_j, together with the angle θ_{ij}^l of T_{ij}^l opposite to the edge $\overline{S_i S_j}$, $l = 1, 2$. We also denote by d_{ij}^l the length

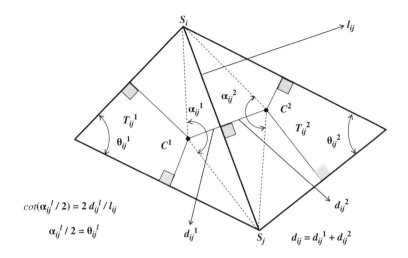

$$cot(\alpha_{ij}^l/2) = 2\,d_{ij}^l/l_{ij}$$

$$\alpha_{ij}^l/2 = \theta_{ij}^l$$

$$d_{ij} = d_{ij}^1 + d_{ij}^2$$

Figure 4.18 Explaining why $a_{i,j} := -d_{ij}/l_{ij} = -(\cot\,\theta_{ij}^1 + \cot\,\theta_{ij}^2)/2$ for $i \neq j$

of the portion of d_{ij} lying inside T_{ij}^l. Then, referring to Figure 4.18, we see that the angle α_{ij}^l satisfies $\cot(\alpha_{ij}^l/2) = 2d_{ij}^l/l_{ij}$. Finally, by a geometric argument to be checked by the reader as Exercise 4.6, α_{ij}^l is found to be equal to $2\theta_{ij}^l$ and we are done. The coincidence of $a_{i,i}$ for both methods also holds and can be handled as a part of Exercise 4.7. Of course, in general the right-side vector $\vec{b}_h$ differs from the one of the $\mathcal{P}_1$ FEM, and only for this reason the Vertex-centred FV scheme (4.41) is not exactly the same as its FE counterpart for the same mesh. However, both ways of handling the datum f can be viewed as just different numerical integrations of $\int_\Omega f\varphi_i\,dxdy$ having the same order of approximation. Otherwise stated, the $\mathcal{P}_1$ FEM and the Vertex-centred FVM for the same mesh yield practically equivalent schemes for solving equation (4.3), and therefore they have the same stability and convergence properties.

We conclude with a few words about the application of the Vertex-centred FVM to solve equation (4.4). It was implicit in the relations of equation (4.41) that homogeneous or inhomogeneous Dirichlet boundary conditions are simply taken into account by assigning to u_j for $j > N_h$ the prescribed value of u at a vertex S_j lying on Γ_0. Thus, the only issue that remains to be clarified concerns the equations derived for a vertex belonging to Γ_1. In Figure 4.19, we illustrate a CV V_i corresponding to a vertex $S_i \in \Gamma_1$. According to our assumptions on the mesh in case Γ_1 is not empty, V_i has two edges contained in Γ_1. Naturally enough, over these segments the **flux** $\partial_\nu u_h$ is to be taken equal to zero instead of being approximated like on the inner edges of V_i. Then, the corresponding equation writes

$$\sum_{j\,|\,S_j \text{ vertex of } \Pi_i} d_{ij}(u_i - u_j)/l_{ij} = \int_{V_i} f\,dxdy \ \text{ for } i = 1, 2, \ldots, N_h,$$

where for an edge $\overline{S_iS_j}$ contained in Γ_1 d_{ij} is the length of the segment of the corresponding perpendicular bisector lying in the sole triangle having such an edge. In Figure 4.19, this segment is $\overline{CD}$ for triangle T'. Here again, it is possible to establish (cf. Exercise 4.7) that the resulting matrix is nothing but the FE matrix A_h for equation (4.4).

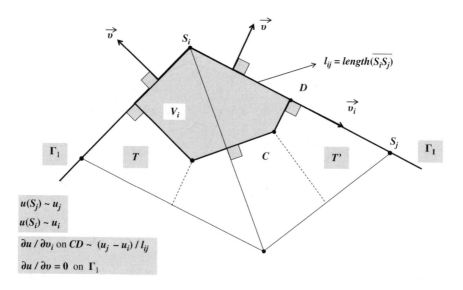

$u(S_j) \sim u_j$

$u(S_i) \sim u_i$

$\partial u / \partial v_i$ on $CD \sim (u_j - u_i) / l_{ij}$

$\partial u / \partial v = 0$ on Γ_1

Figure 4.19   Control volume V_i related to a vertex S_i located on Γ_1

4.4.2   The Cell-centred FVM: Focus on Flux Computations

In this subsection, we consider types of cells very popular among FV users, namely, triangular cells and rectangular cells.

If we want to avoid some technical complications inherent to the case of triangular cells, we have to assume that all the triangles are of the **strict acute type**. This means that their angles are all strictly smaller than $\pi/2$. We observe, however, that it is not always easy to construct this type of triangulation. Fortunately, the Cell-centred FVM can also be defined upon **Delaunay triangulations**, which allow angles greater or equal to $\pi/2$, under the condition that, for every internal edge, the sum of two angles facing it is less than π. Moreover, it is required that no angle facing an edge contained in Γ is obtuse. We refer to Figure 4.20 for

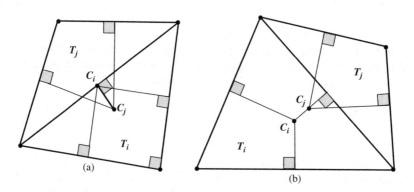

Figure 4.20   Incompatible (A) and compatible (B) configurations with a Delaunay triangulation

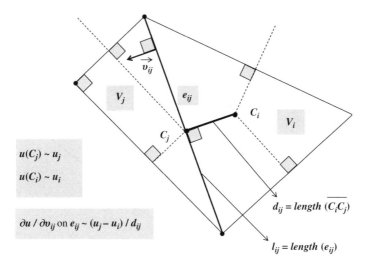

Figure 4.21 Approximation of the flux across an edge common to two triangular cells

self-explanatory illustrations, and to reference [155] or [76] for details on the construction of Delaunay triangulations, amidst a rather vast literature on the subject.

For a mesh constructed as described in Section 4.3, the CVs are the triangles of $\mathcal{T}_h$ themselves. So if we number the elements like in the previous section from T_1 up to T_{M_h}, we approximate the balance relation (equation (4.39), or yet (4.40)) with $V_i = T_i$. Now the question is how to define a discrete flux of a function u_h approximating u, which is constant in each triangle.

Denoting by u_i the value of u_h in the triangular cell V_i, the natural thing to do is to approximate the flux (i.e. the outer normal derivative $\partial u/\partial \nu_{ij}$) across the edge e_{ij} common to another CV V_j, by $(u_j - u_i)/m_{ij}$, where m_{ij} is the length of a segment perpendicular to this edge between two 'reasonable' points attached to V_i and V_j, respectively. For a Delaunay triangulation, these points can be chosen to be the circumcentres C_i and C_j of these triangles. Notice that in the case of acute triangles, they necessarily lie in their respective interiors. Referring to Figure 4.22 in this case, $m_{ij} = d_{ij}$. In contrast, as suggested in Figure 4.21, there is a problem if both angles opposite to an inner edge e_{ij} equal $\pi/2$ because C_i and C_j will coincide. The situation becomes even more troublesome if these angles are obtuse, for in this case the two centres lie outside V_i and V_j. This explains the assumption that the mesh be of the strict acute type.

Dirichlet boundary conditions on the whole Γ (resp. on Γ_0) can be taken into account by simply setting $u_i = 0$ if V_i is a triangular cell having an edge on Γ (resp. on Γ_0). However, the best thing to do in terms of accuracy is to add values $u_j = 0$, for $j > M_h$, formally at a point D_j of a boundary CV V_i, lying on an edge e_{ij} contained in Γ_0 (or in Γ according to the boundary conditions). This is illustrated in Figure 4.22, where indications are given about the special treatment of the flux in such a CV (cf. [68]). In practical terms, this requires precise definitions of D_j together with the lengths l_{ij} and d_{ij} for an edge e_{ij} contained in Γ_0 (resp. Γ). While D_j is the intersection of e_{ij} with the perpendicular bisector of V_i related to this edge, d_{ij} is the distance between the circumcentre C_i of V_i and D_j; l_{ij} in turn still denotes the length

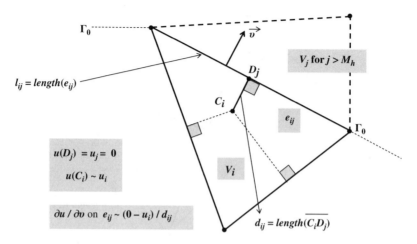

$$\Gamma_0$$

$$\vec{v}$$

$$V_j \text{ for } j > M_h$$

$$D_j$$

$$l_{ij} = length(e_{ij})$$

$$C_i$$

$$e_{ij}$$

$$\Gamma_0$$

$$u(D_j) = u_j = 0$$

$$u(C_i) \sim u_i$$

$$V_i$$

$$\partial u / \partial v \text{ on } e_{ij} \sim (0 - u_i) / d_{ij}$$

$$d_{ij} = length(\overline{C_i D_j})$$

Figure 4.22   Approximation of the flux across an edge located on Γ_0

of e_{ij}. It is convenient to associate with D_j and u_j a fictitious CV V_j for $j > M_h$, which is illustrated in Figure 4.22 with thick dashed lines. In Subsection 5.1.5, we will be more specific about this technique.

Admitting that such V_js were introduced for $j > M_h$, using the above definitions, we obtain equations very similar to equation (4.41), by interchanging l_{ij} with d_{ij}, namely,

The triangle-centred FV scheme for the Poisson equation

$$\sum_{j \,|\, V_j \cap V_i \neq \emptyset} l_{ij}(u_i - u_j)/d_{ij} = \int_{V_i} f \, dxdy \quad \text{for } i = 1, 2, \ldots, M_h. \qquad (4.43)$$

Notice that if an edge is contained in a non-empty Γ_1, we set to zero the mean flux (i.e. the outer normal derivative) across this edge in equation (4.43). We should also comment that the above definition of d_{ij} for an edge e_{ij} contained in Γ_0 or Γ is the reason for ruling out obtuse angles facing this edge. Finally, it is worthwhile pointing out that the choice $m_{ij} = d_{ij}$ means that the CVs for the Vertex-centred FVM belong to the **dual partition** of the domain with respect to the Cell-centred (primal) partition.

Under certain conditions, the Cell-centred FV scheme to solve equation (4.3) or (4.4) can be derived in the light of its relationship to a popular **mixed FEM**, known as the **lowest order Raviart–Thomas mixed FEM** (RT_0) **method**). Rigorously, such a relationship was first established by Baranger–Maître–Oudin in papers published in the 1990s (see. e.g. [19]), for both triangular and tetrahedral meshes. The arguments employed in such works involve rather complicated technicalities. Furthermore, they are fairly complex from the conceptual point of view, although they lead to very practical equivalence results between both discretisation methods. However, at this introductory level, we feel the reader should focus on principles, rather than on lengthy or fastidious calculations. That is why we shall restrict our argumentation about such an equivalence to the framework of problems posed in a rectangular domain $\Omega = (0, L_x) \times (0, L_y)$. Nevertheless, in Chapter 5 we will be more specific about this technique as applied to triangular meshes (cf. Example 5.3).

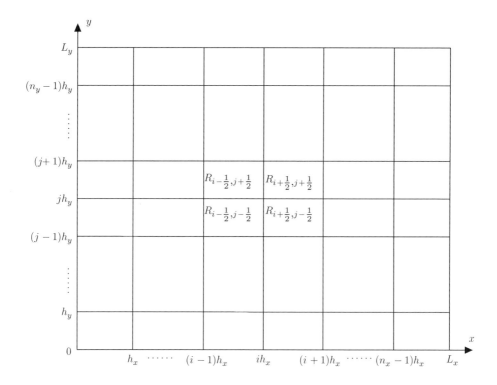

Figure 4.23   A uniform rectangular mesh for the Cell-centred FVM

Assume that rectangle Ω is partitioned into $n_x \times n_y$ equal rectangles of edge sizes $h_x = L_x/n_x$ and $h_y = L_y/n_y$, where n_x and n_y are given integers, as illustrated in Figure 4.23. The rectangular cells are denoted by $R_{q,s}$ where the numbers q and s of the double subscript are both of the form $k - 1/2$, k being a natural number suitably adjusted according to the vertex coordinates, as indicated in self-explanatory Figure 4.23. We denote the resulting mesh of Ω by $\mathcal{R}_h$, setting $h = \max[h_x, h_y]$.

Before going into details, we move back to the general case, in order to introduce the **mixed formulation** of the Poisson problem. This is expressed in terms of two fields, namely, the unknown u itself and the variable $\vec{p} = \mathbf{grad}\ u$, both represented with different types of interpolation. Just to make ideas clearer, we consider the case of Dirichlet boundary conditions only. This means that the problem (4.3) is formulated in the form of a system of first-order PDEs, namely,

$$\left\{ \begin{array}{l} \vec{p} = \mathbf{grad}\ u \\ -div\ \vec{p} = f \end{array} \right\} \ \text{in } \Omega \tag{4.44}$$
$$\left. \begin{array}{l} u = 0 \ \text{on } \Gamma. \end{array} \right.$$

In order to apply the FEM to solve equation (4.44), first of all we have to recast it in a suitable variational form. For this purpose, we introduce the space

$$\mathbf{Q} = \{\vec{q} = [q_x, q_y]^T / q_x, q_y \in L^2(\Omega) \text{ and } div\ \vec{q} \in L^2(\Omega)\}$$

and denote the space $L^2(\Omega)$ by U. Then, we set the

Mixed variational formulation of the Poisson equation

> Find $\vec{p} \in \mathbf{Q}$ and $u \in U$ such that
> $(\vec{p}|\vec{q}) + (u \,|\, div \ \vec{q})_0 = 0 \ \forall \vec{q} \in \mathbf{Q}$ (4.45)
> $(div \ \vec{p}|v)_0 = -(f|v)_0 \ \forall v \in U.$

While the second equation of (4.45) is obviously an equivalent weak form of the second equation of (4.44), the first one is equivalent to the first and third relations of (4.44) all together. Indeed, assuming that $u \in H^1(\Omega)$, we can apply equation (4.11) on the left side, thereby obtaining,

$$\int_{\Omega} (\vec{p}|\vec{q}) \ dx dy = -\oint_{\Gamma} (\vec{q}(s)|\vec{\nu}(s)) u(s) \ ds + \int_{\Omega} (\mathbf{grad} \ u|\vec{q}) \ dx dy \ \forall \vec{q} \in \mathbf{Q}. \qquad (4.46)$$

Taking $\vec{q}$ such that $(\vec{q}|\vec{\nu}) = 0$ on Γ, we derive

$$\int_{\Omega} (\vec{p}|\vec{q}) \ dx dy = \int_{\Omega} (\mathbf{grad} \ u|\vec{q}) \ dx dy \ \forall \vec{q} \in \mathbf{Q}.$$

According to reference [87], this relation must also hold for all $\vec{q}$ having both components in $L^2(\Omega)$ only, and hence necessarily $\vec{p} = \mathbf{grad} \ u$. Since we already know that $\vec{p} = \mathbf{grad} \ u$, we can get rid of the integrals in Ω in (4.46). Hence, we are only left $-\oint_{\Gamma} (\vec{q}(s)|\vec{\nu}(s)) u(s) \ ds$ for all $\vec{q} \in \mathbf{Q}$. Now taking $\vec{q} \in \mathbf{Q}$ such that $(\vec{q}|\vec{\nu}) = u$ on Γ, we readily obtain $u = 0$ on Γ and hence equation (4.45) implies (4.44). Conversely, the same relation (4.11) allows us to reconstruct equation (4.45) starting from (4.44), and therefore both problems are equivalent.

One might ask why it could be advantageous to use the mixed formulation (4.45), rather than the standard Galerkin formulation, to solve the Poisson equation. The explanation can be found at a discrete level and in a context a little wider than this. First of all, if one is interested in representing the solution gradient at least as well as the solution itself, it is wise to introduce this variable explicitly in the formulation and to interpolate this field with polynomial functions finer than (incomplete) piecewise constant. In this respect, it should be noted that in the case of the $\mathcal{P}_1$ FEM the solution gradient belongs to a subset of the space of piecewise constant functions, which is a rather poor approximation. Moreover, in many applications, one has to solve the generalised Poisson problem $-div \ \mathcal{K}\mathbf{grad} \ u = f$ where $\mathcal{K}$ is a 2×2 second-order symmetric tensor that may depend on x and y. This dependence can be such that the values of $\mathcal{K}$ vary dramatically in Ω. For instance, this is the case of flow in highly heterogeneous and/or anisotropic porous media, in which $\mathcal{K}$ is the permeability tensor, u being the *Darcy pressure*. However, physically the flux $\mathcal{K}\mathbf{grad} \ u$ cannot have jumps in the normal direction across a surface separating layers of two different constituents of the porous medium. Therefore, the kind of interpolation of this variable should have some continuity properties–or at least its normal component on inter-element boundaries–otherwise this flux conservation property will be lost. This was actually the main motivation for the introduction of the mixed formulation, which in this case is

$$\left\{ \begin{array}{l} \vec{p} = \mathcal{K}\mathbf{grad} \ u \\ -div \ \vec{p} = f \\ u = 0 \ \text{on} \ \Gamma. \ \blacksquare \end{array} \right\} \ \text{in} \ \Omega$$

The mixed FEM RT_0 was first introduced for triangles (cf. [161]), and in this case it is some-times denoted by $RT_0(T)$. Its version for rectangles denoted by $RT_0(R)$ that we will exploit

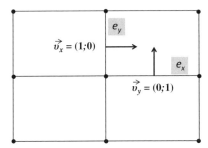

Figure 4.24 Orientation of normal vectors across the cell edges

hereafter is based on the representations of $\vec{p}$ and u by $\vec{p}_h$ and u_h of the following form. At the level of each rectangle in $\mathcal{R}_h$, u_h is constant exactly as in the case of the Cell-centred FVM. On the other hand, in a given rectangle $R \in \mathcal{R}_h$ $\vec{p}_h$ is a field of the form $[a_x^R x + b_x^R, a_y^R y + b_y^R]^T$, $a_x^R, b_x^R, a_y^R, b_y^R$ being real coefficients. Referring to Figure 4.24, we require the equality of the mean (normal) fluxes $\int_e (\vec{p}_h | \vec{v}_e) ds / length(e)$ across an edge e common to two neighbouring rectangles, oriented in the same manner for both, as indicated for an edge e_x such that $\vec{v}_e = \vec{v}_y = (0; 1)$ (horizontal edge) or a vertical edge e_y such that $\vec{v}_e = \vec{v}_x = (1; 0)$ (vertical edge). Let us denote by U_h the space of constant functions in each $R \in \mathcal{R}_h$ and by $\mathbf{Q}_h$ the space of fields satisfying the above properties assigned to $\vec{p}_h$. Then, the mixed problem associated with the $RT_0(R)$ method to solve equation (4.45) is the

Raviart–Thomas analog of Poisson's equation mixed formulation

$$
\begin{array}{|l|}
\hline
\text{Find } \vec{p}_h \in \mathbf{Q}_h \text{ and } u_h \in U_h \text{such that} \\
(\vec{p}_h | \vec{q})_0 + (u_h | div \ \vec{q})_0 = 0 \ \forall \vec{q} \in \mathbf{Q}_h \\
(div \ \vec{p}_h | v)_0 = -(f | v)_0 \ \forall v \in U_h. \\
\hline
\end{array}
\qquad (4.47)
$$

The fact that problem (4.47) has a unique solution was first proven in reference [189]. As for the practical solution of equation (4.47), we apply the usual strategy for the FEM. Noticing that the dimension of $\mathbf{Q}_h$ equals the number of edges in the mesh, that is, $L_h = 2n_x n_y + n_x + n_y$, and the one of U_h equals $M_h = n_x \times n_y$, this means that after numbering the edges from one through L_h we first exhibit **canonical bases** for the spaces $\mathbf{Q}_h$ and U_h, say fields $\vec{\eta}_k$ for $k = 1, 2, \ldots, L_h$ and functions χ_m for $m = 1, 2, \ldots, M_h$, where $\vec{\eta}_k$ is the basis field of $\mathbf{Q}_h$ associated with the kth edge e_k of the mesh and χ_m is the basis function associated with the mth rectangle R_m of the mesh, in the following manner:

$$
\begin{cases}
\int_{e_l} (\vec{\eta}_k | \vec{v}_l) \ ds / length(e_l) = \delta_{kl} \text{ for } l = 1, 2 \cdots, L_h \\
\chi_m = \delta_{mn} \text{in } R_n \text{ for } n = 1, 2, \ldots, M_h.
\end{cases}
$$

where $\vec{v}_l$ is the unit normal vector on e_l (i.e. either $\vec{v}_x$ or $\vec{v}_y$). While on the one hand the analytical expression of χ_m is obvious, the one of $\vec{\eta}_k$ requires more attention (cf. Exercise 4.8). However, this task is more easily carried out at the element level, if we renumber the rectangles and edges using two subscripts, like in the FDM, as seen below.

In order to derive a uniquely solvable SLAE equivalent to equation (4.47), we expand $\vec{p}_h$ and u_h in terms of such bases, that is, $\vec{p}_h = \sum_{l=1}^{L_h} p_l \vec{\eta}_l$ and $u_h = \sum_{n=1}^{M_h} u_n \chi_n$. Then, taking successively $\vec{q} = \vec{\eta}_k$ and $v = \chi_m$, similarly to Section 4.3, we conclude that the coefficient p_l corresponds to the mean value of the total normal flux along the lth edge, that is, $\int_{e_l} (\vec{p}_h|\vec{\nu}_l) ds / length(e_l)$. u_n in turn is nothing but the value of u_h in rectangle R_n, exactly like in the Cell-centred FVM. Let us merge both types of unknowns into a $(L_h + M_h)$-component vector $\vec{c}_h = [c_1, \ldots, c_{L_h}, c_{L_h+1}, \ldots, c_{L_h+M_h}]^T$, whose first L_h components are those of the vector $\underline{\vec{p}}_h := [p_1, p_2, \ldots, p_{L_h}]^T$, and the M_h last ones are those of the vector $\vec{u}_h := [u_1, u_2, \ldots, u_{M_h}]^T$. Then, (4.47) is easily seen to be equivalent to a SLAE $\mathcal{A}_h \vec{c}_h = \vec{g}_h$, where $\vec{g}_h$ is a $L_h + M_h$ component vector whose first L_h components are zero and whose $(L_h + m)$th component equals $-\int_\Omega f \chi_m \, dx dy$. The square matrix $\mathcal{A}_h$ in turn is a **block matrix** of the form:

$$
\begin{array}{cc}
L_h \text{ columns} \quad M_h \text{ columns} & \\
\mathcal{A}_h = \left[\begin{array}{c|c} B_h & C_h \\ \hline C_h^T & \mathcal{O} \end{array} \right] & \begin{array}{l} L_h \text{rows} \\ \\ M_h \text{rows} \end{array}
\end{array}
\tag{4.48}
$$

where $\mathcal{O}$ is the null $M_h \times M_h$ matrix, B_h is a $L_h \times L_h$ matrix whose entry b_{kl} is $(\vec{\eta}_l|\vec{\eta}_k)_0$ and C_h is the $L_h \times M_h$ matrix whose entry c_{kn} is $(\chi_n|div \, \vec{\eta}_k)_0$.

Of course, the system $\mathcal{A}_h \vec{c}_h = \vec{g}_h$ has a unique solution if and only if the only possible solution to the SLAE $\mathcal{A}_h \vec{c}_h = \vec{0}_{L_h+M_h}$ is $\vec{0}_{L_h+M_h}$, where $\vec{0}_s$ represents the null vector with s components. Pre-multiplying $\mathcal{A}_h \vec{c}_h$ by $\vec{c}_h^T$, we come up with $\underline{\vec{p}}_h^T B_h \underline{\vec{p}}_h + \vec{p}_h^T C_h \vec{u}_h = 0$ and $\vec{u}_h^T C_h^T \underline{\vec{p}}_h = 0$. The latter relation can be rewritten as $\underline{\vec{p}}_h^T C_h \vec{u}_h = 0$, and hence the former is reduced to $\underline{\vec{p}}_h^T B_h \underline{\vec{p}}_h = 0$. The fact that the symmetric matrix B_h is positive-definite (to be checked as Exercise 4.8) implies that $\underline{\vec{p}}_h = \vec{0}_{L_h}$. It follows from the first L_h equations of the SLAE $\mathcal{A}_h \vec{c}_h = \vec{0}_{L_h+M_h}$ that $C_h \vec{u}_h = \vec{0}_{L_h}$. Then, $\vec{u}_h = \vec{0}_{M_h}$ if the rank of C_h is M_h. Notice that this is only possible if $M_h \leq L_h$, which is thus a necessary condition for the mixed problem to have a unique solution. Recalling that $L_h = 2n_x n_y + n_x + n_y$ and $M_h = n_x n_y$, this condition is largely fulfilled. Notice that the condition that the rank of C_h be M_h is also necessary for existence and uniqueness of a solution to the mixed problem. Indeed, if it is violated there exists $\vec{u}_h \in \mathfrak{R}^{M_h} \neq \vec{0}_{M_h}$ such that $C_h \vec{u}_h = \vec{0}_{L_h}$. The **rank condition**, however, is not sufficient to ensure neither the uniform stability nor the convergence of the Raviart–Thomas mixed method of the lowest order for rectangles. This is an issue that requires some knowledge lying beyond the scope of this book. Those readers who wish to gain more insight on mixed variational problems and their approximation can consult reference [37][12].

We next push further the study of the Cell-centred FVM for rectangles and its connections with the Raviart–Thomas mixed FE of the lowest order. First, we observe that if B_h were a diagonal matrix, say $B_h = D_h = diag\{d_k\}$, we could easily eliminate the unknowns p_k from the system. Indeed, assuming this is the case, the first L_h equations in the SLAE yield the relation $D_h \underline{\vec{p}}_h + C_h \vec{u}_h = \vec{0}$. Hence, $p_k = -d_k^{-1} \sum_{n=1}^{M_h} c_{kn} u_n$, or yet in matrix form $\underline{\vec{p}}_h =$

---

[12] In mixed formulations, the second unknown field plays the role of a **Lagrange multiplier** of a condition satisfied by the first unknown field. In the case addressed above, this corresponds to the condition $div \, \vec{p} + f = 0$ in weak form.

$D_h^{-1} C_h \vec{u}_h$. Now, plugging the latter relation into the last M_h equations of the system, we come up with the SLAE $A_h \vec{u}_h = \vec{b}_h$ where $\vec{b}_h = [b_1, b_2, \ldots, b_{M_h}]^T$ with $b_m = -\int_{R_m} f \, dx dy$, $m = 1, 2, \ldots, M_h$, and A_h is the $M_h \times M_h$ matrix given by $C_h^T D_h^{-1} C_h$. Unfortunately, if we integrate exactly the coefficients $(\vec{\eta}_l | \vec{\eta}_k)_0$ of matrix B_h, the result is not a diagonal matrix. Then, following reference [106], we resort to the trapezoidal numerical quadrature formula, in each rectangle, which appears to be the simplest reliable way to obtain a diagonal approximation of B_h. By such a formula, the integral in a rectangle R with vertices S_r, $r = 1, 2, 3, 4$ of a continuous function g is approximated by

$$I_R(g) = \sum_{r=1}^{4} g(S_r) area(R)/4.$$

Before pursuing, it is necessary to determine the basis fields $\vec{\eta}_k$ at the element level. Referring to Figure 4.25, here we use the notation $R_{i-1/2, j-1/2}$ for the rectangles of $\mathcal{R}_h$, for $i = 1, 2, \ldots, n_x$ and $j = 1, 2, \ldots, n_y$. The horizontal edges are denoted by $e_{i-1/2, j}$ for $i = 1, 2, \ldots, n_x$ and $j = 0, 1, \ldots, n_y$, and the vertical edges by $e_{i, j-1/2}$ for $i = 0, 1, \ldots, n_x$ and $j = 1, 2, \ldots, n_y$.

The restrictions to rectangle $R_{i-1/2, j-1/2}$ of the four basis (shape) fields that do not vanish identically on it, that is, those attached to the edges $e_{i-1/2, j-1}, e_{i-1/2, j}, e_{i-1, j-1/2}$ and $e_{i, j-1/2}$, are denoted respectively by $\vec{\eta}_{i-1/2, j-1}, \vec{\eta}_{i-1/2, j}, \vec{\eta}_{i-1, j-1/2}$ and $\vec{\eta}_{i, j-1/2}$. Then, as the reader may check without any difficulty, changing variables in $R_{i-1/2, j-1/2}$ into $x_i = x - (i-1)h_x$ and $y_j = y - (j-1)h_y$, we have

$$\begin{cases} \vec{\eta}_{i, j-1/2} = \left[\dfrac{x_i}{h_x}, 0 \right]^T ; \ \vec{\eta}_{i-1\, j-1/2} = \left[\dfrac{h_x - x_i}{h_x}, 0 \right]^T ; \\[3mm] \vec{\eta}_{i-1/2, j} = \left[0, \dfrac{y_j}{h_y} \right]^T ; \ \vec{\eta}_{i-1/2\, j-1} = \left[0, \dfrac{h_y - y_j}{h_y} \right]^T . \end{cases} \quad (4.49)$$

It is as easy to verify that $I_{R_{i-1/2, j-1/2}}(\vec{p}, \vec{q})$ equals zero if $\vec{p}$ and $\vec{q}$ are two distinct fields out of the four fields defined in equation (4.49), and $I_{R_{i-1/2, j-1/2}}(\vec{p}, \vec{p})$ equals $h_x h_y / 2$ if $\vec{p}$ is any

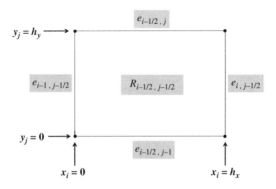

Figure 4.25   Rectangular cell and edge numbering for the Cell-centred FVM

one of them. Consequently, the thus-modified matrix B_h will be effectively a diagonal matrix D_h. Actually, taking into account that each interior edge is shared by two elements in $\mathcal{R}_h$, the pertaining diagonal coefficient d_k equals $h_x h_y$. In the case of a boundary edge, the corresponding coefficient is equal to $h_x h_y/2$.

Now, in order to proceed to the elimination of the unknowns p_k in the SLAE, we first determine C_h. Clearly enough, the only nonzero entries in the kth row of this matrix are the c_{kn}s corresponding to a rectangle R_n that has e_k as an edge. The corresponding value is easily found to be $-h_y$ if R_n (resp. $-h_x$) lies ahead of e_k in the increasing sense of coordinate x (resp. y), and $+h_y$ (resp. $+h_x$) otherwise. Then, using the notation $u_{i-1/2,j-1/2}$ for the approximate value of u in the element $R_{i-1/2,j-1/2}$, and denoting by $p_{i-1/2,j}$ and $p_{i,j-1/2}$ the mean fluxes on edges $e_{i-1/2,j}$ and $e_{i,j-1/2}$, respectively, as illustrated in Figure 4.25, we come up with

$$\begin{cases} p_{i-1/2,j} = (u_{i-1/2,j+1/2} - u_{i-1/2,j-1/2})/h_y \\ \quad \text{for } i = 1, 2, \ldots, n_x \text{ and } j = 1, 2, \ldots, n_y - 1, \\[2ex] p_{i,j-1/2} = (u_{i+1/2,j-1/2} - u_{i-1/2,j-1/2})/h_x \\ \quad \text{for } i = 1, 2, \ldots, n_x - 1 \text{ and } j = 1, 2, \ldots, n_y, \\[2ex] p_{i-1/2,0} = 2u_{i-1/2,1/2}/h_y & \text{for } i = 1, 2, \ldots, n_x, \\ p_{i-1/2,n_y} = -2u_{i-1/2,n_y-1/2}/h_y & \text{for } i = 1, 2, \ldots, n_x, \\ p_{0,j-1/2} = 2u_{1/2,j-1/2}/h_x & \text{for } j = 1, 2, \ldots, n_y, \\ p_{n_x,j-1/2} = -2u_{n_x-1/2,j-1/2}/h_x & \text{for } j = 1, 2, \ldots, n_y. \end{cases} \tag{4.50}$$

We still denote by $\vec{p}_h$ and u_h the new approximations of the unknowns in the mixed formulation and by $\underline{\vec{p}}_h$ and $\vec{u}_h$ the corresponding coefficient vectors. In doing so, we complete the elimination of the flux unknowns by plugging equation (4.50) into the second equation in (4.47), that is, the relation $C_h^T \underline{\vec{p}}_h = \vec{b}_h$. In order to work this out, we remind that the latter relation is equivalent to taking $v = \chi_m$ in the former, for m successively equal to $1, 2, \ldots, M_h$, or yet $v = \chi_{i-1/2,j-1/2}$, using a self-explanatory two-subscript notation corresponding to the adopted element numbering. This relation reduces to $\int_{R_{i-1/2,j-1/2}} div\,\vec{p}_h\,dxdy = -\int_{R_{i-1/2,j-1/2}} f\,dxdy$, to which we apply the divergence theorem. Observing that the normal vector is an outer one along the upper and right edges of the elements and an inner one along the lower and left edges, this directly leads to the following equations for $i = 1, 2, \ldots, n_x$ and $j = 1, 2, \ldots, n_y$:

$$+ h_x(p_{i-1/2,j} - p_{i-1/2,j-1}) + h_y(p_{i,j-1/2} - p_{i-1,j-1/2}) = -\int_{R_{i-1/2,j-1/2}} f\,dxdy. \tag{4.51}$$

Finally, combining equations (4.50) and (4.51), and using the one point quadrature formula $\int_R f\,dxdy \simeq f(C_R)h_x h_y$ where C_R is the centre of R, after straightforward calculations we derive our **Cell-centred FV scheme** in connection with a uniform mesh of a rectangle. More precisely, the scheme allows us to find approximations $u_{i-1/2,j-1/2}$ of u in $R_{i-1/2,j-1/2}$, for

$i = 1, 2, \ldots, n_x$ and $j = 1, 2, \ldots, n_y$, satisfying one equation per cell. For the $(n_x - 1) \times (n_y - 1)$ inner cells, these equations give rise to the

Inner cell-centred FV analog of the Poisson equation in a rectangle

$$\left[\frac{2}{h_x^2} + \frac{2}{h_y^2} \right] u_{i-1/2,j-1/2} - \frac{u_{i-3/2,j-1/2} + u_{i+1/2,j-1/2}}{h_x^2} - \frac{u_{i-1/2,j-3/2} + u_{i-1/2,j+1/2}}{h_y^2}$$
$$= f([i - 1/2]h_x, [j - 1/2]h_y), \text{ for } i = 2, \ldots, n_x - 1, j = 2, \ldots, n_y - 1;$$
$$(4.52)$$

For the boundary cells different from corner cells and the corner cells themselves, similar equations are obtained. It is up to the reader to write them down as Exercise 4.9, for the different types of boundary cells, if she or he wishes to consolidate her or his comprehension of the above concepts. However, instead of exhibiting all the equations in the form of equation (4.52), we shall express the very same relations in terms of mean fluxes across the edges. This procedure, besides allowing for a unification of the equations, makes explicit such quantities, whose control is one of the guiding principles of the FVM. Let us see how to achieve this.

First, we introduce auxiliary boundary values,

$$\begin{cases} u_{i-1/2,-1/2} = u_{i-1/2,n_y+1/2} = u_{-1/2,j-1/2} = u_{n_x+1/2,j-1/2} = 0 \\ \text{for } i = 1, 2, \ldots, n_x \text{ and } j = 1, 2, \ldots, n_y, \end{cases} \tag{4.53}$$

and define associated flux values across edges of the mesh contained in Γ, that is, $e_{i-1/2,0}$, $e_{i-1/2,n_y+1}$, $e_{0,j-1/2}$ and $e_{n_x+1,j-1/2}$. Keeping in mind such definitions, we set up the

Numerical mean fluxes across the boundary Γ

$$F_{i-1/2,0}^{-} = \frac{2(u_{i-1/2,-1/2} - u_{i-1/2,1/2})}{h_y}, \qquad i = 1, \ldots, n_x;$$

$$F_{i-1/2,n_y}^{+} = \frac{2(u_{i-1/2,n_y+1/2} - u_{i-1/2,n_y-1/2})}{h_y}, \quad i = 1, \ldots, n_x;$$

$$\tag{4.54}$$

$$F_{0,j-1/2}^{-} = \frac{2(u_{-1/2,j-1/2} - u_{1/2,j-1/2})}{h_x}, \qquad j = 1, \ldots, n_y;$$

$$F_{n_x,j-1/2}^{+} = \frac{2(u_{n_x+1/2,j-1/2} - u_{n_x-1/2,j-1/2})}{h_x}, \quad j = 1, \ldots, n_y.$$

The relations in equation (4.54) are aimed at approximating the outer normal derivative of u along the corresponding boundary edge, for instance $e_{i-1/2,0}$, by calling attention to the fact that the FV approximation of u in the adjacent cell $R_{i-1/2,1/2}$ varies from $u_{i-1/2,1/2}$ to zero (i.e. $u_{i-1/2,-1/2}$) in the outward direction, along a segment of length equal to $h_y/2$. A similar argument applies to any other boundary edge, as illustrated in Figure 4.26 for the

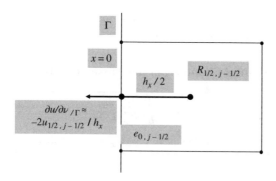

Figure 4.26   Flux computation along an edge $e_{0,j-1/2}$ contained in Γ

edge $e_{0,j-1/2}$. In this figure, we depict the flux calculation, assuming that values not nec-essarily zero $u_{-1/2,j-1/2} = u(0, [j-1/2]h_y)$ for $j = 1, 2, \ldots, n_y$ are prescribed along the edge of Γ given by $x = 0$. Now we complete the above definitions by letting $F^-_{i-1/2,j-1}$, $F^+_{i-1/2,j}$, $F^-_{i-1,j-1/2}$ and $F^+_{i,j-1/2}$ be the numerical fluxes outward CV $R_{i-1/2,j-1/2}$ across the edges $e_{i-1/2,j-1}$, $e_{i-1/2,j}$, $e_{i-1,j-1/2}$ and $e_{i,j-1/2}$, respectively, as long as they are not contained in Γ:

Numerical mean fluxes across inner edges of a rectangular mesh

$$
\begin{aligned}
F^-_{i-1/2,j-1} &= \frac{u_{i-1/2,j-3/2} - u_{i-1/2,j-1/2}}{h_y}, & i = 1, \ldots, n_x, j = 2, \ldots, n_y; \\[2mm]
F^+_{i-1/2,j} &= \frac{u_{i-1/2,j+1/2} - u_{i-1/2,j-1/2}}{h_y}, & i = 1, \ldots, n_x, j = 1, \ldots, n_y - 1; \\[2mm]
F^-_{i-1,j-1/2} &= \frac{u_{i-3/2,j-1/2} - u_{i-1/2,j-1/2}}{h_x}, & j = 1, \ldots, n_y, i = 2, \ldots, n_x; \\[2mm]
F^+_{i,j-1/2} &= \frac{u_{i+1/2,j-1/2} - u_{i-1/2,j-1/2}}{h_x}, & j = 1, \ldots, n_y, i = 1, \ldots, n_x - 1.
\end{aligned}
$$

$$(4.55)$$

The notation for the fluxes carries a pair of subscripts referring to the edge and a sign super-script. The latter identifies the CV they are referred to. While on the one hand the superscript is superfluous in the case of equation (4.54), this is not the case of the fluxes across inner edges defined in equation (4.55). Indeed, a quantity defined by the latter relations is an out-ward flux for the cell under consideration, while being an inward flux for a neighbouring cell.

Remark 4.11 *The author admits that the above notation is quite cumbersome and could be simplified. This was not done here because he believes that it is advisable to stick as much as possible to the presentation of the FVM in one-dimensional space, given in Chapter 1. Nevertheless, the Cell-centred FVM is further developed in Chapters 5 and 6,* *where notations more conforming to simpler ones usually encountered in the literature are adopted.* ■

   With the above definitions, we come up with the following system of equations to solve equation (4.3) in a rectangular domain with a uniform $n_x \times n_y$ mesh by the

Cell-centred FV scheme with rectangular cells in terms of fluxes

$$
\begin{aligned}
&\text{Find } u_{i-1/2,j-1/2}\text{satisfying for } i = 1, 2, \ldots, n_x, j = 1, 2, \ldots, n_y : \\[4pt]
&h_x(F^-_{i-1/2,j-1} + F^+_{i-1/2,j}) + h_y(F^-_{i-1,j-1/2} + F^+_{i,j-1/2}) \\
&= h_x h_y f([i - 1/2]h_x, [j - 1/2]h_y).
\end{aligned}
\tag{4.56}
$$

Clearly enough, dividing both sides of the above relation by $h_x h_y$, and using equations (4.53), (4.54) and (4.55), we obtain equations of the same form as equation (4.52).

   By arguments very close to those already exploited in this chapter and in Chapter 1, one can prove that SLAE (equation (4.52)) has a unique solution. However, we postpone the proof of this result, for in Chapter 5 we demonstrate it as an outcome of the scheme's stability.

   If, instead of the zero boundary values in equation (4.53), inhomogeneous Dirichlet boundary conditions $u = g$ on Γ must be taken into account, we can easily modify the FV scheme (4.56) in order to accommodate them. It turns out that in this case, too, the Cell-centred FV scheme can be derived from the underlying RT_0 mixed FEM of the lowest order, through numerical integration. Exercise 4.10 addresses precisely this issue.

   Summarising, the Cell-centred FVM for the Poisson problem with Dirichlet boundary conditions can be derived from the RT_0 method assorted with a specific quadrature rule to approximate the L^2-inner product of two flux variables $\vec{p}$ and $\vec{q}$ in the elements, as long as we assign zero fluxes in equation (4.40) along edges contained in Γ_1. As a consequence, the stability and convergence results that are known to hold for the thus-modified RT_0 method (i.e. with numerical quadrature) extend to the Cell-centred FVM in the case of both rectangles and triangles (cf. [19] and [106]). In particular, the Cell-centred FVM is first-order convergent in $L^2(\Omega)$, which will be proved in Chapter 5.

Remark 4.12 *This rather poor result can be improved to second-order convergence at practically no additional cost, thanks to a technique of the author described in reference [173] and illustrated in reference [175].* ■

Remark 4.13 *An interesting point about RT_0 elements is the **superconvergence** observed at the element centroids: The values of u_h at these points converge to the exact ones with order two and not one, even if the mesh is distorted. Examples given in reference [174] duly certify this. However, to the best of the author's knowledge, at this writing no formal proofs have yet been given to such a property.* ■

   On the other hand, the flux variable $\vec{p}$ (here the solution gradient), besides having desirable continuity properties, is approximated with order one in the L^2-norm, together with its divergence (here the Laplacian of u) [161]. This is certainly an advantage as compared to the $\mathcal{P}_1$

FEM. Notice that the variable $\vec{p}$ can be easily recovered from the FV solution using relations (4.50).

To conclude, we would like to point out that, likewise the one-dimensional case, stability and convergence results in the maximum norm cannot be worked out in the same manner as for the FDM, even if uniform meshes are employed. The reader might wish to sketch this study, in order to practice the knowledge acquired through the material presented in this section. In doing so, she or he would find out that the rectangular-centred FVM is not even first-order consistent in a FD sense. In Subsection 5.1.5, we will apply a technique of analysis well adapted to FV schemes, which will circumvent such a drawback.

4.5   SLAE Resolution

In this section, we give some hints on the solution of large SLAEs resulting from the discretisation of a boundary value PDE in two space variables. A particular emphasis is given to the right way to number unknowns and/or or cells, in the aim of optimising the solution process. Since here again this is a vast subject, we do not attempt to be exhaustive.

First of all, we should point out that the optimal way to number entities such as grid points, nodes, elements or cells heavily depends on the method employed to solve the SLAE resulting from the discretisation. As pointed out in Subsection 4.2.3, natural numberings, such as those employed in the one-dimensional case, lead to better computational performances for most solution methods. Nevertheless, the concept of natural numbering for two-dimensional or higher dimensional problems is somewhat more fuzzy. Generally speaking, we can say that the best way to number nodes and elements is to follow strips or slices of the domain roughly orthogonal to a line along which the domain's extension is the longest. For instance, if the domain is a rectangle and the mesh is uniform, the node and element numbers should increase in the sense of the longest edges, and within each strip of nodes and elements parallel to the shortest edge, the numbers should increase one by one in the same sense. In this way, the width of the band containing nonzero coefficients of the system matrix is reduced to a minimum, thereby minimising storage requirements. Just to illustrate the above described procedure, we display in Figure 4.27 the numbering of the nodes for a uniform 4×2 mesh of a rectangle, and linear elements, assuming that Γ_0 corresponds to the left vertical edge of the domain. Recalling the procedure suggested in the previous section, the boundary nodes on Γ_0 and Γ_1 are numbered separately for the latter correspond to unknown values. Incidentally, the numbering of the elements does not play an important role for most solution methods, except perhaps for the **frontal method** introduced by Irons [104], in which Gauss eliminations are gradually performed as one sweeps the mesh, in order to assemble the matrix and right-side vector.

We display in Array 4.2 the pattern of the system matrix corresponding to the mesh of Figure 4.27, where the letter z represents a generic a priori nonzero coefficient. As the reader may check, the matrix bandwidth is seven, but since it is symmetric only the upper (resp. lower) band needs to be stored for an upper (resp. a lower) bandwidth equal to four.

Notice that the above numbering procedure is based on the consideration that neither a Cholesky decomposition nor Gaussian elimination without pivoting (cf. [158]) creates nonzero entries outside the band of a banded matrix, that is, the matrix positions filled in with zero entries only, before the process starts. Therefore, storage requirements for the application of both direct methods are known a priori.

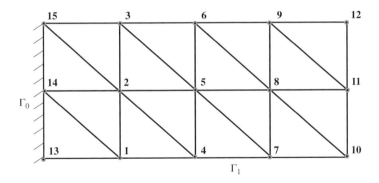

Figure 4.27 An example of optimal node numbering for linear elements

$$A_h = \begin{bmatrix}
z & z & 0 & z & 0 & 0 & 0 & 0 & 0 & 0 & 0 & 0 \\
z & z & z & z & z & 0 & 0 & 0 & 0 & 0 & 0 & 0 \\
0 & z & z & 0 & z & z & 0 & 0 & 0 & 0 & 0 & 0 \\
z & z & 0 & z & z & 0 & z & 0 & 0 & 0 & 0 & 0 \\
0 & z & z & z & z & z & z & z & 0 & 0 & 0 & 0 \\
0 & 0 & z & 0 & z & z & z & z & z & 0 & 0 & 0 \\
0 & 0 & 0 & z & z & 0 & z & z & 0 & z & 0 & 0 \\
0 & 0 & 0 & 0 & z & z & z & z & z & z & z & 0 \\
0 & 0 & 0 & 0 & 0 & z & 0 & z & z & 0 & z & z \\
0 & 0 & 0 & 0 & 0 & 0 & z & z & 0 & z & z & 0 \\
0 & 0 & 0 & 0 & 0 & 0 & 0 & z & z & z & z & z \\
0 & 0 & 0 & 0 & 0 & 0 & 0 & 0 & z & 0 & z & z
\end{bmatrix}$$

(4.57)

Array 4.2

There are algorithms aimed at numbering nodes and elements in an optimal way for arbitrary meshes, in such a way that bandwidths are reduced to a minimum. These procedures are very useful whenever one is dealing with domains of irregular shape. In this case, in principle, the largest column (resp. row) number occupied by a nonzero matrix coefficient varies with the row (resp. column) number. This kind of matrix is called a **skyline matrix**, for which a **skyline storage** is more appropriate (see e.g. [4]). This is nothing but a sort of band storage, in which the zero entries not lying within a distance from the diagonal to the last nonzero entry are disregarded in each row (resp. column). Actually, users can take good advantage of a skyline profile, since akin to a matrix band, it is preserved by a Cholesky or a Crout decomposition without pivoting. Moreover, the effort of coding Cholesky's or Crout's method for skyline matrices is about the same as for banded matrices.

We should also point out that the concept of optimal numbering may be different if iterative solution methods are employed. Nevertheless, otherwise specific rules are prescribed, in general the minimum fixed or variable bandwidth criterion should provide satisfactory results. Further details on FE or FV mesh generation and pertaining numbering aspects can be found in reference [76].

Incidentally, the matrix resulting from the discretisation of the Poisson equation by the FEM besides being symmetric is positive-definite, according to an argument that can be adapted to the FDM and the FVM, and to the FEM itself for solving (P_1). This was seen in detail in Subsection 4.3.3, in connection with the Ritz method. As a consequence, the successful application of several methods to solve the underlying SLAE is guaranteed. For instance, we can solve it for $\vec{u}_h$ by Cholesky's method and the conjugate gradient method already recalled in Section 1.6. The latter method converges faster in case the **pre-conditioning** technique is employed. We know that if we apply standard Cholesky's decomposition to A_h, in principle the nonzero entries of the resulting upper triangular banded matrix U_h will fill the whole band having the same width as the band of A_h. However, we saw that even inside the band the rate of sparsity can be very large. The principle of the incomplete Cholesky's decomposition consists of skipping the computation of a coefficient $u_{i,j}$ of U_h inside its band, as long as $a_{i,j} = 0$. This leads to an approximation $\tilde{U}_h$ of U_h which hopefully satisfies $\tilde{U}_h^T \tilde{U} \simeq A_h$. Actually, the existence of $\tilde{U}_h$ is guaranteed in some cases, for instance, if matrix A_h is diagonally dominant [63]. This happens if the triangulation is of the acute type, which can be checked as Exercise 4.11.

Now, let us define $\vec{d}_h := [\tilde{U}_h^T]^{-1}\vec{b}_h$. Clearly, we have $[\tilde{U}_h^T]^{-1}A_h\vec{u}_h = \vec{d}_h$, and hence if we set $\vec{w}_h = \tilde{U}_h\vec{u}_h$, the system to solve for $\vec{w}_h$ is $[\tilde{U}_h^T]^{-1}U_h^T U_h \tilde{U}_h^{-1}\vec{w}_h = \vec{d}_h$. As long as U_h and $\tilde{U}_h$ are not so different from each other, the matrix $\tilde{I}_h := [\tilde{U}_h^T]^{-1}U_h^T U_h \tilde{U}_h^{-1}$ of the above system is very close to the identity matrix. Actually, whatever the initial guess, the conjugate gradient method applied to this modified SLAE will converge in much less iterations than in the case of the original system, as illustrated in Example 7.5. Notice that the final value of $\vec{u}_h$ is obtained by simple back substitution (i.e. from the relation $\vec{u}_h = [\tilde{U}_h]^{-1}\vec{w}_h$). Moreover, at each iteration a few linear systems with triangular matrices $\tilde{U}_h$ and $[\tilde{U}_h]^T$ have to be solved in order to compute residuals and quantities alike related to system $\tilde{I}_h\vec{z}_h = \vec{d}_h$. We refer to the preliminary section on linear algebra for further details.

4.5.1   Example 4.1: A Crout Solver for Banded Matrices

We supply below a FORTRAN 95 code for solving SLAEs $A\vec{u} = \vec{b}$ by Crout's method with partial pivoting, with an $n \times n$ invertible banded matrix A stemming from FD, FE or FV discretisations. Hopefully, this will represent a modest but useful practical material to beginner readers not willing to start coding activities in numerics from scratch.

We recall that Crout's method works for any non-singular system matrix which may be otherwise arbitrary. Of course, this means that the method is applicable to the case of symmetric positive-definite matrices, such as those considered in this chapter's previous sections.

Some explanations on the code are given prior to its instruction listing, but the reader is expected to read it through in order to make sure to have acquired all the necessary understanding of the solution process by this method.

As pointed out in the preliminary section on linear algebra, this method is particularly efficient when several SLAEs have to be solved with the same invertible matrix. This possibility is taken into account in the program by means of a key: a zero value of the key indicates that a first SLAE is being solved, in which case the Crout decomposition is carried out, followed by the solution itself by forth and back substitution. In case additional systems are being solved, the value of the key is set to one and the decomposition step is skipped. We recall that A is decomposed into the product of two matrices L and U, where L is lower triangular with a

$$\begin{bmatrix}
0 & 0 & 0 & a_{1,1} & a_{1,2} & a_{1,3} & a_{1,4} \\
0 & 0 & a_{2,1} & a_{2,2} & a_{2,3} & a_{2,4} & a_{2,5} \\
0 & a_{3,1} & a_{3,2} & a_{3,3} & a_{3,4} & a_{3,5} & a_{3,6} \\
a_{4,1} & a_{4,2} & a_{4,3} & a_{4,4} & a_{4,5} & a_{4,6} & a_{4,7} \\
a_{5,2} & a_{5,3} & a_{5,4} & a_{5,5} & a_{5,6} & a_{5,7} & a_{5,8} \\
a_{6,3} & a_{6,4} & a_{6,5} & a_{6,6} & a_{6,7} & a_{6,8} & a_{6,9} \\
a_{7,4} & a_{7,5} & a_{7,6} & a_{7,7} & a_{7,8} & a_{7,9} & a_{7,10} \\
a_{8,5} & a_{8,6} & a_{8,7} & a_{8,8} & a_{8,9} & a_{8,10} & a_{8,11} \\
a_{9,6} & a_{9,7} & a_{9,8} & a_{9,9} & a_{9,10} & a_{9,11} & a_{9,12} \\
a_{10,7} & a_{10,8} & a_{10,9} & a_{10,10} & a_{10,11} & a_{10,12} & 0 \\
a_{11,8} & a_{11,9} & a_{11,10} & a_{11,11} & a_{11,12} & 0 & 0 \\
a_{12,9} & a_{12,10} & a_{12,11} & a_{12,12} & 0 & 0 & 0.
\end{bmatrix}$$

Array 4.3

unit diagonal and U is upper triangular. The half bandwidth of A is equal to m assumed to be much less than n. We also assume that partial pivoting is performed along the Gaussian elimination steps. In view of this, A in its initial form will be stored as an $n \times (2m - 1)$ array, whose first (resp. last) $m - 1$ columns are reserved for the matrix lower (resp. upper band) and the mth column for the diagonal of A. In order to conform A to such a storage pattern, at initial stage an upper left (resp. lower right) triangular portion of the storing array is filled in with $m(m - 1)/2$ zeros from the first through the $(m - 1)$th (resp. from the $(n - m + 2)$th through the nth) row. In self-explanatory Array 4.3 supplied hereafter, we illustrate the storage of a 12×12 matrix with a half bandwidth equal to four.

If no partial pivoting occurs, as the Gaussian elimination goes on, the entries of L will occupy the corresponding positions of $a_{i,j}$ for $i > j$ and the entries of U will occupy the corresponding positions of $a_{i,j}$ for $j \geq i$. However, if partial pivoting takes place, the bandwidth may need to be enlarged to the right by $m - 1$ columns. This can be understood by reporting, for instance to Array 4.3. Assume that at the first elimination the pivot is found to be $a_{4,1}$. The fourth row will be interchanged with the first row, and hence the new first row will have nonzero entries up to the 10th column in the storing array. Then, at the end of the first step, in principle, the first four rows of the true matrix will have nonzero entries up to its seventh column. The fact that three additional columns may be necessary in the storing array to carry out the remaining steps with partial pivoting can be checked by inspection. It follows that, in case partial pivoting is implemented, the storage array must have n rows and $3m - 2$ columns. After the Gaussian eliminations are finished, L will still be stored in the first $m - 1$ columns, while U will occupy the last $2m - 1$ columns of the storing array. Finally, referring to the preliminary section on linear algebra, the permutation matrix P can be simply stored in an integer n-component array whose lth entry corresponds to the number of the row interchanged with the lth row at the lth Gaussian elimination step.

We presume that these explanations allow for full understanding of the FORTRAN code given next. Note that this code can be used as an auxiliary procedure in an FD, FE or FV PDE solution package.

Besides the integer pointer key, the following variables and arrays are used: c for both the initial system banded matrix A and its factors L and U; v for both the right-side vector $\vec{b}$ and the solution $\vec{u}$; $ipiv$ for storing the number of the equation interchanged by partial pivoting, with the lth equation at the lth Gaussian elimination; n is the number of equations and unknowns; and lb is the half bandwidth of A (i.e. a third bandwidth at the end).

As the author believes, the meaning of the variables and arrays in subroutine CROUT other than c, v, $ipiv$, n, lb and key is self-explanatory. The reader is strongly encouraged to read the following code through, in order to make sure to have understood the whole procedure.

```
      SUBROUTINE CROUT( c, v, ipiv, n, lb, key)
      IMPLICIT NONE
      INTEGER :: m, lr, kr, i, j, jpiv, jfinal, jm, ir
      INTEGER,          INTENT(IN) :: n, lb, key
      INTEGER,          INTENT(IN OUT), DIMEN-
SION(n)              :: ipiv
      DOUBLE PRECISION, INTENT(IN OUT), DIMENSION(n,3*(lb-1)+1) :: c
      DOUBLE PRECISION, INTENT(IN OUT), DIMENSION(n)           :: v
      DOUBLE PRECISION :: t
!---------- ARGUMENT EXPLANATION -----------------------------
! array c inputs matrix A and outputs triangular factors L and U;
! array v inputs the right side b and outputs the solution u;
! array ipiv stores partial pivoting's line permutations;
! n is the number of unknowns of the linear system;
! lb is the half band width of matrix A;
! key=0 if A does not enter in factorised form LU and key=1 other-
wise.
! Remark: ipiv(l)=1 for all l if key=0.
!-----------------------------------------------------------
      m = 2*lb-1
      IF ( key /= 1) Then,
         lr = lb
         DO kr = 1, lr-1
            DO i = 1, lr-kr
               DO j = 2, m
                  c(kr,j-1) = c(kr,j)
               ENDDO
               c(kr,m) = 0.D0
            ENDDO
         ENDDO
         ipiv(n) = n
         DO i = 1, n-1
            ipiv(i) = i
            DO kr = i+1, lr
               IF ( DABS( c(kr,1)) > DABS( c(ipiv(i),1))) Then,
                  ipiv(i) = kr
               END IF
            ENDDO
            jpiv = ipiv(i)
            IF ( jpiv /= i) Then,
               DO j = 1, m
                  t        = c(i,j)
                  c(i,j)   = c(jpiv,j)
```

```
                    c(jpiv,j) = t
                ENDDO
            END IF
            DO kr = i+1, lr
                t = c(kr,1)
                c(i,m+kr-i) = t/c(i,1)
                DO j = 2, m
                    c(kr,j-1) = c(kr,j) - t*c(i,j)/c(i,1)
                END DO
                c(kr,m) = 0.D0
            ENDDO
            IF ( lr /= n) Then,
                lr = lr + 1
            END IF
        END DO
    END IF
    DO i = 1, n-1
        jfinal = i + lb - 1
        IF ( jfinal > n) Then,
            jfinal = n
        END IF
        t           = v(i)
        v(i)        = v(ipiv(i))
        v(ipiv(i)) = t
        DO kr = i+1, jfinal
            v(kr) = v(kr) - c(i,m+kr-i)*v(i)
        END DO
    END DO
    v(n) = v(n)/c(n,1)
    jm = 2
    DO ir = 1, n-1
        kr = n - ir
        DO j = 2, jm
            v(kr) = v(kr) - c(kr,j)*v(kr-1+j)
        END DO
        v(kr) = v(kr)/c(kr,1)
        IF ( jm /= m) Then,
            jm = jm + 1
        END IF
    END DO
END SUBROUTINE CROUT
```

4.5.2   *Example 4.2: Iterative Solution of Equivalent FD–FE–FV SLAEs*

To close this chapter, let us do some more number crunching in a rectangular domain $\Omega = (0, L_x) \times (0, L_y)$.

As long as the right-side datum is a continuous function, under certain additional conditions specified hereafter, the Five-point FDM is equivalent to the $\mathcal{P}_1$ FEM and the Vertex-centred

FVM to solve the Poisson equation. More specifically, it is assumed that the nodes of the underlying triangular mesh coincide with the FD grid points, and moreover that suitable quadrature rules are employed to compute the right side in the case of the FEM and the FVM. For the sake of simplicity, we take the case of homogeneous Dirichlet boundary conditions as a model, and consider only uniform grids, together with meshes constructed by subdividing each rectangle having four neighbouring grid points as vertices into two right triangles by its diagonal parallel to the diagonal d^+ of Ω with a positive slope. In Figure 5.14, we illustrate the resulting mesh, in case Ω is a unit square.

We saw in Subsection 4.3.1 that the matrix coefficient $a_{i,j}$ for both the $\mathcal{P}_1$ FEM and the Vertex-centred FVM associated with two nodes S_i and S_j, which are end-points of a mesh edge e, is given by $(\cot\theta^1_{i,j} + \cot\theta^2_{i,j})/2$, where $\theta^1_{i,j}$ and $\theta^2_{i,j}$ are the angles opposite to this edge in the triangles T_1 and T_2, respectively, whose intersection is e. Thus, at least for meshes of a rectangular Ω constructed as above, $a_{i,j} = 0$ for both methods provided the pair of nodes $(S_i; S_j)$ is such that $\overline{S_i S_j}$ is parallel to d^+. This means that the sparsity structure of the matrices for the Five-point FDM, the $\mathcal{P}_1$ FEM and the Vertex-centred FVM is the same. Actually, except for different scalings, the SLAEs for all the three methods are identical, as long as suitable numerical integration formulae are used to compute the right-side vectors. To begin with, we consider the right-side vectors $\vec{b}_h$. We know that for the FDM, this vector is given by $[f_1, f_2, \ldots, f_{N_h}]$ where $N_h = (n_x - 1)(n_y - 1)$, n_x and n_y being the number of subdivisions of L_x and L_y, respectively, used to construct the grid or mesh. Let $h_x = L_x/n_x$ and $h_y = L_y/n_y$. If we apply the one-point quadrature rule $f(S_i)area(V_i) \simeq \int_{V_i} f\, dxdy$, where V_i is the CV about node S_i, since the area of V_i equals $h_x h_y$ for all V_i (please check!), the right-side vector for the FVM is the same as for the FDM times $h_x h_y$. In the case of the FEM, we assume that the trapezoidal rule $[g(S^T_1) + g(S^T_2) + g(S^T_3)]area(T)/3$ is employed to approximate the integral of a continuous function g in any mesh triangle T with vertices S^T_r, $r = 1,2,3$. Applying this formula to the function $f\varphi_i$ in all the triangles T having S_i as a vertex, φ_i being the shape function of the $\mathcal{P}_1$ FEM associated with node S_i, it is easily seen that the result is the total area of these triangles multiplied by $f(S_i)$ and divided by 3. Referring to Figure 5.11, it turns out that there are exactly six such triangles with area equal to $h_x h_y/2$, and hence the final value of the corresponding component of $\vec{b}_h$ for the FEM is also equal to $h_x h_y$.

All that is left to do is establishing that the nonzero entries for the $\mathcal{P}_1$ FEM and the Vertex-centred FV matrix are the same as the corresponding entries of the FD matrix multiplied by $h_x h_y$. Taking any mesh edge parallel to the y-axis with end-points S_i and S_j, applying the same co-tangent formula as above, the mean value of the pair of pertaining co-tangents is readily seen to be equal to h_y/h_x. Similarly, for any edge parallel to the x-axis with end-points S_i and S_j the corresponding entry $a_{i,j}$ is h_x/h_y. Dividing these matrix coefficients by the scaling factor $h_x h_y$, and recalling the form of the matrix for the FDM studied in Subsection 4.2.1, the reader can easily conclude as Exercise 4.13 that the matrices and right-side vector coincide for the three methods considered in this example.

Next, we endeavour to solve the Poisson equation $-\Delta u = 1$ with homogeneous Dirichlet boundary conditions in a unit square with successively refined uniform grids, and hence with the $\mathcal{P}_1$ FEM and the Vertex-centred FVM. Notice that in this example, the integrals of the right sides for the FEM and the FVM can be computed exactly. The solution u is not computable exactly by an analytic expression in this case, and hence we will carry out a sensitivity analysis. This consists of comparisons of the solutions successively determined for $n_x \times n_y$ grids or meshes with $n_x = n_y = n = 10 \times 2^j$ for $j = 1,2,3,\ldots,m$ with another numerical solution

$i \to$	10	11	12	13	14	15	16	17	18	19
$x \to$	0.50	0.55	0.60	0.65	0.70	0.75	0.80	0.85	0.90	0.95
j y										
10 0.50	0.29399	0.29149	0.28395	0.27118	0.25293	0.22881	0.19834	0.16094	0.11592	0.06254
11 0.55	0.29149	0.28902	0.28156	0.26893	0.25087	0.22699	0.19681	0.15975	0.11510	0.06213
12 0.60	0.28395	0.28156	0.27433	0.26211	0.24462	0.22148	0.19218	0.15613	0.11262	0.06086
13 0.65	0.27118	0.26893	0.26211	0.25057	0.23404	0.21211	0.18430	0.14997	0.10839	0.05871
14 0.70	0.25293	0.25087	0.24462	0.23404	0.21884	0.19864	0.17293	0.14107	0.10226	0.05558
15 0.75	0.22881	0.22699	0.22148	0.21211	0.19864	0.18069	0.15773	0.12911	0.09399	0.05135
16 0.80	0.19834	0.19681	0.19218	0.18430	0.17293	0.15773	0.13819	0.11366	0.08324	0.04582
17 0.85	0.16094	0.15975	0.15613	0.14997	0.14107	0.12911	0.11366	0.09408	0.06951	0.03870
18 0.90	0.11592	0.11510	0.11262	0.10839	0.10226	0.09399	0.08324	0.06951	0.05200	0.02949
19 0.95	0.06254	0.06213	0.06086	0.05871	0.05558	0.05135	0.04582	0.03870	0.02949	0.01724

Values at the coarsest grid points (mesh nodes) directly computed thereupon

Array 4.4

$i \to$	10	11	12	13	14	15	16	17	18	19
$x \to$	0.50	0.55	0.60	0.65	0.70	0.75	0.80	0.85	0.90	0.95
j y										
10 0.50	0.29407	0.29157	0.28403	0.27127	0.25302	0.22890	0.19843	0.16101	0.11598	0.06257
11 0.55	0.29157	0.28911	0.28164	0.26902	0.25096	0.22708	0.19690	0.15982	0.11516	0.06216
12 0.60	0.28403	0.28164	0.27442	0.26221	0.24472	0.22158	0.19227	0.15621	0.11268	0.06090
13 0.65	0.27127	0.26902	0.26221	0.25068	0.23414	0.21222	0.18441	0.15007	0.10846	0.05875
14 0.70	0.25302	0.25096	0.24472	0.23414	0.21896	0.19877	0.17306	0.14118	0.10234	0.05563
15 0.75	0.22890	0.22708	0.22158	0.21222	0.19877	0.18082	0.15787	0.12924	0.09410	0.05141
16 0.80	0.19843	0.19690	0.19227	0.18441	0.17306	0.15787	0.13835	0.11381	0.08338	0.04591
17 0.85	0.16101	0.15982	0.15621	0.15007	0.14118	0.12924	0.11381	0.09425	0.06967	0.03881
18 0.90	0.11598	0.11516	0.11268	0.10846	0.10234	0.09410	0.08338	0.06967	0.05217	0.02963
19 0.95	0.06257	0.06216	0.06090	0.05875	0.05563	0.05141	0.04591	0.03881	0.02963	0.01741

Coarse grid (mesh) values computed with the finer grid (mesh)

Array 4.5

$i \to$	10	11	12	13	14	15	16	17	18	19
$x \to$	0.50	0.55	0.60	0.65	0.70	0.75	0.80	0.85	0.90	0.95
j y										
10 0.50	0.29409	0.29160	0.28406	0.27130	0.25305	0.22893	0.19846	0.16104	0.11599	0.06258
11 0.55	0.29160	0.28913	0.28167	0.26905	0.25099	0.22712	0.19693	0.15985	0.11518	0.06217
12 0.60	0.28406	0.28167	0.27445	0.26224	0.24475	0.22161	0.19231	0.15624	0.11270	0.06091
13 0.65	0.27130	0.26905	0.26224	0.25071	0.23418	0.21226	0.18444	0.15010	0.10849	0.05876
14 0.70	0.25305	0.25099	0.24475	0.23418	0.21899	0.19881	0.17310	0.14122	0.10237	0.05564
15 0.75	0.22893	0.22712	0.22161	0.21226	0.19881	0.18086	0.15792	0.12929	0.09413	0.05143
16 0.80	0.19846	0.19693	0.19231	0.18444	0.17310	0.15792	0.13840	0.11386	0.08342	0.04593
17 0.85	0.16104	0.15985	0.15624	0.15010	0.14122	0.12929	0.11386	0.09431	0.06972	0.03885
18 0.90	0.11599	0.11518	0.11270	0.10849	0.10237	0.09413	0.08342	0.06972	0.05223	0.02968
19 0.95	0.06258	0.06217	0.06091	0.05876	0.05564	0.05143	0.04593	0.03885	0.02968	0.01746

Extrapolated values at the coarsest grid points (mesh nodes)

Array 4.6

theoretically much more accurate. For instance, one might take $n = 10 \times 2^l$ for l significantly greater than m. However, we will not do this, for our main purpose here is to illustrate the performance of iterative methods. Instead, we will solve the problem with only two nested equally spaced grids or meshes in both Cartesian directions, with mesh sizes $h = 1/20$ and $h = 1/40 \,(h_x = h_y = h)$, Then we apply Richardson's extrapolation (cf. [158]) combining the two results at the points of the coarsest grid (mesh nodes), in order to obtain a more accurate solution at those points.

Similarly to the presentation of the FDM, for convenience we use the same double subscript numbering of the unknowns as in the presentation of the FDM. In doing so, it can be easily checked that the SLAE to solve writes, for $n = 20$ or $n = 40$:

$$4u_{i,j} - u_{i,j-1} - u_{i,j+1} - u_{i-1,j} - u_{i+1,j} = h^2 \text{ for } i = 1, \ldots, n-1 \text{ and } j = 1, \ldots, n-1$$

with $u_{0,j} = u_{n,j} = u_{i,0} = u_{i,n} = 0$ for all i, j. However, owing to symmetry we can restrict the computations to the domain $(1/2, 1) \times (1/2, 1)$, for example by correcting the above equations for $i = n/2$ or $j = n/2$. More precisely, enforcing $u_{i,n/2-1} = u_{i,n/2+1}$ and $u_{n/2-1,j} = u_{n/2+1,j}$ for $i, j \geq n/2$, the equations become

$$4u_{i,n/2} - 2u_{i,n/2+1} - u_{i-1,n/2} - u_{i+1,n/2} = h^2 \text{ for } i = n/2+1, \ldots, n-1;$$

$$4u_{n/2,j} - 2u_{n/2+1,j} - u_{n/2,j-1} - u_{n/2,j+1} = h^2 \text{ for } j = n/2+1, \ldots, n-1;$$

$$4u_{n/2,n/2} - 2u_{n/2+1,n/2} - 2u_{n/2,n/2+1} = h^2.$$

Richardson's extrapolation is applied, assuming that the exact solution is smooth enough for the three methods to be of the second order in the point wise sense. This property will be formally confirmed in the next chapter (cf. Subsection 5.1.1). More precisely, denoting by $u^1_{i,j}$ and $u^2_{l,m}$ the approximations of $u(i/20, j/20)$ for $i, j = 1, \ldots, 19$ and of $u(l/40, m/40)$ for $l, m = 1, \ldots, 39$ respectively, obtained for $n = 20$ and $n = 40$, the following formula yields the extrapolated approximations $u^e_{i,j}$ at the grid point (mesh node) $(i/20; j/20)$, for $i, j = 1, \ldots, 19$:

$$u^e_{i,j} = (4u^2_{2i,2j} - u^1_{i,j})/3.$$

We selected the Gauss–Seidel method to solve the corresponding SLAEs, starting from a zero initial guess, and decided to sweep the grid points (nodes) by letting i vary from $n/2$ through $n - 1$ for each value of j from $n/2$ through $n - 1$. In doing so, taking into account the corrections for i or j equal to $n/2$, and incorporating the boundary conditions in an obvious manner, at the kth iteration we have for $j = n/2, \ldots, n-1$ and for each j, $i = n/2, \ldots, n-1$,

$$u^k_{i,j} = [u^k_{i,j-1} + u^{k-1}_{i,j+1} + u^k_{i-1,j} + u^{k-1}_{i+1,j} + h^2]/4.$$

In Arrays 4.4, 4.5 and 4.6, we display the approximate values at the coarse grid points (coarse mesh nodes), obtained for $n = 20$, for $n = 40$ and by Richardson's extrapolation, respectively, after running a FORTRAN 95 code. The iterations went on until the maximum absolute value of the relative increments between two successive iterations of all the unknowns–that is $|u^k_{i,j} - u^{k-1}_{i,j}|/|u^k_{i,j}|$–was less than a given tolerance, which was fixed to 10^{-5}. We found out that the total number of iterations necessary to satisfy the stop criterion was 325 for $n = 20$ and 1067 for $n = 40$. This is in agreement with the expected behaviour of the Gauss–Seidel method in this case: the more approximation points, the slower the convergence (cf. [199]).

As one can infer from the above tables, the values obtained with the two nested grids come closer to the extrapolated solution within an acceptable degree. More precisely, a three to four-digit accuracy is attained for meshes not so fine, to a great extent thanks to the rather good performance of the Gauss–Seidel method.

4.6   Exercises

4.1   Prove the validity of equation (4.8) for any bounded continuous function v in $\overline{\Omega}$ that vanishes on Γ, assuming that Ω is sufficiently smooth. Specify a possibly small value of the constant C_P depending only on Ω (hint: consider a rectangular domain R strictly containing Ω and the prolongation of v to R by zero outside Ω).

4.2   Under the same regularity assumptions on v in Exercise 4.1, prove the validity of equation (4.8) for any v, in case Ω is a rectangle and v vanishes on one of its edges.

4.3   Assume that the transition points between Γ_0 and Γ_1, if any, are mesh nodes. Show that any function in V_h (resp. V_{h0}) belongs to $H^1(\Omega)$ and vanishes on Γ_0 (resp. on Γ), that is, that V_h (resp. V_{h0}) is a subspace of V (resp. V_0).

4.4   Prove that the set of shape functions φ_j for $j = 1, 2, \ldots, K_h$ forms a partition of unity in Ω.

4.5   Check the equivalence between equations (4.36) and (4.37).

4.6   In Figure 4.18, two mesh triangles T_{ij}^1 and T_{ij}^2 assumed to be acute and having a common edge $\overline{S_i S_j}$ are represented, together with pertaining angles θ_{ij}^l and α_{ij}^l for $l = 1, 2$. Check that α_{ij}^l equals $2\theta_{ij}^l$.

4.7   We refer to Figure 4.19 and to the flux approximation in CVs involving triangles that have an edge on Γ_1. Check that the matrix for the Vertex-centred FVM corresponding to the Dirichlet-Neumann problem (equation (4.4)) is the same as for the $\mathcal{P}_1$ FEM.

4.8   Show that the $\vec{\eta}_k$s for $1 \leq k \leq L_h$ can be uniquely defined, so as to have the property specified in Subsection 4.4.2, without resorting to equation (4.49). Conclude that they are linearly independent and that matrix B_h in equation (4.48) is symmetric positive-definite (hint: first verify that whenever the four mean flux values across the edges of a rectangle R vanish, the field $\vec{p}_h$ of the form $(a_x^R x + b_x^R; a_y^R + b_y^R)$ in R having such mean fluxes vanishes identically in R).

4.9   Derive the equations for the rectangle-centred FVM corresponding to the boundary cells different from corner cells, and to the corner cells themselves.

4.10   Starting from the underlying RT_0 mixed FEM for rectangles, use the mid-point numerical quadrature formula to modify the equations of the Cell-centred FV scheme of Subsection 4.3.2 only for the boundary cells, in order to accommodate inhomogeneous Dirichlet boundary conditions $u = g_0$ on Γ. Asume that g_0 is regular enough to be the trace on Γ of a function in $H^1(\Omega)$–for instance, a continuous function everywhere on Γ (hint: first check that, in this case, the term $\int_\Gamma g_0(\vec{q}|\vec{\nu})ds$ must be added to the right side of the appropriate equation of the mixed variational formulation).

4.11   Check that if the triangulation is of the acute type, matrix A_h for the $\mathcal{P}_1$ FEM is diagonally dominant (hint: use the property of Exercise 4.4).

4.12   Study the FORTRAN subroutine supplied in Example 4.1, until all its steps and instructions are fully understood. Run the program and recover the solution $[1, 1, 1, 1, 1, 1]^T$ for the matrix and right-side vector given by $a_{i,i} = 3$ for $1 \leq i \leq 6$, $a_{i,i+2} = -2$ for $1 \leq i \leq 4$, $a_{i-2,i} = -1$ for $3 \leq i \leq 6$ and $a_{i,j} = 0$ otherwise; $b_1 = b_2 = 1$, $b_3 = b_4 = 0$ and $b_5 = b_6 = 2$.

4.13   Show that if the numerical integration technique specified in Example 4.2 is employed, the Five-point FDM and the Vertex-centred FVM coincide for every non-uniform rectangular grid (mesh), constructed in order to solve the Poisson problem described therein. Akin to Example 4.2, the triangular mesh is constructed from the partition of Ω into rectangular cells underlying the FD grid, in such a way that each rectangular cell is subdivided into two triangles by its diagonal with a positive slope.

5

Analyses in Two Space Variables

> Young man, in mathematics
> you don't understand things.
> You just get used to them.
>
> John Von Neumann

The purpose of this chapter is to carry out rigorous and complete reliability studies of numerical methods for solving PDEs posed in two-dimensional spatial domains with time dependence or not. The systematic employed in this aim is the one already exploited to derive stability, consistency and convergence results for a model ODE in Chapter 2 and also in Chapter 3 for time-dependent problems in one space variable. Intrinsically, the problems we deal with in this chapter are more complex than those addressed in both aforementioned chapters. In particular, in contrast to the one-dimensional case, the assumption that the **family of meshes** have suitable **regularity properties** becomes mandatory for the **convergence of the FEM in energy norms**. Nevertheless, an attempt is made to render the analyses as clear as possible, through the application of the very same principles, in a unified and structured manner.

Chapter outline: Section 5.1 is devoted to the convergence study of the four discretisation methods to solve the Poisson equation presented in Chapter 4. In Subsection 5.1.1, we derive, by standard arguments, results known to hold for the FDM for several decades. As far as the FEM and the FVM are concerned, novel techniques are developed, allowing us to obtain well-known results in Subsections 5.1.2 and 5.1.5, respectively, though using the same recipe that proved effective for the FDM. Numerical experiments are reported in Subsections 5.1.3, 5.1.4 and 5.1.6, in support of these theoretical results. In Section 5.2, analyses are conducted for the FDM and the FEM to solve the heat equation in two space variables. In Subsection 5.2.1, convergence results in the maximum norm are obtained for a combination of the Backward or the Forward Euler schemes for the time integration with the Five-point FD scheme for the space discretisation. In Subsection 5.2.2, a study of the Crank–Nicolson scheme is conducted, combined with a $\mathcal{P}_1$ FE space discretisation. Several results are obtained in spatial mean-square norms assorted with maximum or mean-squares norms in time. We conclude in Subsection

Numerical Methods for Partial Differential Equations:
An Introduction Finite Differences, Finite Elements and Finite Volumes, First Edition. Vitorino Ruas.
© 2016 John Wiley & Sons, Ltd. Published 2016 by John Wiley & Sons, Ltd.
Companion Website: www.wiley.com/go/ruas/numericalmethodsforpartial

5.2.3 with some remarks on the convergence of the FEM and FVM in the pointwise sense.

5.1 Methods for the Poisson Equation

5.1.1 Convergence of the Five-point FDM

In this subsection, we study the reliability of the Five-point FD scheme for the Poisson equation in two space variables presented in Chapter 4. We follow the same guidelines as in Chapter 2 for the Three-point FD scheme to solve two-point boundary value problems. For this reason, here the study will be conducted a little more briefly, while highlighting the aspects of the analysis inherent to the two-dimensional case.

We confine ourselves to the case of inhomogeneous Dirichlet boundary conditions. More precisely, we study the following Poisson equation:

Given bounded functions f and g defined everywhere in Ω and on Γ, respectively, we wish to find u such that

$$\begin{cases} -\Delta u = f \text{ in } \Omega, \\ u = g \text{ on } \Gamma. \end{cases} \tag{5.1}$$

First of all, recalling equation (4.5), we specify the local truncation error associated with a non-uniform rectangular grid for

The Five-point FD scheme for solving equation (5.1)

> Find $u_{i,j}$ for $1 \leq i \leq n_x - 1$ and $1 \leq j \leq n_y - 1$
> satisfying for i and j in the same range:
> $-\Delta_h[u_{i,j}] = f_{i,j}$
> where $\Delta_h[u_{i,j}] := \delta_x^2(u_{i,j}) + \delta_y^2(u_{i,j})$
> and $f_{i,j} := f(x_i, y_j)$, with
> $u_{i,0} = g(ih_x, 0);\ u_{i,n_y} = g(ih_x, L_y)$ for $0 \leq i \leq n_x$ and
> $u_{0,j} = g(0, jh_y);\ u_{n_x,j} = g(L_x, jh_y)$ for $0 \leq j \leq n_y$. $\tag{5.2}$

Recalling that $(x_i; y_j)$ are the coordinates of a generic grid-point $M_{i,j}$, as long as the solution u is three times continuously differentiable in the set $\overline{\Omega}$, the application of operator Δ_h defined by (4.5) to the $u(x_i, y_j)$'s instead of the $u_{i,j}$'s gives for $1 \leq i \leq n_x - 1$ and $1 \leq j \leq n_y - 1$,

$$\begin{cases} \Delta_h[u(x_i, y_j)] - \Delta[u(x_i, y_j)] = & \dfrac{h_{i+1}^2 \partial_{xxx} u(\xi_{i+1}, y_j) - h_i^2 \partial_{xxx}(\xi_i, y_j)}{6h_{i+1/2}} \\ & + \dfrac{k_{j+1}^2 \partial_{yyy} u(x_i, \chi_{j+1}) - k_j^2 \partial_{yyy}(x_i, \chi_j)}{6k_{j+1/2}}, \end{cases} \tag{5.3}$$

for suitable ξ_i and χ_j fulfilling $x_{i-1} \leq \xi_i \leq x_i$, $i = 1, \ldots, n_x$ and $y_{j-1} \leq \chi_j \leq y_j$, $j = 1, \ldots, n_y$. The reader can easily verify the validity of the above expression using standard Taylor expansions of u about point (x_i, y_j).

Now we define the **grid size** $h = \max[\max\limits_{1 \leq i \leq n_x} h_i, \max\limits_{1 \leq j \leq n_y} k_j]$ bound to be arbitrarily small.

We assume that there exists a **uniformity constant** $c > 0$ such that $h_i/h > c$ and $k_j/h > c$

for every i, j and for every n_x, n_y, thereby characterising a **quasi-uniform family of grids**. $R_{i,j}(u)$ being the residual defined below in connection with equation (5.3), it holds true the

Upper bound for the local truncation error of FD scheme (5.2)

Defining $R_{i,j}(u) := -\Delta_h[u(x_i, y_j)] - f(x_i, y_j)$ for $1 \leq i \leq n_x, 1 \leq j \leq n_y$,
$|R_{i,j}(u)| \leq C_D h \max_{(x,y)\in\overline{\Omega}} [|\partial_{xxx} u(x, y)| + |\partial_{yyy} u(x, y)|] \; \forall i, j$ $\qquad$ (5.4)

The constant C_D can be taken equal to c^{-1} multiplied by a fixed real number κ independent of h and u. While $1/3$ appears to be the actual value of κ, it is possible to do better as far as C_D is concerned, by noting that $h_{i+1/2} \geq \max[h_i, h_{i+1}]/2$ and $k_{j+1/2} \geq \max[k_j, k_{j+1}]/2$ (please check!).

Estimate (5.4) means that the order of consistency of the FD scheme (5.2) with a non-uniform grid to solve equation (5.1) is one. It is not a difficult matter to establish that in case the grid is uniform (i.e. $h_i = h_x$ for all i and $k_j = h_y$ for all j), provided the function u is four times continuously differentiable in Ω and its fourth-order derivatives are bounded in $\overline{\Omega}$, under the same assumption on the grids, the local truncation error can be bounded by an expression of the form $C_U h^2 \max_{(x,y)\in\overline{\Omega}} [|\partial_{xxxx} u(x, y)| + |\partial_{yyyy} u(x, y)|]$, where C_U is a constant independent of u and h.

This means that in this case, the method is second-order consistent.

Next, we consider the stability of scheme (5.2) in the maximum norm. First, we recall that the Poisson equation satisfies a maximum–minimum principle (see e.g. [109]), namely,

$$\max_{(x,y)\in\overline{\Omega}} u(x, y) \leq \max_{(x,y)\in\Gamma} u(x, y) \text{ if } f(x, y) \leq 0 \; \forall(x, y) \in \Omega,$$

$$\min_{(x,y)\in\overline{\Omega}} u(x, y) \geq \min_{(x,y)\in\Gamma} u(x, y) \text{ if } f(x, y) \geq 0 \; \forall(x, y) \in \Omega,$$

Similarly, the FD scheme (5.2) satisfies the following:

Discrete maximum–minimum principle for scheme (5.2)

$$\max_{(i;j) \mid (x_i, y_j)\in\overline{\Omega}} u_{i,j} \leq \max_{(i;j) \mid (x_i, y_j)\in\Gamma} u_{i,j} \text{ if } f_{i,j} \leq 0 \; \forall(x_i, y_j) \in \Omega,$$
$$\text{and} \qquad (5.5)$$
$$\min_{(i;j) \mid (x_i, y_j)\in\overline{\Omega}} u_{i,j} \geq \min_{(i;j) \mid (x_i, y_j)\in\Gamma} u_{i,j} \text{ if } f_{i,j} \geq 0 \; \forall(x_i, y_j) \in \Omega.$$

As a consequence, in case $f = 0$, it holds

Discrete maximum–minimum principle for the Laplace equation

If $\Delta_h[u_{i,j}] = 0$ for $1 \leq i \leq n_x$ and $1 \leq j \leq n_y$ then,
$$\max_{(i;j) \mid (x_i, y_j)\in\overline{\Omega}} u_{i,j} \leq \max_{(i;j) \mid (x_i, y_j)\in\Gamma} u_{i,j} \text{ and} \qquad (5.6)$$
$$\min_{(i;j) \mid (x_i, y_j)\in\overline{\Omega}} u_{i,j} \geq \min_{(i;j) \mid (x_i, y_j)\in\Gamma} u_{i,j}$$

Equation (5.5) can be established by means of arguments in all similar to those exploited below to prove equation (5.6).

Assume that the maximum (respectively minimum) value of $u_{i,j}$ is attained for a pair $(i_0; j_0)$ with $0 < i_0 < n_x$ and $0 < j_0 < n_y$. From equation (5.2), it is easy to see that

$$u_{i_0,j_0} = \alpha_1 u_{i_0-1,j_0} + \alpha_2 u_{i_0+1,j_0} + \alpha_3 u_{i_0,j_0-1} + \alpha_4 u_{i_0,j_0+1}, \tag{5.7}$$

where $0 < \alpha_k < 1$ for $k = 1, 2, 3, 4$ and $\sum_{k=1}^{4} \alpha_k = 1$. Thus, the value u_{i_0,j_0} is less (resp. greater) than or equal to the maximum (resp. minimum) of the four approximate values of u at the grid points specified in equation (5.7). But since the value on the left side of equation (5.7) is the maximum (resp. minimum) of all the $u_{i,j}$s by assumption, all the four values on the right side of equation (5.7) must be equal to u_{i_0,j_0}. If one of the corresponding grid points lie on Γ, the result is proved. In case not, we centre the scheme at one of the corresponding grid points, to conclude the same regarding its four neighbours. The process goes on successively in the same way until we necessarily reach a grid point on Γ. It follows that either all the values $u_{i,j}$ are equal, or the maximum (respectively minimum) is attained at a boundary point. Whatever the case, the result follows. From equation (5.6), we conclude that the following stability result holds true for the above FD scheme (4.6) applied to the Laplace equation (see e.g. [69]):

$$\max_{0 \le i \le n_x, 0 \le j \le n_y} |u_{i,j}| \le \max_{(i;j)|(x_i;y_j) \in \Gamma} |g(x_i, y_j)|,$$

The counterpart in the case where f does not vanish identically is

Stability inequality for scheme (5.2)

$$\boxed{\max_{0 \le i \le n_x, 0 \le j \le n_y} |u_{i,j}| \le \max_{(i;j) \,|\, (x_i;y_j) \in \Gamma} |g(x_i, y_j)| + C \max_{0 < i < n_x, 0 < j < n_y} |f_{i,j}|,} \tag{5.8}$$

where C is a constant depending on Ω but neither on h nor on f. The proof of equation (5.8) is quite similar to the one of equation (2.3): we apply equation (5.5) to an auxiliary problem involving the function $\phi(x, y) = (x/L_x)^2/2$. If ϕ is the solution to equation (4.1), then the right side must be equal to $-L_x^{-2}$. Moreover, the maximum (resp. minimum) of the corresponding boundary value is $1/2$ (resp. 0). Applying the scheme (2.2) to the exact values $\phi_{i,j} = \phi(x_i, y_j)$, since ϕ does not depend on y, recalling equation (4.5) after a few calculations, we obtain

$$-L_x^2 \Delta_h(\phi_{i,j} - 1/2) = -L_x^2 \delta_x^2(\phi_{i,j} - 1/2) = -\frac{x_{i+1} + x_i}{h_i + h_{i+1}} + \frac{x_i + x_{i-1}}{h_i + h_{i-1}} = -1 \tag{5.9}$$

The reader is advised to check the validity of equation (5.9), so that she or he makes sure to have properly understood FD schemes with a non-uniform grid.

Next, setting $F = \max_{(x;y) \in \overline{\Omega}} |f(x, y)|$, we define the $(n_x - 1)(n_y - 1)$-component vectors $\vec{w}^+$ and $\vec{w}^-$ by $w_{i,j}^+ = +u_{i,j} + FL_x^2[\phi_{i,j} - 1/2]$ and $w_{i,j}^- = -u_{i,j} + FL_x^2[\phi_{i,j} - 1/2]$, $i = 1, 2, \cdots, n_x, j = 1, 2, \ldots, n_y$. Then, owing to equations (5.9) and (4.6), we have

$$\left\{ \frac{2(2w_{i,j}^{\pm} - w_{i-1,j}^{\pm} - w_{i+1,j}^{\pm})}{h_i(h_i + h_{i+1})} + \frac{2(2w_{i,j}^{\pm} - w_{i,j-1}^{\pm} - w_{i,j+1}^{\pm})}{k_j(k_j + k_{j+1})} = \pm f_{i,j} - F \le 0. \right. \tag{5.10}$$

Let $g_{max} = \max_{(x;y) \in \Gamma} g(x, y)$, $g_{\min} = \min_{(x;y) \in \Gamma} g(x, y)$ and $g_{abs} = \max[|g_{max}|, |g_{\min}|]$. By virtue of equation (5.5), and since $0 \le \phi(x, y) \le 1/2$ for all $(x; y) \in \overline{\Omega}$, equation (5.10) implies that $w_{i,j}^+ \leq \max\limits_{(i;j)}[u_{i,j}] = g_{max} \leq g_{abs}$ and $w_{i,j}^- \leq \max\limits_{(i;j)}[-u_{i,j}] = -g_{min} \leq g_{abs}$ for every $(i;j)$. However, $\pm u_{i,j} = w_{i,j}^\pm + FL_x^2[1/2 - \phi_{i,j}] \leq w_{i,j}^\pm + FL_x^2/2 \leq g_{abs} + FL_x^2/2$, since $\phi(x,y) \geq 0$ for every $(x;y)$ and $w_{i,j}^\pm \leq g_{abs}$. It immediately follows that we have stability in the sense of equation (5.8) with $C = L_x^2/2$ (of course, we could have obtained $C = L_y^2/2$ as well).

Finally, in the same way as in the case of the FD scheme studied in Chapter 2, the stability result (5.8), together with the first-order estimate for the local truncation error (equation (5.4)), directly leads to an error estimate for the FD scheme with a non-uniform grid defined by equation (4.6), to approximate (4.3) in a rectangle $(0, L_x) \times (0, L_y)$: Provided the third order derivatives of u are continuous in $\overline{\Omega}$, we derive the

Pointwise error estimate for FD scheme (5.2)

$$\max\limits_{(i;j)|\; 0<i<n_x,0<j<n_y} |u_{i,j} - u(x_i, y_j)|$$
$$\leq CC_D h[\max\limits_{(x,y)\in\overline{\Omega}} |\partial_{xxx} u(x,y)| + \max\limits_{(x,y)\in\overline{\Omega}} |\partial_{yyy} u(x,y)|]. \tag{5.11}$$

Equation (5.11) means that under the underlying assumption on the differentiability of u, convergence holds in the following sense:

For points (x,y) that remain grid points, say (x_i, y_j), with i and j varying according to a sequence of successively refined grids, the corresponding sequence of approximations $u_{i,j}$ tends to $u(x,y)$ as h goes to zero.

By the same arguments, an estimate qualitatively equivalent to equation (5.11) applies to the modified FD scheme using fictitious grid points described in Subsection 4.2.2, in order to take into account Neumann boundary conditions. However, in order to attain second-order consistency and hence convergence in the case of uniform grids, it is wise to improve the modification of the Five-point FD scheme at points belonging to the portion of Γ, where the outer normal derivative of u is prescribed. One of the possible modifications and the study of its properties are proposed as Exercises 5.1 and 5.2.

5.1.2   Convergence of the $\mathcal{P}_1$ FEM

Now we endeavour to derive convergence results for the $\mathcal{P}_1$ FEM introduced in Section 4.3, as applied to the equations (4.3) or (4.4), for the sake of conciseness. The analysis is confined to convergence in a suitable mean-square sense like in Chapter 2. The extensions of the results established below to the case of pure homogeneous Neumann and inhomogeneous Dirichlet or Neumann boundary conditions are the object of Exercises 5.3, 6.1 and 6.2.

Hereafter, we use the terms 'mesh' and 'triangulation' as synonyms.

Before going into the convergence results, we recall that owing to the Friedrichs–Poincaré inequality (equation (4.8)), the semi-norm $|\cdot|_{1,2}$ of $H^1(\Omega)$ is a natural norm for the problem under study. Hence, for the sake of simplicity, in contrast to the one-dimensional case, we will consider the convergence of the FEM in the sense of this norm, instead of $\|\cdot\|_{1,2}$, without any essential loss. This means that we must ensure both the stability in terms of $|\cdot|_{1,2}$, and the consistency of the method.

Here, we could follow the same script as in Subsection 2.3.2. This means considering a problem more general than equation (4.18) for spaces V_{h0} or V_h, namely,

$$a(u_h, v) = F(v) + G(\mathbf{grad}\ v)\ \forall v \in V_{h0}\ \text{or}\ v \in V_h, \tag{5.12}$$

where $G(\vec{q}) = (\vec{g}|\vec{q})_0$, and $\vec{g} = (g_1; g_2)$ satisfies $g_1, g_2 \in L^2(\Omega)$. Then, using arguments entirely analogous to those developed below, we would come up with the (extended) stability result:

$$|u_h|_{1,2} \le C_P \parallel f \parallel_{0,2} + \parallel \vec{g} \parallel_{0,2}. \tag{5.13}$$

where C_P is the constant of equation (4.8) depending only on Ω.

However, in this section, we prefer to study the stability and the convergence of this method in the classical manner employed in most books on the subject. In this approach, stability is a consequence of the fact that

$$a(u_h, u_h) = |u_h|_{1,2}^2 \tag{5.14}$$

Owing to the Cauchy–Schwarz inequality followed by equation (4.8), we readily obtain

$$F(u_h) \le \parallel f \parallel_{0,2} \parallel u_h \parallel_{0,2} \le C_P \parallel f \parallel_{0,2} |u_h|_{1,2} \tag{5.15}$$

Combining equations (5.14) and (5.15), we immediately establish the

Stability inequality for the $\mathcal{P}_1$ FEM in the gradient mean–square norm

$$\boxed{|u_h|_{1,2} \le C_P \parallel f \parallel_{0,2}.} \tag{5.16}$$

Still in the classical approach, the consistency step directly leads to convergence without resorting to equation (5.16), and not even to equation (5.13). Here, we shall adopt a strategy which is halfway between such an approach and the above one. Before going into the analysis, we observe that we are assuming exact computation of the integral $\int_{\Omega} fv\ dxdy$ for every $v \in V_h$. The case where numerical quadrature is employed to compute such a term can be addressed very much like in the study of the vertex centred FVM in Part A of Subsection 5.1.5.

Like in Chapter 2, it is handy to introduce an interpolating function $\tilde{u}_h$ of u belonging to W_h, assuming that u is continuous in $\overline{\Omega}$. Unfortunately, in the two-dimensional case, a function $u \in H^1(\Omega)$ needs to be neither continuous nor bounded (cf. [182]) in Ω, and therefore in general such interpolate is not defined for a function u in this space. Nevertheless, according to the **Sobolev embedding theorem** (see e.g. [1]), whenever all the second-order partial derivatives of u belong to $L^2(\Omega)$, then $u \in C^0(\overline{\Omega})$, which suffices to define the interpolating function $\tilde{u}_h \in W_h$ satisfying $\tilde{u}_h(S) = u(S)$ at every node S of $\mathcal{T}_h$. For this reason, we assume henceforth that $u \in H^2(\Omega)$, where $H^2(\Omega)$ is the Sobolev space consisting of functions in $L^2(\Omega)$ whose partial derivatives of all order up to two also belong to $L^2(\Omega)$. Notice that this is always true whenever Ω is a convex polygon (cf. [92]), but even if it is not so the reader should keep in mind that nothing prevents such a regularity of u from holding for certain problem data.

As we saw in Section 4.3, $\tilde{u}_h \in V_h$ if $u \in V$ since Ω is a polygon by assumption. Moreover, thanks to the mesh positioning vis-à-vis the transition points between Γ_0 and Γ_1, we also have $\tilde{u}_h \in V_h$ whenever $u \in V$ (cf. Subsection 4.3.3).

Now, we note that $a(u_h - \tilde{u}_h, v) = F(v) - a(\tilde{u}_h, v) = F(v) + a(u - \tilde{u}_h, v) - a(u, v)$ for every v in V_{h0} (resp. V_h). Since $a(u, v) = F(v)$, we have

$$a(u_h - \tilde{u}_h, v) = a(u - \tilde{u}_h, v) \ \forall v \in V_h \ (\text{resp.} V_{h0}). \tag{5.17}$$

Taking $v = u_h - \tilde{u}_h$, recalling that $a(v, v) = |v|^2_{1,2}$ and noting that $a(w, v) \leq |w|_{1,2}|v|_{1,2}$ for all w and v in $H^1(\Omega)$, from equation (5.17) we immediately obtain $|u_h - \tilde{u}_h|_{1,2} \leq |u - \tilde{u}_h|_{1,2}$, or, using the triangle inequality,

$$|u_h - u|_{1,2} \leq 2|u - \tilde{u}_h|_{1,2} \tag{5.18}$$

In Chapter 2, we estimated the interpolation error in the one-dimensional case for a function u such that $u'' \in L^2(0, L)$, using some properties of integrals. In two or higher dimensions this is still possible, although in this case, in most books on mathematical aspects of the FEM, somewhat more abstract arguments are employed. Unfortunately, these lie beyond the scope of this book (see e.g. [45], [31] and [66]). Since here we have to resort to multiple integrals, clearly enough the calculations become more intricate than in a one-dimensional setting. Nevertheless, we attempt to carry them out as smoothly and as close as possible to the case of one space variable, while highlighting essential aspects inherent to interpolation error in two dimensions.

Noting that $|\tilde{u}_h - u|^2_{1,2} = \sum_{T \in \mathcal{T}_h} \int_T |\mathbf{grad}(\tilde{u}_h - u)(x, y)|^2 dx dy$, we only need to estimate a term of this sum for an arbitrary triangle T. Let h_T be the maximum edge length of triangle T (cf. Figure 5.2). We further define $h := \max_{T \in \mathcal{T}_h} h_T$, referred to as the **mesh size**. For convenience, we change the notations used in Chapter 4 to represent the vertices of T into $S_i = (x_i; y_i)$, $i = 1, 2, 3$, and set $u_i := u(S_i)$. Now for an arbitrary point $P = (x; y) \in T$, s_i is the abscissa along the axis $\overrightarrow{S_iP}$ with origin at S_i and directed towards P. Setting $r_i = |\overrightarrow{S_iP}|$, similarly to Section 2, we would like to use an expansion of $u(S_i)$ in terms of both $u(P)$ and first- and second-order derivatives of u along $\overrightarrow{S_iP}$. However, in contrast to the one-dimensional case, point values of the solution u or of its first-order derivatives, or yet line integrals of second-order derivatives of u may not be defined. In order to overcome these restrictions, we temporarily assume that u is twice continuously differentiable in Ω and that its second-order derivatives are bounded in $\overline{\Omega}$. Such a function space is usually denoted by $W^{2,\infty}(\Omega)$ (see e.g. [1]). As seen below, the final result requires only that $u \in H^2(\Omega)$. Hence, we can legitimately conjecture that the argument remains valid, by requiring only such a lower regularity of u (see also Remark 5.3).

Under the assumption that $u \in W^{2,\infty}(\Omega)$, first we observe that

$$u_i = u(P) - r_i \frac{\partial u}{\partial s_i}(P) + \int_0^{r_i} \frac{\partial^2 u}{\partial s_i^2}[x(s_i), y(s_i)]s_i ds_i, \tag{5.19}$$

which can be easily checked. Next, we multiply both sides of equation (5.19) with the barycentric coordinate λ_i^T at point P, and sum up the resulting relations from $i = 1$ through $i = 3$. Noting that $\tilde{u}_h$ in T equals $\sum_{i=1}^3 u_i \lambda_i^T$, and recalling that $\sum_{i=1}^3 \lambda_i^T \equiv 1$, this gives, for every point $P \in T$:

$$[\tilde{u}_h - u](P) = -\sum_{i=1}^3 \lambda_i^T(P) \left\{ r_i \frac{\partial u}{\partial s_i}(P) - \int_0^{r_i} \frac{\partial^2 u}{\partial s_i^2}[x(s_i), y(s_i)]s_i ds_i \right\}. \tag{5.20}$$

Now, we note that

$$\sum_{i=1}^{3} \lambda_i^T(P) r_i \frac{\partial u}{\partial s_i}(P) = \sum_{i=1}^{3} \lambda_i^T(P)(\mathbf{grad}\ u(P)|\overrightarrow{S_i P}).$$

Then, denoting by $\mathbf{x}$ and $\mathbf{x}_i$ the position vectors of P and S_i respectively, this easily gives

$$\sum_{i=1}^{3} \lambda_i^T(P) r_i \frac{\partial u}{\partial s_i}(P) = -\left(\mathbf{grad}\ u(P)\Big|\sum_{i=1}^{3} \mathbf{x}_i \lambda_i^T(P)\right) + (\mathbf{grad}\ u(P)|\mathbf{x}). \qquad (5.21)$$

But owing to Property 4 of λ_i^T, $\sum_{i=1}^{3} \mathbf{x}_i \lambda_i^T(P) \equiv \mathbf{x}$, and hence equation (5.21) yields

$$\sum_{i=1}^{3} \lambda_i^T(P) r_i \frac{\partial u}{\partial s_i}(P) = 0. \qquad (5.22)$$

On the other hand, since the integrand on the right side of equation (5.19) is independent of P, the derivatives of the corresponding integral with respect to either x or y are determined by differentiating only its upper limit r_i. Then,

$$\mathbf{grad} \int_0^{r_i} \frac{\partial^2 u}{\partial s_i^2}[x(s_i), y(s_i)]s_i ds_i = \left\{\frac{(x - x_i; y - y_i)}{r_i} \frac{\partial^2 u}{\partial s_i^2}[x(s_i), y(s_i)]s_i\right\}_{/s_i = r_i}$$

or

$$\mathbf{grad} \int_0^{r_i} \frac{\partial^2 u}{\partial s_i^2}[x(s_i), y(s_i)]s_i ds_i = (x - x_i; y - y_i)\frac{\partial^2 u}{\partial s_i^2}(P). \qquad (5.23)$$

Putting together equations (5.20), (5.22) and (5.23), we have for all $P \in T$:

$$\mathbf{grad}[\tilde{u}_h - u](P) = \sum_{i=1}^{3} \lambda_i^T(P)(x - x_i; y - y_i)\frac{\partial^2 u}{\partial s_i^2}(P)$$

$$+ \sum_{i=1}^{3} \mathbf{grad}\lambda_i^T \int_0^{r_i} \frac{\partial^2 u}{\partial s_i^2}[x(s_i), y(s_i)]s_i ds_i. \qquad (5.24)$$

From equation (5.24), we easily derive

$$\int_T |\mathbf{grad}(\tilde{u}_h - u)(x, y)|^2 dx dy \leq 2[E_1(u, T) + E_2(u, T)] \qquad (5.25)$$

where

$$\begin{cases} E_1(u, T) := \int_T \left[\sum_{i=1}^{3} \lambda_i^T(x, y)|(x - x_i; y - y_i)| \left|\frac{\partial^2 u}{\partial s_i^2}(x, y)\right|\right]^2 dx dy \\ E_2(u, T) := \int_T \left[\sum_{i=1}^{3} |\mathbf{grad}\lambda_i^T| \left|\int_0^{r_i} \frac{\partial^2 u}{\partial s_i^2}[x(s_i), y(s_i)]s_i ds_i\right|\right]^2 dx dy. \end{cases} \qquad (5.26)$$

At this point, we recall that the second-order derivatives of u can be grouped in the form of a 2×2 second-order symmetric tensor $H(u)$ called

The Hessian of a function u

$$H(u) := \begin{bmatrix} \dfrac{\partial^2 u}{\partial x^2} & \dfrac{\partial^2 u}{\partial x \partial y} \\ \dfrac{\partial^2 u}{\partial x \partial y} & \dfrac{\partial^2 u}{\partial y^2} \end{bmatrix} \tag{5.27}$$

Denoting by $\vec{f}_i$ the unit vector in the direction of $\overrightarrow{S_i P}$, we also recall that $\dfrac{\partial^2 u}{\partial s_i^2} = (H(u) \vec{f}_i | \vec{f}_i)$.
Thus, applying a couple of times the Cauchy–Schwarz inequality, we obtain for $i = 1, 2, 3$:

$$\left| \frac{\partial^2 u}{\partial s_i^2} \right| \leq |H(u)\vec{f}_i| \leq \left[\left(\frac{\partial^2 u}{\partial x^2} \right)^2 + 2 \left(\frac{\partial^2 u}{\partial x \partial y} \right)^2 + \left(\frac{\partial^2 u}{\partial y^2} \right)^2 \right]^{1/2}. \tag{5.28}$$

For convenience, we introduce the following notation:

$$\| H(u) \|_{0,T} := \left\{ \int_T \left[\left(\frac{\partial^2 u}{\partial x^2} \right)^2 + 2 \left(\frac{\partial^2 u}{\partial x \partial y} \right)^2 + \left(\frac{\partial^2 u}{\partial y^2} \right)^2 \right] dx dy \right\}^{1/2}. \tag{5.29}$$

Making use of the above tools, estimating the term $E_1(u, T)$ is not so difficult. Indeed, applying again the Cauchy–Schwarz inequality, we obtain

$$E_1(u, T) \leq \int_T \left\{ \sum_{i=1}^3 [\lambda_i^T(x, y)]^2 |\overrightarrow{S_i P}|^2 \right\} \left\{ \sum_{i=1}^3 \left[\frac{\partial^2 u}{\partial s_i^2}(x, y) \right]^2 \right\} dx dy \tag{5.30}$$

Noticing that $|\overrightarrow{S_i P}| \leq h_T \ \forall P \in T$, and $0 \leq \lambda_i^T \leq 1$ for all i, we further derive

$$E_1(u, T) \leq 3 h_T^2 \sum_{i=1}^3 \int_T \left[\frac{\partial^2 u}{\partial s_i^2}(x, y) \right]^2 dx dy. \tag{5.31}$$

Finally, using equations (5.28) and (5.29), we come up with

$$E_1(u, T) \leq 9 h_T^2 \| H(u) \|_{0,T}^2. \tag{5.32}$$

Now, we switch to the estimate of $E_2(u, T)$. Referring to Figure 5.1, it is convenient to employ a local polar coordinate systems (r_i, φ_i) with origin S_i for each term of the summation over i in the expression of $E_2(u, T)$. The azimuthal coordinate φ_i is measured in the counterclockwise sense by sweeping the interior of T starting from the appropriate edge of T. Let e_i be the edge of T opposite to S_i and l_i be its length. Furthermore, θ_i represents the angle of T with vertex S_i and $d_i(\varphi_i)$ is the length of the (straight) segment joining S_i to e_i along the line defined by a given φ_i, $0 \leq \varphi_i \leq \theta_i$.

Reporting to Subsection 4.3.4, we know that the gradient of λ_i^T is orthogonal to e_i, and $|\mathbf{grad} \ \lambda_i^T| = 1/h_i$, where h_i is (the length of) the height of T through S_i. It follows that

$$E_2(u, T) \leq \int_T \left\{ \sum_{i=1}^3 h_i^{-2} \left| \int_0^{r_i} \frac{\partial^2 u}{\partial s_i^2}[x(s_i), y(s_i)] s_i ds_i \right| \right\}^2 dx dy. \tag{5.33}$$

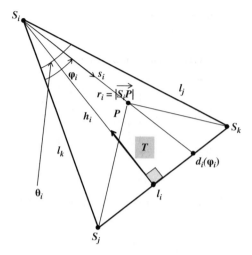

Figure 5.1   Triangle T and pertaining lengths, angles and polar coordinate system $(S_i; r_i, \varphi_i)$

Then, using the (Cauchy–Schwarz) inequality $\int_0^r g(s)ds \leq \sqrt{r}\{\int_0^r [g(s)]^2 ds\}^{1/2}$, we obtain

$$E_2(u,T) \leq \int_T \left\{ \sum_{i=1}^3 \frac{r_i}{h_i^2} \int_0^{r_i} \left| \frac{\partial^2 u}{\partial s_i^2}[x(s_i), y(s_i)] \right|^2 s_i^2 ds_i \right\} dxdy. \tag{5.34}$$

Next we switch to the local coordinate systems for each value of i, to easily derive from equation (5.34),

$$E_2(u,T) \leq \sum_{i=1}^3 \left\{ \int_0^{\theta_i} \int_0^{d_i(\varphi_i)} \frac{r_i}{h_i^2} \left[\int_0^{r_i} \left| \frac{\partial^2 u}{\partial s_i^2}(s_i, \varphi_i) \right|^2 s_i^2 ds_i \right] r_i dr_i d\varphi_i \right\} \tag{5.35}$$

Working out trivial upper bounds of the right side of equation (5.35), we further obtain

$$E_2(u,T) \leq \sum_{i=1}^3 \frac{a_i^2}{h_i^2} \left\{ \int_0^{\theta_i} \int_0^{d_i(\varphi_i)} \left[\int_0^{d_i(\varphi_i)} \left| \frac{\partial^2 u}{\partial s_i^2}(s_i, \varphi_i) \right|^2 s_i ds_i \right] r_i dr_i d\varphi_i \right\}. \tag{5.36}$$

where a_i stands for the length of the longest edge of T intersecting at S_i. Since the integral with respect to s_i in equation (5.36) does not depend on r_i, this inequality further yields

$$E_2(u,T) \leq \sum_{i=1}^3 \frac{a_i^4}{2h_i^2} \int_0^{\theta_i} \int_0^{d_i(\varphi_i)} \left| \frac{\partial^2 u}{\partial s_i^2}(s_i, \varphi_i) \right|^2 s_i ds_i d\varphi_i. \tag{5.37}$$

The double integral in equation (5.37) is an area integral sweeping the whole T. Recalling equations (5.28) and (5.29), this means that

$$E_2(u,T) \leq \| H(u) \|_{0,T}^2 \sum_{i=1}^3 \frac{a_i^4}{2h_i^2} \tag{5.38}$$

Now denoting by ζ_i the smallest angle of T whose vertex is not S_i, it is easy to see that $h_i = a_i \sin \zeta_i$. Then from equation (5.38), we obtain

$$E_2(u, T) \leq \| H(u) \|_{0,T}^2 \sum_{i=1}^{3} \frac{a_i^2}{2(\sin \zeta_i)^2}. \tag{5.39}$$

Finally, putting together equations (5.25), (5.32) and (5.39), setting $\theta_T = \min_{1 \leq i \leq 3} \theta_i$, and noting that $\sin \theta_i \geq \sin \theta_T$ for $i = 1, 2, 3$, we come up with

$$\int_T \| \mathbf{grad}(u - \tilde{u}_h) \|_{0,2}^2 \leq h_T^2 [18 + 3(\sin \theta_T)^{-2}] \| H(u) \|_{0,T}^2. \tag{5.40}$$

Now let $\theta_{\min} > 0$ be the smallest angle of all mesh elements. Noting that $\max_{T \in \mathcal{T}_h} h_T$ is the mesh size h, setting $c_h = \sin \theta_{\min}$, and recalling equation (5.18), it is readily seen from equation (5.40) that

$$|u - u_h|_{1,2} \leq C_h h \| H(u) \|_{0,2} \tag{5.41}$$

where $C_h = 2\sqrt{18 + 3c_h^{-2}}$.

In principle, the constant C_h depends on the mesh; and hence, unless we know more about this dependence, equation (5.41) does not allow us to conclude the convergence rate of the method, and not even convergence at all. However, if we assume that we are computing with a family of meshes whose mesh sizes decrease indefinitely in such a manner that the smallest angle of all the elements of a mesh in this family are uniformly bounded below for every mesh, say by θ_0, then $C_h \leq 2\sqrt{18 + 3(\sin \theta_0)^{-2}}$. In this case, the method is first-order convergent in the norm $| \cdot |_{1,2}$.

Remark 5.1 *Method's first order of consistency lies behind the above analysis. Indeed, if u_h is replaced by $\tilde{u}_h$ in equation (4.14) (respectively equation (4.15)), the **variational residual** $a(\tilde{u}_h, v) - F(v)$ is bounded above by $C_h h \| H(u) \|_{0,2} |v|_{1,2}/2 \, \forall v \in V_h$ (respectively V_{h0}). The reader may check this assertion as Exercise 5.4.∎*

Remark 5.2 *It is also worth highlighting an interesting facet to error estimate (5.41): if the solution of the Poisson problem happens to be linear in the whole domain, then $u_h \equiv u$. This is surely the least to be expected of an approximation method based on (piecewise) linear functions.∎*

Remark 5.3 *Proving the validity of relations such as inequality (4.8) and identity (5.19) making use of expressions defined only for smooth functions is admissible under certain conditions. The indispensable requirement is that the final result be well defined for the target class of less smooth functions. The principle lying behind such a technique is the so-called density argument: If a relation defined for functions in a certain class E is valid for every function in a sub-class $F \subset E$ containing functions that can be as close as one wishes to any function in E in an appropriate norm, then it also holds for every function in E. We refer to any book on elementary functional analysis for more details and examples.∎*

Remark 5.4 *In Johnson's celebrated book [110], an estimate of the error of interpolation in the maximum norm, of functions in $W^{2,\infty}(\Omega)$ and their gradient, with functions in FE subspaces is given. Similarly to the case of equation (5.41), the proof is based only on elementary* *geometry and integral and differential calculus, so that no concepts of functional analysis are needed. These estimates in reference [110] are of the same nature as equation (5.41) by changing the mean-square norms into maximum norms. However, in contrast, they apply only to functions in $W^{2,\infty}(\Omega)$, and hence no density argument is necessary. Notice that such esti-mates in L^∞-norms could be used for the purpose of the error estimates for the $\mathcal{P}_1$ FEM, assuming that u belongs indeed to $W^{2,\infty}(\Omega)$, and this is the purpose of Exercise 5.5.■*

In order to be more specific about the geometric conditions to be fulfilled so as to guarantee the convergence of the FEM, we must exploit the concepts of regular and quasi-uniform families of meshes. Like in the analysis leading to equation (5.41), this can be done by exploiting the ratio between the length of an element edge and the corresponding height. However, in a more classical way (see e.g. [45]), we shall rather classify meshes through the ratio between the square of the maximum edge length and the area of an element. Clearly enough, using elementary geometry, it is possible to find equivalent conditions associated with these two different ratios, and the reader is encouraged to derive them.

Referring to Figure 5.2 we define ρ_T to be the radius of the circle inscribed in $T \in \mathcal{T}_h$, and referring to Figure 5.1 we let $p := \sum_{i=1}^3 l_i$. Since $2h_T \le p \le 3h_T$ and $area(T) = p\rho_T/2$, we have $2h_T/(3\rho_T) \le h_T^2/area(T) \le h_T/\rho_T$. For this reason, we can equivalently work with the ratios $h_T^2/area(T)$ and h_T/ρ_T. Since most authors use the latter ratio, we will do the same.

Now supposedly a family of triangulations $\{\mathcal{T}_h\}_h$ is employed to solve a corresponding sequence of problems (4.18) for $V_n = V_h$ or $V_n = V_{h0}$, with $h \le h_0, \forall \mathcal{T}_h$. We assume that such a family is

A regular family of meshes $\{\mathcal{T}_h\}_h$

There exists a constant $c_0 > 0$ such that $\frac{\rho_T}{h_T} \ge c_0 \ \forall T \in \mathcal{T}_h$ and $\forall h$.   (5.42)

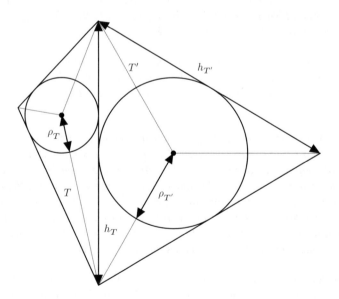

Figure 5.2   Maximum edge lengths and radii of inscribed circles for triangles T, T'

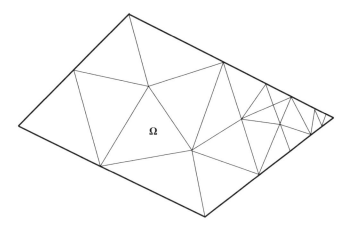

Figure 5.3   A typical mesh in a regular family not suited to a quasi-uniform family

where c_0 is called the **regularity constant of the family**. By an argument from elementary geometry, we can interpret this property as follows (the reader is advised to justify the assertion as Exercise 5.6): there exists an angle $\theta_0 > 0$ such that the angles of the triangles remain bounded below and above by θ_0 and $\pi - 2\theta_0$, respectively, for every mesh in the family of triangulations in use, however fine it is. Notice that, even if they satisfy equation (5.42), the triangles within a given mesh may have very different sizes as illustrated in Figure 5.3. That is why a more strict definition of mesh regularity is often employed (cf. [94]). This gives rise to the definition of a family of triangulations sometimes quite abusively called *uniform*, since in this case neither the sizes nor the shapes of the mesh elements can differ so much from each other. Here, we will rather call it,

A quasi-uniform family of meshes $\{\mathcal{T}_h\}_h$

There exists a constant $c > 0$ such that $\frac{\rho_T}{h} \geq c \; \forall T \in \mathcal{T}_h$ and $\forall h$.

$$(5.43)$$

where c is the **uniformity constant** of the family. It should be understood that in any regular family of triangulations (and hence in any quasi-uniform family) h can be as close to zero as one wishes.

Remark 5.5 *Similarly to the case of FE meshes, a regular family of FD grids could be defined, but it is not so meaningful. This is because any local refinement necessarily impacts far-away grid portions, where stretched patterns will appear.* ∎

Anyway, as long as the family of triangulations in use is regular, we can assert that $\mathcal{C}_h$ depends on the constant c_0 and Ω but not on h [45]. For this reason, one had better use the notation $\mathcal{C}$ for this constant in this case. Then definitively, provided the family of triangulations $\{\mathcal{T}_h\}_h$ is regular and the solution of the Poisson equation belongs to $H^2(\Omega)$, the following result applies to the $\mathcal{P}_1$ FEM:

Error estimate for the $\mathcal{P}_1$ FEM with a regular $\{\mathcal{T}_h\}_h$

> There exists a mesh-independent constant $\mathcal{C} > 0$ such that
> $$|u - u_h|_{1,2} \leq \mathcal{C}h \parallel H(u) \parallel_{0,2}.$$
(5.44)

Since supposedly the regular family of triangulations in use allows for h to tend to zero, equation (5.44) implies that under the above assumptions on u, the sequence of FE approximations $\{u_h\}_h$ converges to u in the $|\cdot|_{1,2}$ norm. The corresponding convergence rate is one in view of the power of h in the error estimate.

Notice that, like in Chapter 2, thanks to the Friedrichs–Poincaré inequality, a qualitatively identical estimate applies to the error measured in the L^2-norm. However, here again, this is not satisfactory, at least for equation (4.3), in which case the $\mathcal{P}_1$ FEM is actually second-order convergent in this norm, as long as Ω is convex. This result is due to a classical **duality argument** known as **the Aubin–Nitsche trick**, whose presentation as such does not quite fit the scope of this book. The interested reader can find more details about it in the text-books on mathematics of FE quoted above.

Here, we endeavour to simplify the argument by adopting an approach analogous to the one we used to prove the error estimate (2.51).

First, we recall that the solution w of the Poisson equation $-\Delta w = g$ with homogeneous Dirichlet boundary conditions $w = 0$ on Γ does belong to $H^2(\Omega)$ for any $g \in L^2(\Omega)$ if Ω is convex (see e.g. [92]). It is also possible to prove that in this case there exists a constant C_H depending only on Ω such that [92]

$$\parallel H(w) \parallel_{0,2} \leq C_H \parallel \Delta w \parallel_{0,2}.$$
(5.45)

In order to be more convincing about the validity of equation (5.45), we demonstrate that, as long as we consider functions vanishing on Γ sufficiently smooth, we can take $C_H = 1$ for a rectangular Ω. Indeed, let w have continuous partial derivatives up to the second order in $\overline{\Omega} = [0, L_x] \times [0, L_y]$, and have third-order derivatives in $L^2(\Omega)$. Clearly enough, the first-order tangential derivatives of w along Γ (i.e., the derivative with respect to x (resp. to y) along the edges given by $y = 0$ or $y = L_y$ (respectively by $x = 0$ or $x = L_x$)), vanish identically. Then, by First Green's identity, we have

$$\int_\Omega |\Delta w|^2 dxdy = \oint_\Gamma (\Delta \, w\vec{\nu}|\mathbf{grad} \, w)ds - \int_\Omega (\mathbf{grad}\Delta w|\mathbf{grad} \, w) \, dxdy.$$

Next, we note that $\mathbf{grad}\Delta w \equiv \Delta \mathbf{grad} \, w$. Hence, applying the same Green's identity the other way around, we obtain (please check!):

$$\int_\Omega |\Delta w|^2 dxdy = \oint_\Gamma (\Delta \, w\vec{\nu}|\mathbf{grad} \, w) \, ds - \oint_\Gamma (H(w)\vec{\nu}|\mathbf{grad} \, w)ds + \int_\Omega |H(w)|^2 dxdy.$$

Now, owing to the vanishing tangential derivatives of w up to the second order, both integrals along Γ in the above expression cancel out (checking this assertion is proposed to the reader as Exercise 5.7). This establishes the claimed result.

Remark 5.6 *The identity $\parallel \Delta w \parallel_{0,2} = \parallel H(w) \parallel_{0,2}$ for a rectangular Ω conforms to the inequality $\parallel H(w) \parallel_{0,2} \leq \parallel \Delta w \parallel_{0,2}$ for $w \in H^2(\Omega) \cap H_0^1(\Omega)$ that holds for some kinds of domains (cf. [34]).*■

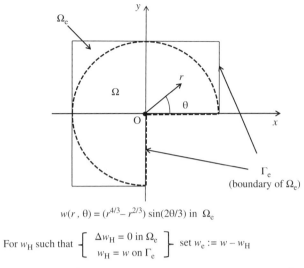

$$w(r, \theta) = (r^{4/3} - r^{2/3}) \sin(2\theta/3) \text{ in } \Omega_e$$

For w_H such that $\left\{ \begin{array}{l} \Delta w_H = 0 \text{ in } \Omega_e \\ w_H = w \text{ on } \Gamma_e \end{array} \right\}$ set $w_e := w - w_H$

w_e lies in $H_0^1(\Omega_e)$, Δw_e lies in $L^2(\Omega_e)$, but the components of $H(w_e)$ do not lie in $L^2(\Omega_e)$

Figure 5.4   Defining $w_e \in H_0^1(\Omega_e)$ s.t. $\Delta w_e \in L^2(\Omega)$ but $H(w_e) \notin L^{2 \times 2}(\Omega_e)$ in hexahedron Ω_e

Remark 5.7 *Unfortunately, for general convex polygons Ω, the argument exploited above does not apply to all functions $w \in H_0^1(\Omega)$ whose Laplacian belongs to $L^2(\Omega)$, let alone non-convex polygons, for which equation (5.45) does not even hold [92]. An example of the violation of equation (5.45) for the non-convex domain defined in polar coordinates by $\Omega = \{(r; \theta) \,|\, r < 1 \text{ and } 0 < \theta < 3\pi/2\}$ is the function $w(r, \theta) := (r^{4/3} - r^{2/3}) \sin(2\theta/3)$. As one can easily check, $w \in H_0^1(\Omega)$ and $\Delta w = 4r^{-2/3} \sin(2\theta/3) \in L^2(\Omega)$. However, no second-order derivative of w belongs to $L^2(\Omega)$, and hence equation (5.45) cannot hold. Referring to Figure 5.4, one can extend the function w to the complement of Ω with respect to the non-convex polygon $\Omega_e := \{(x; y) \,|-1 < x, y < 1 \text{ and } xy > 0 \text{ if } x > 0\}$, and then construct another function $w_e \in H_0^1(\Omega_e)$ whose Laplacian belongs to $L^2(\Omega_e)$, although its second-order derivatives do not.$\blacksquare$*

Thanks to equation (5.45), and using ingredients analogous to those in Subsection 2.4.2, the following

Error estimate for the $\mathcal{P}_1$ FEM in the L^2-norm when Ω is convex

> There exists a mesh-independent constant $\mathcal{C}_0 > 0$ such that
> $$\| u - u_h \|_{0,2} \leq \mathcal{C}_0 h^2 \, \| H(u) \|_{0,2}.$$

(5.46)

can be proved. The validity of equation (5.46) can be established as Exercise 5.8.

Remark 5.8 *The assumption that the family of triangulations be regular is not really necessary to derive estimates (5.44) and (5.46) according to reference [107], but only the upper bound $\pi - 2\theta_0$ for the angles, as illustrated by an example in reference [182]. However, in* *practice in most situations, it is not recommended to use triangles that become too flat owing to the lack of a lower bound for their angles, since this may bring about unnecessary computational effort.* ■

5.1.3   Example 5.1: Solving the Poisson Equation with Neumann Boundary Conditions

In this example, discrete counterparts of the Poisson equation with Neumann boundary conditions are investigated. Although the general considerations on the numerical resolution of such a problem are basically the same for all discretisation methods, for the sake of conciseness we confine ourselves to the $\mathcal{P}_1$ FEM. The problem to solve is: Given continuous functions f and g_1 defined in a bounded domain Ω of $\mathfrak{R}^2$ and on its boundary Γ, respectively, find a function $u \in H^1(\Omega)$ such that

$$\begin{cases} -\Delta u = f \text{ in } \Omega \\ \partial_\nu u = g_1 \text{ on } \Gamma. \end{cases} \tag{5.47}$$

First, we note that the Neumann problem (5.47) for the Laplacian operator in any bounded domain Ω is only solvable if $\int_\Omega f \, dx dy + \oint_\Gamma g_1 \, ds = 0$. This is because, by the divergence theorem,

$$\int_\Omega f \, dx dy = -\int_\Omega \Delta u \, dx dy = -\oint_\Gamma \partial_\nu u \, ds = -\oint_\Gamma g_1 \, ds.$$

Assuming that the above condition is fulfilled, the solution u to equation (5.47) is not unique, for $u + d$ is also a solution for any constant d. If one wishes to uniquely determine u, it is necessary to fix the additive constant d up to which it is defined. This can be achieved by enforcing $\int_\Omega u \, dx dy = 0$. We can also prescribe $u(P) = 0$ at a certain point of $P \in \Omega$, among other possibilities.

In this example, we take $g_1 \equiv 0$ and $\Omega = (0, 1) \times (0, 1)$. The equivalent variational form of equation (5.47) is

$$\int_\Omega (\mathbf{grad} \ u | \mathbf{grad} \ v) dx dy = \int_\Omega fv \, dx dy \ \forall v \in H^1(\Omega).$$

The discrete counterpart of this problem is obtained by replacing u with u_h belonging to the $\mathcal{P}_1$ FE space W_h and $H^1(\Omega)$ with W_h. The compatibility condition analogous to $\int_\Omega f \, dx dy = 0$ ensuring the existence of u_h is equivalent to

$$\sum_{i=1}^{K_h} \int_\Omega (\mathbf{grad} \ u_h | \mathbf{grad} \ \varphi_i) dx dy = 0,$$

where K_h is the total number of nodes (i.e. of $\mathcal{P}_1$ shape functions φ_i), according to the following argument. First, we note that this relation holds true since the shape functions φ_i form a partition of unity (please check!), and hence the sum of their gradients is the null vector everywhere in Ω. On the other hand, by this property of the φ_i's, the SLAE's right side also fulfils $\sum_{i=1}^{K_h} \int_\Omega f \varphi_i dx dy = \int_\Omega f \, dx dy = 0$ by assumption.

For the reason to be specified hereafter, a well-balanced choice of the arbitrary additive constant d_h, up to which u_h is defined, consists of prescribing $\int_{\Omega_h} u_h \, dx dy = 0$. However, in an FE solution procedure, this relation couples all the unknowns in the underlying SLAE, and

Table 5.1 Absolute errors in three different norms for $\mathcal{P}_1$ FE solutions of equation (5.47)

$h \quad \rightarrow$	1/4	1/8	1/16	1/32	1/64
$H^1 \quad \rightarrow$	$0.34482 \times 10^{+0}$	$0.17939 \times 10^{+0}$	0.90802×10^{-1}	0.45572×10^{-1}	0.22811×10^{-1}
$L^2 \quad \rightarrow$	0.27599×10^{-1}	0.73573×10^{-2}	0.18751×10^{-2}	0.47135×10^{-3}	0.11803×10^{-3}
$L^\infty \quad \rightarrow$	0.49925×10^{-1}	0.21113×10^{-1}	0.75295×10^{-2}	0.24529×10^{-2}	0.75607×10^{-3}

hence the matrix band structure is lost. Furthermore, we have to eliminate one of the remaining equations in the linear system. Since the choice of this equation is essentially arbitrary, the SLAE becomes lopsided anyway. That is why we shall rather prescribe $u_h(P) = 0$ at a given mesh node P. Notice that it is easy to determine a posteriori another solution, say $\acute{u}_h$, satisfying the zero integral condition. Indeed, it suffices to take $d_h := -\int_\Omega u_h dx dy / area(\Omega)$ and set $\acute{u}_h := u_h + d_h$ to have $\int_\Omega \acute{u}_h = 0$.

Now assume for simplicity that in the adopted systematic of node numbering, P is the K_h-th node. This means that φ_{K_h} yields the K_h-th equation in the original linear system. Hence, by prescribing $u(P) = 0$, we reduce the problem to solve into a SLAE with $N_h := K_h - 1$ unknowns and N_h equations. But then the remaining N_h shape functions no longer form a partition of unity, which means that the compatibility issue is thus resolved.

In our computations, we took the manufactured solution $u(x, y) = (1 - x^2)^2 + (1 - y^2)^2$ having a zero normal derivative on the boundary of the unit square Ω, satisfying $u(1, 1) = 0$. We also prescribed $u_h(1, 1) = 0$, but do not compare u with u_h in the sense of $L^2(\Omega)$ or $L^\infty(\Omega)$. In the comparisons in those norms, we rather take $\acute{u}_h := u_h + d_h$ instead of u_h, where d_h is determined by enforcing $\int_\Omega [u - \acute{u}_h] dx dy = 0$. This is enough to ensure second-order convergence of $\acute{u}_h$ to u in $L^2(\Omega)$, according to variants of the convergence analyses conducted in Subsection 5.1.2 (cf. Exercise 5.30).

The numerical solution was carried out with uniform meshes constructed in the same way as in Example 4.2. In Table 5.1, we give the absolute errors in the (semi-)norm of $H^1(\Omega)$, in the norm of $L^2(\Omega)$ and in the L^∞-norm restricted to the nodes. Each column corresponds to a value of $h = 1/2^m$ for $m = 2, 3, 4, 5, 6$. The numerical values indicate that from each mesh level to the next finer level, roughly the errors in the sense of H^1 decrease by a factor of two, while the errors in the sense of L^2 decrease by a factor of four. This behaviour suggests the method's first- and second-order convergence in these norms, respectively, like in the case of Dirichlet boundary conditions studied in Subsection 5.2.1. Establishing these properties is proposed to the reader as Exercise 5.30. Notice that the approximations in the discrete L^∞-norm appear to decrease by a factor close to 3.2.

5.1.4 Example 5.2: Convergence of the $\mathcal{P}_1$ FEM to Non-smooth Solutions

The aim of this example is to show that, whenever it is not possible to ensure that the Hessian of the solution to a Poisson equation in a bounded two-dimensional domain Ω belongs to $[L^2(\Omega)]^{2 \times 2}$, the order of convergence of the sequence of solutions obtained by the $\mathcal{P}_1$ FEM with a family of nested quasi-uniform meshes is strictly less than one in the space $H^1(\Omega)$ and strictly less than two in $L^2(\Omega)$. The study is purely numerical, and the author would like to acknowledge that the corresponding results obtained with **FreeFem++** software were supplied by Dr Sílvia Barbeiro from the University of Coimbra, Portugal [22].

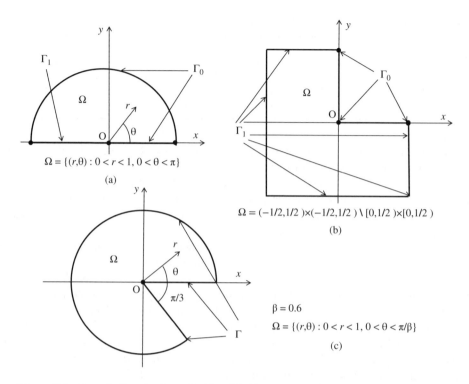

$\Omega = \{(r,\theta) : 0 < r < 1, 0 < \theta < \pi\}$

(a)

$\Omega = (-1/2,1/2\,)\times(-1/2,1/2\,) \setminus [0,1/2\,)\times[0,1/2\,)$

(b)

$\beta = 0.6$

$\Omega = \{(r,\theta) : 0 < r < 1, 0 < \theta < \pi/\beta\}$

(c)

Figure 5.5   Domain Ω and subsets Γ_0, Γ_1 of Γ: (a) test case 1; (b) test case 2; (c) test case 3

Three problems were solved: the first one with mixed Dirichlet–Neumann boundary conditions in a convex domain, and the second and third ones with pure Dirichlet boundary conditions in non-convex domains. In all cases, the exact solution belongs to a space strictly contained in $H^1(\Omega)$, but strictly containing $H^2(\Omega)$.

Test case 1: The homogeneous mixed Dirichlet–Neumann problem was solved in the domain defined in polar coordinates (r, θ) by $\Omega = \{(r, \theta) : 0 < r < 1; \; 0 < \theta < \pi\}$. The boundary portion Γ_1 of Γ is the segment given by $-1 < x < 0$ and $y = 0$, Γ_0 being the complementary portion of Γ (cf. Figure 5.5a). The right side is the function $f = 6r^{1/2}\sin(\theta/2) \in L^2(\Omega)$. However, the solution $u \in H^1(\Omega)$ given by $u(r, \theta) = (1 - r^2)r^{1/2}\sin(\theta/2)$ does not belong to $H^2(\Omega)$, as the reader can verify by calculating the L^2-norm of the two terms whose sum is the Laplacian of u in polar coordinates, namely, $r^{-1}\partial(r\partial u/\partial r)/\partial r$ and $r^{-2}\partial^2 u/\partial\theta^2$. In Figure 5.6a, we show log-log plots of the errors of the numerical solution as h decreases, measured in both the L^2-norm $\|\cdot\|_{0,2}$ and the H^1-(semi-)norm $|\cdot|_{1,2}$ for meshes successively refined respecting the minimum angle criterion. The slopes of the straight lines best fitting the error decrease in the least-square sense, indicate that for this problem the P_1 FEM converges in $H^1(\Omega)$ and $L^2(\Omega)$ with orders roughly equal to 0.51874 and 1.0043, respectively. These results suggest that the expected regularity of the solution u, halfway between $H^2(\Omega)$ and $H^1(\Omega)$, dictates such rather low convergence rates. Actually, this poorer regularity can be explained by the fact that a transition point between Γ_0 and Γ_1 lies away from a boundary corner (cf. [92]).

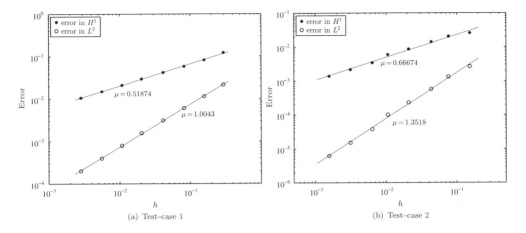

(a) Test–case 1   (b) Test–case 2

Figure 5.6   Errors in the L^2-norm and in the H^1-(semi-)norm for test cases 1 and 2

Test case 2: This time, a Laplace equation with inhomogeneous mixed Dirichlet–Neumann boundary conditions was solved in the domain defined in Cartesian coordinates by $\Omega = \{(x,y) := (-1/2, 1/2) \times (-1/2, 1/2) \setminus [0, 1/2) \times [0, 1/2)\}$. The boundary portion Γ_0 of Γ is the union of the segments given by $0 \leq x \leq 1/2$; $y = 0$ and $0 \leq y \leq 1/2$; $x = 0$, Γ_1 being the complementary set of Γ_0 (cf. Figure 5.5b). We are given a harmonic solution u defined in polar coordinates by $u(r, \theta) = r^{2/3} \sin(2\theta/3 - \pi/3)$, the inhomogeneous boundary conditions being adjusted accordingly. This function belongs to $H^1(\Omega)$ but does not lie in $H^2(\Omega)$, since the integral of the square of both $r^{-1}\partial(r\partial u/\partial r)/\partial r$ and $r^{-2}\partial^2 u/\partial\theta^2$ are unbounded in Ω. In Figure 5.6b, we supply log-log plots of the errors of the numerical solution as h decreases, measured in both the L^2-norm $\|\cdot\|_{0,2}$ and the H^1-semi-norm $|\cdot|_{1,2}$ for successively refined uniform meshes. Now the slopes of the straight lines best fitting the error decrease in the least-squares sense, suggest that for this problem, the $\mathcal{P}_1$ FEM converges in $H^1(\Omega)$ and $L^2(\Omega)$ with orders roughly equal to 0.66674 and 1.3518, respectively. These convergence rates, a little better than in the case of test case 1, are in good agreement with the expected intermediate regularity of the solution, when the transition between Γ_0 and Γ_1 occurs at boundary corners (cf. [92]).

Test case 3: This test was inspired by a problem proposed in reference [34]. The Poisson equation with homogeneous Dirichlet boundary conditions was solved in the domain defined in polar coordinates by $\Omega = \{(r, \theta) : 0 < r < 1, 0 < \theta < \pi/\beta\}$, where β is a real number in the interval $(1/2, 1)$ (cf. Figure 5.5c). For a right side $f = (4\beta + 4)r^\beta \sin(\beta\theta) \in L^2(\Omega)$, the solution is given by $u(r, \theta) = (1 - r^2)r^\beta \sin(\beta\theta)$, which belongs to $H^1(\Omega)$ for $1/2 < \beta < 1$. But since both $r^{-1}\partial u/\partial r$ and $\partial^2 u/\partial r^2$ behave like $r^{\beta-2}$ near $r = 0$, for $1/2 < \beta < 1$, u cannot be in $H^2(\Omega)$. Notice, however, that the space u belongs to approaches $H^2(\Omega)$ as β tends to one (cf. [92]). Log-log plots of the errors of the numerical solution as h decreases are displayed in Figures 5.7a and 5.7b for five different values of β. On the left we show errors measured in the H^1-(semi-)norm, and on the right errors measured in the L^2-norm for meshes successively refined respecting the minimum angle criterion. The decrease of the fitted slopes of the error in the least-squares sense indicates that for this problem, the $\mathcal{P}_1$ FEM converges in $H^1(\Omega)$ and $L^2(\Omega)$ with minimum orders for $\beta = 0.51$, slightly greater than 0.5 and 1, respectively. The

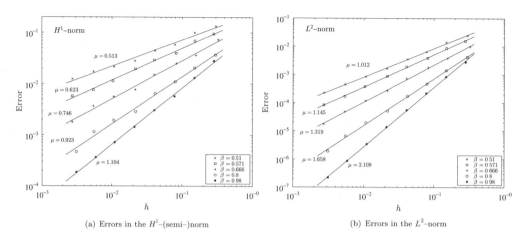

(a) Errors in the H^1-(semi-)norm   (b) Errors in the L^2-norm

Figure 5.7   Errors for test case 3 with different values of β

maximum orders in turn exceed a little bit 1 and 2, respectively, for $\beta = 0.98$. These results are also in good agreement with the increasing regularity of u as β approaches 1 (cf. [92]).

As a final comment on this example, let us add that the singular behaviour of exact solutions about certain points, such as those in the above test cases, motivated the proposal of the XFEM (for extended FEM, cf. [141]). The idea is to modify the standard FE shape functions by enriching the corresponding space with functions mimicking the solution singularity, though only locally. In such a way, optimal approximation orders can be recovered. For further details, the author refers the reader to references [23] and references therein among many others.

5.1.5   Convergence of the FVM

In this subsection, we address the convergence of both FV schemes considered in detail in Section 4.4. The analysis corresponding to the Vertex-centred FVM can be viewed as a variant of the one conducted in the previous subsection.

On the other hand, we give a new convergence analysis of the Cell-centred FVM in the sense of $L^2(\Omega)$. Although the study is limited to rectangular cells, it can be easily extended to the case of arbitrary triangular cells of the strict acute type. For the sake of brevity, we refrain from studying the latter case here and refer to reference [68] for the analysis of the FVM for cells of arbitrary polygonal shape.

A. The Vertex-centred FVM:

The shortest path leading to convergence results for this method is to mimic as much as possible the analysis carried out in the previous subsection. This is because for a given triangulation, the matrix of the SLAE corresponding to this method is the same as for the $\mathcal{P}_1$ FEM, as we saw in Subsection 4.4.1. This means that the difference between the FE and FV solutions can only be due to the way the right side $\int_\Omega fv\, dxdy$ is handled, v being either a piecewise linear function or a constant function in every CV. That is why we first consider different possibilities to approximate the ith component b_i of the right-side vector in connection with such an integral, by assuming that f is continuous in $\overline{\Omega}$.

In the case of the FVM, $f(S_i)area(V_i)$ is the most usual approximation to the integral $\int_{V_i} f\,dxdy$ (i.e. the ith component b_i of the right-side vector). On the other hand, we note that in the case of the FEM, both the matrix and the right-side vector are assembled element by element. Hence, a practical way to approximate the integral of fv in Ω is to use in each element T the quadrature formula $\int_T fv\,dxdy \approx f(G_T)v(G_T)area(T)$, where G_T is the centroid of T and v is a $\mathcal{P}_1$ FE test function. Such a one-point quadrature formula is not so rough, for it does not affect the method's final order of approximation in the norm of $H^1(\Omega)$ (checking this assertion is the object of Exercise 5.9). However, it is as simple to select the trapezoidal rule extended to a triangular domain, to compute the right side. This corresponds to taking $\int_T fv\,dxdy \approx \sum_{r=1}^3 f(S_r^T)v(S_r^T)area(T)/3$, the S_r^T's being the vertices of T. We recall that the component b_i is given by the integral of $f\varphi_i$ in Ω, where in any triangle T containing the ith mesh vertex S_i, φ_i equals the barycentric coordinate of T related to S_i. Then, clearly enough, b_i is replaced by the summation of terms of the form $f(S_i)area(T)/3$, over such elements T, with $S_i \equiv S_r^T$ for a certain r, $1 \leq r \leq 3$. In this case, either approximation of the exact right side does not downgrade the order of convergence of the FEM (cf. Exercise 5.10). It turns out that such a way to determine the right-side vector for the FEM could also be well-suited to the Vertex-centred FVM. More precisely, this is true in the case of meshes having particular uniformity properties; for example, meshes for which the area of each V_i happens to be equal to the area of the polygons $\tilde{V}_i$ about S_i, whose vertices are the mid-points of the edges containing S_i and the centroids of the elements having S_i as a vertex. In Figure 5.8, both types of CV are displayed for the uniform mesh of a rectangular domain consisting of equal right triangles with edge length h.

In this particular example, the areas of both types of CV are the same, except for those associated with corner nodes. In fact, the areas of both V_i and $\tilde{V}_i$ equal h^2 for any node in the

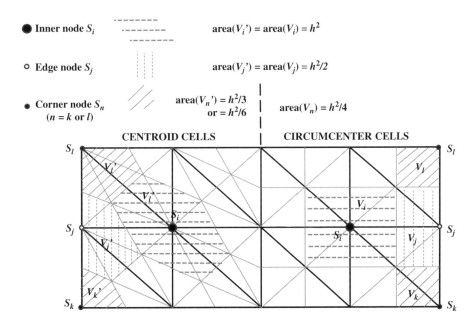

Figure 5.8   Cells $\tilde{V}_i$ and V_i based on centroids and circumcentres of a triangular FE mesh

interior of the domain, which to the least perfectly suits the case of homogeneous Dirichlet boundary conditions. However, in the general case these areas are different. Therefore, we will rather modify the weights of $1/3$ in the above trapezoidal rule for triangles, in such a way that the right sides of both methods coincide.

The crucial issue is to evaluate how the differences between both areas influence the FE approximate solution in the general case. However, even if we are able to establish that it is negligible regarding the final convergence properties, this does not mean that we can simply extend to the case of the Vertex-centred FVM the convergence results thus derived for the $\mathcal{P}_1$ FEM. This is because, although the corresponding SLAEs will coincide in this case, both methods are conceptually different. However, we can attempt to do so by mixing up concepts as we do below, although we should warn the reader about the fact that it is not advisable to adopt such a strategy in a systematic way.

For several reasons, it is necessary to use tools fitting better the FVM in the study of these methods. For instance, in contrast to the FEM, the FV approximate solution u_h has got no gradient in this operator's strict sense. The way out could be deriving error estimates in the mean-square sense for the solution rather than its gradient, or for a flux-based discrete gradient. Here, we will work with the discrete gradient of piecewise constant functions v, and by these means we will bring the FEM and the FVM closer to each other. The discrete gradient can be viewed as the two-dimensional analog of the discrete derivative v'^h, v being a piecewise constant function of a single variable (cf. Subsection 1.4.3).

Referring to Figure 4.16, we define the (discrete) first-order derivative of the FV solution u_h in $T \cap V_i$ in the direction $\overrightarrow{S_i S_j}$ by $D_{T,i,j} u_h := (u_j - u_i)/l_{ij}$, where $l_{ij} = |\overrightarrow{S_i S_j}|$. Now if S_k is the third vertex of T, analogously the (discrete) first-order derivative of u_h in $T \cap V_i$ in the direction $\overrightarrow{S_i S_k}$ is given by $D_{T,i,k} u_h := (u_k - u_i)/l_{ik}$, where $l_{ik} = |\overrightarrow{S_i S_k}|$. Referring to Figure 5.9, let us define the unit vectors $\vec{e}_{im} := \overrightarrow{S_i S_m}/|\overrightarrow{S_i S_m}|$ for $m = j$ or $m = k$, together with the direct orthonormal basis $\mathcal{B}_{T.i} = \{\vec{e}_{im}, \vec{f}_{im}\}$ of $\Re^2$ for $m = j$ or $m = k$, m being chosen in such a way that $(\vec{f}_{im}|\vec{e}_{il}) < 0$ with $l = k$ or $l = j$, $l \neq m$ ($m = k$ in Figure 5.9).

By definition, the **discrete gradient** of u_h denoted by $\mathbf{grad}_h u_h$ restricted to $T \cap V_i$ is given by $\mathcal{M}_{T,i}[D_{T,i,m} u_h, D_{T,i,l} u_h]^T$, where $\mathcal{M}_{T,i}$ is the inverse of the matrix transforming the basis $\{\vec{e}_{im}, \vec{e}_{il}\}$ into $\mathcal{B}_{T,i}$. The Jacobian of the inverse transformation matrix $\mathcal{M}_{T,i}$ is

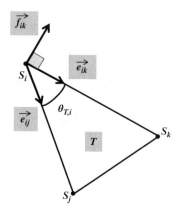

Figure 5.9   Orthonormal basis $[\vec{e}_{ik}, \vec{f}_{ik}]$ associated with edge $\overline{S_i S_k}$ of triangle T

$-[\sin\theta_{T,i}]^{-1}$, where $\theta_{T,i}$ is the angle of T whose vertex is S_i (cf. Figure 5.9). The underlying covariant components of $\mathbf{grad}_h u_h$ expressed in the direct oblique basis $\{\vec{e}_{im}, \vec{e}_{il}\}$ are $(D_{T,i,m}u_h, D_{T,i,l}u_h)$. Then, assuming again that u is continuous in $\overline{\Omega}$ and denoting by $\bar{u}_h$ the piecewise constant function whose value in each V_i is $u(S_i)$, by a straightforward calculation using elementary tensor calculus[1], we obtain

$$sin^2\theta_{T,i}|[\mathbf{grad}_h(u_h - \bar{u}_h)]|_{(V_i\cap T)}|^2 = [D_{T,i,j}(u_h - \bar{u}_h)]^2$$
$$+ [D_{T,i,k}(u_h - \bar{u}_h)]^2 + 2cos\theta_{T,i}D_{T,i,j}(u_h - \bar{u}_h)D_{T,i,k}(u_h - \bar{u}_h). \qquad (5.48)$$

Now, let us assume that the nodal values of the FV solution u_h coincide with those of the $\mathcal{P}_1$ FE solution obtained with the same mesh. Notice that this will be the case if we use a modified trapezoidal rule to compute $\int_T fv\,dxdy$ with $v = \lambda_r^T$, $r = 1, 2, 3$ in every triangle T, in such a way that the ith component of the right-side vector $\int_\Omega f\varphi_i dxdy$ is approximated by $f(S_i)area(V_i)$ (cf. (4.21)). This can be accomplished by associating the weights $area(V_i \cap T)/area(T)$ to $[f\varphi_i](S_i)$ instead of $1/3$, which does no harm to the order of convergence of the $\mathcal{P}_1$ FEM in the $|\cdot|_{1,2}$-norm, as long as $f \in C^1(\overline{\Omega})$. The result we need to support this assertion is an estimate for the difference between $\int_\Omega fv\,dxdy$ and its approximation $I_h(fv)$ for v in V_h or in V_{h0}, given by the modified trapezoidal rule, namely, the

Quadrature rule equalising $\mathcal{P}_1$ FEM and Vertex-centred FVM right sides

$$\boxed{\int_\Omega fv\,dxdy \approx I_h(fv) := \sum_{i=1}^{N_h} f(S_i)v(S_i)area(V_i).} \qquad (5.49)$$

In order to accomplish this, at first we note that $I_h(fv) = \sum_{T\in\mathcal{T}_h}\sum_{r=1}^3 f(S_r^T)v(S_r^T)area(V_r^T)$, V_r^T being the subset $V_i \cap T$ if S_r^T is node S_i. Let us estimate

$$E^T(fv) := \int_T fv\,dxdy - \sum_{r=1}^3 f(S_r^T)v(S_r^T)area(V_r^T).$$

for an arbitrary mesh triangle T. We have

$$\begin{cases} E^T(fv) \le E_1^T(fv) + E_2^T(fv), \text{ where} \\ E_1^T(fv) := \left|\sum_{r=1}^3 \left[\int_{V_r^T} fv(S_r^T)dxdy - f(S_r^T)v(S_r^T)area(V_r^T)\right]\right| \\ E_2^T(fv) := \left|\int_T fv\,dxdy - \sum_{r=1}^3 \int_{V_r^T} fv(S_r^T)dxdy\right|. \end{cases} \qquad (5.50)$$

In order to estimate $E_1^T(fv)$, we first observe that by an elementary Taylor expansion this term can be bounded as follows:

$$E_1^T(fv) \le h\,\|\,\mathbf{grad}\,f\|_{0,\infty} \sum_{r=1}^3 |v(S_r^T)|area(V_r^T).$$

Since the mesh is of the acute type, $area(V_r^T) \le area(T)$ for $r = 1, 2, 3$. Moreover, $\sum_{r=1}^3 |v(S_r^T)| \le \left[3\sum_{r=1}^3 v(S_r^T)^2\right]^{1/2}$. Thus, taking $p_r = q_r = v(S_r^T)$ in equation (4.32), we

---
[1] For a quick refreshment on pertaining concepts, the reader can consult reference [205].

conclude that $\int_T v^2 dxdy \geq area(T) \sum_{r=1}^{3} v(S_r^T)^2/12$. It easily follows that

$$E_1^T(fv) \leq 6h\sqrt{area(T)} \parallel \mathbf{grad}\ f \parallel_{0,\infty} \left[\int_T v^2 dxdy\right]^{1/2} \tag{5.51}$$

As for $E_2^T(fv)$, we proceed as follows. First, we use the identity $\sum_{s=1}^{3} \lambda_s^T \equiv 1$ and the fact that $v = \sum_{s=1}^{3} v(S_s^T)\lambda_s^T$, to derive

$$E_2^T(fv) = \left| \sum_{r=1}^{3} \int_{V_r^T} f \sum_{s=1}^{3} \left[v(S_s^T)\lambda_s^T - v(S_r^T)\lambda_s^T \right] dxdy \right| \tag{5.52}$$

Now, since v is a linear function in T for every pair $(r;s)$, it trivially holds that $|v(S_s^T) - v(S_r^T)| \leq h|\mathbf{grad}\ v|$. Plugging this into equation (5.52) and using again the fact that $\sum_{s=1}^{3} \lambda_s^T \equiv 1$ we readily obtain

$$E_2^T(fv) \leq h \left| \sum_{r=1}^{3} \int_{V_r^T} |f| |\mathbf{grad}\ v| dxdy \right| = h \int_T |f| |\mathbf{grad}\ v| dxdy. \tag{5.53}$$

The remainder of the task is now only a matter of summing up over the mesh triangles the bounds of $E_1^T(fv)$ and $E_2^T(fv)$ given by equations (5.51) and (5.53), and then applying the Cauchy–Schwarz inequality to the resulting summations. In this manner, we have estimated the additional error of the $\mathcal{P}_1$ FEM due to numerical integration of the right side using a modified trapezoidal rule, so as to make it coincide with the Vertex-centred FVM. More specifically, recalling the constant C_P of equation (4.8), for all v in V_h or in V_{h0}, according to the problem under study, we have established an

Error estimate for the quadrature rule (5.49) to compute the $\mathcal{P}_1$FEM right side

$$\left| \int_\Omega fv\ dxdy - I_h(fv) \right| \leq h[6C_P\sqrt{area(\Omega)} \parallel \mathbf{grad}\ f \parallel_{0,\infty} + \parallel f \parallel_{0,2}] |v|_{1,2}. \tag{5.54}$$

Taking equation (5.54) for granted, the reader is expected to complete the necessary calculations leading to the desired error estimate as Exercise 5.11. More precisely, recalling that $\tilde{u}_h$ denotes the $\mathcal{P}_1$-interpolate of u upon mesh $\mathcal{T}_h$, she or he should apply the same arguments as in the previous subsection to derive

$$|\underline{u}_h - \tilde{u}_h|_{1,2} \leq \underline{\mathcal{C}}_h h[\parallel H(u) \parallel_{0,2} + \parallel \mathbf{grad}\ f \parallel_{0,\infty}]. \tag{5.55}$$

where $\underline{\mathcal{C}}_h$ has the same nature as $\mathcal{C}_h$, but in contrast to the latter the former constant takes into account the additional error in the approximation of $(f|v)_0$ given by equation (5.54) (cf. Exercise 5.11).

Now for an integer m, $1 \leq m \leq 5$, let $[m]_3 := m$ if $m \leq 3$ and $[m]_3 := m - 3$ if $m > 3$. Setting $j = [i + 1]_3$ and $k = [i + 2]_3$ for $1 \leq i \leq 3$ and $\hat{u}_h := \underline{u}_h - \tilde{u}_h$, we claim that (cf. Exercise 5.12):

$$
|[\mathbf{grad}\hat{u}_h]_{|T}|^2 = \frac{1}{3}\sum_{i=1}^{3}\frac{1}{sin^2\theta_{T,i}} \times \left\{ \left[\frac{\hat{u}_h(S_j) - \hat{u}_h(S_i)}{\ell_{ij}}\right]^2 + \left[\frac{\hat{u}_h(S_k) - \hat{u}_h(S_i)}{\ell_{ik}}\right]^2 \right.
$$

$$
\left. -2cos\theta_{T,i}\left[\frac{\hat{u}_h(S_j) - \hat{u}_h(S_i)}{\ell_{ij}}\right]\left[\frac{\hat{u}_h(S_k) - \hat{u}_h(S_i)}{\ell_{ik}}\right]\right\}. \tag{5.56}
$$

Actually, each term in the above summation has the same value, for it simply expresses $|\mathbf{grad}[\underline{u}_h - \tilde{u}_h]_{|T}|^2$ in one out of three different bases, namely, $\mathcal{B}^{T,i}$ for $i = 1, 2, 3$. Thus, recalling the definition of $D_{T,i,m}$, for $m = j$ or $m = k$, comparing equation (5.56) with (5.48), and taking into account that $T = \cup_{i=1}^{3}V_i \cap T$ and the V_i's do not overlap, by a straightforward argument we come up with

$$
\int_T |\mathbf{grad}_h(u_h - \bar{u}_h)|^2 \, dxdy = \int_T |\mathbf{grad}(\underline{u}_h - \tilde{u}_h)|^2 \, dxdy \; \forall T \in \mathcal{T}_h. \tag{5.57}
$$

It follows from equations (5.55) and (5.57) that the following estimate in the mean-square sense of discrete gradients holds:

$$
\| \, \mathbf{grad}_h(u_h - \bar{u}_h) \, \|_{0,2} \leq \underline{\mathcal{C}}h[\| \, H(u) \, \|_{0,2} + \| \, \mathbf{grad}\,f \, \|_{0,\infty}], \tag{5.58}
$$

provided we are using a regular family of meshes, $\mathcal{C}$ being a mesh-independent constant. Otherwise stated, – (5.58) means that, as h goes to zero, the discrete gradient of the FV solution converges linearly in the norm of $L^2(\Omega)$ to the corresponding discrete gradient of the interpolate of u at the mesh nodes, which is constant in the corresponding CVs.

The interpretation of such a convergence result may be unclear in practice. That is why many authors (cf. [68]) derive error estimates in the L^2-norm from equation (5.58), by exploiting the fact that for a suitable mesh-independent constant C_D the following:

Discrete Friedrichs–Poincaré inequality for the Vertex-centred FVM

$$
\boxed{\| \, v \|_{0,2} \leq C_D \, \| \, \mathbf{grad}_h v \|_{0,2},} \tag{5.59}
$$

holds, for any function v which is piecewise constant in the CVs and vanishes in the CVs having a non-empty intersection with Γ_0.

The proof of inequality (5.59) can be carried out in full rather easily thanks to the fact that, provided all the meshes in use are of the acute type, there exists a quantifiable mesh independent constant C_V, such that for every function $\underline{v} \in V_h$ we have

$$
\left[\sum_{k=1}^{N_h} [\underline{v}(S_k)]^2 area(V_k)\right]^{1/2} \leq C_V \, \| \, \underline{v} \|_{0,2}. \tag{5.60}
$$

It is handier to prove inequality (5.60) by considering at first a generic triangle $T \in \mathcal{T}_h$. For $r = 1, 2, 3$, let V_{k_r} be the three CVs having a non-empty intersection with T, and S_{k_r} be the corresponding mesh nodes. Recalling equation (4.32), let us take $p_r = q_r = \underline{v}(S_{k_r})$ for $r = 1, 2, 3$. For this choice, we clearly have

$$\int_T v(x, y)^2 dx dy \geq \sum_{r=1}^{3} [\underline{v}(S_{k_r})]^2 area(T)/12. \tag{5.61}$$

Since for meshes of the acute type, a CV V_k about a vertex S_k has empty intersections only with mesh triangles that do not contain S_k, we necessarily have

$$\sum_{T|\ S_k \in T} area(T) \geq area(V_k).$$

Thus, summing up over $T \in \mathcal{T}_h$ both sides of relation (5.61), we easily obtain equation (5.60), with $C_V = \sqrt{12}$.

Notice that any function in V_h belongs to $H^1(\Omega)$ and vanishes on Γ_0. Hence, by the Friedrichs–Poincaré inequality, we have

$$\| \underline{v} \|_{0,2} \leq C_P |\underline{v}|_{1,2} \ \forall v \in V_h. \tag{5.62}$$

On the other hand, letting v be the piecewise constant function, whose value in each CV V_k is $\underline{v}(S_k)$ for an arbitrary function $\underline{v} \in V_h$, we can write

$$\| v \|_{0,2} = \left[\sum_{k=1}^{N_h} [\underline{v}(S_k)]^2 area(V_k) \right]^{1/2}. \tag{5.63}$$

Thus by virtue of equation (5.60),

$$\| v \|_{0,2} \leq C_V \| \underline{v} \|_{0,2}. \tag{5.64}$$

On the other hand, as a direct by-product of the calculations leading to equation (5.57), we have (please check!),

$$|\underline{v}|_{1,2} = \| \mathbf{grad}_h v \|_{0,2}. \tag{5.65}$$

Thus combining equations (5.64), (5.62) and (5.65) we have proved equation (5.59) with $C_D = C_P C_V$.

Remark 5.9 *A discrete Friedrichs–Poincaré inequality associated with the Cell-centred FVM for rectangular meshes is proved hereafter, without resorting to equation (4.8).* ∎

The application of equation (5.59) in our case directly leads to first-order convergence results in the L^2 norm. Indeed, setting $\overline{C} = \underline{C} C_D$ we first note that

$$\| u_h - \bar{u}_h \|_{0,2} \leq \overline{C} h [\| H(u) \|_{0,2} + \| \mathbf{grad} \ f \|_{0,\infty}], \tag{5.66}$$

Next, observing that

$$\| u_h - u \|_{0,2} \leq \| u_h - \bar{u}_h \|_{0,2} + \| \bar{u}_h - u \|_{0,2}, \tag{5.67}$$

all that is left to do is estimating an interpolation error in $L^2(\Omega)$, namely, the L^2-norm of $u - \bar{u}_h$. Here, the polar coordinate trick similar to the one employed in the previous subsection does not lead to an optimal estimate of $\| u - \bar{u}_h \|_{0,2}$. As we should point out, the classical interpolation theory (cf. [45]) guarantees that, if the family of meshes in use is regular, there exists a constant C_2 such that

$$\| u - \bar{u}_h \|_{0,2} \leq C_2 h |u|_{1,2} \tag{5.68}$$

However, the proof of this result requires some mathematical tools that are not supposed to be mastered by most of this book's readers.

Remark 5.10 *For the applications in view in this chapter, the standard result on the interpolation error can be stated as follows (cf. [45]): Let ω be a bounded subset of $\mathfrak{R}^2$ and h_ω be the maximum distance between two points in $\overline{\omega}$ (i.e. the diameter of ω). Let also ρ_ω be the maximum radius of the circles contained in $\overline{\omega}$. If g is a function in $H^l(\omega)$ for $l = 1$ or $l = 2$, and g_h is an approximation of g such that $g_h \equiv g$ whenever g in ω is a polynomial of degree less than or equal to k, for $k = 0$ or $k = 1$, then there exists a constant C_{kl} such that*

$$\| g - g_h \|_{l,2} \leq C_{kl} h_\omega^{k+1} \rho_\omega^{-l} |g|_{k+1,2,\omega}, 0 \leq l \leq k,$$

*where $|\cdot|_{1,2,\omega}^2 := \int_\omega |\mathbf{grad}(\cdot)|^2 dx dy$ and $|\cdot|_{2,2,\omega}^2 := \int_\omega |H(\cdot)|^2 dx dy$. A relevant application of this result is the approximation of $\int_\Omega fv\, dx dy$ for a constant function v in each CV in the case of the Vertex-centred FVM, or piecewise linear in the case of the $\mathcal{P}_1$ FEM, by $\int_\Omega f_h v_h\, dx dy$, where f_h is the constant function in each CV V_i, given by $f_h := \sum_{i=1}^{K_h} \frac{1}{area(V_i)} \left[\int_{V_i} f\, dx dy \right] \chi_i$, χ_i being the **characteristic function** of V_i [2], with an analogous definition of v_h. According to the above estimate, as long as $f \in H^1(\Omega)$, we have $\| f - f_h \|_{0,2} \leq C_{00} h |f|_{1,2}$.* ∎

Nonetheless, in an attempt to circumvent equation (5.68), here we shall rather require a little bit more regularity of u, by assuming that $|\mathbf{grad}\, u| \in L^\infty(\Omega)$. This assumption is not superfluous for, in contrast to the one-dimensional case, a function of the H^1-class, such as first-order derivatives of functions in $H^2(\Omega)$, need not to be bounded in a two-dimensional domain (a famous counterexample given by Strang and Fix [182] is the function $\log [\log (x^2 + y^2)]$, when Ω contains the unit disk). Notice, however, that in most practical applications, the gradient of the solution to the Poisson equation is a physical quantity that cannot be unbounded, and hence our assumption is far from unreasonable.

Let then P be an arbitrary point in V_i. Since u is continuous, we know that

$$u(P) - u(S_i) = \int_{S_i}^P (\mathbf{grad}\, u(x(s), y(s)) | \vec{e_i})\, ds$$

where $\vec{e_i}$ is the unit vector of the axis $\overrightarrow{S_i P}$.

Hence, for all P in V_i, $|u(P) - u(S_i)| \leq \| |\mathbf{grad}\, u| \|_{0,\infty} h$, which yields

$$\int_{V_i} |u(x,y) - u(S_i)|^2\, dx dy \leq area(V_i)\, \| |\mathbf{grad}\, u| \|_{0,\infty}^2 h^2.$$

Summing up over i, we come up with

$$\| u - \bar{u}_h \|_{0,2} \leq \sqrt{area(\Omega)}\, \| |\mathbf{grad}\, u| \|_{0,\infty} h \tag{5.69}$$

---

[2] For those not well acquainted with this concept, $\chi_i(\mathbf{x}) = 1$ if $\mathbf{x} \in V_i$ and $\chi_i(\mathbf{x}) = 0$ otherwise.

Remark 5.11 *The reader should bear in mind that by using the 'brutal' technique to derive equation (5.69), we gave up the search for optimality, in terms of both order of convergence and exact solution regularity. Notice, however, that our main goal here is to show once more that, in most cases, the rigorous study of convergence to smooth solutions of discretisation methods for linear PDEs (and very often for non-linear ones too) is a task that can be successfully achieved with the help of basic mathematical tools, and nothing else.*∎

Finally, provided we are using a regular family of meshes, and as long as $u \in H^2(\Omega)$ and $|\mathbf{grad}\, u| \in L^\infty(\Omega)$, combining equations (5.66), (5.67) and (5.69), we derive the

Error estimate for the Vertex-centred FVM

$$\| u_h - u \|_{0,2} \leq [\sqrt{area(\Omega)} \; \| |\mathbf{grad}\, u| \|_{0,\infty} + \overline{C} \; \| H(u) \|_{0,2}]h. \tag{5.70}$$

Notice that, in contrast to the $\mathcal{P}_1$ FEM which represents exactly linear solutions, in principle equation (5.70) is the best we can hope for, since the FVM is capable of reproducing only constant solutions exactly. Nevertheless, this point deserves a comment; in this respect, we refer to part C of this subsection, 'Final Comment on the FVM'.

B. Convergence of-the Cell-centred FVM:

To begin with, we recall that the Cell-centred FVM was described in Subsection 4.4.2 for a rectangular domain Ω and equal rectangular cells. Confining ourselves here again to the case of homogeneous Dirichlet boundary conditions, we first generalise the method to the case of a mesh consisting of rectangles of different sizes. The corresponding mesh is commonly called a **Cartesian mesh**.

Among the necessary adaptations to this new environment, we first modify equations (4.54), (4.55) and (4.56) into the relations satisfied by the unknowns $u_{i-1/2,j-1/2}$, namely, the approximations of u at the centres $(x_{i-1/2}, y_{j-1/2})$ of variable rectangular cells $R_{i-1/2,j-1/2}$, for $i = 1, \ldots, n_x$ and $j = 1, \ldots, n_y$, where $R_{i-1/2,j-1/2} = (x_{i-1}, x_i) \times (y_{j-1}, y_j)$, with $0 = x_0 < x_1 < x_2 < \cdots < x_{n_x-1} < x_{n_x} = L_x$ and $0 = y_0 < y_1 < y_2 < \cdots < y_{n_y-1} < y_{n_y} = L_y$. In order to take into account homogeneous Dirichlet boundary conditions, we set $u_{-1/2,j-1/2} = u_{n_x+1/2,j-1/2} = 0$ for $j = 1, 2, \ldots, n_y$ and $u_{i-1/2,-1/2} = u_{i-1/2,n_y+1/2} = 0$ for $i = 1, 2, \ldots, n_x$. We denote the mesh spacings by $h_i := x_i - x_{i-1}$ for $i = 1, 2, \ldots, n_x$ and $k_j := y_j - y_{j-1}$ for $j = 1, 2, \ldots, n_y$, and introduce intermediate spacings $h_{i-1/2}$ for $i = 1, \ldots, n_x + 1$ and $k_{j-1/2}$ for $j = 1, \ldots, n_y + 1$ defined by

$$h_{1/2} = h_1/2; \; h_{n_x+1/2} = h_{n_x}/2;$$

$$k_{1/2} = k_1/2; \; k_{n_y+1/2} = k_{n_y}/2;$$

$$h_{i+1/2} = (h_i + h_{i+1})/2 \text{ for } i = 1, 2, \ldots, n_x - 1;$$

$$k_{j+1/2} = (k_j + k_{j+1})/2 \text{ for } j = 1, 2, \ldots, n_y - 1.$$

Keeping the edge notation of Figure 4.25, the flux $\partial u / \partial \nu$ from $R_{i-1/2,j-1/2}$ outwards, for $i = 1, 2, \ldots, n_x$ and $j = 1, 2, \ldots, n_y$, is approximated by the following quantities:

- Flux across edge $e_{i-1,j-1/2} : F_{i-1,j-1/2}^- := (u_{i-3/2,j-1/2} - u_{i-1/2,j-1/2})/h_{i-1/2};$
- Flux across edge $e_{i,j-1/2} : F_{i,j-1/2}^+ := (u_{i+1/2,j-1/2} - u_{i-1/2,j-1/2})/h_{i+1/2};$

- Flux across edge $e_{i-1/2,j-1}$: $F_{i-1/2,j-1}^- := (u_{i-1/2,j-3/2} - u_{i-1/2,j-1/2})/k_{j-1/2}$;
- Flux across edge $e_{i-1/2,j}$: $F_{i-1/2,j}^+ := (u_{i-1/2,j+1/2} - u_{i-1/2,j-1/2})/k_{j+1/2}$.

Then, letting $A_{i,j}$ be the area of $R_{i-1/2,j-1/2}$, in contrast to scheme (4.56) for a uniform rectangular mesh we work with the right side given by $\tilde{f}_{i-1/2,j-1/2} = \int_{R_{i-1/2,j-1/2}} f(x,y)\,dxdy/A_{i,j}$. This is because the underlying analysis is a little simpler. We note, however, that in practice it may be impossible to compute integrals of f exactly. Thus, the use of numerical integration is necessary. The one-point quadrature formula yielding a right side $f_{i-1/2,j-1/2} := f(x_{i-1/2}, y_{j-1/2})$ is sufficient to avoid lowering of the method's order of convergence. The formal proof of this result is proposed to the reader as Exercise 5.13.

Now noting that the length of both $e_{i-1,j-1/2}$ and $e_{i,j-1/2}$ equals k_j, and the length of both $e_{i-1/2,j-1}$ and $e_{i-1/2,j}$ equals h_i, the FVM for solving equation (4.1) with homogeneous Dirichlet boundary conditions[3] in flux formulation according to the above definitions is

The Cell-centred FV scheme for rectangular cells of variable size.

$$
\boxed{
\begin{array}{l}
\text{Find } u_{i-1/2,j-1/2} \text{ for } i = 1, 2, \ldots, n_x \text{ and } j = 1.2, \ldots, n_y \text{ such that} \\
-k_j[F_{i-1,j-1/2}^- + F_{i,j-1/2}^+] - h_i[F_{i-1/2,j-1}^- + F_{i-1/2,j}^+] = A_{i,j}\tilde{f}_{i-1/2,j-1/2}
\end{array}
}
\tag{5.71}
$$

We recall that our goal is to prove stability and convergence results for scheme (5.71) in a suitable mean-square sense. In this aim, we first define a function u_h (almost) everywhere in $\overline{\Omega}$ associated with the solution of equation (5.71) as follows. At every point, $P \in \Omega$, $u_h(P) = u_{i-1/2,j-1/2}$ if P belongs to $R_{i-1/2,j-1/2}$. Moreover, u_h takes zero values at the mid-point of all boundary edges, and consequently at all points belonging to Γ.

Before undertaking this study, it is necessary to establish a discrete analog of the Friedrichs–Poincaré inequality, associated with this FV discretisation. This inequality has the same nature as the one that holds for piecewise constant functions associated with triangle-based Vertex-centred FV discretisations. The inequality considered here applies to piecewise constant functions v_h in each rectangular CV, that by definition take zero values at the mid-point of boundary edges. Let us denote the set of such functions by $\overline{V}_h$.

Similarly to the FV solution u_h, the constant value of v_h in CV $R_{i-1/2,j-1/2}$ is denoted by $v_{i-1/2,j-1/2}$ and by definition $v_{0,j-1/2} = v_{n_x,j-1/2} = 0$ for $j = 1, 2, \ldots, n_y$ and $v_{i-1/2,0} = v_{i-1/2,n_y} = 0$ for $i = 1, 2, \ldots, n_x$. In order to prove a discrete Friedrichs–Poincaré inequality applying to an arbitrary function $v_h \in \overline{V}_h$, we first have to define its **discrete gradient grad$_h v_h$**. The natural choice here is to let each component of **grad$_h v_h$** be constant in each cell of the corresponding **semi-dual mesh** $\mathcal{D}_{xh}$ or $\mathcal{D}_{yh}$. The former is the set of $(n_x + 1)n_y$ rectangles $(x_{i-1/2}, x_{i+1/2}) \times (y_{j-1}, y_j)$ for $i = 0, 1, \ldots, n_x$ and $j = 1, \ldots, n_y$, where $x_{-1/2} := x_0 = 0$; $x_{n_x+1/2} := x_{n_x} = L_x$, and the latter is the set of $n_x(n_y + 1)$ rectangles $(x_{i-1}, x_i) \times (y_{j-1/2}, y_{j+1/2})$ for $i = 1, \ldots, n_x$ and $j = 0, 1, \ldots, n_y$, where $y_{-1/2} := y_0 = 0$; $y_{n_y+1/2} := y_{n_y} = L_y$. Let us denote the rectangle $(x_{i-1/2}, x_{i+1/2}) \times (y_{j-1}, y_j)$ in $\mathcal{D}_{xh}$ by $X_{i,j}$ for $i = 0, 1, \ldots, n_x$ and $j = 1, \ldots, n_y$ and the rectangle $(x_{i-1}, x_i) \times (y_{j-1/2}, y_{j+1/2})$ in $\mathcal{D}_{yh}$ by $Y_{i,j}$. Notice that by construction $area(X_{i,j}) = h_{i+1/2}k_j$ and $area(Y_{i,j}) = h_i k_{j+1/2}$. We refer to Figure 5.10 for an illustration of some cells in these semi-dual meshes. Incidentally, we

---

[3] For the case of Neumann and inhomogeneous Dirichlet boundary conditions, we refer to reference [68].

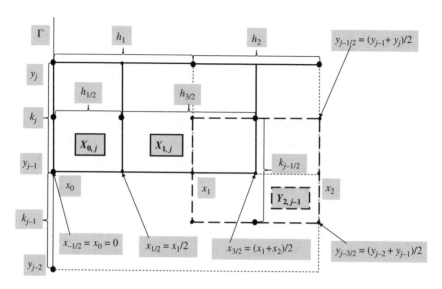

Figure 5.10 Cells $X_{0,j}$ and $X_{1,j}$ of mesh $\mathcal{D}_{xh}$ and (dashed) cell $Y_{2,j-1}$ of mesh $\mathcal{D}_{yh}$

recall that $x_{i-1/2} = (x_{i-1} + x_i)/2$ for $i = 1, 2, \ldots, n_x$ and $y_{j-1/2} = (y_{j-1} + y_j)/2$ for $j = 1, 2, \ldots, n_y$.

By definition, we have

$$
\begin{cases}
\mathbf{grad}_h v_h = (\partial_{xh}[v_h]; \ \partial_{yh}[v_h]) \text{ with} \\
\text{In } X_{ij} : \ \partial_{xh}[v_h] = \partial_{xh}[v_h]_{ij} := \dfrac{v_{i+1/2,j-1/2} - v_{i-1/2,j-1/2}}{h_{i+1/2}}, \\
i = 0, 1, \ldots, n_x, j = 1, \ldots, n_y \\
\text{In } Y_{ij} : \ \partial_{yh}[v_h] = \partial_{yh}[v_h]_{ij} := \dfrac{v_{i-1/2,j+1/2} - v_{i-1/2,j-1/2}}{k_{j+1/2}}, \\
j = 0, 1, \ldots, n_y, i = 1, \ldots, n_x.
\end{cases}
\tag{5.72}
$$

Next, we state

The discrete Friedrichs–Poincaré inequality for a rectangular FV mesh

> There exists a constant $C_D(\Omega)$ depending only on Ω such that
> $\| v_h \|_{0,2} \le C_D(\Omega) \| \mathbf{grad}_h v_h \|_{0,2} \ \forall v_h \in \overline{V}_h.$ (5.73)

In order to demonstrate equation (5.73), we first write down the expressions of both norms involved in this inequality, namely,

$$
\| v_h \|_{0,2}^2 = \sum_{i=1}^{n_x} h_i \sum_{j=1}^{n_y} k_j v_{i-1/2,j-1/2}^2.
\tag{5.74}
$$

$$\begin{cases} \| \operatorname{grad}_h v_h \|_{0,2}^2 = \| \partial_{xh}[v_h] \|_{0,2}^2 + \| \partial_{yh}[v_h] \|_{0,2}^2 \\ \text{with} \\ \| \partial_{xh}[v_h] \|_{0,2}^2 = \sum_{i=0}^{n_x} \sum_{j=1}^{n_y} k_j \frac{(v_{i+1/2,j-1/2} - v_{i-1/2,j-1/2})^2}{h_{i+1/2}} \\ \| \partial_{yh}[v_h] \|_{0,2}^2 = \sum_{j=0}^{n_y} \sum_{i=1}^{n_x} h_i \frac{(v_{i-1/2,j+1/2} - v_{i-1/2,j-1/2})^2}{k_{j+1/2}}. \end{cases} \tag{5.75}$$

Notice that equation (5.74) results from summations over elements of the (primal) mesh. On the other hand, equation (5.75) is the outcome of summations over the elements of the semi-dual meshes $\mathcal{D}_{xh}$ and $\mathcal{D}_{yh}$.

Now, we write for $j = 1, 2, \ldots, n_y$ and $i = 1, 2, \ldots, n_x$ the obvious identity:

$$v_{i-1/2,j-1/2} \equiv \sum_{m=0}^{i-1} \frac{v_{m+1/2,j-1/2} - v_{m-1/2,j-1/2}}{\sqrt{h_{m+1/2}}} \sqrt{h_{m+1/2}} \tag{5.76}$$

Taking the square of both sides of equation (5.76) and applying the Cauchy–Schwarz inequality to the resulting right side, we easily obtain

$$v_{i-1/2,j-1/2}^2 \leq \left[\sum_{m=0}^{i-1} \frac{(v_{m+1/2,j-1/2} - v_{m-1/2,j-1/2})^2}{h_{m+1/2}} \right] \sum_{m=0}^{i-1} h_{m+1/2} \tag{5.77}$$

Since $\sum_{m=0}^{i-1} h_{m+1/2} \leq L_x$ for all $i \leq n_x$, equation (5.77) further yields

$$v_{i-1/2,j-1/2}^2 \leq L_x \left[\sum_{m=0}^{n_x} \frac{(v_{m+1/2,j-1/2} - v_{m-1/2,j-1/2})^2}{h_{m+1/2}} \right] \forall i \leq n_x, i > 0. \tag{5.78}$$

Then, multiplying both sides of equation (5.78) with h_i and summing up the resulting relations from $i = 1$ through $i = n_x$, since $\sum_{i=1}^{n_x} h_i = L_x$, we readily obtain

$$\sum_{i=1}^{n_x} h_i v_{i-1/2,j-1/2}^2 \leq L_x^2 \left[\sum_{m=0}^{n_x} \frac{(v_{m+1/2,j-1/2} - v_{m-1/2,j-1/2})^2}{h_{m+1/2}} \right]. \tag{5.79}$$

Recalling equation (5.72), we observe that each term in the summation on the right side of equation (5.79) is nothing but $\{\partial_{xh}[v_h]_{mj}\}^2$ multiplied by $h_{m+1/2}$. Therefore,

$$\sum_{i=1}^{n_x} h_i v_{i-1/2,j-1/2}^2 \leq L_x^2 \left[\sum_{m=0}^{n_x} h_{m+1/2} \{\partial_{xh}[v_h]_{mj}\}^2 \right]. \tag{5.80}$$

Then multiplying both sides of equation (5.80) by k_j, summing up from $j = 1$ through $j = n_y$ and recalling equations (5.74)–(5.75), we come up with

$$\| v_h \|_{0,2}^2 \leq L_x^2 \| \partial_{xh}[v_h] \|_{0,2}^2. \tag{5.81}$$

Adding $L_x^2 \| \partial_{yh}[v_h] \|_{0,2}^2$ to the right side of equation (5.81) we establish equation (5.73) with $C_D(\Omega) = L_x$ (notice that we could have obtained $C_D(\Omega) = L_y$ as well).

Next, we endeavour to prove the convergence to the exact solution in the L^2-norm, of the FV solution, under certain mesh conditions.

Once more, we need a preliminary stability result applying to equation (5.71), in terms of the norms under consideration. Stability inequalities in the mean-square sense can be obtained for this scheme, by means of the following procedure.

First, we replace the fluxes of types F^+ and F^- in equation (5.71), by their four possible values given on the list supplied at the beginning of part B of this subsection, thereby obtaining,

$$
- k_j \left(\frac{u_{i-3/2,j-1/2} - u_{i-1/2,j-1/2}}{h_{i-1/2}} + \frac{u_{i+1/2,j-1/2} - u_{i-1/2,j-1/2}}{h_{i+1/2}} \right)
$$
$$
- h_i \left(\frac{u_{i-1/2,j-3/2} - u_{i-1/2,j-1/2}}{k_{j-1/2}} + \frac{u_{i-1/2,j+1/2} - u_{i-1/2,j-1/2}}{k_{j+1/2}} \right) = A_{i,j} \tilde{f}_{i-1/2,j-1/2}.
$$
$$(5.82)$$

Then, we multiply both sides of each equation in (5.82) by the corresponding value $u_{i-1/2,j-1/2}$ and sum up the resulting relations from $i = 1$ through $i = n_x$ and from $j = 1$ through $j = n_y$. Setting $B_{i,j} := area(X_{i,j})$ and $D_{i,j} := area(Y_{i,j})$ and denoting by f_h the piecewise constant function whose value in $R_{i-1/2,j-1/2}$ is $\tilde{f}_{i-1/2,j-1/2}$, we derive

$$
\sum_{i=1}^{n_x} \sum_{j=1}^{n_y} u_{i-1/2,j-1/2} \left[B_{i-1,j} \frac{u_{i-1/2,j-1/2} - u_{i-3/2,j-1/2}}{h_{i-1/2}^2} + B_{i,j} \frac{u_{i-1/2,j-1/2} - u_{i+1/2,j-1/2}}{h_{i+1/2}^2} \right.
$$
$$
\left. + D_{i,j-1} \frac{u_{i-1/2,j-1/2} - u_{i-1/2,j-3/2}}{k_{j-1/2}^2} + D_{i,j} \frac{u_{i-1/2,j-1/2} - u_{i-1/2,j+1/2}}{k_{j+1/2}^2} \right] = \int_\Omega f_h u_h \, dx dy.
$$
$$(5.83)$$

as the reader can easily check. The summations on the left side of equation (5.83) can be rearranged so as to separate the term associated with $\partial_{xh}[u_h]$ from the term associated with $\partial_{yh}[u_h]$. Such manipulations together with equation (5.73) lead to

A first set of stability inequalities for the Cell-centred FVM

$$
\boxed{
\begin{aligned}
&\| \mathbf{grad}_h u_h \|_{0,2} \leq C_D(\Omega) \, \| f_h \|_{0,2}, \\
&\| u_h \|_{0,2} \leq [C_D(\Omega)]^2 \, \| f_h \|_{0,2}.
\end{aligned}
}
$$
$$(5.84)$$

We decline to develop the calculations yielding equation (5.84) starting from equation (5.83), and instead we encourage the reader to do it (cf. Exercise 5.14). This is because we derive below another stability inequality for a scheme more general than equation (5.71). Nevertheless, it is not superfluous to recall that equation (5.84) suffices to guarantee the existence and uniqueness of a solution to equation (5.71).

In principle, taking equation (5.84) for granted, the next step would be the study of the consistency of equation (5.71) by replacing each value $u_{i-1/2,j-1/2}$ by $\tilde{u}_{i-1/2,j-1/2} := u(x_{i-1/2}, y_{j-1.2})$, and checking that the corresponding order is strictly positive. Unfortunately, the stability inequalities (equation (5.84)) are not well suited to the convergence analysis in view. This is because the resulting local truncation error is only an

$O(h^0)$, as the reader may check (cf. Exercise 5.15). Similarly to the case of the FEM, the way out is to consider a problem more general than equation (5.71), incorporating a term on the right side that mimics the operations involving the unknown function on the left side. In the case under study, the only operation of this kind is the calculation of a flux across the boundary of a given CV. Quite naturally, this suggests us to set up a problem with the same kind of data. More precisely, we mean a problem in which a datum of the form $\oint_{\partial R} \mathcal{F}_{R,h} ds$ is added to the right-side of the original one. $\mathcal{F}_{R,h}$ is actually a (numerical) flux across the boundary ∂R of a given cell R $(R = R_{i-1/2,j-1/2}$ in the case under study). We assume that $\mathcal{F}_{R,h}$ is bounded on ∂R for every R.

It is important to note that, like the solution fluxes, the right-side fluxes must be **conservative**. More precisely, they have to satisfy a balance relation, by enforcing the same constraint as on the left side: on every edge e of the mesh common to two CVs, say, R and S, we must have $\mathcal{F}_{R,h} + \mathcal{F}_{S,h} = 0$. Furthermore, owing to the wider framework the following study applies to, it is advisable to undertake it by using notations simpler and more general at a time, than those we have adopted so far. As a guide, the reader might think of the notation employed for the FEM on arbitrary triangular meshes.

First, we denote by $\mathcal{E}$ the set of edges in the FV mesh. Let R be an arbitrary CV and $\mathcal{E}_R$ be the set of edges belonging to ∂R. If S is a CV sharing an edge with R, we denote by e_{RS} such an edge. The length of e_{RS} is denoted by l_{RS}. If R has one or more edges contained in Γ, for each one of them we introduce a fictitious CV S' lying outside Ω, and sharing one of those edges with R, denoted by $e_{RS'}$. The form of such fictitious CVs plays no role.

As pointed out at the beginning of Section 4.4, an admissible FV mesh must be designed in such a way that each CV R has a representative point $\mathbf{x}_R$ satisfying the **positive transmissivity property** (cf. [68]). In abridged terms, this means that for every pair of CVs R and S having a common edge e_{RS} the segment joining $\mathbf{x}_R$ and $\mathbf{x}_S$ is orthogonal to e_{RS} and has a strictly positive length d_{RS}. In all the sequel, we assume that $\mathbf{x}_R$ lies in the interior of R for every R. In case S' is a fictitious CV, by definition $\mathbf{x}_{S'}$ is the intersection of the perpendicular to Γ through $\mathbf{x}_R$, with Γ itself. Notice that in this case, $d_{RS'}$ is the distance from $\mathbf{x}_R$ to Γ. To each CV R corresponds a constant value of the numerical solution u_h denoted by u_R. If the CV is a fictitious one, say S', the corresponding value $u_{S'}$ equals zero. Furthermore, for each non-fictitious CV R, the constant value u_R in R is viewed as an approximation of $u(\mathbf{x}_R)$. In order to unify the writing in the expressions that follow, S will represent a generic CV, whether fictitious or not.

Remark 5.12 *Just to keep ideas clear, we recall that for the rectangular domain $(0, L_x) \times (0, L_y)$, the mesh $\mathcal{T}_h$ we are using is the union of $R_{i-1/2,j-1/2}$ for $i = 1, \ldots, n_x$ and $j = 1, \ldots, n_y$. For each rectangle, $R \in \mathcal{T}_h$, $\mathbf{x}_R$ is the centre of R, l_{RS} equals either h_i or k_j for certain i and j, and d_{RS} is either $h_{i-1/2}$ or $k_{j-1/2}$ for suitable values of i and j. Furthermore, if $R = R_{i-1/2,j-1/2}$, $u_R = u_{i-1/2,j-1/2}$.* ∎

Now, setting $A_R = area(R)$, defining $\tilde{f}_R := \int_R f\, dxdy / A_R$ and associating a function f_h to the f_R's in the same manner as for $\tilde{f}_{i-1/2,j-1/2}$, the more general problem writes

$$\text{For every } R \in \mathcal{T}_h, \quad \sum_{S \cap R \neq \emptyset} l_{RS} \frac{u_R - u_S}{d_{RS}} = A_R \tilde{f}_R + \oint_{\partial R} \mathcal{F}_{R,h} ds \qquad (5.85)$$

Like in the case of equation (5.71), the stability of equation (5.85) in the mean-square sense can be established by multiplying both sides of it with u_R and then summing up over the elements of the mesh. Similarly to equation (5.83), this gives

$$\sum_{R\in\mathcal{T}_h} u_R \left[\sum_{S\cap R\neq\emptyset} l_{RS} \frac{u_R - u_S}{d_{RS}} \right] = \int_\Omega f_h u_h dxdy + \sum_{R\in\mathcal{T}_h} u_R \left[\sum_{S\cap R\neq\emptyset} \int_{e_{RS}} \mathcal{F}_{R,h} ds \right] \tag{5.86}$$

Notice that, if $S\cap R\neq\emptyset$, the terms $u_S l_{SR} \dfrac{u_S - u_R}{d_{SR}} = -u_S l_{RS} \dfrac{u_R - u_S}{d_{RS}}$ and $u_S \int_{e_{SR}} \mathcal{F}_{S,h} ds = -u_S \int_{e_{RS}} \mathcal{F}_{R,h} ds$ implicitly appear in both the left and right-side summations respectively. Thus, we may rewrite the summations over the CVs in equation (5.86), in the form of summations over the edges. In a further attempt to simplify the notation in the four equations that follow, the reader should bear in mind that R and S generically represent a pair of CVs with a common edge e. Then, setting $l_e = length(e)$ and $d_e = d_{RS}$, this gives

$$\sum_{e\in\mathcal{E}_h} l_e \frac{(u_R - u_S)^2}{d_e} = \int_\Omega f_h u_h dxdy + \sum_{e\in\mathcal{E}_h} (u_R - u_S) \int_e \mathcal{F}_{e,h} ds \tag{5.87}$$

where $\mathcal{F}_{e,h}$ was chosen to be $\mathcal{F}_{R,h} = -\mathcal{F}_{S,h}$.

Setting $\mathcal{F}_{max} := \max_{e\in\mathcal{E}} |\mathcal{F}_{e,h}|$, we next apply the Cauchy–Schwarz inequality to the inner product of f_h and u_h in equation (5.87). After some trivial manipulations, we come up with the upper bound,

$$\sum_{e\in\mathcal{E}_h} l_e \frac{(u_R - u_S)^2}{d_e} \leq \| f_h \|_{0,2} \| u_h \|_{0,2} + \mathcal{F}_{max} \sum_{e\in\mathcal{E}_h} \frac{\sqrt{l_e}|u_R - u_S|}{\sqrt{d_e}} \sqrt{l_e d_e} \tag{5.88}$$

Applying again the Cauchy–Schwarz inequality to the right-side summation over $\mathcal{E}_h$, we derive

$$\sum_{e\in\mathcal{E}_h} l_e d_e \left[\frac{u_R - u_S}{d_e} \right]^2 \leq \| f_h \|_{0,2} \| u_h \|_{0,2} + \mathcal{F}_{max} \sqrt{\sum_{e\in\mathcal{E}_h} l_e d_e \left(\frac{u_R - u_S}{d_e} \right)^2} \sqrt{\sum_{e\in\mathcal{E}_h} l_e d_e} \tag{5.89}$$

Referring to Figure 5.11, we note that, if e is not contained in Γ, $l_e d_e$ is twice the area of the convex quadrilateral Q_e whose vertices are $\mathbf{x}_R$, $\mathbf{x}_S$ and the end-points of e, and otherwise $l_e d_e$ is twice the area of the triangle Q'_e whose vertices are $\mathbf{x}_R$ and the end-points of the edge common to R and a fictitious CV S'.

By inspection, we find out that the whole polygon Ω is the union over $\mathcal{E}_h$ of the quadrilaterals Q_e and the triangles Q'_e. Since all these elements are disjoint, we have $\sum_{e\in\mathcal{E}_h} l_e d_e = 2area(\Omega)$.

For the sake of conciseness, we resort to the notation of reference [68], namely,

$$\| u_h \|_{1,\mathcal{T}} := \left[\sum_{e\in\mathcal{E}_h} l_e d_e \left(\frac{u_R - u_S}{d_e} \right)^2 \right]^{1/2},$$

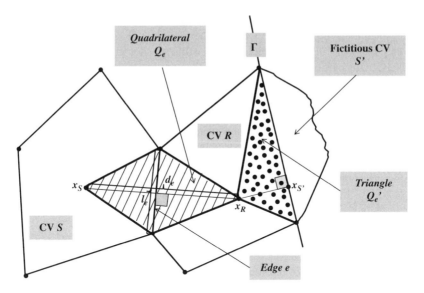

Figure 5.11 CVs R, S with a common edge e and dashed quadrilateral Q_e; fictitious CV S' having a common edge with R and dotted triangle Q'_e

In doing so, we apply the inequality $ab \leq (a^2 + b^2)/2$ for $a, b \in \mathfrak{R}$, to the second term on the right side of equation (5.89), to easily obtain

$$\| u_h \|_{1,\mathcal{T}}^2 \leq 2 \| f_h \|_{0,2} \| u_h \|_{0,2} + 2area(\Omega)\mathcal{F}_{max}^2. \tag{5.90}$$

In reference [68], the validity of a discrete Friedrichs–Poincaré inequality is proved in the general framework of equation (5.90). More precisely, existence is established of a mesh-independent constant $C_P(\Omega)$, such that $\| v\|_{0,2} \leq C_P(\Omega) \| v\|_{1,\mathcal{T}}$ for every piecewise constant function v in the CVs, vanishing at the representative points of the fictitious boundary cells. The curious reader would eventually be tempted to consult reference [68] in order to check the proof. In doing so, she or he would figure out that the convergence results we are about to prove are as general as can be. Since here we are considering only Cartesian meshes, we switch back to the pertaining notations.

By inspection, we note that in this case $\| u_h \|_{1,\mathcal{T}}$ is nothing but $\| \mathbf{grad}_h u_h\|_{0,2}$. Therefore, using equation (5.73) and Young's inequality with $\delta = 1/2$, from equation (5.90), we easily derive

$$\| \mathbf{grad}_h u_h \|_{0,2}^2 \leq 4[C_D(\Omega)]^2 \| f_h \|_{0,2}^2 + 4L_x L_y \mathcal{F}_{max}^2. \tag{5.91}$$

Setting $L := \max[L_x, L_y]$ and using the discrete Friedrichs–Poincaré inequality, we obtain two

Stability inequalities for the rectangular Cell-centred FV scheme

$$\begin{aligned} \| \mathbf{grad}_h u_h\|_{0,2} &\leq 2L(\| f_h\|_{0,2} + \mathcal{F}_{max}) \\ \| u_h\|_{0,2} &\leq 2L^2(\| f_h\|_{0,2} + \mathcal{F}_{max}). \end{aligned} \tag{5.92}$$

Before pursuing the convergence study of scheme (5.71), we should stress again that the main lines of the underlying proof apply to FV schemes with polygonal CVs of arbitrary shape, as long as they form a family of **restricted admissible meshes** in the sense of reference [68]. More precisely, in the particular case of scheme (5.71), setting $diameter(R_{i,j}) := \max[h_i, k_j]$ we would require the existence of a constant $\beta > 0$ such that for all the meshes considered to solve equation (4.1) $\min[h_i, k_j] \geq \beta \, diameter(R_{i,j})$ holds $\forall i, j$. This definition conforms to the general case, as a FV counterpart of the concept of a regular family of FE meshes. In the case of a rectangular (structured) mesh, however, we had better ask for more, for the above condition is of limited use, as pointed out in Remark 5.5. Although for proving the convergence of the FVM in the L^2-norm hereafter this is not necessary, we make the following assumption: There exists a constant $\gamma > 0$ such that for all meshes in the family,

$$h_i \geq \gamma h \ \forall i \ \text{and} \ k_j \geq \gamma h \ \forall j, \ \text{with} \ h = \max[h_{max}, k_{max}], h_{max} = \max_{1 \leq i \leq n_x} h_i, k_{max} = \max_{1 \leq j \leq n_y} k_j.$$

Notice that this is the FV counterpart of a quasi-uniform family of FE meshes, which implicitly allow for h to go to zero.

In contrast to Theorem 9.4 of reference [68], where less regularity is required, we assume that u is twice continuously differentiable in $\overline{\Omega}$. Clearly enough, such an assumption allows for a simplification of the analysis. Incidentally, for general meshes, the convergence of the FVM relies upon a property, stating the **consistency of the numerical fluxes** [68, 202]. Let us see this concept in detail in the case of a Cartesian mesh.

For every CV R, let $\tilde{u}_R := u(\mathbf{x}_R)$, $\bar{u}_R = u_R - \tilde{u}_R$ and $f_R := [\int_R f(x, y) \, dx dy]/A_R$. Furthermore, we denote by $\vec{\nu}_R$ the unit outer normal vector to ∂R.

Similarly to u_h, we denote by $\tilde{u}_h$ the constant function in each CV $S \in \mathcal{T}_h$, such that $\tilde{u}_h(\mathbf{x}_S) = \tilde{u}_S$, and $\tilde{u}_h = 0$ everywhere on Γ. We also set $\bar{u}_h := u_h - \tilde{u}_h$. Now, for every edge $e \in \mathcal{E}_R$, we define the **mean flux consistency error** along e from CV R outwards, by

$$\bar{E}_R(e) := \frac{1}{l_e} \int_e \partial_{\nu_R} u \, ds - \frac{\tilde{u}_S - \tilde{u}_R}{d_e}. \tag{5.93}$$

S being the CV (eventually a fictitious one) sharing edge e with R (the reader should bear in mind that in the case under study, $\vec{\nu}_R$ is either $\pm(1; 0)$ or $\pm(0; 1)$).

Now we note that Rolle's theorem ensures the existence of a point $\tilde{\mathbf{x}}_e$ on the segment joining $\mathbf{x}_R$ and $\mathbf{x}_S$ such that

$$(\tilde{u}_S - \tilde{u}_R)/d_e = [\partial_{\nu_R} u](\tilde{\mathbf{x}}_e). \tag{5.94}$$

Moreover, by the mean value theorem for integrals, there exists a point $\hat{\mathbf{x}}_e \in e$ such that

$$\int_e \partial_{\nu_R} u \, ds = l_e \, [\partial_{\nu_R} u](\hat{\mathbf{x}}_e). \tag{5.95}$$

It follows from equations (5.93), (5.94) and (5.95) that

$$\bar{E}_R(e) = [\partial_{\nu_R} u](\hat{\mathbf{x}}_e) - [\partial_{\nu_R} u](\tilde{\mathbf{x}}_e). \tag{5.96}$$

Now we apply again Rolle's theorem to the above expression, this time to the difference of partial derivatives in brackets. Referring to Figure 5.12, the notation $\partial_{\nu_R \sigma_e} u$ is used to represent the partial derivative in the direction of $\vec{\sigma}_e$ of the partial derivative of u in the direction of $\vec{\nu}_R$, where $\vec{\sigma}_e$ is the unit vector along the segment with length $\bar{d}_e$, joining $\tilde{\mathbf{x}}_e$ and $\hat{\mathbf{x}}_e$, oriented from

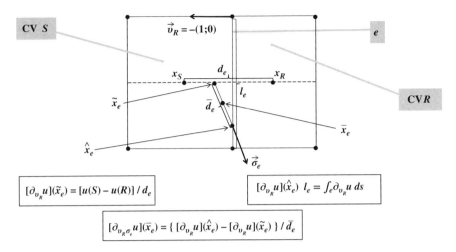

Figure 5.12   Remarkable points, lengths and unit vectors pertaining to the pair of CVs R and S with a common edge e

the former towards the latter. In doing so, for a certain point $\overline{\mathbf{x}}_e$ belonging to this segment, we can write

$$\bar{E}_R(e) = \bar{d}_e \; [\partial_{\nu_R\sigma_e} u](\overline{\mathbf{x}}_e) \leq h \parallel H(u) \parallel_{0,\infty} / \sqrt{2}, \tag{5.97}$$

where $\parallel H(u) \parallel_{0,\infty}$ is the maximum in $\overline{\Omega}$ of $|H(u)| := [|\partial_{xx}u|^2 + |\partial_{yy}u|^2 + |\partial_{xy}u|^2]^{1/2}$. Indeed, referring again to Figure 5.12, it is not difficult to figure out that $\bar{d}_e \leq h/\sqrt{2}$. In short, equation (5.97) tells us that the maximum value of the mean flux consistency error over the edges of $\mathcal{E}_h$ cannot exceed $h \parallel H(u) \parallel_{0,\infty} / \sqrt{2}$.

To conclude the convergence study, all we have to do is to replace in equation (5.85) the u_R's with the $\bar{u}_R$'s. Taking into account the scheme's definition and recalling equation (5.94), we note that the expression of the resulting residual at the level of a CV R is

$$\mathcal{G}_R(u) := \int_R f(x, y) \, dxdy - \sum_{e \in \mathcal{E}_R} l_e \, [\partial_{\nu_R} u](\widetilde{\mathbf{x}}_e). \tag{5.98}$$

From the true balance equation in R and equation (5.95), it follows that the residual $\mathcal{G}_R(u)$ is given by

$$\mathcal{G}_R(u) = \sum_{e \in \mathcal{E}_R} l_e \{[\partial_{\nu_R} u](\widehat{\mathbf{x}}_e) - [\partial_{\nu_R} u](\widetilde{\mathbf{x}}_e)\}. \tag{5.99}$$

Then, recalling equation (5.96), equation (5.99) implies that

$$\mathcal{G}_R(u) = \sum_{e \in \mathcal{E}_R} l_e \bar{E}_R(e) \tag{5.100}$$

Equation (5.100) means that $\bar{u}_h := u_h - \tilde{u}_h$ is the solution of equation (5.85) for $f_h = 0$ and for fluxes $\mathcal{F}_{R,h}$, whose restriction to each edge e of ∂R is the constant $\bar{E}_R(e)$ given by equation (5.96).

Thus, taking into account equations (5.92) and (5.97), $\bar{u}_h$ is readily seen to satisfy

$$\| \bar{u}_h \|_{0,2} \leq Ch \parallel H(u) \parallel_{0,\infty} \text{ with } C = \sqrt{2}L^2. \tag{5.101}$$

Finally, similarly to equation (5.69), we can assert that

$$\| u - \tilde{u}_h \|_{0,2} \leq h\sqrt{area(\Omega)} \parallel \mathbf{grad}\ u \parallel_{0,\infty}. \tag{5.102}$$

where $\parallel \mathbf{grad}\ u \parallel_{0,\infty}$ is to be understood as the maximum in $\overline{\Omega}$ of $|\mathbf{grad}\ u|$. Since $area(\Omega) \leq L^2$, combining equations (5.101) and (5.102), and using the triangle inequality, we obtain the

Error estimate for the rectangular Cell-centred FVM

$$\boxed{\| u - u_h \|_{0,2} \leq h\ L[\sqrt{2}L \parallel H(u) \parallel_{0,\infty} + \parallel \mathbf{grad}\ u \parallel_{0,\infty}].} \tag{5.103}$$

This establishes the error estimate we were searching for, together with the corresponding first-order convergence of u_h to u in $L^2(\Omega)$.

C. Final Comments on the FVM:

As already pointed out in Part A of this subsection, generally speaking first-order convergence is what we can hope for the scheme (5.71), since we are approximating u by piecewise constant functions. Actually, using equation (5.92), we can prove a qualitatively equivalent result for $\parallel \mathbf{grad}\ u - \mathbf{grad}_h u_h \parallel_{0,2}$, under the same assumptions on u and on the mesh. In this respect, we refer to reference [68]. Furthermore, it is noteworthy that completely equivalent results hold for both u and its gradient, in the case of general polygonal meshes, as long as they form a restrictive admissible partition of Ω (cf. [68]). The arguments this result is based upon are not so different from those we employed in Part B of this subsection, although they involve some additional technicalities. For instance, a more general definition of the discrete gradient is necessary in this case. The basic idea is to define it in terms of the fluxes across edges with arbitrary slopes. The outcome comes closer to the convergence results known to hold for the lowest order Raviart–Thomas mixed element, in view of the connections between both methods (cf. Subsection 4.4.2).

On the other hand, it is possible to prove that the order of convergence of u_h to u in the maximum norm is almost linear. More precisely, error estimates in terms of an $O(h|\log h|)$ are proved in reference [68] (cf. Corollary 9.1), by requiring the same regularity of u and under some additional assumption on the mesh. However, the arguments leading to such a result lie beyond the scope of this book. Notice that, in contrast, an error estimate in the maximum norm in $O(h^2|\log h|)$ holds for the $\mathcal{P}_1$ FE solution of equation (4.1) or equations alike in a convex polygon. This result is due to Nitsche [144] for equation (4.3), and to Scott [180] for the equation $-\Delta u + u = f$ with Neumann boundary conditions.

In view of the above comments, one might ask oneself why so many practitioners prefer the FVM to the FEM, at least in contemporary **CFD - computational fluid dynamics (CFD)**[4]. From the author's point of view, the original philosophy of FVM creators and first users (see e.g. [193] and [177]), well reported in subsequent literature on CFD (see e.g. [98], [72] and [137]), could be the explanation. According to it, the approximations should be based

---

[4] CFD is the branch of applied Science devoted to the numerical simulation of fluid flow, encompassing mass and heat transfer.

on piecewise constant functions in CVs. In this manner, the balance of relevant physical quantities within or across the interfaces of non-overlapping cells, such as heat and mass fluxes would become easier to control. Whatever the reason, it is undeniable that the FVM is widely accepted in the various technological fields in which CFD is an indispensable tool, such as the automotive and aerospace industries.

Going back to the Vertex-centred FVM studied in Part A of this subsection, we should report a more recent trend among users to assimilate it to a modified FEM. Some authors call this technique the **finite volume-finite element method** (cf. [68]), or the **finite volume element method** (cf. [39]). More specifically, such a method is based on the approximation of the solution by a continuous piecewise linear function (called a **trial function**), while the **test functions** v remain piecewise constant in the CVs. This requires the use of a particular variational formulation akin to the one-dimensional case (cf. Subsection 1.4.3), but on the other hand the new approach gives rise to enhanced approximation properties. For instance, one might expect second-order convergence in the L^2-norm for the thus-modified FVM, instead of the first-order ones (equation (5.70)). However, it should be pointed out that at this writing the FVM is still a subject of active research, and the object of several proposals to improve its performance, even in the method's most basic aspects. In this respect the author refers to references [147] and [43], among many other works, including one of his own [59].

5.1.6   *Example 5.3: Triangle-Centred FVM versus RT_0 Mixed FEM*

In Subsection 4.4.2, we studied the connection between the Cell-centred FVM and the Raviart–Thomas mixed FE of the lowest order for rectangles, known as the RT_0 method. Although we did not elaborate on this issue, a similar relationship exists between the Cell-centred FVM and the version of the RT_0 method for triangles. In this example, we assess the performances of both methods by comparing them in the solution of the Poisson equation with manufactured solutions. Before presenting numerical results, let us complete the description of the RT_0 method for triangles, together with the technique by means of which the corresponding Cell-centred FVM can be derived. First, we recall the problem (4.47) associated with a mixed formulation of the Poisson equation $-\Delta u = f$ with homogeneous Dirichlet boundary conditions discretised by the FEM:

$$\begin{cases} \text{Find } \vec{p}_h \in \mathbf{Q}_h \text{ and } u_h \in U_h \text{ such that} \\ (\vec{p}_h | \vec{q})_0 + (u_h \mid div\ \vec{q})_0 = 0\ \forall \vec{q} \in \mathbf{Q}_h \\ (div\ \vec{p}_h \mid v)_0 = -(f|v)_0\ \forall v \in U_h. \end{cases}$$

where $\vec{p}_h$ is an approximation of **grad** u in a FE space $\mathbf{Q}_h$ and u_h is an approximation of u in the FE space U_h, both being spaces associated with $\mathcal{T}_h$.

Referring to Figure 5.13, where a typical triangle T in a mesh $\mathcal{T}_h$ of the **strict acute type** is represented, we denote by $(x_i^T; y_i^T)$ the coordinates of vertex S_i^T of T and by e_i^T the opposite edge. The (length of the) height of T with respect to S_i^T is denoted by h_i^T, the length of e_i^T by l_i^T, and the unit outer normal vectors to e_i^T by $\vec{v}_i^T$, $i = 1, 2, 3$. $\mathbf{Q}_h$ is defined as the space of vector fields $\vec{q}$ whose restriction to every $T \in \mathcal{T}_h$ is of the form $a^T \mathbf{x} + \mathbf{b}^T$, where $\mathbf{x}$ is the position vector, a^T is a real coefficient and $\mathbf{b}^T$ is a real vector $[b_x^T, b_y^T]^T$. We require that the mean value of the normal flux $(\vec{q}|\vec{v}_e)$ across a mesh edge e with unit normal vector $\vec{v}_e$ oriented in a unique manner for an inner edge, is the same from both sides

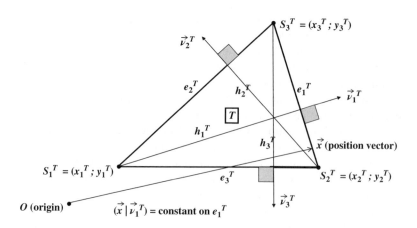

Figure 5.13   A triangle T and entities used to define the Raviart–Thomas space $\mathbf{Q}_h$

of e $\forall \vec{q} \in \mathbf{Q}_h$. Coefficients a^T, b_x^T and b_y^T are determined in such a way that $(\vec{q}_{|T}|\vec{\nu}_i^T)$ has prescribed values for $i = 1, 2, 3$. The three coefficients can be uniquely defined (cf. [162]) by noticing that $(\vec{x}|\vec{\nu}_i^T)$ is constant along e_i^T as the modulus of the position vector's orthogonal projection over $\vec{\nu}_i^T$. Therefore, $\vec{q} \in \mathbf{Q}_h$ is also constant along any mesh edge. U_h in turn is defined to be the space of functions v whose restriction to every $T \in \mathcal{T}_h$ is constant. This combination of polynomial interpolations of the primal variable u_h and the flux (dual) variable $\vec{p}_h$ is not chosen by chance. Besides balancing orders of approximation of u and $\vec{p}$ in the natural norms, this choice brings about the satisfactory fulfilment of a stability condition inherent to mixed formulations. It is known as the **Babuška–Brezzi condition** (cf. [11] and [36]), or the **inf-sup condition**, whose rigorous statement lies beyond the scope of this book.

The local basis field $\vec{\eta}_i^T$ satisfying $(\vec{\eta}_i^T|\vec{\nu}_j^T)/l_j^T = \delta_{ij}$ for space $\mathbf{Q}_h$ is easily found to be $\vec{\eta}_i^T = (x - x_i^T, y - y_i^T)/h_i^T$ (cf. Exercise 5.17), while the basis (shape) functions of U_h are the characteristic functions of the triangles.

Letting L_h be the number of edges and M_h be the number of triangles of $\mathcal{T}_h$, if we number the edges from one through L_h and elements from one through M_h, at a global level, the shape functions of $\mathbf{Q}_h$ are denoted by $\vec{\eta}_l$ for $l = 1, \ldots, L_h$ and those of U_h are χ_m for $m = 1, \ldots, M_h$. Like in Subsection 4.4.2, let $\vec{p}_h := \sum_{l=1}^{L_h} p_l \vec{\eta}_l$ and $u_h = \sum_{n=1}^{M_h} u_n \chi_n$, and the vector of $L_h + M_h$ unknowns be $\vec{c}_h = [c_1, \ldots, c_{L_h}, c_{L_h+1}, \ldots, c_{L_h+M_h}]^T$, whose first L_h components are the p_ls and the M_h last ones are the u_ns. Then, the above mixed system is seen to be equivalent to the SLAE $\mathcal{A}_h \vec{c}_h = \vec{g}_h$, where $\vec{g}_h$ is an $L_h + M_h$ component vector whose first L_h components are zero and whose $(L_h + m)$-th component equals $-\int_\Omega f \chi_m \, dx dy$. $\mathcal{A}_h$ is a block matrix of the form (4.48) where B_h and C_h are $L_h \times L_h$ and $L_h \times M_h$ matrices obtained in the same manner as in Section 4.3.2 with respect to the bases $\{\vec{\eta}_l\}_{l=1}^{L_h}$ and $\{\chi_m\}_{m=1}^{M_h}$.

Recalling that $(\vec{p}_h|\vec{q})_0 = \sum_{T \in \mathcal{T}} \int_T (\vec{p}_h(\mathbf{x})|\vec{q}(\mathbf{x})) dx dy$, we apply a quadrature formula of the form,

$$\int_T (\vec{p}_h(\mathbf{x})|\vec{q}(\mathbf{x})) dx dy \simeq \sum_{i=1}^{3} \alpha_i^T \int_{e_i^T} (\vec{p}_h|\vec{\nu}_i^T) ds \int_{e_i^T} (\vec{q}|\vec{\nu}_i^T) ds.$$

in every $T \in \mathcal{T}_h$, where the α_i^T's are suitably chosen real coefficients. The matrix D_h approximating B_h through the application of such a quadrature formula is diagonal, owing to the properties of the flux shape functions $\vec{\eta}_l$. Letting θ_i^T be the angle of T at vertex S_i^T, according to references [19] and [106], if $\alpha_i^T = \cot \ \theta_i^T /2$, the thus-modified mixed formulation derived from the standard one by replacing B_h with D_h will yield first-order approximations of $\vec{p}$, u and $div \ \vec{p}$ in the L^2-norm, in the same way as the original mixed formulation [162]. After elimination of the flux unknown vector $\vec{p}_h = [p_1, p_2, \ldots, p_{L_h}]^T$, the resulting system is $A_h \vec{u}_h = \vec{b}_h$, where $A_h := C_h^T D_h^{-1} C_h$ and $\vec{b}_h$ is the M_h-component real vector whose mth component is $-\int_\Omega f \chi_m \ dxdy$. More than this, such a system is identical to the one corresponding to the Cell-centred FVM with the same triangular mesh [19].

Two Poisson equations were solved with both the RT_0 method and the Cell-centred FVM. In order to observe the convergence of the thus-generated approximation with both methods of u, **grad** u and Δu, we computed with meshes consisting of $2J^2$ triangles for $J = 2^j$, with $j = 2, 3, 4, 5, 6$. Sticking to the principles of the FVM, rigorously the approximation of **grad** u is constant in the influence domain Q_e of each mesh edge e. As illustrated in Figure 5.14, if e is an inner edge, Q_e is the convex quadrilateral whose vertices are the centroids of the triangles T_{1e} and T_{2e} having e as a common edge and the end-points of e. Then, the approximation of the gradient of u in Q_e is the difference between the values of u_h in T_{1e} and T_{2e} divided by the distance d_e between the centroids C_{1e} and C_{2e} of both triangles multiplied by the unit normal vector to e pointing outwards T_{2e}. If e is a boundary edge, a similar construction applies, but in this case the influence domain Q_e is the triangle whose vertices are the end-points of e

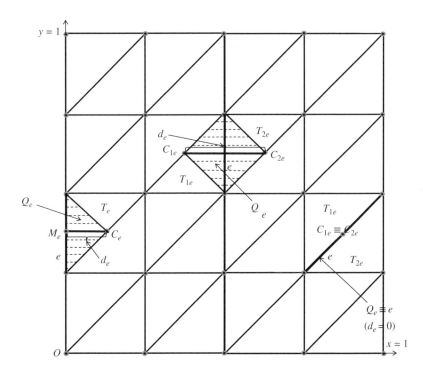

Figure 5.14 Mesh for test case 1 ($J = 4$) and domain Q_e for an edge e

and the centroid of the sole triangle T_e having e as an edge. Then, the approximation of the gradient of u in Q_e is the difference between the value of u_h in T_e and prescribed value of u at M_e, namely, the mid-point of e, divided by the distance d_e between the centroid C_e of T_e and M_e (cf. Figure 5.14). However, $d_e = 0$ if an edge e has only right angles as opposite angles (see Remark 5.13). In this case, as recommended in reference [68], we replace both triangles T_{1e} and T_{2e} by their union, which becomes a quadrilateral CV. Its representative point is the common centroid of both right triangles, namely, the mid-point of e. If e is a boundary edge having an opposite right angle of T_e, the problem is more delicate, but fortunately this does not occur in the examples given below. The approximation of Δu in turn is computed in each CV as the sum of (outgoing) fluxes across its three or four edges divided by its area, while for the RT_0 method it equals the constant $div\ \vec{p}_h$ in each element.

Remark 5.13 *It is interesting to observe that if the Cell-centred FVM is implemented using the mixed RT_0 method as an auxiliary tool, the case of meshes containing pairs of right triangles $(T_{1e}; T_{2e})$ with right angles opposite to their common edge e can be treated with the Cell-centred FVM without assembling them into a quadrilateral CV. Notice that this situation occurs in the important case of uniform meshes of rectangular domains. Letting C_i be the circumcentre of T_{ie} for $i = 1, 2$, in this case $C_{1e} = C_{2e}$ (i.e. the mid-point of e), and the mean flux $(u_{T_{1e}} - u_{T_{2e}})/d_e$ becomes singular for $d_e = 0$. Thus, rigorously $u_{T_{1e}} = u_{T_{2e}}$ must hold. However, it is not necessary to bother about this if the mixed FE approach is used, for it will naturally lead to the right value of this quotient, as seen next in test case 1.∎*

Test case 1: Ω is the unit square $(0, 1) \times (0, 1)$ and $u(x, y) = (x - x^2)(y - y^2)/4$, in which case $f(x, y) = (x - x^2 + y - y^2)/2$. Uniform meshes used in the computations were constructed by first subdividing Ω into J^2 equal squares, each one of them being subdivided in turn into two triangles by its diagonal parallel to the line $x = y$. The mesh is illustrated in Figure 5.14 for $J = 4$. We used the three edge mid-point quadrature formula, which yields the exact integrals of quadratic functions in the triangles.

By means of log-log plots, in Figure 5.15a we show the evolution of the errors of u_h for these two methods, both in the norm of $L^2(\Omega)$ and in the sense of the maximum absolute value

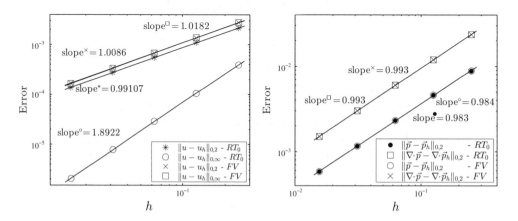

Figure 5.15 Approximation errors of u, $\vec{p} = \nabla u \equiv \mathbf{grad}\ u$ and $\nabla \cdot \vec{p} \equiv div\ \vec{p} = \Delta u$ for test case 1

at the triangle centroids. We call this norm the *discrete L^∞-norm*. First-order convergence in $L^2(\Omega)$ is confirmed for both methods, but RT_0 is a little more accurate than the Cell-centred FVM. It is interesting to observe the super-convergence at the triangle centroids of order not far from two in the case of the RT_0 mixed method only. Similarly, Figure 5.15b displays expected first-order L^2-errors of $\vec{p}_h$ and $div\ \vec{p}_h$, and errors of practically the same magnitudes for both methods. On the other hand, although we do not supply error plots for the flux variable $\vec{p}_h$ in the discrete L^∞-norm, we observed a convergence rate close to the golden number ϕ for both the RT_0 mixed FEM and the Cell-centred FVM, the former being slightly more accurate. However, we should emphasise again that the enhanced accuracy of RT_0 as compared to the Cell-centred FVM is achieved at the price of solving a coupled linear SLAE, while in the case of the latter $\vec{p}_h$ is computed by a simple post-processing of u_h.

Test case 2: Ω is the unit disk with centre at the origin and $u(x, y) = r^2 \log\ r/4$ where $r = \sqrt{x^2 + y^2}$, in which case $f(x, y) = -\log\ r - 1$. Owing to symmetry, only the quarter of Ω given by $x > 0$ and $y > 0$ was taken into account, by applying homogeneous (symmetry) boundary conditions on the edges given by $x = 0$ and $y = 0$. Setting $\vec{p}_h = [p_{hx}, p_{hy}]^T$, in the case of the RT_0 element this means prescribing $p_{hx} = 0$ and $p_{hy} = 0$ respectively. A small subset of a quasi-uniform family of meshes with $2J^2$ triangles was used in the computations. The meshes were generated by mapping the nodes of meshes used in test case 1 onto the quarter disk's triangular mesh. This mapping is based on subdivisions of the polar coordinates r and θ, instead of subdivisions of the Cartesian coordinates x and y, following a procedure described in reference [166]. The thus-generated triangles are of the strict acute type as illustrated in Figure 5.16.

The same quadrature formula as in the previous test case was used to compute the integral of f in the triangles. In doing so, we avoided the singularity of f at the origin.

One of the purposes of the numerical experiments in this test case is to detect how the approximation of the boundary of Ω by a sequence of polygons converging to it affects the orders of convergence of the methods being evaluated. By watching the log-log plots in Figure 5.17a and 5.17b, we infer that the most conspicuous conclusions for Test-case 1 also apply here. Furthermore, they suggest that the boundary approximation by a polygon does not erode the

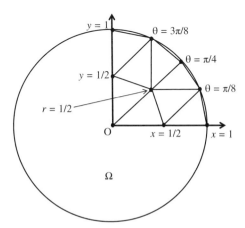

Figure 5.16   Mesh of a quarter of unit disk with $2J^2$ triangles for test case 2 with $J = 2$

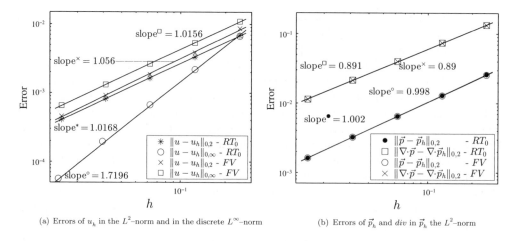

(a) Errors of u_h in the L^2–norm and in the discrete L^∞–norm   (b) Errors of $\vec{p}_h$ and $div\ \vec{p}_h$ the L^2–norm

Figure 5.17   Approximation errors of u, $\vec{p} = \nabla u \equiv \mathbf{grad}\ u$ and $\nabla \cdot \vec{p} \equiv div\ \vec{p} = \Delta u$ for test case 2

first-order convergence in the L^2-norm, known to hold for both u_h and $\vec{p}_h$ in the case of a polygonal domain. However, the observed convergence rate for the RT_0 method in the discrete L^∞-norm decreases to ca. 1.72 and a slightly lower convergence rate of $div\ \vec{p}_h$ to Δu close to 0.89 can be reported for both methods. Finally, the golden number convergence rate for the discrete L^∞-norm of $\vec{p} - \vec{p}_h$ can no longer be notified. Linear convergence rates were observed for both methods instead, but here again we refrain from showing error plots.

Summarising, like in test case 1 the RT_0 method for triangles shows to be globally more accurate than the Cell-centred FVM, and the boundary approximation do influence errors committed with both methods, at least in certain senses.

5.2   Time Integration Schemes for the Heat Equation

In this section, we consider extensions of the analyses conducted in Chapter 3 for the heat equation in one-dimensional space. For the sake of conciseness, we shall often use in this section the *dot symbol* to express dependence on space variables x and y, that is, on position $\mathbf{x}$, in the form of either $(\cdot)$ or $(\cdot, t)$.

More specifically, we address here the convergence analysis of space–time discretisations of the following equation for given f, u_0 and g, namely,

The two-dimensional heat equation

$$
\begin{aligned}
&\partial_t u - p\Delta\ u = f \ \text{in}\ \Omega \times (0, \Theta], \\
&u(\mathbf{x}, 0) = u_0(\mathbf{x}) \quad \forall \mathbf{x} \in \Omega, \\
&u(\mathbf{x}, t) = g(\mathbf{x}, t) \quad \forall (\mathbf{x}, t) \in \Gamma \times (0, \Theta].
\end{aligned}
\tag{5.104}
$$

Among other interpretations, equation (5.104) expresses the distribution of the temperature u during a time interval $(0, \Theta]$, in a medium with constant heat conductivity p occupying a region Ω of the plane, subject to a source f, and prescribed temperature g on the boundary of Ω, given a temperature distribution u_0 at time $t = 0$. This equation has a unique solution. For this property of equation (5.104), among others such as the **maximum principle**, we refer to reference [77].

Since the tools exploited in Chapter 3 combined with those developed in the previous section are all that is needed to study numerical schemes to solve equation (5.104), we will proceed to their convergence analysis in more abridged form.

5.2.1 Pointwise Convergence of Five-point FD Schemes

To begin with, we consider explicit and implicit Euler schemes to solve equation (5.104) combined with a space discretisation based on FD, assuming that Ω is a rectangle. Referring to Figure 4.2, for simplicity we take a uniform grid with spacings h_x and h_y. Then, given an integer $l > 1$ and the associated time step $\tau = \Theta/l$, and setting $u_{i,j}^0 = u_0(ih_x, jh_y)$ for $i = 1, \ldots, n_x$, and $j = 1, \ldots, n_y$ and $u_{0,j}^k = g(0, jh_y, k\tau)$, $u_{i,0}^k = g(ih_x, 0, k\tau)$, $u_{n_x,j}^k = g(L_x, jh_y, k\tau)$, $u_{i,n_y}^k = g(ih_x, L_y, k\tau)$, and $f_{i,j}^k = f(ih_x, jh_y, k\tau)$, we determine successively for $k = 1, 2, \ldots, l$, approximations $u_{i,j}^k$ of $u(ih_x, jh_y, k\tau)$ for $1 \leq i \leq n_x - 1$ and $1 \leq j \leq n_y - 1$,

either explicitly by **the Forward Euler scheme (FES)**

$$u_{i,j}^k = u_{i,j}^{k-1} + \tau[p\Delta_h(u_{i,j}^{k-1}) + f_{i,j}^{k-1}]; \qquad (5.105)$$

or implicitly by **the Backward Euler scheme (BES)**

$$u_{i,j}^k - \tau p\Delta_h(u_{i,j}^k) = u_{i,j}^{k-1} + \tau f_{i,j}^k. \qquad (5.106)$$

We refer to equations (4.5) and (4.6) for the definition of Δ_h with constant spacings $h_{i-1/2} = h_x$ for $1 \leq i \leq n_x$ and $k_{j-1/2} = h_y$ for $1 \leq j \leq n_y$. Notice that both equations (5.105) and (5.106) extend to the case of variable grid spacings h_i and k_j in a straightforward manner.

The schemes can be conveniently and equivalently rewritten in matrix form by introducing an $m \times m$ matrix A_h^+ with $m = (n_x + 1)(n_y + 1) - 4$, once we have suitably numbered the values $u_{i,j}^k$ from one through m, thereby defining a vector $\vec{u}_h^k$ of $\Re^m$. Naturally enough, the numbering of the given values $u_{i,j}^{k-1}$ and $f_{i,j}^k$ follows the same systematic. This leads to a vector $\vec{b}_h^k$ of $\Re^m$ consisting of given values. The reader is encouraged to determine matrix A_h^+ as Exercise 5.18, in case the numbering of the n unknowns with $n = (n_x - 1)(n_y - 1)$ follows the systematic suggested in Chapter 4, and the relevant known boundary values (i.e. all of them except $u_{0,0}^k$, $u_{n_x,0}^k$, u_{0,n_y}^k and u_{n_x,n_y}^k) are numbered from $n + 1$ through m in the counterclockwise sense, starting from $u_{1,0}^k$. On the other hand, all the entries in the last $m - n$ rows of A_h^+ are zero. As a consequence, the entry of the right-side vector $\vec{b}_h^k$ at such positions is just the value of g at the corresponding boundary points. We further define two vectors $\vec{f}_{Fh}^k$ and $\vec{f}_{Bh}^k$ of $\Re^m$ whose first n entries are respectively the $f_{i,j}^{k-1}$'s or the $f_{i,j}^k$'s, and the last $m - n$ entries are the corresponding ones of the vector $[\vec{u}_h^k - \vec{u}_h^{k-1}]/\tau$. In doing so, we may write

The FES in matrix form

$$\vec{u}_h^k = \vec{b}_h^k \text{ with } \vec{b}_h^k = (I - p\tau A_h)\vec{u}_h^{k-1} + \tau \vec{f}_{Fh}^k,$$

The BES in matrix form

$$(I + pA_h)\vec{u}_h^k = \vec{b}_h^k \text{ with } \vec{b}_h^k = \vec{u}_h^{k-1} + \tau \vec{f}_{Bh}^k.$$

The last $m - n$ scalar equations in each one of the above matrix relationships are trivially satisfied. They were added just to complete a SLAE having exactly m equations and m unknowns instead of n. As for the rth equation with $1 \leq r \leq n$, we observe that the entries a_{rs} of A_h^+ satisfy $\sum_{s=1}^m a_{rs} = 0$ and $a_{rs} \leq 0$ for $s \neq r$. This is because $a_{rr} = 2(h_x^{-2} + h_y^{-2})$, and there are only four nonzero off-diagonal matrix coefficients a_{rs}, whose values are either h_x^{-2} or h_y^{-2}, each one of them corresponding to a pair of such coefficients.

Let us examine the stability of both schemes in the vector $l^\infty -$ norm.

A. Pointwise Stability of the Euler Schemes:

Provided for every $r \leq n$ we have $\tau p a_{rr} \leq 1$, all the coefficients of the matrix $I - p\tau A_h$ are non-negative. Furthermore, the sum of all the coefficients in the rth row of this matrix is exactly equal to one for every $r \leq n$. Thus, using the same arguments as in Section 3.3, a sufficient condition for stability of an explicit scheme states that there must be a bound for the time step in terms of the grid sizes. In the case of equation (5.105), this bound is easily found to be $\tau \leq \dfrac{h_x^2 h_y^2}{2p(h_x^2 + h_y^2)}$. This condition must be expressed in terms of the **minimum grid size** in order to cover the case of non-uniform grids. For a uniform grid, the resulting lower bound for the right side of the above inequality gives rise to the

Stability condition for the Forward Euler scheme

$$\tau \leq \frac{h_{min}^2}{4p} \text{ where } h_{min} := \min[h_x, h_y] \tag{5.107}$$

On the other hand, since for every $r \leq n$ the coefficient a_{rr} of matrix A_h equals minus the sum of the off-diagonal coefficients in the same row, we can adapt the arguments exploiting equation (2.6) employed in Section 3.3, in order to study the stability of the Backward Euler scheme for the one-dimensional heat equation. More precisely, we first replace in equation (2.6) p_i with a_{rr} and the pair (p_i^+, p_i^-) with the set of coefficients $-a_{rs}$ for $r \neq s$. Guessing that the remaining adaptations leading to the desired stability result are not so difficult to carry out, we leave them to the reader as Exercise 5.19.

We conclude that the same stability result holds for both Euler schemes, although without stability condition (5.107) in the case of the implicit scheme. In short, we have

Stability inequalities for the Euler schemes with a uniform grid

If for the FES τ fulfils (5.107), $\forall k \leq l$ it holds for both the FES and the BES:
$$\| \vec{u}_h^k \|_\infty \leq \max \Big[\max_{\substack{(x,y,t) \in \\ \Gamma \times [0,\Theta]}} |g(x,y,t)|, \| \vec{u}_h^0 \|_\infty \Big] + \Theta \max_{\substack{1 \leq i \leq n_x; 1 \leq j \leq n_y \\ 0 \leq q \leq k}} |f(ih_x, jh_y, q\tau)|.$$

$$\tag{5.108}$$

Remark 5.14 *It turns out that condition (5.107) is not really necessary for stability of the Forward Euler scheme if $h_x \neq h_y$. Similarly to Section 3.3, this can be checked by means of the Von Neumann criterion (cf. Exercise 5.20).* ∎

B. Convergence of the Euler Schemes:

First, we turn our attention to the consistency of both Euler schemes, assuming that u has bounded partial derivatives in time and space up to the second and fourth order, respectively. We do this rather briefly, for it follows from straightforward extensions of the related material presented in Chapter 3 for the heat equation, combined with the results for the Poisson equation in the previous subsection.

Using the heat equation at time $k'\tau$ with $k' = k - 1$ or at time $k\tau$, replacing $u_{i,j}^k$ by $\tilde{u}_{i,j}^k :=$ $u(ih_x, jh_y, k\tau)$ in equation (5.105) or in (5.106), we come up with the following residuals:

Local truncation error for the Forward Euler scheme

$$
\begin{aligned}
r_{i,j}^{F,k}(u) &= \tau^2 \partial_{tt} u(ih_x, jh_y, [k - \underline{\theta}_{ijk}]\tau)/2 \\
&+ \tau p\{h_x^2[\partial_{xxxx} u([i + \mu_{ijk}^+]h_x, jh_y, k'\tau) + \partial_{xxxx} u([i - \mu_{ijk}^-]h_x, jh_y, k'\tau)] \\
&+ h_y^2[\partial_{yyyy} u(ih_x, [j + \nu_{ijk}^+]h_y, k'\tau) + \partial_{yyyy} u(ih_x, [j - \nu_{ijk}^-]h_y, k'\tau)]\}/24.
\end{aligned}
\tag{5.109}
$$

Local truncation error for the Backward Euler scheme

$$
\begin{aligned}
r_{i,j}^{B,k}(u) &= \tau^2 \partial_{tt} u(ih_x, jh_y, [k - \overline{\theta}_{ijk}]\tau)/2 \\
&\tau p\{h_x^2[\partial_{xxxx} u([i + \overline{\mu}_{ijk}^+]h_x, jh_y, k\tau) + \partial_{xxxx} u([i - \overline{\mu}_{ijk}^-]h_x, jh_y, k\tau)] \\
&+ h_y^2[\partial_{yyyy} u(ih_x, [j + \overline{\nu}_{ijk}^+]h_y, k\tau) + \partial_{yyyy} u(ih_x, [j - \overline{\nu}_{ijk}^-]h_y, k\tau)]\}/24.
\end{aligned}
\tag{5.110}
$$

where the symbols $\underline{\theta}, \mu^+, \mu^-, \nu^+, \nu^-, \overline{\theta}, \overline{\mu}^+, \overline{\mu}^-, \overline{\nu}^+$ and $\overline{\nu}^-$ with subscript ijk are used to represent real numbers in the interval $[0, 1]$.

Finally, combining equation (5.108) with (5.109) and (5.110), and using the same arguments as in Section 3.4, we immediately conclude the validity of the

Pointwise error estimates for the Euler schemes

$$
\begin{aligned}
&\text{Under condition (5.107) for the FES and whatever } \tau \text{ for the BES,} \\
&\text{For } k, 1 \le k \le l \text{ it holds } \forall\, (i, j) \in \{1, 2, \dots, n_x\} \times \{1, 2, \dots, n_y\}. \\
&|u_{i,j}^k - \tilde{u}_{i,j}^k| \le \frac{\Theta}{2} \left\{ \tau \, \| \partial_{tt} u \|_{0,\infty} + \frac{ph^2}{6} [\| \partial_{xxxx} u \|_{0,\infty} + \| \partial_{yyyy} u \|_{0,\infty}] \right\}
\end{aligned}
\tag{5.111}
$$

Summarising, equation (5.111) tells us that, if we reduce the time step τ and refine the FD grid indefinitely, in such a way that h goes to zero, then for every point $(x, y) \in \Omega$ and for every $t \in (0, \Theta]$, there will be a sequence of numerical approximations provided by the Forward or the Backward Euler scheme converging to $u(x, y, t)$, under the stability condition (5.107) for the former. Moreover, as long as (x, y, t) corresponds to grid points for all the levels of grid refinement, from given values of n_x, n_y, l on, we can assert that the sequence of approximations provided by the Euler schemes will converge to $u(x, y, t)$ as fast as $\tau + h^2$ goes to zero. Notice that in the case of the Forward Euler scheme, necessarily the global error is an $O(h^2)$, whereas for the Backward Euler scheme the global of convergence rate will depend on the way the user decides to relate τ and h.

To conclude, it should be stressed that the Forward Euler scheme is suboptimally consistent if we do not use equally spaced grids. This is because time discretisation errors will be an

$O(h^2)$, while the scheme will only first - order consistent, owing to space local truncation errors. On the other hand, if we use non-uniform grids and choose $\tau = O(h)$, the Backward Euler scheme will yield optimal global $O(h)$ errors, though at the price of solving a SLAE at every time step. Notice that the use of Cholesky's method is recommended, for the system matrix is positive-definite and remains fixed.

5.2.2   *Convergence of $\mathcal{P}_1$ FE Schemes in the Mean-square Sense*

Here, we proceed to the most widespread way of studying the reliability of the FEM, as a space discretisation method to solve linear parabolic PDEs, based on H^1- or L^2-norms in space and L^∞-norm in time. This is because such an analysis is natural and leads to optimal results (see e.g. [190]). As we should note, the pioneering work in this connection is generally acknowledged to be the one of Douglas and Dupont, dating back to 1970 [64]. Like in other cases treated so far, the idea is to mimic some properties of the equation being solved numerically. In the present study, we will deal with the solution stability with respect to the data in the sense of mean-square (energy) norms. In order to carry it out, we legitimate beforehand all the operations that follow involving the solution u, thereby postponing the statement of the assumptions that validate them. In particular, several calculations in the sequel strongly rely on Fubini's theorem for multiple integrals [79].

For the sake of brevity, we confine ourselves to the rather academic case where $g \equiv 0$. The general case can be treated in a similar way, though at the price of even lengthier manipulations. Multiplying both sides of this equation by u and integrating in Ω, we obtain

$$\frac{1}{2}\int_\Omega \frac{\partial u^2}{\partial t} dxdy - p\int_\Omega u\Delta\, u\, dxdy = \int_\Omega fu\, dxdy \; \forall t \in (0, \Theta]. \tag{5.112}$$

Applying First Green's identity to equation (5.112), we further establish that

$$\frac{\partial}{\partial t}\left[\int_\Omega \frac{u^2}{2} dxdy\right] + p\int_\Omega |\mathbf{grad}\, u|^2 dxdy = \int_\Omega fu\, dxdy \; \forall t \in (0, \Theta]. \tag{5.113}$$

Integrating both sides of **equation** (5.113) with respect to time from 0 up to an arbitrary $t > 0$, we obtain $\forall t \in (0, \Theta]$,

$$\frac{1}{2}\int_\Omega u^2(\cdot, t)dxdy + p\int_0^t\left[\int_\Omega |\mathbf{grad}\, u|^2 dxdy\right] ds = \int_0^t\left[\int_\Omega fu\, dxdy\right] ds + \frac{1}{2}\int_\Omega u_0^2 dxdy. \tag{5.114}$$

Next, we rewrite the product fu as $fu = [fC_P\sqrt{1/p}][u\sqrt{p}/C_P]$, where C_P is the constant of the Friedrichs–Poincaré inequality (4.8). This trivially leads to the inequality $fu \le [C_P^2 f^2/p + pu^2/C_P^2]/2$. Applying such a result to equation (5.114), together with equation (4.8) to $u(\cdot, s)$ for $s \in (0, t]$, the reader may check without any difficulty the continuous dependence of the solution of equation (5.104) on the data u_0 and f, for $g \equiv 0$, which is a

First bound for the solution of equation (5.104) with $g \equiv 0$ in energy norms

$$\begin{aligned} &\int_\Omega u^2(\cdot, t)dxdy + p\int_0^t[\int_\Omega |\mathbf{grad}\, u|^2 dxdy]ds \\ &\le \frac{C_P^2}{p}\int_0^t\left[\int_\Omega f^2 dxdy\right] ds + \int_\Omega u_0^2 dxdy \; \forall t \in (0, \Theta]. \end{aligned} \tag{5.115}$$

Let us now assume that $\partial_t u$ belongs to $H^1(\Omega)$ for every $t \in [0, \Theta]$. Since necessarily $\partial_t u$ vanishes on Γ for all t, multiplying both sides of equation (5.104) with this function and integrating the resulting relation in Ω, we obtain

$$\int_\Omega [\partial_t u]^2 dxdy + \frac{p}{2} \int_\Omega \partial_t |\mathbf{grad}\ u|^2 dxdy = \int_\Omega f \partial_t u \ dxdy \ \forall t \in (0, \Theta]. \tag{5.116}$$

Next, we integrate both sides of equation (5.116) in time from 0 up to an arbitrary $t > 0$, thereby obtaining $\forall t \in (0, \Theta]$,

$$\int_0^t \left\{ \int_\Omega [\partial_t u]^2 dxdy \right\} ds + \frac{p}{2} \int_\Omega |\mathbf{grad}\ u(\cdot, t)|^2 dxdy = \frac{p}{2} \int_\Omega |\mathbf{grad}\ u_0|^2$$
$$+ \int_0^t \left[\int_\Omega f \partial_t u \ dxdy \right] ds.$$

Finally, using for the integral of $f\partial_t u$ a trick of the same type as for the right side of equation (5.114), after some elementary operations, we derive a

Second bound for the solution of equation (5.104) with $g \equiv 0$ in energy norms

$$\int_0^t \left\{ \int_\Omega [\partial_t u]^2 dxdy \right\} ds + p \int_\Omega |\mathbf{grad}\ u(\cdot, t)|^2 dxdy$$
$$\leq p \int_\Omega |\mathbf{grad}\ u_0|^2 dxdy + \int_0^t \left[\int_\Omega f^2 dxdy \right] ds \ \forall t \in (0, \Theta]. \tag{5.117}$$

Visible interpretations of inequalities (5.115) and (5.117) are as follows. First of all, whenever $f \equiv 0$, the H^1-semi-norm of $u(\cdot, t)$ will be bounded above by the H^1-semi-norm of u_0 at every time t. Moreover, if f does not vanish, the growth of the H^1-norm of $u(\cdot, t)$ with t will be controlled by the growth of the mean-square norms of f involved in both inequalities. Somehow, properties (5.115) and (5.117) should be reproduced by a numerical solution method, otherwise its reliability would be questionable. As seen below, standard FE schemes perfectly do the job.

Explicit and implicit Euler schemes combined with a FE space discretisation can certainly be applied to solve equation (5.104) numerically. However, we have thoroughly studied such methods for parabolic equations in Chapter 3 and in the previous subsection. That is why we will only consider here the Crank–Nicolson scheme, whose analysis incorporates some new aspects. The Euler schemes can be treated by combining the underlying arguments with those previously employed. For this reason, their study is left as Exercises 5.21 and 5.22.

To begin with, still assuming that $g \equiv 0$, under the assumption that $u_0 \in L^2(\Omega)$ and $f(\cdot, t) \in L^2(\Omega)$ for every $t \in (0, \Theta]$, we write equation (5.104) in the following:

Standard Galerkin variational form of the heat equation

Find $u(\cdot, t)$ such that $u(\cdot, 0) = u_0(\cdot)$ in Ω, and for every $t \in (0, \Theta]$,
$u(\cdot, t) \in H_0^1(\Omega)$ and $\partial_t u(\cdot, t) \in L^2(\Omega)$, satisfying $\forall t \in (0, \Theta]$:
$\int_\Omega [\partial_t u \ v + p(\mathbf{grad}\ u|\mathbf{grad}\ v)] dxdy = \int_\Omega fv \ dxdy \ \forall v \in H_0^1(\Omega).$ $\tag{5.118}$

The equivalence of equations (5.118) and (5.104) with $g \equiv 0$ can be easily established.

Let $u_{0,h}$ be the **Ritz projection** of u_0 onto V_{h0}, that is, the unique function in V_{h0} defined by $(\mathbf{grad}\ u_{0,h}|\mathbf{grad}\ v)_0 = (\mathbf{grad}\ u_0|\mathbf{grad}\ v)_0 \ \forall v \in V_{h0}$. We next perform a semi-discretisation in space of equation (5.118), thereby posing a discrete counterpart of this problem, namely,

$$\begin{cases} \text{For every } t \in (0, \Theta] \text{ find } u_h(\cdot, t) \in V_{h0} \text{ with } u_h(\cdot, 0) = u_{0,h}(\cdot) \text{ such that} \\ \int_\Omega [\partial_t u_h\ v + p(\mathbf{grad}\ u_h|\mathbf{grad}\ v)]dxdy = \int_\Omega fv\ dxdy\ \forall v \in V_{h0}. \end{cases} \quad (5.119)$$

Recalling the basis $\mathcal{B} = \{\varphi_j\}_{j=1}^{N_h}$ of V_{h0}, for every t we may write $u_h(\cdot, t) = \sum_{j=1}^{N_h} u_j(t)\varphi_j(\cdot)$ for suitable functions $u_j(t)$. Then, akin to Subsection 3.2.4, equation (5.119) is easily seen to be equivalent to a linear system of ODEs. That is where the time discretisation scheme comes into play. Let again $\tau = \Theta/l$ be a given time step, and u_h^k be an approximation of u_h (and hence of u) at time $t = k\tau$, also belonging to V_{h0}. In the case of the Crank–Nicolson scheme, we approximate the time derivative of u_h at time $(k - 1/2)\tau$ by the FD $(u_h^k - u_h^{k-1})/\tau$.

Before describing the solution scheme, we introduce the following notation: For a real number q in $[0, l]$, and a function ϕ defined (almost) everywhere in a given space–time domain $\omega \times (0, t)$, we denote by ϕ^q the function defined (almost) everywhere in ω by $\phi^q(\cdot) := \phi(\cdot, q\tau)$. In doing so, u_h^k is successively determined by

The FE Crank–Nicolson scheme to solve the heat equation

For $k = 1, 2, \ldots, l$ find $u_h^k \in V_{h0}$ such that $u_h^0 = u_{0,h}$ and $\forall v \in V_{h0}$:

$$\int_\Omega \left\{ (u_h^k - u_h^{k-1})v + \frac{\tau p}{2}(\mathbf{grad}[u_h^k + u_h^{k-1}]|\mathbf{grad}\ v) \right\} dxdy = \frac{\tau}{2} \int_\Omega (f^k + f^{k-1})v\ dxdy.$$

$$(5.120)$$

Since, for every k, we have $u_h^k = \sum_{j=1}^{N_h} u_j^k \varphi_j$, u_j^k being the approximations of u at the mesh node S_j at time $k\tau$, problem (5.120) is nothing but a SLAE having the same set of unknowns. The fact that it has a unique solution can be established as Exercise 5.23.

Incidentally, we note that numerical integration is required in most cases to compute the integral of fv in Ω. However, at this stage we assume that the reader is aware enough about the way such an additional approximation should be handled. That is why, for the sake of brevity, we decline to consider numerical integration here, and propose the corresponding analysis as Exercise 5.24.

Let us switch to the notation $(\mathcal{F}|\mathcal{G})_0$ for the standard L^2-inner product of two functions or vector fields $\mathcal{F}$ and $\mathcal{G}$. For the purpose of studying the convergence of the Crank–Nicolson scheme, we first consider problem (5.121) more general than equation (5.120), in which the integral in Ω of $(f^k + f^{k-1})v/2$ is replaced by the linear functional $d^{k-1/2}(v)$ defined for any $v \in V_{h0}$. The functional $d^{k-1/2}$ is assumed to be bounded in the sense that $d^{k-1/2}(v) \leq D_{k-1/2} \| v \|_{0,2} \ \forall v \in L^2(\Omega)$, where $D^{k-1/2}$ is a constant. We wish to

$$\begin{cases} \text{Find } u_h^k \in V_{h0} \text{ such that } u_h^0 = u_{0,h} \text{ and } \forall v \in V_{h0}: \\ ([u_h^k - u_h^{k-1}]|v)_0 + \frac{\tau p}{2}(\mathbf{grad}[u_h^k + u_h^{k-1}] \mid \mathbf{grad}\ v)_0 = \tau d^{k-1/2}(v) \end{cases} \quad (5.121)$$

Taking $v = u_h^k + u_h^{k-1}$, we come up with

$$\| u_h^k \|_{0,2}^2 - \| u_h^{k-1} \|_{0,2}^2 + \frac{\tau p}{2} \| \mathbf{grad}[u_h^k + u_h^{k-1}] \|_{0,2}^2 = \tau d^{k-1/2}(u_h^k + u_h^{k-1}). \quad (5.122)$$

Now we manipulate the right side of equation (5.122), using the properties of functional $d^{k-1/2}$ combined with the Friedrichs–Poincaré inequality, to obtain

$$\| u_h^k \|_{0,2}^2 + \frac{\tau p}{2} \| \mathbf{grad}[u_h^k + u_h^{k-1}] \|_{0,2}^2 \leq \| u_h^{k-1} \|_{0,2}^2 + \tau C_P D_{k-1/2} \| \mathbf{grad}[u_h^k + u_h^{k-1}] \|_{0,2}. \tag{5.123}$$

Next, applying Young's inequality with $a = \sqrt{\tau} C_P D_{k-1/2}$, $b = \sqrt{\tau} \| \mathbf{grad}(u_h^k + u_h^{k-1}) \|_{0,2}$ and $\delta = 2/p$, to the right side of equation (5.123), we derive

$$\| u_h^k \|_{0,2}^2 + \frac{\tau p}{4} \| \mathbf{grad}[u_h^k + u_h^{k-1}] \|_{0,2}^2 \leq \| u_h^{k-1} \|_{0,2}^2 + \frac{\tau}{p} C_P^2 D_{k-1/2}^2. \tag{5.124}$$

Finally, adding up both sides of equation (5.124) from $k = 1$ up to an arbitrary n, we obtain

A first stability inequality for the FE Crank–Nicolson scheme

$$\forall n \leq l : \| u_h^n \|_{0,2}^2 + \sum_{k=1}^{n} \frac{\tau p}{4} \| \mathbf{grad}[u_h^k + u_h^{k-1}] \|_{0,2}^2 \leq \| u_h^0 \|_{0,2}^2 + \frac{C_P^2 \tau}{p} \sum_{k=1}^{n} D_{k-1/2}^2. \tag{5.125}$$

Equation (5.125) ensures that **the Crank–Nicolson scheme is unconditionally stable**. Once this stability result is established, the next step consists of determining the scheme's order of consistency. In this aim, in contrast to the studies conducted so far, we will not replace the approximate values by interpolates of u, but rather use the **Ritz projection**. As the reader can certainly figure out by following carefully the next steps, this is not a real change, owing to the well known estimates of the difference between both functions, as one can infer from equation (5.14) for example. The Ritz projection $u_{m,h}$ of $u(\cdot, m\tau)$ onto V_{h0}, for $m = 1, 2, \ldots, l$, is the unique solution of the problem,

$$(\mathbf{grad}\ u_{m,h}(\cdot)|\mathbf{grad}\ v)_0 = (\mathbf{grad}\ u(\cdot, m\tau)|\mathbf{grad}\ v)_0 \ \forall v \in V_{h0}.$$

According to the results of Subsection 5.1.2, provided $u(\cdot, t)$ belongs to $H^2(\Omega)$ for every $t \in [0, \Theta]$, and the family of triangulations in use is regular, we have

$$|u_{m,h}(\cdot) - u(\cdot, m\tau)|_{1,2} \leq Ch \| H[u(\cdot, m\tau)] \|_{0,2}. \tag{5.126}$$

Moreover, according to equation (5.46), as long as Ω is convex, the estimate

$$\| u_{m,h}(\cdot) - u(\cdot, m\tau) \|_{0,2} \leq C_0 h^2 \| H[u(\cdot, m\tau)] \|_{0,2}. \tag{5.127}$$

holds true, under the same conditions as equation (5.126), C and C_0 being two mesh-independent constants. Henceforth, we shall assume that the conditions for equation (5.127) to hold are fulfilled.

Hereafter, we will use the following identity, obtained by straightforward calculations:

$$\begin{cases} [u^k - u^{k-1}](\cdot) = \frac{\tau}{2} \left[(\partial_t u)^k + (\partial_t u)^{k-1} \right](\cdot) + E(u, [k - 1/2]\tau)(\cdot) \\ \text{where } E(u, [k - 1/2]\tau)(\cdot) := \frac{1}{2} \int_{(k-1)\tau}^{k\tau} (t - k\tau)[t - (k - 1)\tau][\partial_{ttt} u](\cdot, t) dt. \end{cases} \tag{5.128}$$

Now, we replace u_h^m with $u_{m,h}$ for $m = k - 1$ and $m = k$ on the left side of equation (5.120). From the definition of $u_{m,h}$ for every m and by linearity, this leads to the following relation for every $v \in V_{h0}$:

$$([u_{k,h} - u_{k-1,h}]|\, v)_0 + \frac{\tau p}{2}(\mathbf{grad}[u_{k,h} + u_{k-1,h}]|\, \mathbf{grad}\, v)_0$$

$$= ([u_{k,h} - u_{k-1,h}]|\, v)_0 + \frac{\tau p}{2} + (\mathbf{grad}[u^k + u^{k-1}]|\, \mathbf{grad}\, v)_0 \qquad (5.129)$$

Notice that, owing to equation (5.128), we have $\forall v \in V_{h0}$,

$$([u_{k,h} - u_{k-1,h}]|\, v)_0 = ([u_{k,h} - u^k] - [u_{k-1,h} - u^{k-1}]|\, v)_0$$

$$+ \frac{\tau}{2}([(\partial_t u)^k + (\partial_t u)^{k-1}]|\, v)_0 + \int_\Omega (E(u, [k-1/2]\tau)|\, v)dxdy \qquad (5.130)$$

Thus, plugging equation (5.130) into the right side of equation (5.129), we obtain

$$(u_{k,h} - u_{k-1,h}|\, v)_0 + \frac{\tau p}{2}(\mathbf{grad}[u_{k,h} + u_{k-1,h}]|\mathbf{grad}\, v)_0 =$$

$$([u_{k,h} - u^k]|\, v)_0 - ([u_{k-1,h} - u^{k-1}]|\, v)_0 + \int_\Omega (E(u, [k-1/2]\tau)|\, v)dxdy \qquad (5.131)$$

$$+ \frac{\tau}{2}[([(\partial_t u)^k + (\partial_t u)^{k-1}]|\, v)_0 + p(\mathbf{grad}[\, u^k + u^{k-1}]|\mathbf{grad}\, v)_0].$$

Now, leaving aside the solution of equation (5.119), we redefine the function u_h in $\Omega \times (0, \Theta]$ as the Ritz projection of $u(\cdot, t)$ onto V_{h0} at every time t. Notice that $u_h(\cdot, k\tau) = u_{k,h}(\cdot)$ in Ω for every k, and clearly enough $[(u_{k,h} - u^k) - (u_{k-1,h} - u^{k-1})](\cdot) = \int_{(k-1)\tau}^{k\tau} \partial_t[u_h - u](\cdot, t)dt$. Then setting

$$d^{k-1/2}(v) := \tau^{-1}\left(E(u, [k - 1/2]\tau) + \int_{(k-1)\tau}^{k\tau} \partial_t[u_h - u](t)dt \,|\, v\right)_0, \qquad (5.132)$$

$\forall v \in L^2(\Omega)$, we apply equation (5.118) at times $t = k\tau$ and $t = (k-1)\tau$ to equation (5.131), thereby obtaining

$$(u_{k,h} - u_{k-1,h}|v)_0 + \frac{\tau p}{2}(\mathbf{grad}[u_{k,h} + u_{k-1,h}]|\mathbf{grad}v)_0 = \frac{\tau}{2}([f^k + f^{k-1}]|v)_0 + \tau d^{k-1/2}(v).$$

$$(5.133)$$

Subtracting both sides of equations (5.133) and (5.120), we derive the equations satisfied by the difference $\bar{u}_h^k := u_{k,h} - u_h^k$, for values of k ranging from 1 to l, namely,

$$(\bar{u}_h^k - \bar{u}_h^{k-1}|\, v)_0 + \frac{\tau p}{2}(\mathbf{grad}[\bar{u}_h^k + \bar{u}_h^{k-1}]|\mathbf{grad}\, v)_0 = \tau d^{k-1/2}(v). \qquad (5.134)$$

Now, noticing that $u^k - u_h^k = u^k - u_{k,h} + \bar{u}_h^k$ and taking into account equation (5.134), the consistency of the Crank–Nicolson scheme (without specification of the order) will be ensured, if we establish that $\forall v$, $d^{k-1/2}(v)$ (or yet $D_{k-1/2}$) tends to zero, as τ and h go to zero. While on the one hand this is quite obvious as far as $\tau^{-1}E(u, [k - 1/2]\tau)$ is concerned, on the other hand handling the term $\tau^{-1}\int_{(k-1)\tau}^{k\tau} \partial_t[u_h - u](t)dt$ is more delicate. The key to the problem is a property of the Ritz projection that can be stated as follows: Assume that w is a function defined in $\Omega \times (0, \Theta]$ satisfying $w(\cdot, t) \in H_0^1(\Omega)$ and having a well-defined partial derivative with respect to t such that $\partial_t w(\cdot, t) \in H^1(\Omega)$ for every t. Since this implies that $\partial_t w$ vanishes on Γ for every t, we may consider the Ritz projection $\mathcal{D}w_{t,h}$ of $\partial_t w(\cdot, t)$ onto V_{h0} for every t. The Ritz projection $w_h(\cdot, t)$ of $w(\cdot, t)$ onto V_{h0} is related to $\mathcal{D}w_{t,h}(\cdot)$ for every t as follows:

$$
\begin{cases}
\text{If } \forall v \in V_{h0} \text{ and } \forall t \in [0, \Theta], \ w_h \text{ and } \mathcal{D}w_{t,h} \text{ satisfy} \\
([\mathbf{grad}\ w_h](\cdot, t)|[\mathbf{grad}\ v](\cdot)\ )_0 = ([\mathbf{grad}\ w](\cdot, t)|[\mathbf{grad}\ v](\cdot))_0 \\
\text{and} \\
([\mathbf{grad}\mathcal{D}w_{t,h}](\cdot)|[\mathbf{grad}\ v](\cdot)\ )_0 = ([\mathbf{grad}\ \partial_t w](\cdot, t)|[\mathbf{grad}\ v](\cdot))_0 \\
\text{then } \mathcal{D}w_{t,h}(\cdot) = \partial_t w_h(\cdot, t) \text{ for every } t \in [0, \Theta].
\end{cases}
\tag{5.135}
$$

The validity of equation (5.135) can be justified by simply interchanging the order of differentiation with respect to t in the expression of the above inner products (i.e. $\int_\Omega (\mathbf{grad}\ \partial_t u|\mathbf{grad}\ v) dx dy = \int_\Omega (\partial_t \mathbf{grad}\ u|\mathbf{grad}\ v) dx dy)$, provided the function u is sufficiently smooth. Notice that this is the case of w_h owing to our assumptions on w.

Let us assume that $\| \partial_t u(\cdot, t) \|_{1,2}$ is bounded in $[0, \Theta]$. The reader may check as Exercise 5.25 that, in this case, $\| \partial_t u_h(\cdot, t) \|_{1,2}$ is also bounded in $[0, \Theta]$ (cf. [40]). Then, we may write

$$
\left\| \tau^{-1} \int_{(k-1)\tau}^{k\tau} \partial_t[u_h - u](\cdot, t) dt \right\|_{0,2} \leq \max_{t \in [0,\Theta]} \| \partial_t[u_h - u](\cdot, t) \|_{0,2}.
\tag{5.136}
$$

It follows from equations (5.135), (5.136) and (5.127) that, as long as all the second-order partial derivatives of $\partial_t u$ with respect to x and y belong to $L^2(\Omega)$ for every $t \in [0, \Theta]$ and the function of t given by $\| H(\partial_t u[\cdot, t]) \|_{0,2}$ belongs to $L^\infty(0, \Theta)$, the following estimate holds:

$$
\left\| \tau^{-1} \int_{(k-1)\tau}^{k\tau} \partial_t[u_h - u](\cdot, t) dt \right\|_{0,2} \leq C_0 h^2 \max_{t \in [0,\Theta]} \| H(\partial_t u[\cdot, t]) \|_{0,2}.
\tag{5.137}
$$

Summarising the results obtained so far for the Crank–Nicolson scheme (5.120), we can assert that it is stable in the sense of equation (5.125) and consistent. This allows us to conclude that it converges in a sense related to equation (5.125) as both h and τ go to zero, provided the family of meshes in use is regular. In this aim we have to make a few assumptions on the regularity of u. So far we have implicitly assumed that $\partial_{ttt} u$ is well-defined in $\Omega \times (0, \Theta)$ and $\int_0^\Theta \| \partial_{ttt} u \|_{0,2}^2 \, dt$ is finite, and moreover that $H(\partial_t u)$ is well-defined in $\Omega \times [0, \Theta]$ and $\| H(\partial_t u[\cdot, t]) \|_{0,2} \in L^\infty(0, \Theta)$. Actually the latter assumption can be weakened (cf. Exercise 5.27).

Now taking into account equations (5.134), (5.128) and (5.132), we apply (5.125) to the $\bar{u}_h^k$'s.

Noticing that $\bar{u}_h^0 = 0$, after a series of manipulations using more particularly the Cauchy–Schwarz inequality for time integration, a combination of the above results allows us to write

$$
\begin{cases}
\forall n \leq l: \| \bar{u}_h^n \|_{0,2}^2 + \sum_{k=1}^{n} \frac{\tau p}{4} \| \mathbf{grad}[\bar{u}_h^k + \bar{u}_h^{k-1}] \|_{0,2}^2 \\
\leq C_1^2 \tau^4 \int_0^{n\tau} \| \partial_{ttt} u \|_{0,2}^2 \, dt + C_2^2 h^4 \int_0^{n\tau} \| H(\partial_t u) \|_{0,2}^2 \, dt.
\end{cases}
\tag{5.138}
$$

for suitable constants C_1 and C_2 depending on p and C_P (cf. (4.8)). The reader is encouraged to work out the details leading to estimate equation (5.138) as Exercises 5.26 and 5.27.

Since $\| u_{n,h} - u^n \|_{0,2} \le C_0 h^2 \| H(u^n) \|_{0,2}$, from equation (5.138), we readily derive

A first error estimate for the FE Crank–Nicolson scheme

For every $n \le l$: $\| u_h^n - u^n \|_{0,2} \le$

$$
C_1 \tau^2 \left[\int_0^{\Theta} \| \partial_{ttt} u \|_{0,2}^2 \, dt \right]^{1/2} + C_2 h^2 \left[\int_0^{\Theta} \| H(\partial_t u) \|_{0,2}^2 \, dt \right]^{1/2} + C_0 h^2 \| H(u^n) \|_{0,2}.
$$
(5.139)

Equation (5.139) means that the Crank–Nicolson scheme (5.120) is second-order convergent in time and space in the maximum norm in t of the norm of $L^2(\Omega)$. Moreover, provided τ varies as an $O(h)$ the scheme is globally second-order convergent in terms of h. Actually, this is the optimal choice of τ, since the time step is not limited by whatever stability condition. Equation (5.125) also indicates that a result similar to equation (5.139) holds for the error of the solution gradient. However, in this case, the error in time is no longer in the pointwise sense, but rather in the mean-square sense. Notice that such an error is only an $O(\tau^2) + O(h)$, since the approximation of the solution gradient in space in L^2 is of the first order. Furthermore, there is another problem here: estimate (5.140) given below bounds above the error with respect to an approximation of the norm of $L^2(0, l\tau)$ obtained by applying a modified one-point quadrature rule in each time interval $([k-1]\tau, k\tau)$, and not the error with respect to the exact L^2-norm; the modification consists of using mean values $[u^k + u^{k-1}]/2$, instead of $u^{k-1/2}$. More precisely, as the reader might check without any particular difficulty, using equation (5.126), we derive the following estimate for every n, $1 \le n \le l$:

$$
\left\{ \sum_{k=1}^{n} \tau p \left\| \frac{\mathbf{grad}[u_h^k + u_h^{k-1}] - \mathbf{grad}[u^k + u^{k-1}]}{2} \right\|_{0,2}^2 \right\}^{1/2} \le C_1 \, \tau^2 \left[\int_0^{\Theta} \| \partial_{ttt} u \|_{0,2}^2 \, dt \right]^{1/2}
$$

$$
+ C_2 h^2 \left[\int_0^{\Theta} \| H(\partial_t u) \|_{0,2}^2 \, dt \right]^{1/2} + Ch \left[\sum_{k=1}^{n} \tau p \, \frac{H(u^k) + H(u^{k-1})}{2} \right]_{0,2}^{2} \Bigg]^{1/2}.
$$
(5.140)

In short, equation (5.140) has to be improved, so that it possibly becomes an equivalent estimate for

$$
G_h(u) := \left\{ \int_0^{\Theta} \| \mathbf{grad}[u(\cdot, t) - u_{h,\tau}(\cdot, t)] \|_{0,2}^2 \, dt \right\}^{1/2},
$$
(5.141)

from the qualitative point of view, $u_{h,\tau}(\cdot, t)$ being the time-independent function in each time interval $([k-1]\tau, k\tau)$, coinciding with $[u_h^k + u_h^{k-1}]/2$ therein at every point of Ω, for $k = 1, 2, \ldots, l$. Let us give some hints on the way $G_h(u)$ can be estimated under the reasonable assumption that $\partial_t \mathbf{grad} \, u$ belongs to $L^2[\Omega \times (0, \Theta)]$.

For convenience, we introduce the function $u_\tau(\cdot, t)$, whose value in the whole time interval $([k-1]\tau, k\tau)$ equals $[u^k + u^{k-1}]/2$ everywhere in Ω, for $k = 1, 2, \ldots, l$. Next setting $\bar{u}_{h,\tau} := u_\tau - u_{h,\tau}$, $\bar{u}_\tau := u - u_\tau$, the right side of equation (5.140) is seen to provide an error bound for $[\int_0^{n\tau} \| \mathbf{grad} \, \bar{u}_{h,\tau} \|_{0,2}^2]^{1/2}$, and thus for the norm of $\mathbf{grad} \, \bar{u}_{h,\tau}$ in $L^2[\Omega \times (0, \Theta)]$ taking $n = l$.

Now we note that, owing to the triangle inequality,

$$
G_h(u) \le \left[\int_0^\Theta \| \mathbf{grad}\ \bar{u}_{h,\tau} \|_{0,2}^2\ dt \right]^{1/2} + \left[\int_0^\Theta \| \mathbf{grad}\ \bar{u}_\tau \|_{0,2}^2\ dt \right]^{1/2}.
$$

On the other hand, by straightforward calculations, we obtain

$$
\begin{cases}
\mathbf{grad}\ \bar{u}_\tau(\cdot, t) = \vec{g}_{k-1/2}(u)(\cdot, t)\ \forall t \in ([k-1]\tau, k\tau), \\
\text{where } \vec{g}_{k-1/2}(u)(\cdot, t) := \frac{1}{2} \int_{(k-1)\tau}^t \partial_t \mathbf{grad}\ u(\cdot, s)ds - \frac{1}{2} \int_t^{k\tau} \partial_t \mathbf{grad}\ u(\cdot, s)ds.
\end{cases} \tag{5.142}
$$

It follows that

$$
\int_0^\Theta \int_\Omega |\mathbf{grad}\ \bar{u}_\tau(\cdot, t)|^2 d(\cdot)\ dt = \sum_{k=1}^l \int_{(k-1)\tau}^{k\tau} \int_\Omega |\vec{g}_{k-1/2}(u)(\cdot, t)|^2 d(\cdot)\ dt. \tag{5.143}
$$

Applying the inequality $\vec{a} - \vec{b} \le |\vec{a}| + |\vec{b}|,\ \forall \vec{a}, \vec{b} \in \Re^2$ to $\vec{g}_{k-1/2}(u)$ given by equation (5.142), and then the Cauchy–Schwarz inequality to the integrals in $([k-1]\tau, t)$ and $(t, k\tau)$, we easily derive

$$
\sum_{k=1}^l \int_{(k-1)\tau}^{k\tau} \int_\Omega |\vec{g}_{k-1/2}(u)(\cdot, t)|^2 d(\cdot) dt \le
$$

$$
\frac{1}{2} \sum_{k=1}^l \int_{(k-1)\tau}^{k\tau} [|k\tau - t| + |t - (k-1)\tau|] dt \int_{(k-1)\tau}^{k\tau} \int_\Omega |\partial_t \mathbf{grad}\ u(\cdot, t)|^2 d(\cdot) dt. \tag{5.144}
$$

Thus, from equations (5.143) and (5.144), trivial calculations and upper bounds lead to

$$
\int_0^\Theta \int_\Omega |\mathbf{grad}\ \bar{u}_\tau(x, y, t)|^2 dxdydt \le \frac{\tau^2}{2} \int_0^\Theta \int_\Omega |\partial_t \mathbf{grad}\ u(x, y, t)|^2 dxdydt. \tag{5.145}
$$

Now, the remainder of the estimate of $G_h(u)$ becomes a rather easy task, and is left to the reader as Exercise 5.28. More precisely, a final estimate in terms of $O(h) + O(\tau)$, of the error in the norm of $L^2[\Omega \times (0, \Theta)]$ (i.e. $\left\{ \int_0^\Theta \int_\Omega |\mathbf{grad}[u - u_{h,\tau}]|^2 dxdy\ dt \right\}^{1/2}$) should be found, under the regularity assumptions on u that we have specified.

One can expect to obtain qualitatively equivalent pointwise error estimates in time for the solution gradient, starting from another stability result applying to equation (5.120) analogous to equation (5.117). Let us see how this works.

First, we take $v = [u_h^k - u_h^{k-1}]/\tau \in V_{h0}$ in equation (5.120), which easily gives

$$
\tau \left\| \frac{u_h^k - u_h^{k-1}}{\tau} \right\|_{0,2}^2 + \frac{p}{2} \|\mathbf{grad}\ u_h^k\|_{0,2}^2 = \frac{p}{2} \|\mathbf{grad}\ u_h^{k-1}\|_{0,2}^2 + \tau \left| d^{k-1/2} \left(\frac{u_h^k - u_h^{k-1}}{\tau} \right) \right|. \tag{5.146}
$$

Then, we apply Young's inequality with $\delta = 1$ to the right side of equation (5.146), to obtain

$$
\tau \left\| \frac{u_h^k - u_h^{k-1}}{\tau} \right\|_{0,2}^2 + p\|\mathbf{grad}\ u_h^k\|_{0,2}^2 \le p\|\mathbf{grad}\ u_h^{k-1}\|_{0,2}^2 + \tau D_{k-1/2}^2. \tag{5.147}
$$

Finally, adding up both sides of equation (5.147) from $k = 0$ through $k = n$, $n \leq l$, we establish a discrete analog of equation (5.117), namely,

A second stability inequality for the FE Crank–Nicolson scheme

$$\forall n : \tau \sum_{k=1}^{n} \left\| \frac{u_h^k - u_h^{k-1}}{\tau} \right\|_{0,2}^2 + p\|\mathbf{grad}\ u_h^n\|_{0,2}^2 \leq p\|\mathbf{grad}\ u_h^0\|_{0,2}^2 + \tau \sum_{k=1}^{n} D_{k-1/2}^2. \quad (5.148)$$

As far as consistency is concerned, we recall the expression (5.132) of the functional $d^{k-1/2}$ in terms of $E(u, [k - 1/2]\tau)$ given by equation (5.128). Thus, by mimicking the steps leading to equation (5.139), the remainder of the analysis is straightforward. This leads to

A second error estimate for the FE Crank–Nicolson scheme

For every $n \leq l$: $\| \mathbf{grad}\ [u^n - u_h^n] \|_{0,2} \leq$

$$C_2\tau^2 \left[\int_0^\Theta \| \partial_{ttt}u \|_{0,2}^2\ dt \right]^{1/2} + \frac{C_2}{\sqrt{p}}h^2 \left[\int_0^\Theta \| H(\partial_t u) \|_{0,2}^2\ dt \right]^{1/2} + Ch\ \| H(u^n)\|_{0,2}, \quad (5.149)$$

C being the constant independent of h in equation (5.44).

As one might expect, under the previous assumptions on u, the error of the solution gradient in the pointwise sense in time and in the mean-square sense in space of the Crank–Nicolson scheme is of the second order in terms of τ and of the first order in terms of h.

To conclude, we should point out that in many applications, the time derivative of the solution is an important physical quantity. For this reason, controlling the numerical error of $\partial_t u$ in some sense could be a must. As a by-product of equation (5.148), it is possible to obtain error estimates for the time derivative of u in the underlying natural norm. Besides exploiting a stability result like in the other cases, we have to estimate the square root of the deviation,

$$\int_0^\Theta \left[\int_\Omega |\partial_t(u - \tilde{u})|^2 dx dy \right] dt = \int_\Omega \sum_{k=1}^{l} \left[\int_{(k-1)\tau}^{k\tau} \left| \partial_t u - \frac{u^k - u^{k-1}}{\tau} \right|^2 dt \right] dx dy,$$

where $\tilde{u}$ is a function varying linearly with t in each interval $([k - 1]\tau, k\tau)$ such that $\tilde{u}(\cdot, k\tau) = u^k(\cdot)$ for $k = 0, 1, \ldots, l$. The result is an additional $O(\tau)$ term under the assumption that $\partial_{tt}u$ is square integrable in $\Omega \times (0, \Theta)$. Hence, the order of approximation of the solution time derivative in the norm of $L^2[\Omega \times (0, \Theta)]$ is one with respect to τ and two with respect to h. The detailed calculations leading to such an error estimate, including the specification of the constants appearing therein, are proposed to the reader as Exercise 5.29.

5.2.3 Pointwise Behaviour of FE and FV Schemes: An Overview

To close this chapter, we address a few words about the pointwise behaviour of FE and FV schemes in the framework of parabolic problems. This completes our comments on this issue made in Part C of Subsection 5.1.5.

To begin with, we underline that in the two-dimensional case, both discretisation methods are designed to operate with arbitrary meshes. The terms **structured** or **unstructured** are commonly used to qualify a mesh, whether or not its pattern follows a fixed number of roughly parallel lines. For example, the mesh of Figure 4.11 can be viewed as an unstructured triangular mesh. In contrast, the non-uniform cartesian mesh associated with the FD grid considered in Subsection 5.1.1 is a **structured mesh**. Since both the FEM and the FVM are not subject to any preliminary structuring, in current applications unstructured triangular meshes are often handled when using these methods to solve a two-dimensional boundary value problem. However, as we have seen so far, the convergence results derived for both methods are mostly in the mean square sense. This is because they are based on integral formulations, and hence it seems more natural to establish their convergence in this sense. Of course, a priori, nothing prevents one from doing so in the pointwise sense, akin to Chapter 2 in the one-dimensional case. However, obtaining such convergence results in connection with unstructured meshes for two-dimensional or higher dimensional problems, is much more laborious, as already pointed out in Subsection 5.2.3. Nevertheless, we would like to give a quick overview of the results in the pointwise sense that are known to hold for the P_1 FEM and the Cell-centred FVM, as applied to parabolic problems.

As far as stability is concerned, the pointwise analysis carried out for the FDM in Subsection 5.1.1 can be adapted to FE or FV discretisations and equation (5.104), provided some geometric conditions are fulfilled. More specifically, the same kind of pointwise stability for the FDM holds, as long as Delaunay triangulations are used and no obtuse angle faces a boundary edge. As a matter of fact, under this assumption all the off-diagonal coefficients of the underlying matrix A_h are negative, and the sum of the coefficients in every row of A_h is non-negative. An outcome of this property, eventually combined with a condition relating the mesh and time steps in the case of an explicit scheme, is a numerical behaviour qualitatively comparable to the one inherent to the Five-point FDM.

On the other hand, we recall that the error in the mean-square sense of the Ritz projection was the key to the consistency analysis of Subsection 5.2.2. However, if one is dealing with the maximum norm and unstructured meshes, this is more tricky, owing to limitations for using Taylor expansions in this case. In short, this means that the same difficulties encountered in the study of equation (4.1) must be overcome.

We should stress that, since the 1970s and up to the 2000s, deriving optimal convergence results in maximum norms, not only for the P_1 FEM (and consequently for related FVM) but also for more general FVM, as applied to elliptic or parabolic equations, has been the object of rather sophisticated mathematical analyses (see e.g. [144], [145], [180], [168], [68], [81] and [42]). That is why in this book we will not go beyond the mere quotation of the most representative known results in the framework of elliptic problems (cf. Part C of Subsection 5.1.5). The interested reader will find complete studies and analyses in the works quoted above and in references therein.

5.3   Exercises

5.1   Consider the Laplace equation $\Delta u = 0$ in the rectangle $(0, L_x) \times (0, L_y)$ with inhomogeneous Dirichlet boundary conditions $u = g_0$ everywhere except along edge $y = 0$, where the inhomogeneous Neumann boundary condition $\partial_y u = -g_1$ is enforced, g_1 being a continuous function. We wish to solve this problem by the Five-point FDM with a uniform $n_x \times n_y$ grid, and in this aim we propose the following use of the Laplace equation at $M_{i,0}$, in order to take into account the boundary condition at point $M_{i,0}$, for $i = 1, \ldots, n_x - 1$:

$$[u_{i,1} - (1 + h_y^2/h_x^2)u_{i,0} + h_y^2(u_{i+1,0} + u_{i-1,0})/(2h_x^2)]/h_y = -g_1(M_{i,0}).$$

Show that the local truncation error in the approximation of the Neumann boundary condition at $M_{i,0}$, when $u_{k,0}$ is replaced by $u(M_{k,0})$ for $k = i - 1$, $k = i$ or $k = i + 1$ and $u_{i,1}$ by $u(M_{i,1})$, is an $O(h_y^2 + h_y h_x^2)$.

5.2 Inspired by the technique indicated in Exercise 5.1, we assume that the second derivative g_{1xx} of g_1 with respect to x is well-defined and continuous everywhere on Γ_1. Akin to the case of equation (1.34), we define fictitious approximations $u_{i,-1}$ at points $M_{i,-1}$ of coordinates $(ih_x; -h_y)$ by $u_{i,-1} = u_{i,1} + 2h_y g_1(M_{i,0}) - h_y^3 g_{1xx}(M_{i,0})/3$. Then, we apply the Five-point FD scheme as follows: $\dfrac{2u_{i,j} - u_{i+1,j} - u_{i-1,j}}{h_x^2} + \dfrac{2u_{i,j} - u_{i,j+1} - u_{i,j-1}}{h_y^2} = 0$ for $i = 1, \ldots, n_x - 1$, and $j = 0, 1, \ldots, n_y - 1$, with $u_{0,j} = g_0(M_{0,j})$ and $u_{n_x,j} = g_0(M_{n_x,j})$ for $j = 1, \ldots, n_y - 1$ and $u_{i,n_y} = g_0(M_{i,n_y})$ for $i = 1, \ldots, n_x - 1$. Determine the local truncation error of this scheme at $M_{i,0}$ for $i = 1, \ldots, n_x - 1$. Infer that the method is second-order convergent in terms of $h = max[h_x, h_y]$.

5.3 Following the steps leading to the error estimate in the semi-norm $|\cdot|_{1,2}$ in Subsection 5.1.2, extend them in order to derive equivalent error estimates for $\| u - u_h \|_{1,2}$, where u is the solution of the equation $-\Delta u + u = f \in L^2(\Omega)$ with homogeneous Neumann boundary conditions, and u_h is its $\mathcal{P}_1$ FE approximation. Akin to Section 5.1.2, assume that u belongs to $W^{2,\infty}(\Omega)$, and consider only the case of exact integrations.

5.4 Replace u_h by $\tilde{u}_h$ in equation (4.15), and define the variational residual $R_h(\tilde{u}_h, v) := a(\tilde{u}_h, v) - F(v) \; \forall v \in V_h$. Show that, provided u is sufficiently smooth, R_h is bounded above by $C_h h \| H(u) \|_{0,2} |v|_{1,2}/2 \; \forall v \in V_h$.

5.5 Let Ω be a polygon, u be a function in $W^{2,\infty}(\Omega)$, and $\tilde{u}_h$ be its $\mathcal{P}_1$ interpolate at the nodes of a partition $\mathcal{T}_h$ of Ω into triangles, belonging to a regular family of triangulations. Find an expression for the mesh-independent constant C for which the following error bound holds:

$$\| u - \tilde{u}_h \|_{0,\infty} + h \| \mathbf{grad}[u - \tilde{u}_h] \|_{0,\infty} \leq Ch^2 \| H(u) \|_{0,\infty}.$$

Although other definitions are commonly found in the literature, work with the norms

$$\| \mathbf{grad} \cdot \|_{0,\infty} := \| |\mathbf{grad} \cdot| \|_{0,\infty} \text{ and } \| H(\cdot) \|_{0,\infty} := \| |H(\cdot)| \|_{0,\infty},$$

where $|\cdot|$ is the Euclidean norm of N-component fields, N being 2 and 4, respectively.

5.6 Let $\{\mathcal{T}_h\}_h$ be a regular family of triangulations of a polygon Ω. Show that there exists an angle $\theta_0 > 0$ such that the angles of all triangles in $\mathcal{T}_h$ remain bounded below and above by θ_0 and $\pi - 2\theta_0$, respectively, whatever the mesh in the family.

5.7 Let Ω be the rectangle $(0, L_x) \times (0, L_y)$ with boundary Γ and unit outer normal vector $\vec{\nu}$, and w be a function in $W^{2,\infty}(\Omega) \cap H_0^1(\Omega)$ having third-order derivatives in $L^2(\Omega)$

– that is, $w \in H^3(\Omega)$ (cf. Adams, Sobolev spaces, 1975). Γ being the boundary of Ω, show that $\oint_\Gamma ([\Delta w] \vec{\nu} | \mathbf{grad} \ w) \ ds = \oint_\Gamma (H(w) \vec{\nu} | \mathbf{grad} \ w) ds$.

5.8 Prove the validity of equation (5.46) under the assumption that Ω is a convex polygon and that a regular family of triangulations of Ω is in use. In this aim, admit the following well-known result: as long as Ω is of such a shape, whenever $w \in H_0^1(\Omega)$ and $\Delta w \in L^2(\Omega)$, then $w \in H^2(\Omega)$, and there exists a constant $C(\Omega)$ depending only on Ω such that $\| H(w) \|_{0,2} \leq C(\Omega) \| \Delta w \|_{0,2} \ \forall w \in H^2(\Omega) \cap H_0^1(\Omega)$.

5.9 Let f be a continuous function in $\overline{\Omega}$, Ω being a polygon. In each element T of a triangulation $\mathcal{T}_h$ of Ω, $\int_T fv \ dxdy$ is approximated by $J_T(fv) := f(G_T)v(G_T)area(T)$, where G_T is the centroid of T and v is a function in the $\mathcal{P}_1$ FE space V_h. Assuming that f has a bounded gradient in $\overline{\Omega}$, show that the use of this one-point quadrature formula does not affect $\mathcal{P}_1$ FEM's final order of approximation of the solution to equation (4.15) (or equation (4.14)) in the seminorm of $H^1(\Omega)$ (hint: use the fact that $\int_T v \ dxdy = area(T)v(G_T) \ \forall v \in V_h$ and $\forall T \in \mathcal{T}_h$, and estimate the additional term $\sum_{T \in \mathcal{T}_h} \left[\int_T fv \ dxdy - J_T(fv) \right]$ in the variational residual, before applying the appropriate stability inequality).

5.10 Under the same assumptions as in Exercise 5.9, replace the quadrature formula $J_T(fv)$ in each element T of a triangulation of Ω by the trapezoidal rule, namely, $K_T(fv) := \sum_{i=1}^3 [fv](S_r^T)area(T)/3$, where S_r^T for $r = 1, 2, 3$ are the vertices of T and v is a function in the $\mathcal{P}_1$ FE space V_{h0}. Show that the same conclusion of Exercise 5.9 holds.

5.11 Starting from equation (5.54), derive equation (5.55). Give an expression for the constant $\mathcal{C}_h$ in equation (5.55) in terms of the minimum angle θ_0 of the mesh triangles in a regular family of partitions.

5.12 Check the validity of identity (5.56).

5.13 In the rectangular Cell-centred FVM, approximate the right side by the one-point quadrature formula $\int_R f \ dxdy \simeq f_R area(R)$, with $R = R_{i-1/2,j-1/2}$, assuming that f is continuous, where $f_R = f(x_{i-1/2}, y_{j-1/2})$. Prove that this does not cause any lowering of the method's order of convergence, as long as f is sufficiently smooth.

5.14 Derive equation (5.84) starting from equation (5.83).

5.15 Show that the local truncation error of scheme (5.71) is an $O(h^0)$. Instead of $\tilde{f}_{i-1/2,j-1/2}$, take $f_{i-1/2,j-1/2} = f(x_{i-1/2}, y_{j-1/2})$.

5.16 Prove that equation (2.54) holds for scheme (1.27), assuming that p is constant and both q and f are continuous, with $q(x) \geq \beta > 0 \ \forall x \in [0, L]$. Give an expression for the constant $C_c(u)$ (hint: use the fact that both u' and u'' belong to $C[0, L]$).

5.17 Check that the local basis field $\vec{\eta}_i^T$ satisfying $\int_{e_j^T} (\vec{\eta}_i^T | \vec{\nu}_j) dxdy/l_j^T = \delta_{ij}$ for the flux space $\mathbf{Q}_h$ associated with the RT_0 mixed FE for a triangle T, whose vertices are $S_i^T = (x_i^T; y_i^T)$ for $i = 1, 2, 3$, is given by $\vec{\eta}_i^T(x, y) = [x - x_i^T, y - y_i^T]^T/h_i^T$; h_i^T is the (length of the) height of T related to S_i^T, l_i^T is the length of edge e_i^T of T opposite to S_i^T, and $\vec{\nu}_i$ is the unit outer normal vector along e_i^T, $i = 1, 2, 3$.

5.18 Specify the nonzero coefficients of an $n \times m$ matrix A_h for the two-dimensional Forward Euler scheme with $n = (n_x - 1)(n_y - 1)$ and $m = (n_x + 1)(n_y + 1) - 4$, assuming that the numbering of the n unknowns follows the system suggested in Chapter 4 for the Five-point FDM, and the $m - n$ relevant prescribed boundary values are numbered from $n + 1$ through m in the counterclockwise sense, starting from $u_{1,0}^k$. Sketch the form of A_h by representing its nonzero entries by an x and its zero entries by a 0. Take significant but small values of n_x and n_y.

5.19 Starting from equation (2.6), establish in detail the pointwise stability inequality (5.108) for the two-dimensional FD Backward Euler scheme.

5.20 Check by means of a Von Neumann analysis, analogous to the one carried out in Section 3.3, that the condition $2p\tau[h_x^{-2} + h_y^{-2}] \leq 1$ is necessary for the stability of the Forward Euler scheme. Check also that it is sufficient for stability, as a relaxed condition (equation (5.107)).

5.21 The purpose of this exercise is to study the stability of the FE-FES in the particular case of homogeneous Dirichlet boundary conditions. More precisely, letting u_h^0 be the Ritz projection of $u_0 \in H_0^1(\Omega)$ onto V_{h0}, determine for $k = 1, 2, \ldots, l$ $u_h^k \in V_{h0}$ such that, $(u_h^k - u_h^{k-1}|v)_0 + \tau p(\mathbf{grad}\, u_h^{k-1}|\mathbf{grad}\, v)_0 = \tau(f^{k-1}|v)_0 \ \forall v \in V_{h0}$.
Consider exact integration, but take $f \equiv 0$. Assuming that the family of meshes in use is quasi-uniform, first establish the following **inverse inequality**: There exists a mesh-independent constant C_I such that,

$$\| \mathbf{grad}\, v\|_{0,2} \leq C_I h^{-1} \| v\|_{0,2} \text{ holds } \forall v \in V_{h0}$$

Prove that under a condition on τ and h involving constant C_I, the following stability inequality holds:

$$\| u_h^m \|_{0,2}^2 \leq 2 \| u_h^0 \|_{0,2}^2 + \tau p \| \mathbf{grad}\, u_h^0 \|_{0,2}^2 \text{ for } 1 \leq m \leq l.$$

5.22 Still consider only homogeneous Dirichlet boundary conditions and that exact integration is implemented, but assume no longer that $f \equiv 0$ and that the family of meshes is quasi-uniform. Taking u_h^0 like in exercise 5.21, we set up the FE-BES, for $k = 1, 2, \ldots, l$:
Find $u_h^k \in V_{h0}$ such that $(u_h^k - u_h^{k-1}|v)_0 + \tau p(\mathbf{grad}\, u_h^k|\mathbf{grad}\, v)_0 = \tau(f^k|v)_0$ $\forall v \in V_{h0}$.
Derive the stability inequality below by specifying the constant C_B:

$$\| u_h^m \|_{0,2}^2 + \tau p \sum_{k=1}^m \| \mathbf{grad}\, u_h^k \|_{0,2}^2 \leq \| u_h^0 \|_{0,2}^2 + C_B \tau \sum_{k=1}^m \| f^k \|_{0,2}^2 \text{ for } 1 \leq m \leq l.$$

5.23 Show that equation (5.120) has a unique solution at every step.

5.24 Assume that f is continuous in $\overline{\Omega} \times [0, \Theta]$ and that the Crank–Nicolson scheme studied in Subsection 5.2.2 is modified by using the trapezoidal quadrature rule to compute the integrals of fv in every mesh triangle, instead of exact integration. Modify all the error estimates obtained therein in order to take into account this additional approximation. Specify the required regularity of f in a maximum norm, in order to avoid qualitative erosion in the error estimate (5.139).

5.25 Let $u_h(\cdot, t)$ be the Ritz projection (w.r. to the inner product $(\mathbf{grad} \cdot |\mathbf{grad}\cdot)_0$ of $H_0^1(\Omega)$) of $u(\cdot, t)$ onto V_{h0} for every t, where u is the solution to the heat equation. Show that if $|\partial_t u(\cdot, t)|_{1,2}$ is bounded in $[0, \Theta]$, then so is $\| \partial_t u_h(\cdot, t) \|_{1,2}$.

5.26 Equation (5.138) results from the application of equation (5.125) to $\bar{u}_h^k = u_{k,h} - u_h^k$, after a combination of equations (5.128), (5.132) and (5.134), $u_{k,h}$ being the Ritz projection of $u(\cdot, k\tau)$ onto V_{h0}, and u_h^k being the solution provided by the Crank–Nicolson scheme at the kth time step. Express the constant C_1 in estimate (5.138) in terms of problem data.

5.27 Check that the term $\max_{t \in [0, n\tau]} \| H(\partial_t u[\cdot, t]) \|_{0,2}^2$ can be replaced by $\int_0^{n\tau} \| H(\partial_t u) \|_{0,2}^2\, dt$ in equation (5.137). Then give an estimate for the constant C_2 in equation (5.138).

5.28 Starting from equation (5.144), estimate $G_h(u)$ given by equation (5.141).

5.29 Derive an error estimate establishing that the order of approximation of the solution time derivative in the norm of $L^2[\Omega \times (0, \Theta)]$ is one with respect to τ and two with respect to h. Specify the constants appearing therein.

5.30 This exercise is related to Example 5.1, and here we consider only the case where Ω is a convex polygon. Let u_h be a $\mathcal{P}_1$-FE solution of equation (5.47) with $g_1 \equiv 0$ and $f \in L^2(\Omega)$ satisfying $\int_\Omega f\, dxdy = 0$, by further prescribing $\int_\Omega u\, dxdy = 0$. The additive constant d_h up to which u_h is defined is fixed by enforcing $\int_\Omega u_h dxdy = 0$. We assume that any solution to the Poisson equation $-\Delta v = g$ in Ω with homogeneous Neumann boundary conditions belongs to $H^2(\Omega)$ for every $g \in L^2(\Omega)$ satisfying $\int_\Omega g\, dxdy = 0$, and that moreover there exists a constant C_N depending only on Ω such that $\| H(v) \|_{0,2} \leq C_N \| g \|_{0,2}$, as long as the additional condition $\int_\Omega v\, dxdy = 0$ is fulfilled. Show that an error estimate in $O(h^2)$ for $u - u_h$ holds in $L^2(\Omega)$.

6

Extensions

Truth is ever to be found in simplicity, and not in the multiplicity and confusion of things.

Isaac Newton

So far, we have confined ourselves to the simplest versions of the FDM, FEM and FVM in both one and two space variables and mostly to homogeneous Dirichlet and Neumann boundary conditions. This chapter brings about several important complements to the contents of Chapters 1 through 5 in both respects. More precisely, one of the purposes of this chapter is to present some other versions of the FEM that are supposedly more accurate, thereby showing how universal and versatile the method can be. The treatment of more general boundary conditions than those considered so far is also addressed, extended to the case of problems posed in three-dimensional spatial domains.

Chapter outline: An outline of the chapter is as follows. In Section 6.1, we consider FEMs based on polynomials of degree greater than one, to solve second-order elliptic equations. We start the presentation in Subsection 6.1.1 by considering the one-dimensional case. Then, in Subsections 6.1.2 and 6.1.3 we introduce the bilinear FEM for quadrialteral meshes and the quadratic FEM for triangular meshes. In Subsection 6.1.4 the case of curved domains is thoroughly discussed, and thereby the concept of **isoparametric finite elements** is illustrated. In Subsection 6.1.5 several numerical examples are given and in Subsection 6.1.6, we complete the directions suggested in Chapter 4 for implementing the $\mathcal{P}_1$ FEM in the case of more general boundary conditions, extending them also to quadratic FEs. In Section 6.2 we address essential aspects of the solution of the Poisson equation in a three-dimensional bounded domain by the FDM, the FVM and the FEM. We study in more detail the three-dimensional versions of the FDM in Subsection 6.2.1 and of the $\mathcal{P}_1$ FEM in Subsection 6.2.2. Subsection 6.2.3 is devoted to the practical implementation of the latter method, and a MATLAB code for this purpose is supplied in Subsection 6.2.4.

Numerical Methods for Partial Differential Equations:
An Introduction Finite Differences, Finite Elements and Finite Volumes, First Edition. Vitoriano Ruas.
© 2016 John Wiley & Sons, Ltd. Published 2016 by John Wiley & Sons, Ltd.
Companion Website: www.wiley.com/go/ruas/numericalmethodsforpartial

6.1   Lagrange FEM of Degree Greater than One

One of the reasons why the FEM became so effective and widespread in countless scientific and technological applications is the possibility to attain enhanced accuracy in a rational and simple manner at a time. One of the means to achieve this is by increasing the degree of the local polynomial representation the method is based upon. Far from being exhaustive on this issue, in this section we intend to illustrate how this refinement can be enforced in a few settings that we selected, in accordance with their relevance.

It is important to underline that, for second-order boundary value problems, a FE solution needs to have no unknown (or **degree of freedom** as some authors prefer) other than function values. In the literature, these methods are referred to as Lagrange FEM, owing to their connections to Lagrange interpolation (see e.g. [158]). The $\mathcal{P}_1$ FEM is the simplest example of a Lagrange FEM. This is in contrast to Hermite FEM, for example, which incorporated derivatives of the solution as degrees of freedom. Incidentally, in Chapter 7 we study some Hermite methods for fourth-order PDEs.

6.1.1   The $\mathcal{P}_k$ FEM in One-dimension Space for $k > 1$

There is certainly no better framework to introduce a numerical method for PDEs than an ODE posed in a real interval. Keeping pace with the steps we followed in the book's first chapters, we show in this subsection how the $\mathcal{P}_1$ FEM extends to the so-called $\mathcal{P}_k$ FEM for an integer $k > 1$, in order to solve equation (P$_1$) or problems alike posed in terms of a single space variable. More precisely, we consider below some important variants of this equation, namely,

For $0 < \alpha \leq p(x) \leq A$ and $0 \leq q(x) \leq B \ \forall x \in [0, L]$:

The homogeneous Dirichlet problem

$$\boxed{\begin{aligned} -(pu')' + qu &= f \\ u(0) = u(L) &= 0, \end{aligned}} \tag{6.1}$$

The homogeneous Robin problem

$$\boxed{\begin{aligned} -(pu')' + qu &= f \\ \gamma u(0) - [pu'](0) &= 0 \\ \gamma u(L) + [pu'](L) &= 0 \end{aligned}} \left.\right\} \gamma > 0 \tag{6.2}$$

For $0 < \alpha \leq p(x) \leq A$ and $0 < \beta \leq q(x) \leq B \ \forall x \in [0, L]$:

The homogeneous Neumann problem

$$\boxed{\begin{aligned} -(pu')' + qu &= f \\ u'(0) = u'(L) &= 0 \end{aligned}} \tag{6.3}$$

Under the assumption that $f \in L^2(0, L)$, one can easily establish the equivalence between these three problems respectively with

The Galerkin variational formulation of equation (6.1)

$$
\begin{aligned}
&\text{Find } u \in H_0^1(0, L) \text{ such that} \quad \forall v \in H_0^1(0, L) \\
&\int_0^L pu'v' \, dx + \int_0^L quv \, dx = \int_0^L fv \, dx;
\end{aligned}
\tag{6.4}
$$

The Galerkin variational formulation of equation (6.2)

$$
\begin{aligned}
&\text{Find } \quad u \in H^1(0, L) \text{ such that} \quad \forall v \in H^1(0, L) \\
&\int_0^L pu'v' \, dx + \int_0^L quv \, dx + \gamma[u(0)v(0) + u(L)v(L)] = \int_0^L fv \, dx;
\end{aligned}
\tag{6.5}
$$

The Galerkin variational formulation of equation (6.3)

$$
\begin{aligned}
&\text{Find } \quad u \in H^1(0, L) \text{ such that} \quad \forall v \in H^1(0, L) \\
&\int_0^L pu'v' \, dx + \int_0^L quv \, dx = \int_0^L fv \, dx.
\end{aligned}
\tag{6.6}
$$

Using basic knowledge of linear ODEs, it is not difficult to check that all those problems have a unique solution under the respective assumptions on p, q and γ (cf. Exercise 6.3)

For the sake of simplicity, we restrict the presentation hereunder to the case of a uniform mesh of the interval $(0, L)$ with n elements and mesh size $h = L/n$. The reader would certainly agree that the extension to the case of non-equally spaced meshes is straightforward, owing to the systematic employed hereafter.

Clearly enough, whatever the case considered above, one may approximate u by means of the FEM, and in particular by the $\mathcal{P}_1$ FEM. Recalling the shape functions φ_i introduced in Chapter 1, in this case we work with spaces V_h^1 consisting of linear combinations of the following bases:

$$
\{\varphi_i\}_{i=1}^{n-1} \text{ for (6.4)} \quad \text{and} \quad \{\varphi_i\}_{i=0}^{n} \text{ for (6.5) and (6.6)}.
$$

Then, recalling equation (1.21), in all cases we solve the approximate variational problem,

$\mathcal{P}_1$ FE approximation of equations (6.4), (6.5) and (6.6)

$$
\begin{aligned}
&\text{Find } \quad u_h^1 \in V_h^1 \text{ such that } a(u_h^1, v) = F(v) \ \forall v \in V_h^1 \\
&\qquad\qquad\qquad \text{where} \\
&a(u, v) := \int_0^L [pu'v' + quv] dx + \beta[u(L)v(L) + u(0)v(0)] \text{ with} \\
&\qquad \beta = 0 \text{ for (6.4) and (6.6) and} \quad \beta = \gamma \text{ for (6.5),} \\
&\qquad\qquad \text{and } F(v) := \int_0^L fv \, dx.
\end{aligned}
\tag{6.7}
$$

There is no essential difficulty to check that in all cases, equation (6.7) has a unique solution u_h^1. Moreover, like in the case of (P_3), as long as $u \in H^2(0, L)$, the approximation $u_h^1 \in V_h^1$ of u determined by the $\mathcal{P}_1$ FEM satisfies, in all three cases,

$$
\|u - u_h^1\|_{1,2} \le C_1 h \|u''\|_{0,2},
$$

and, if $|p'|$ is bounded,

$$||u - u_h^1||_{0,2} \leq \tilde{C}_{01} h^2 ||u''||_{0,2}$$

where C_1 and C_{01} are constants independent of u and h.

Now let us consider higher order piecewise polynomial approximations of well-posed two-point boundary value problems, such as (P$_3$), (6.4), (6.5) and (6.6).

Let V be the corresponding test function space (i.e., a subspace of either $H_0^1(0, L)$ or $H^1(0, L)$). We define corresponding approximation spaces as follows.

First for a given integer $k > 1$ and an interval T, let $\mathcal{P}_k(T)$ be the set of polynomials of degree less than or equal to k defined in T. Next we define

The space associated with the $\mathcal{P}_k$ FEM for $k \geq 1$

$$W_h^k = \{v | v \in C^0[0, L], \, v_{|T} \in \mathcal{P}_k(T) \text{ for } T = [x_{i-1}, x_i], \, i = 1, \dots, n\} \qquad (6.8)$$

Notice that in all the cases being considered, $v \in W_h^k$ implies that $v \in H^1(0, L)$. Like in the case of the $\mathcal{P}_1$ FEM, the subspace V_h^k of W_h^k that approximates V in each case is the set of those functions $v \in W_h^k$ that satisfy the boundary conditions to be enforced for any function in V. For instance, in approximating equation (6.4) we must have $v(0) = v(L) = 0, \forall v \in V_h^k$, while v does not need to satisfy any boundary condition in the case of equations (6.6) and (6.5).

First, we take $k = 2$. A convenient basis for W_h^2 is the set of **piecewise quadratic shape functions** $\{\varphi_i\}_{i=0}^{2n}$ constructed as follows. An 'even' shape function φ_{2j} equals one at x_j for a certain j, $0 \leq j \leq n$, and vanishes not only at all x_i for $i \neq j$, but also at the mid-points $x_{j-1/2}$ of T_j for $j = 1, 2, \cdots, n$. An 'odd' shape function φ_{2j-1} in turn, for a certain j, $1 \leq j \leq n$, equals one at $x_{j-1/2}$ and vanishes at all the $x_{i-1/2}$s if $i \neq j$ and at all the x_js, for $j = 0, 1, \cdots, n$. Indeed, v being a generic function of W_h^2 we have by definition $v = \sum_{l=0}^{2n} v_l \varphi_l$, where the v_l's are real coefficients. From the above construction, we readily infer that $v(x_{m/2}) = v_m$ for $m = 0, 1, \cdots, 2n$. Referring to the points $x_{m/2}$ as the nodes S_m associated with the space W_h^2, we conclude that the coefficients v_l in the expansion of v are nothing but the nodal values of v, i.e. $v(S_l)$, for $l = 0, 1, \cdots, 2n$. Illustrations of the above shape functions are given in Figure 6.1 and Figure 6.2.

Next we switch to the case of W_h^3 (resp. V_h^3). In this case, we subdivide each element T into three equal intervals by taking intermediate points $x_{j-2/3} := (2x_{j-1} + x_j)/3$ and $x_{j-1/3} := (x_{j-1} + 2x_j)/3$. Then we define **piecewise cubic shape functions** φ_{3j-2}, φ_{3j-1} and φ_{3j} for $j = 1, 2, \cdots, n$ together with φ_0, vanishing at all the points $x_{m/3}$ for $m = 0, 1, \cdots, 3n$ but one, say $x_{l/3}$, at which it equals one. The corresponding shape function is precisely $\varphi_{l/3}$, $l = 0, 1, \cdots, 3n$. The reader could draw pictures similar to Figures 6.1 and 6.2 for those 'natural' shape functions of W_h^k for $k = 3$.

Now, in order to write analytic expressions of the shape functions for $k > 1$, the use of barycentric coordinates is very handy. Letting T denote a generic mesh element $[x_1^T, x_2^T]$, we recall the one-dimensional barycentric coordinates λ_1^T and λ_2^T given by equation (4.26).

For $k = 2$, we denote by S_{pp}^T the point with x coordinate x_p^T, $p = 1$ or $p = 2$ and by S_{12}^T the mid-point of T, that is, the point whose x coordinate is $(x_1^T + x_2^T)/2$. Then, in each element T we have,

$\mathcal{P}_2$ shape function for the end-points S_{pp}^T

$$\varphi_{pp}^T = (2\lambda_p^T - 1)\lambda_p^T, \text{ for } \quad p = 1, 2$$

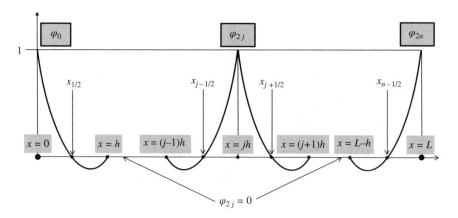

Figure 6.1   Shape functions φ_{2j} for the nodes at $x = jh$, $0 < j < n$, φ_0 and φ_{2n}

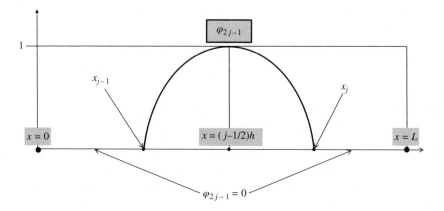

Figure 6.2   Shape functions φ_{2j-1} for the nodes at $x = (j - 1/2)h$, $1 \leq j \leq n$

$\mathcal{P}_2$ shape function for the mid-point S_{12}^T

$$\varphi_{12}^T = 4\lambda_1^T \lambda_2^T.$$

Let us check that the thus constructed shape functions correspond to the definitions of the three φ_m's that do not vanish identically in $T = [x_{j-1}, x_j]$ for a suitable j (cf. Figures 6.1 and 6.2). Indeed, since the barycentric coordinates are linear functions, the product of two linear expressions involving them is a quadratic function. Noticing that the value of both λ_1^T and λ_2^T at the mid-point of T is $1/2$, and $\lambda_p^T(S_{qq}) = \delta_{pq}$, the remainder of the checking procedure is trivial.

For $k = 3$, we use an analogous notation for the nodes, namely, S_{ppq}^T, which is the point with x coordinate $(2x_p^T + x_q^T)/3$, for p and q equal to either 1 or 2. Notice that p may be equal to q, and in this case S_{ppp}^T is an end-point of T.

As for the corresponding shape functions restricted to an interval T, we have

$\mathcal{P}_3$ shape functions for the end-points S^T_{ppp}

$$\varphi^T_{ppp} = (3\lambda^T_p - 2)(3\lambda^T_p - 1)\lambda^T_p/2, \text{ for } p = 1, 2;$$

$\mathcal{P}_3$ shape functions for the third points S^T_{ppq}

$$\varphi^T_{ppq} = 9\lambda^T_p\lambda^T_q(3\lambda^T_p - 1)/2 \text{ for } 1 \le p, q \le 2, p \ne q.$$

Then the shape function associated with node S_{3j} for a given j is the function φ_{3j} whose restriction to an element T containing S_j, that is S^T_{ppp} for p equal to either 1 or 2, is the function φ^T_{ppp}. The shape function φ_{3j-2} (resp. φ_{3j-1}) associated with node S_{3j-2} (resp. S_{3j-1}) for a certain j has nonzero values only in element $T_j = [x_{j-1}, x_j]$ where it is equal to φ^T_{ppq} for $p = 1$ and $q = 2$ (resp. $p = 2$ and $q = 1$).

In practice, there is seldom need to go beyond $k = 3$, and for this reason our detailed presentation of the Lagrange $\mathcal{P}_k$ FEM in one-dimensional space ends here. However, we believe that the reader has already gathered enough insight on the above procedure to realise how to construct the shape functions associated with $\mathcal{P}_k$ Lagrange FEs in one-dimensional space, for any value of k. Actually, the barycentric coordinates can be used *ipso facto* in higher dimensions, as seen in Subsection 6.1.3 (see also [210]).

Once we have defined the basis of V^k_h, we may compute $u^k_h \in V^k_h$, namely, the corresponding Ritz approximation of u, thereby obtaining a SLAE $A_h\vec{u}^k_h = \vec{b}_h$ that is similar to the one generated with piecewise linear elements. For instance, for $k = 2$, in the case of problem (6.4), A_h is the $(2n - 1) \times (2n - 1)$ matrix whose coefficients $a_{i,j}$ are given by $a_{i,j} = \int_0^L (p\varphi'_j\varphi'_i + q\varphi_j\varphi_i)dx$ and $\vec{b}_h$ is the vector of $\mathfrak{R}^{2n-1}$ whose $i - th$ component is $\int_0^L f\varphi_i dx$, while $\vec{u}_h = [u_1, u_2, \cdots, u_{2n-1}]^T$ is the vector of coefficients in the expansion of u_h with respect to the basis $\{\varphi_i\}_{i=1}^{2n-1}$. Clearly enough, $u_j = u_h(S_j)$ for all j.

Once W^k_h is defined as the space spanned by the shape functions associated with the sets $\mathcal{P}_k(T)$ in each mesh element T, we can define the FE approximate space V^k_h to be either W^k_h itself, or the subspace of W^k_h incorporating Dirichlet boundary conditions, according to the problem to solve. Recalling equation (6.7), this gives rise to

The $\mathcal{P}_k$ FE approximation of equations (P$_3$), (6.4), (6.5) and (6.6) for $k \ge 1$

$$\begin{array}{l} \text{Find } u^k_h \in V^k_h \text{ such that} \\ a(u^k_h, v) = F(v) \; \forall v \in V^k_h \end{array} \tag{6.9}$$

Finally, the approximation errors $||u - u^k_h||_{1,2}$ and $||u - u^k_h||_{0,2}$ may be estimated by means of analyses entirely analogous to those carried out in detail for the $\mathcal{P}_1$ FEM applied to (P$_3$).

More specifically, let us assume that the solution of problems equations (P$_3$), (6.4), (6.5) and (6.6) belongs to $u \in H^{k+1}(0, L)$, where for an integer $k > 0$,

$$H^k(0, L) := \{v | v, v', \cdots, v^{(k)} \in L^2(0, L)\}.$$

The problem to solve is approximated by equation (6.9), a being the bilinear form $\int_0^L pu'v'dx + \int_0^L quvdx$ for problems (P$_3$), (6.4) and (6.6) or the same expression modified by the addition of $\gamma[u(0)v(0) + u(L)v(L)]$ for problem (6.5), and $F(v) := \int_0^L fv\,dx$. Then there exists a constant C_k independent of u and h for which the

Error estimate for equation (6.9) in the norm of $H^1(0, L)$

$$\|u - u_h^k\|_{1,2} \le C_k h^k \|u^{(k+1)}\|_{0,2} \tag{6.10}$$

holds. If, in addition to the assumptions we made to validate equation (6.10), we further assume that p is such that the solution to any problem among (P_3), (6.4), (6.5) and (6.6) belongs to $H^2(0, L)$ whenever $f \in L^2(0, L)$, then there exists a constant C_{0k} independent of u and h such that

Error estimate for problem (6.9) in the L^2-norm

$$\|u - u_h^k\|_{0,2} \le C_{0k} h^{k+1} \|u^{(k+1)}\|_{0,2}. \tag{6.11}$$

The proof of estimate (6.10) using Fourier analysis can be found in reference [182] or [170]. Estimate (6.11) in turn follows from the same duality argument exploited in Section 2.2.

Remark 6.1 *An analysis analogous to the one carried out for problem (P_3) indicates that the only assumption on p necessary for equation (6.11) to hold is the boundedness of $|p'|$.* ∎

To conclude this subsection, we observe that both error estimates (6.10) and (6.11) extend with minor modifications to the approximation by the $\mathcal{P}_k$ FEM, of inhomogeneous counterparts of problems (6.4), (6.6), (6.5) and (P_3), that is,

The inhomogeneous Dirichlet problem

$$-(pu')' + qu = f \text{ in } (0, L)$$
$$u(0) = a_1, u(L) = a_2$$

The inhomogeneous Neumann problem

$$-(pu')' + qu = f \text{ in } (0, L)$$
$$-[pu'](0) = b_1, [pu'](L) = b_2$$

The inhomogeneous Robin problem

$$-(pu')' + qu = f \text{ in } (0, L)$$
$$\gamma u(0) - [pu'](0) = b_1, \gamma u(L) + [pu'](L) = b_2$$

The inhomogeneous mixed Dirichlet–Neumann problem

$$-(pu')' + qu = f \text{ in } (0, L)$$
$$u(0) = a_1; [pu'](L) = b_2,$$

for given real numbers a_1, a_2, b_1 and b_2. This remark extends to the same second-order ODE supplemented with several other possible combinations of the above boundary conditions, under appropriate assumptions on p and q. The way to deal with different types of inhomogeneous boundary conditions in a variational context is to be explored as Exercise 6.4, in complement to the material presented in this subsection.

6.1.2 A FEM for Quadrilateral Meshes

In this section, we go back to the solution of equation (4.1) in a polygonal domain Ω with homogeneous Dirichlet–Neumann boundary conditions.

   In some cases, even an irregular polygonal domain Ω can be easily partitioned into a mesh $\mathcal{T}_h$ consisting of convex quadrilaterals generically denoted by Q, respecting compatibility conditions similar to those enforced for triangulations. This means that the intersection (of the closures) of two mesh quadrilaterals is either empty, a common vertex or a common edge, and moreover the transition points between Γ_0 and Γ_1 are mesh vertices. In this case, the popular $\mathcal{Q}_1$ FEM can be used. In Figure 6.3, we illustrate a compatible quadrilateral mesh.

   In contrast to the $\mathcal{P}_k$ FEM, which is based on complete polynomials of degree less than or equal to k, defined in segments or triangles, the $\mathcal{Q}_1$ FEM is constructed upon polynomials of degree less than or equal to one in each space variable. However, unless quadrilateral Q is a rectangle, the underlying solution representation at the element level is not a linear combination of the monomials $1, x, y, xy$, that is, a **bilinear function**. In principle, the solution representation in each $Q \in \mathcal{T}_h$ is performed by means of **isoparametric transformations** onto Q of a **master square** $\hat{Q}$ using bilinear functions in terms of the variables ξ and η of the coordinate system $(\hat{O}; \xi, \eta)$ of the plane of $\hat{Q}$ is referred to. Since the shape functions in Q transforms into bilinear functions of ξ and η in $\hat{Q}$, this type of representation is called **isoparametric**.

   The concept is illustrated in Figure 6.4: A bilinear mapping $\mathbf{F}_Q$ of $\hat{Q}$ onto a current convex quadrilateral Q takes vertex $\hat{S}_i$ of $\hat{Q}$ to vertex S_i^Q of Q for $i = 1, 2, 3, 4$. The existence and uniqueness of $\mathbf{F}_Q$ whenever the area of Q is nonzero are ensured by exhibiting the scalar bilinear components F_{Qx} and F_{Qy} of $\mathbf{F}_Q$ that map the four vertices of $\hat{Q}$ onto the $(x; y)$-coordinates of the corresponding vertices of Q, respectively.

   Next, we determine mathematical expressions for mapping $\mathbf{F}_Q$. For simplicity, we drop the subscript Q related to coordinates and coefficients, keeping in mind however that they are related to a specific quadrilateral Q.

   Let $(x_i; y_i)$ be the Cartesian coordinates of the vertices of Q in a suitable reference frame. The question is whether it is possible or not to uniquely determine functions

$$F_{Qx}(\xi, \eta) = a_1 \xi \eta + a_2 \xi + a_3 \eta + a_4 \text{ and } F_{Qy}(\xi, \eta) = b_1 \xi \eta + b_2 \xi + b_3 \eta + b_4,$$

that is, eight coefficients a_j and b_j for $j = 1, 2, 3, 4$ such that for $i = 1, 2, 3, 4$,

$$x_i = a_1 \xi_i \eta_i + a_2 \xi_i + a_3 \eta_i + a_4 \text{ and } \quad y_i = b_1 \xi_i \eta_i + b_2 \xi_i + b_3 \eta_i + b_4$$

where $(\xi_1; \eta_1) = (0; 0)$, $(\xi_2; \eta_2) = (1; 0)$, $(\xi_3; \eta_3) = (0; 1)$ and $(\xi_4; \eta_4) = (1; 1)$. The answer is yes, and these coefficients are given by

$$a_4 = x_1, a_2 = x_2 - x_1, a_3 = x_3 - x_1, a_1 = x_4 + x_1 - x_2 - x_3;$$
$$b_4 = y_1, b_2 = y_2 - y_1, b_3 = y_3 - y_1, b_1 = y_4 + y_1 - y_2 - y_3.$$

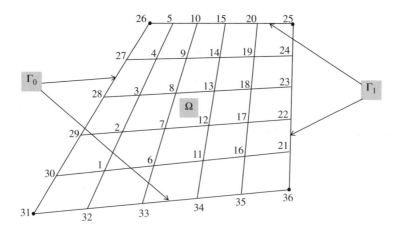

Figure 6.3 A sample FE mesh with convex quadrilaterals and numbered nodes

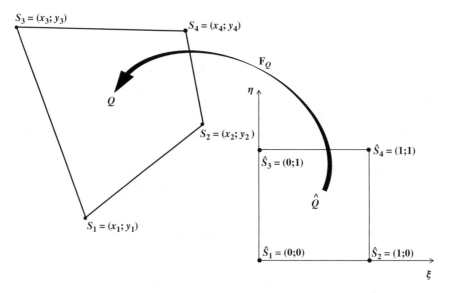

Figure 6.4 Bilinear mapping $\mathbf{F}_Q$ from the master square $\hat{Q}$ onto a quadrilateral Q

It is very important to notice that by construction the restrictions of the thus-defined mapping $\mathbf{F}_Q$ to any edge of $\hat{Q}$ is a linear function of the abscissa along this edge. This is because either ξ or η is constant thereupon. It follows that the image of any edge of $\hat{Q}$ is a straight segment (similarly to the case of a linear mapping from a master triangle onto a current triangle). This argument also shows that the whole Q is represented through the one-to-one mapping $\mathbf{F}_Q$. Indeed, the other way around, to every point of Q with coordinates $(x; y)$ corresponds a single point of $\hat{Q}$ with coordinates $(\xi; \eta)$. This is because Q can be viewed as equipped with a local plane coordinates defined by two systems of straight lines corresponding to the images of the lines of $\hat{Q}$ given by ξ (resp. η) equal to a constant c_ξ (resp. c_η) between zero and one. Figure 6.3 somehow illustrates such systems of lines, for the sample mesh has aligned edges in two different directions. Since all those lines are completely contained in Q and do not intersect each other within the same system of lines, every point of Q with coordinates $(x; y)$ necessarily correspond to only one pair of intersecting lines defined by constants c_ξ and c_η, and hence to a single point of $\hat{Q}$. It is by no means difficult to figure out that the above argument fails in case Q is not convex. We refer to more complete text books on the FEM for further explanations on non linear isoparametric mappings, such as references [210] and [45].

The main problem with functions defined in mesh elements through an isoparametric transformation is the fact that, in general, they are rational functions of x and y. However, their counterparts in $\hat{Q}$ are bilinear functions of ξ and η. For this reason, it is advisable to carry out the element matrix and right-side vector computations in the master element. Let us see how this works.

First, we define a space W_h associated with the quadrilateral mesh $\mathcal{T}_h$ in such a way that every function in W_h is continuous and its restriction to every $Q \in \mathcal{T}_h$ is the image of a bilinear function in $\hat{Q}$ through the unique bilinear mapping $\mathbf{F}_Q$ from $\hat{Q}$ onto Q. This construction is compatible with the continuity requirement. Indeed, in order to fulfil this condition, it suffices to assign the same value at a given mesh node to the mapped functions pertaining to all the quadrilaterals having this node as a common vertex. In doing so, the traces along any mesh edge of the mapped bilinear functions related to two adjacent quadrilaterals are linear functions that have the same values at each one of its end-points by construction. For the same reason, every function in the subspace V_h of W_h made up of functions that vanish at the nodes located in Γ_0 vanishes everywhere thereupon. Recalling the forms a and F defined by equation (4.13), this means that the problem

$$\begin{cases} \text{Find } u_h \in V_h \text{ such that} \\ a(u_h, v) = F(v) \; \forall v \in V_h. \end{cases} \qquad (6.12)$$

to approximate the solution u of equation (4.15) (i.e. the solution of equation (4.3) or (4.4)) makes sense. More than this, we define a basis of V_h consisting of functions $\varphi_i \in V_h$ such that $\varphi(S_j) = \delta_{ij}$ for $1 \leq i, j \leq N_h$, where N_h is the number of mesh vertices (i.e. nodes) not located on Γ_0. In doing so, here again equation (6.12) is equivalent to a SLAE for the N_h unknown values of u_h at the mesh nodes. Notice that the values of u_h at the mesh nodes belonging to Γ_0 are all equal to zero. This SLAE has a unique solution since, at least written as such, problem (6.12) shares the very same stability properties with its $\mathcal{P}_1$ counterpart studied in Chapters 4 and 5. For the sake of conciseness, we do not elaborate any further on the form of the SLAE equivalent to equation (6.12) for it is in all aspects identical to those considered so far. Therefore, in the remainder of this subsection, we concentrate on the computation of element matrices and right-side vectors, since here they are considerably different from what we saw in Chapter 4.

To begin with, like in the case of triangulations, an $M_h \times 4$ incidence matrix $\mathcal{N} = [n_{m,r}]$ for mesh $\mathcal{T}_h$ is supposed to be available, M_h being the number of elements in the mesh. Once we have numbered the elements of $\mathcal{T}_h$ using a convenient systematic, we denote by Q_m the mth element of $\mathcal{T}_h$ and by E^m the associated the element matrix, whose entries are $e_{r,s}^m$ for $1 \leq r, s \leq 4$. The latter are given by

$$e_{r,s}^m = \int_{Q_m} (\mathbf{grad}\varphi_j | \mathbf{grad}\varphi_i) dx dy \quad \text{for} \quad i = n_{m,r}, j = n_{m,s}.$$

Henceforth, we use rather the notation $e^Q_{r,s}$ for entries related to a generic quadrilateral Q.

Let $\hat{\varphi}_s$ be the bilinear function defined in $\hat{Q}$ such that $\hat{\varphi}_s(\hat{S}_r) = \delta_{rs}$ for $1 \leq r, s \leq 4$. Notice that such bilinear functions are uniquely defined, and referring to Figure 6.4 their expressions are found to be

$$\hat{\varphi}_1 = (1 - \xi)(1 - \eta); \quad \hat{\varphi}_2 = \xi(1 - \eta); \quad \hat{\varphi}_3 = (1 - \xi)\eta; \quad \hat{\varphi}_4 = \xi\eta.$$

The restriction of φ_i to a quadrilateral Q having S_i as a vertex (i.e. the rth local vertex S_r^Q) is the image through $\mathbf{F}_Q$ of $\hat{\varphi}_r$. Thus, using the chain rule, we derive for $(x; y) \in Q$: $\quad \partial_x \varphi_i(x, y) = \partial_\xi \hat{\varphi}_r(\xi, \eta)\partial_x \xi + \partial_\eta \hat{\varphi}_r(\xi, \eta)\partial_x \eta \quad$ and $\quad \partial_y \varphi_i(x, y) = \partial_\xi \hat{\varphi}_r(\xi, \eta)\partial_y \xi + \partial_\eta \hat{\varphi}_r(\xi, \eta)\partial_y \eta.$

Now, let $\mathbf{G}_Q = (G_{Q\xi}; G_{Q\eta})(x; y)$ be the uniquely defined inverse of the transformation $\mathbf{F}_Q$, mapping Q onto $\hat{Q}$. Denoting by $\mathcal{G}_Q$, the gradient of $\mathbf{G}_Q$, that is,

$$\mathcal{G}_Q := \begin{bmatrix} \partial_x G_{Q\xi} & \partial_y G_{Q\xi} \\ \partial_x G_{Q\eta} & \partial_y G_{Q\eta} \end{bmatrix},$$

using the chain rule, we may write in vector form,

$$\mathbf{grad}\varphi_i(x, y) = \mathcal{G}_Q^T[F_{Qx}(\xi, \eta), F_{Qy}(\xi, \eta)]\widehat{\mathbf{grad}\hat{\varphi}_r}(\xi, \eta),$$

where $\widehat{\mathbf{grad}\hat{g}}(\xi, \eta)$ represents $(\partial_\xi \hat{g}(\xi, \eta); \partial_\eta \hat{g}(\xi, \eta))$, for any differentiable function $\hat{g}$ defined in $\hat{Q}$.

On the other hand, denoting by J_F^Q the Jacobian of $\mathbf{F}_Q$ (i.e. the determinant of the tensor equal to the gradient of $\mathbf{F}_Q$), from differential calculus in multiple variables, we know that J_F^Q is everywhere strictly positive in $\hat{Q}$ and that for all $(x; y) = \mathbf{F}_Q(\xi, \eta)$, we have $\mathcal{G}_Q(x, y) = \{[J_F^Q]^{-1}\mathcal{M}_Q\}(\xi, \eta)$, where

$$\mathcal{M}_Q := \begin{bmatrix} \partial_\eta F_{Qy} & -\partial_\eta F_{Qx} \\ -\partial_\xi F_{Qy} & \partial_\xi F_{Qx} \end{bmatrix},$$

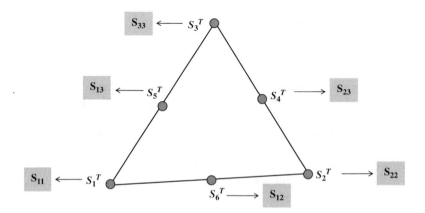

Figure 6.5 A triangle T with six $\mathcal{P}_2$ nodes S_r^T and corresponding local notation S_{pq}

While, on the one hand, the components of the tensor $\mathcal{M}_Q$ are linear functions of ξ and η, on the other hand in principle $[J_F^Q]^{-1}$ is a rational function of these variables (more precisely, the inverse of a linear function, as the reader can easily check).

Summarizing the above calculations, and noting that the area element $dxdy$ at any point $(x; y)$ of Q is given by $J_F^Q(\xi, \eta)d\xi d\eta$, the entry $e_{r,s}^Q$ of the element matrix related to a quadrilateral Q, computed by transformation onto element $\hat{Q}$, can be expressed as follows:

$$e_{r,s}^Q = \int_{\hat{Q}} (\mathcal{M}_Q^T \widehat{\mathbf{grad}}\hat{\varphi}_s | \mathcal{M}_Q^T \widehat{\mathbf{grad}}\hat{\varphi}_r)[J_F^Q]^{-1}d\xi d\eta.$$

However, there is a practical problem here, for in general the integral in the expression of $e_{r,s}^Q$ for isoparametric FEs is either too difficult to evaluate or cannot be computed exactly. Thus, as a rule we must resort to numerical integration, which always involves some risks. The first one to be reported is related to the stability of the FEM. The results of Chapter 5 in this respect clearly extend to the $\mathcal{Q}_1$ FEM, but with exact linear and bilinear forms. If the latter is approximated, we must make sure that stability still holds. In order to concentrate on crucial issues, we address the approximation of the right-side vector by numerical integration a little later.

Let us denote by a^* the approximations of bilinear form a obtained by numerical integration, and by u_h^* the underlying solution. This is the presumably unique function in V_h satisfying $a^*(u_h^*, v) = F^*(v)$, where $F^*(v)$ is an approximation of $\int_\Omega fv \, dxdy \; \forall v \in V_h$. Basically, we must establish the existence of a mesh-independent constant C^* such that $|\tilde{u}_h^*|_{1,2} \leq C^* \| f \|_{0,2}$. Moreover, we have to make sure that numerical integration maintains method's theoretical order of convergence with utopian exact integration. Notice that we have

$$a^*(u_h^*, v) = \sum_{Q \in \mathcal{T}_h} \sum_{r=1}^{4} \sum_{s=1}^{4} u_s^{Q*} e_{r,s}^{Q*} v_r^Q dxdy,$$

where $e_{r,s}^{Q*}$ is the approximation of $e_{r,s}^Q$ obtained by numerical integration in $\hat{Q}$, and u_s^{Q*} is the value of u_h^* at the sth vertex of Q. v_r^Q in turn is the value of test function v at the rth node of Q. This means that numerical integration in $\hat{Q}$ is to be applied to the integral

$$e_{r,s}^Q = \int_{\hat{Q}} (\mathcal{G}_Q^T \widehat{\mathbf{grad}}\hat{\varphi}_s | \mathcal{M}_Q^T \widehat{\mathbf{grad}}\hat{\varphi}_r)d\xi d\eta.$$

Notice that this function is a polynomial of degree four divided by a nonvanishing linear function (please check!).

A generally accepted rule validated by theoretical results (cf. [182]) states that a numerical quadrature formula, applied to the variational approximation of the Poisson equation, should integrate exactly $a^*(u_h^*, v)$ whenever u_h^* is linear in each element. Notice that in this case, the gradient of u_h^* in each Q is constant, say $\vec{g}^Q$. Thus, modifying the above calculations in order to take into account this new information, we find out that the quadrature formula should integrate exactly $(\vec{g}^Q | \mathcal{M}_Q^T \widehat{\mathbf{grad}}\hat{\varphi}_r)$ in $\hat{Q}$, which is no longer a rational function. On the other hand, both the entries of $\mathcal{M}_Q$ and the components of $\widehat{\mathbf{grad}}\hat{\varphi}_r$ are linear functions. It follows that a quadrature formula which is exact for quadratic functions in a square is all we need. The four-point Gaussian quadrature formula specified in the following paragraph fulfils the required condition to ensure stability, together with optimal order of accuracy in both the H^1- and the L^2-norm.

Similar but simpler calculations yield the component f_r^Q of the right-side element vector $\vec{f}^Q$, whose exact value is given by

$$f_r^Q = \int_{\hat{Q}} \hat{f} \hat{\varphi}_r J_F^Q \, d\xi \, d\eta, \quad \text{for} \quad r = 1, 2, 3, 4.$$

where $\hat{f}$ is the function defined in $\hat{Q}$ as the transform of f by change of variables using mapping $\mathbf{F}_Q$. However, even if f is not so complicated, here again the use of numerical quadrature in a rectangle (square) is recommended. Notice that the Q_1 FEM is a first-order method in the norm of $H^1(\Omega)$, for its approximation properties are equivalent to those that hold for the P_1 FEM (cf. [45]). This is because the space Q_1 contains only complete polynomials of degree less than or equal to one. The reader might check this assertion as Exercise 6.5, by transforming a linear function in $\hat{Q}$ onto Q using the bilinear mapping $\mathbf{F}_Q$. In view of this, it is no point using sophisticated quadrature formulae, and basically the one-point Gaussian quadrature rule $\int_{\hat{Q}} \hat{g}(\xi, \eta) \, d\xi \, d\eta \simeq \hat{g}(1/2, 1/2)$ would do, for it integrates exactly bilinear functions in $\hat{Q}$ (please check!). However, if Ω is convex, one can only expect to obtain second-order error estimates in the L^2-norm by a classical duality argument (cf. Chapter 2), if a higher order quadrature formula is employed. In this case, the four-point Gaussian quadrature rule is recommended, which integrates exactly bi-cubic functions in $\hat{Q}$. This formula writes $\int_{\hat{Q}} \hat{g} \, d\xi \, d\eta \simeq \frac{1}{4} \sum_{i=1}^4 \hat{g}(\chi_i, \zeta_i)$, where $(\chi_1; \zeta_1) = (a; a)$, $(\chi_2; \zeta_2) = (b; a)$, $(\chi_3; \zeta_3) = (a; b)$ and $(\chi_4; \zeta_4) = (b; b)$, with $a = (3 - \sqrt{3})/6$ and $b = (3 + \sqrt{3})/6$.

Remark 6.2 *We do not find it appropriate to develop numerical integration in depth in this book, for it is a very technical topic. After all, the methodology we are presenting is limited to FVM and FEM defined upon polynomials of low degree, and in this case selecting the right quadrature formula in each situation is not such an intricate task. For more complete studies on issues pertaining to numerical integration, we refer to Strang and Fix [182], Raviart and Thomas [162] and Ciarlet [45], among many books on mathematics of the FEM.*∎

To conclude, employing arguments of the same nature as those the convergence of the P_1 FEM for equation (4.1) relies upon, the following result can be established for the Q_1 FEM to solve equations (4.3) or (4.4) (cf. [45]):

Assume that for every mesh $\mathcal{T}_h$ in a family $\{\mathcal{T}_h\}_h$ and $\forall Q \in \mathcal{T}_h$, the ratio $\rho(Q)/h(Q)$ is bounded below by a strictly positive mesh-independent constant, $\rho(Q)$ and $h(Q)$ being the radius of the largest circle inscribed in Q and the largest edge of Q, respectively, and moreover there exists a constant c, $c < 1$, such that the absolute value of the cosinus of the four angles of mesh quadrilaterals are less than c for all meshes in the family. Then provided the solution u of equation (4.3) or (4.4) lies in $H^2(\Omega)$, and as long as the quadrature formulae specified above are employed, there exists a mesh-independent constant C_Q such that

$$|u_h^* - u|_{1,2} \leq C_Q h \, \| H(u) \|_{0,2}.$$

where $h := \max_{Q \in \mathcal{T}_h} h(Q)$,

On the other hand an error analysis in the sense of the $L^2(\Omega)$-norm for isoparametric finite elements is much more complex, owing to the compulsory use of numerical integration to handle the underlying bilinear form in the case of a distorted mesh. For this reason we refrain from developing it here.

Remark 6.3 *In contrast to the case of triangulations, a family of convex quadrilateral meshes satisfying both conditions above is called **shape-regular**.*∎

6.1.3   Piecewise Quadratic FEs in Two Space Variables

Even if the the solution of equations (4.3)–(4.4) is smoother than a function in $H^2(\Omega)$, the $\mathcal{Q}_1$ FEM brings about little or no improvement with respect to the $\mathcal{P}_1$ FEM, as far as accuracy is concerned. If one wishes to take advantage of a finer solution regularity to attain more accurate FE approximations, the use of piecewise polynomials of degree higher than one in x and y is mandatory. In the two-dimensional case, there are two families of Lagrange FEs, whose lowest order members are the $\mathcal{P}_1$ FE and the $\mathcal{Q}_1$ FE, respectively. These methods are known as the Lagrange $\mathcal{P}_k$ and the $\mathcal{Q}_k$ FEM, to be used in connection with triangulations and **quadrangulations** of Ω, respectively. Like in the one-dimensional case, k is the highest degree of a monomial involved in the solution representation at the element level for the $\mathcal{P}_k$ FEM. In the case of the $\mathcal{Q}_k$ FEM, k is the highest power of each space variable appearing in the local representation. According to celebrated work on the FEM, such as those already quoted, the sky is the limit. This means that we can go as high as we wish with the values of k, thereby generating methods satisfying all the requirements to yield more and more accurate solutions, as long as the exact solution is sufficiently smooth. In Subsection 7.3.2, we shall elaborate a little more about this. However, if we stick to the logic attached to the numerical methods we have studied so far, better accuracy is searched for through mesh refinement, rather than by method's increasing order. Actually, generally speaking, in a significant majority of cases, practitioners adopt the former strategy to attain this goal. Practical limitations are also a good reason for this choice, since the implementation of numerical methods based on polynomials of high degree is usually a complex task. That is why we confine ourselves to a degree k not greater than three throughout this text, and in this section in particular, to the $\mathcal{P}_2$ FEM for solving equation (4.1).

Without loss of essential results, for the sake of conciseness, in the remainder of this subsection we focus only on equation (4.3).

We still assume that Ω is a polygon and f is continuous and also suppose that a regular family of triangulations $\mathcal{T}_h$ is available, each member of which satisfies the compatibility conditions already specified for the $\mathcal{P}_1$ FEM. Like in the previous subsection, we solve the FE approximate problem (6.12), but this time with spaces W_h and V_h redefined in accordance with the $\mathcal{P}_2$ FEM. $\mathcal{P}_2(T)$ being the set of all polynomials of degree less than or equal to two defined in $T \in \mathcal{T}_h$, we denote by

- W_h, the set of continuous functions whose restriction to each $T \in \mathcal{T}_h$ lies in $\mathcal{P}_2(T)$;
- V_h, the subset of functions in W_h that vanish identically on Γ.

The fact that such constructions are feasible is due to the particular nodal structure inherent to the $\mathcal{P}_2$ FEM: In this method, besides the mesh vertices, all the edge mid-points are included in the set of nodes. It follows that a given mesh triangle T contains six mesh nodes. In Figure 6.5, we illustrate the local numbering of these nodes, that is, S_r^T for $1 \leq r \leq 6$, where S_1^T is arbitrarily chosen. The rest follows the counterclockwise sense, S_{r+3}^T being the edge mid-point opposite to vertex S_r^T. For the purpose of this description, it is convenient to use an alternative two-subscript local notation for nodes S_r^T, as indicated in Figure 6.5: node S_{rr} is S_r^T for $r = 1, 2, 3$, and node S_{pq} is the mid-point of the edge $S_p^T S_q^T$, for $1 \leq p < q \leq 3$. Now, we note that for any set of six values, say α_{pq} for $1 \leq p \leq q \leq 3$, there exists only one function ϕ belonging to $\mathcal{P}_2(T)$ that takes them up at points S_{pq}. Recalling the barycentric coordinates of T given by equation (4.23), the function ϕ is defined as follows

$$\phi := \sum_{p=1}^{3} \sum_{q=p}^{3} \alpha_{pq} \phi_{pq}, \tag{6.13}$$

where

$$\phi_{pp} = \lambda_p^T (2\lambda_p^T - 1) \text{ for } p = 1, 2, 3; \tag{6.14}$$

$$\phi_{pq} = 4\lambda_p^T \lambda_q^T \text{ for } 1 \le p < q \le 3.$$

Indeed, recalling the arguments in Subsection 6.2.1, one can easily check that for $1 \le r, q \le 3$, $\phi_{pp}(S_{qq}) = \delta_{pq}$, and that for $1 \le q < s \le 3$, $\phi_{pp}(S_{qs}) = 0$; moreover, for $1 \le q < s \le 3$, $\phi_{qs}(S_{pp}) = 0$ for $p = 1, 2, 3$ and for $1 \le p < r \le 3$, $\phi_{qs}(S_{pr}) = \delta_{pq}\delta_{rs}$. Actually, we have just established that the six functions defined in equation (6.14) are linearly independent. In short, they form a basis of the space $\mathcal{P}_2(T)$, since the dimension of this space is six, as a space of functions spanned by the six monomials $1, x, y, x^2, xy$ and y^2. In view of the expansion (equation (6.13)) of an arbitrary function $\phi \in \mathcal{P}_2(T)$, the set of six quadratic functions in equation (6.14) is said to be the **canonical basis** of this space related to the vertices and the edge mid-points of T. This property is precisely what allows for the construction of space W_h consisting of piecewise quadratic functions defined by their values at the vertex and edge mid-points of the mesh. Indeed, the function continuity is ensured according to the following argument. The traces of such a function on any edge common to two mesh triangles coincide at its two end-points and mid-point by construction. Since both traces are quadratic functions of the abscissa along the edge, by an elementary property of parabolae they must coincide everywhere. A similar argument shows that V_h is the subspace of W_h consisting of those functions that vanish at all the mesh vertices and edge mid-points contained in Γ.

If K_h is the total number of $\mathcal{P}_2$ nodes in the mesh conveniently numbered, the dimension of W_h is easily seen to be K_h, and a natural basis of this space is the set of **shape functions** $\{\varphi_i\}_{i=1}^{K_h}$ defined as follows. If a mesh triangle T contains node S_i, the restriction of φ_i to T is the local shape function ϕ_{pq} corresponding to this node, and otherwise φ_i vanishes identically. Clearly enough, φ_i belongs to W_h for every i, and in particular it is a continuous function. Moreover, every function in $w \in W_h$ can be expressed by $w = \sum_{i=1}^{K_h} w_i \varphi_i$, for $w_i = w_h(S_i)$ since $\varphi_i(S_j) = \delta_{ij}$ by construction. In view of this, here again we come up with the SLAE equivalent to equation (6.12) whose unknowns are the values u_j of u_h at the N_h nodes of the mesh not contained in Γ, u_h being the variational approximation of u by the $\mathcal{P}_2$ FEM. The practical computation of the system matrix and right-side vector in a wider context will be addressed in Subsection 6.2.5.

To conclude, we refer to any classical book on the mathematical theory of the FEM such as references [34], [31], [66], among many others, for the proof of the following qualitative results that hold for the $\mathcal{P}_2$ FE approximation u_h of the solution u to equations (4.3)–(4.4).

Assume that all the third-order partial derivatives of u belong to $L^2(\Omega)$. Then, provided $\mathcal{T}_h$ belongs to a regular family of triangulations, there exists a constant $C_{23}(u)$ depending only on the third-order partial derivatives of u and on the family of meshes under consideration, but not on the maximum element size h, such that $|u - u_h|_{1,2} \le C_{23}(u)h^2$. Moreover, in the case of problem (4.3) and assuming that Ω is convex, for another constant C'_{23} with the same characteristics as C_{23}, it holds that $\| u - u_h \|_{0,2} \le C'_{23}h^3$.

The results given above are coherent, in the sense that the P_2 FEM should be a higher order method as compared to both the P_1 and the Q_1 FEM. This is essentially because it is capable of reproducing an exact solution, whenever it is a polynomial of degree less than or equal to two. However, the reader has certainly noticed that higher order is effectively attained, only in case the solution is more regular than functions in $H^2(\Omega)$. This is a problem if Ω is a polygon, since in general the best we can hope for is a solution u belonging to this space under the condition that $\Gamma_0 = \Gamma$ and Ω be convex. Indeed, if any of these two conditions is violated, under normal circumstances u does not even belong to $H^2(\Omega)$ (cf. [92]).

On the other hand, assuming that Γ is smoother than a convex polygon, one can expect that the solution third-order partial derivatives do belong to $L^2(\Omega)$ in case $f \in H^1(\Omega)$. But in this case, in principle Ω must have a curved boundary sufficiently smooth. This implies that Γ must be approximated by a polygon, if we are to use the method as described in this subsection. Unfortunately, such an approximation of the domain erodes the method's accuracy to the point of downgrading its second-order convergence properties. In the next subsection, we will address this issue in more detail, and will present an isoparametric technique aimed at remedying this drawback.

As a conclusion, the P_2 FEM should not be employed if the user cannot be sure to significantly enhance the accuracy of a numerical solution. Indeed, for a mesh not so coarse, computations with this method are substantially more costly than with the P_1 FEM and the FVM, let alone the FDM for a rectangular domain.

6.1.4   The Case of Curved Domains

The geometric versatility of the FEM is one of the main reasons for the great popularity this method has won, since it started being employed in engineering applications several decades ago. So far, we have studied the FEM to solve boundary value problems posed only in intervals, rectangles or general polygons. This was mainly motivated by the simpler presentation of the method these cases allow for. However, absolutely nothing prevents one from applying the classical FEM successfully, for solving problems posed in domains of arbitrary shape, as long as they don't have singularities such as cusps or irregularities alike. Incidentally, we observe in this respect that in some singular cases it is possible to resort to variants of the FEM. For example, deformations of solids with cracks are currently simulated with special versions of the FEM. The **XFEM** [141] plays a prominent role in this class of methods, for it is much better suited than classical FEM to cope with singular stress distributions about a crack.

It turns out that in the case of low-order methods, it does no harm approximating a smooth domain Ω by the polygon Ω_h with boundary Γ_h, whose vertices are the mesh vertices located on Γ (cf. Figure 6.6). More specifically, we mean the P_1 and the Q_1 FEM. This assertion is justified by the arguments given below for the case of Dirichlet boundary conditions. In order not to burden the reader any longer with lengthy mathematical analyses, in this subsection we refer to [182, pp. 197–199] for the case of Neumann boundary conditions.

In the remainder of this subsection, we assume that we are working with a quasi-uniform family of meshes.

Let us first assume that $\Omega_h \subset \Omega$, which always happens if Ω is convex. Recalling the a priori error analysis of Subsection 5.1.2 for the P_1 FEM, it is not difficult to figure out that here the consistency equality (5.17) holds with the same bilinear form a and linear form F, provided the functions $v \in V_h$ are prolongated by zero in the set $\Omega \setminus \Omega_h$. Thus, recalling equation

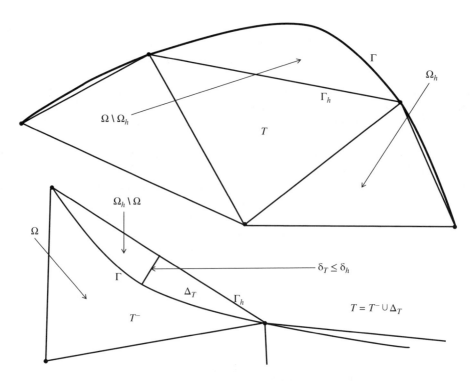

Figure 6.6 Convex and concave portions of a curved Ω and polygon Ω_h near Γ

(5.18) in the case of a curved domain, the estimate of the difference between u and u_h in the norm of $H^1(\Omega_h)$ results from the measure of the difference between u and its $\mathcal{P}_1$ interpolate $\tilde{u}_h$ in the same norm. Thus, this error norm is first bounded by a mesh-independent constant times h multiplied by the norm of $H(u)$ in $L^2(\Omega_h)$. Since $\Omega_h \subset \Omega$, we can augment this bound by taking the norm of $H(u)$ in the whole Ω. But what about the error in the norm of $H^1(\Omega)$? This question makes sense because we are considering that u_h is prolongated by zero in the skin $\Omega \setminus \Omega_h$. For this reason, the additional term is just the L^2-norm of **grad** u in the skin. We assume hereafter that the gradient of u is bounded in Ω, which is reasonable if f is bounded, since Ω is smooth [128]. Then, the term to estimate is easily seen to be bounded above by $\| \textbf{grad } u \|_{0,\infty,\Omega} \sqrt{area(\Omega \setminus \Omega_h)}$. Referring to Figure 6.6, the area of the skin $\Omega \setminus \Omega_h$ is clearly bounded by its maximum chord δ_h times the length of Γ. Since δ_h is bounded above by a mesh-independent constant times h^2, the additional term is still an $O(h)$. As a result, the final estimate of $u - u_h$ in the norm of $H^1(\Omega)$ will be globally an $O(h)$ term multiplied by $\| H(u) \|_{0,2} + \| \textbf{grad } u \|_{0,\infty,\Omega}$.

Now what happens if Ω is not convex? In principle, we no longer have $\Omega_h \subset \Omega$, and therefore a function $v \in V_h$, even prolongated by zero in the convex portions of the skin, is not necessarily in V, owing to its concave portions. Therefore, the consistency equality (5.17) does not apply, and the validity of equation (5.18) is to be questioned. However, it is still possible to show that this does not affect the method's order of convergence, as we do next. For the sake of simplicity, we assume that concave and convex portions of Γ are separated by mesh vertices.

First, we note that for a non-convex Ω, the problem we are solving for u_h is

$$a_h(u_h, v) = F_h(v) \; \forall v \in V_h,$$

where $a_h(u, v) := \int_{\Omega_h} (\mathbf{grad} \; u | \mathbf{grad} \; v) dx dy \; \forall u, v \in V_h$ and $F_h(v) := \int_{\Omega_h} f v \; dx dy \; \forall v \in V_h$, V_h being the space of continuous functions vanishing on Γ_h, whose restriction to every triangle $T \in \mathcal{T}_h$ belongs to $\mathcal{P}_1(T)$. Therefore, now instead of equation (5.17), we have

$$a_h(u_h - \tilde{u}_h, v) = [F_h - F](v) + [F(v) - a(u, v)] + [a - a_h](\tilde{u}_h, v) + [a(u - \tilde{u}_h, v)].$$

Denoting by $\| \cdot \|_{0,h}$ the norm of $L^2(\Omega_h)$ and taking $v = \bar{u}_h := u_h - \tilde{u}_h$, we come up with

$$\| \mathbf{grad} \; \bar{u}_h \|_{0,h}^2 \leq |[F_h - F](\bar{u}_h)| + |F(\bar{u}_h) - a(u, \bar{u}_h)| + |[a - a_h](\tilde{u}_h, \bar{u}_h)| + |a(u - \tilde{u}_h, \bar{u}_h)|$$

In order to shorten the calculations, in some cases the above terms will be estimated by 'brutal force'. We do so because our goal is just showing that a first-order error estimate still holds.

From the boundedness assumption on f, the first term is less than $\|f\|_{0,\infty} \sum_{T \in \mathcal{T}_h} \int_{\Delta_T} |\bar{u}_h|$ $dx dy$, where Δ_T is the intersection of a triangle T with the skin $\Omega_h \setminus \Omega$. But since $\bar{u}_h$ vanishes on Γ_h, its value in each non-empty Δ_T cannot exceed δ_h multiplied by $|\mathbf{grad} \; \bar{u}_h|$, which is constant in T. Therefore, we have

$$\int_{\Delta_T} |\bar{u}_h| dx dy \leq \delta_h \frac{area(\Delta_T)}{\sqrt{area(T)}} \left[\int_T |\mathbf{grad} \; \bar{u}_h|^2 dx dy \right]^{1/2} \qquad (6.15)$$

Taking into account that $area(\Delta_T) \leq area(T)$ and $area(T) \leq h_T^2/2$, where h_T is the length of the largest edge of T, we obtain

$$\int_{\Delta_T} |\bar{u}_h| dx dy \leq \delta_h \sqrt{area(T)} \left[\int_T |\mathbf{grad} \; \bar{u}_h|^2 dx dy \right]^{1/2} \qquad (6.16)$$

Then, applying the Cauchy–Schwarz inequality, equation (6.16) easily yields

$$\sum_{T \in \mathcal{T}_h} \int_{\Delta_T} |\bar{u}_h| dx dy \leq \sqrt{area(\Omega_h)} \delta_h \| \mathbf{grad} \; \bar{u}_h \|_{0,h} \qquad (6.17)$$

Since δ_h is an $O(h^2)$, from equation (6.17) the term $|[F_h - F](\bar{u}_h)|$ is found to be bounded above by $C_{D1} h^2 \|f\|_{0,\infty} \| \mathbf{grad} \; \bar{u}_h \|_{0,h}$, C_{D1} being a mesh-independent constant.

The second term is handled by applying First Green's identity in each $T \in \mathcal{T}_h$ and summing up the individual results over the mesh. Since $-\Delta \; u - f = 0$ in every T, $\tilde{u}_h$ is continuous at inter-element boundaries and the outer normal derivatives of u have opposite signs thereupon, all the terms cancel out except $\int_\Gamma \partial_\nu u \bar{u}_h \; ds$. From the assumption that the gradient of u is bounded in Ω, we have

$$|a(u, \bar{u}_h) - F(\bar{u}_h)| = \left| \int_\Gamma \partial_\nu u \bar{u}_h ds \right| \leq \| \mathbf{grad} \; u \|_{0,\infty} \sum_{T \in \mathcal{T}_h} \int_{\Gamma \cap T} |\bar{u}_h| ds. \qquad (6.18)$$

Taking into account that if $\Gamma \cap T$ is not empty, $\bar{u}_h$ is bounded above by δ_h multiplied by $|\mathbf{grad} \; \bar{u}_h|$, thereupon we can write

$$|a(u, \bar{u}_h) - F(\bar{u}_h)| \leq \delta_h \| \mathbf{grad} \; u \|_{0,\infty} \sum_{T \in \mathcal{T}_h} \int_{\Gamma \cap T} |\mathbf{grad} \; \bar{u}_h| ds.$$

Taking the maximum of $|\mathbf{grad}\ \bar{u}_h|$ over all mesh elements, we further obtain

$$|a(u, \bar{u}_h) - F(\bar{u}_h)| \leq length(\Gamma)\delta_h\ \|\ \mathbf{grad}\ u\|_{0,\infty}\ \|\ \mathbf{grad}\ \bar{u}_h\|_{0,\infty} \tag{6.19}$$

On the other hand, if T_M is an element in which the maximum of $\mathbf{grad}\ \bar{u}_h$ is attained, we clearly have

$$\|\ \mathbf{grad}\ \bar{u}_h\|_{0,\infty} = \max_{T \in \mathcal{T}_h} |\mathbf{grad}\ \bar{u}_h|_{|T} \leq [\sqrt{area(T_M)}]^{-1}\ \|\ \mathbf{grad}\ \bar{u}_h\|_{0,h}. \tag{6.20}$$

Since the family of meshes is quasi-uniform, equation (6.19) implies that there exists a mesh-independent constant C_I such that

$$\|\ \mathbf{grad}\ \bar{u}_h\|_{0,\infty} \leq C_I h^{-1}\ \|\ \mathbf{grad}\ \bar{u}_h\|_{0,h}. \tag{6.21}$$

Remark 6.4 *Equation (6.21) is an example of the so-called **inverse inequalities** in one of its simplest expressions. A similar result applies to many other pairs of norms and to any FE space construction with quasi-uniform families of meshes (see e.g. [45]). Inverse inequalities play a very important role in the literature on the mathematical analysis of the FEM. We refer to Exercise 5.21 for a remarkable aqpplication of inverse inequalities.* ∎

Combining equations (6.19) and (6.21), we conclude that there is a mesh-independent constant C_{D2} such that

$$|F(\bar{u}_h) - a(u, \bar{u}_h)| \leq C_{D2}h\ \|\ \mathbf{grad}\ u\|_{0,\infty}\ \|\ \mathbf{grad}\ \bar{u}_h\|_{0,h}.$$

The third term is just the integral over the skin $\Omega_h \setminus \Omega$ of $(\mathbf{grad}\ \tilde{u}_h|\mathbf{grad}\ \bar{u}_h)$. Since $\tilde{u}_h$ is linear in every element T, by Rolle's theorem, $\partial_x \tilde{u}_h$ (resp. $\partial_y \tilde{u}_h$) is easily found to be bounded above by $\|\ \partial_x u\|_{0,\infty}$ (resp. $\|\ \partial_y u\|_{0,\infty}$). Then, we have

$$|[a - a_h](\tilde{u}_h, \bar{u}_h)| \leq\|\ \mathbf{grad}\ u\|_{0,\infty} \sum_{T \in \mathcal{T}_h} \int_{\Delta_T} |\mathbf{grad}\ \bar{u}_h| dx dy.$$

Since the sum of the areas of Δ_T is the area of the skin $\Omega_h \setminus \Omega$ (cf. Figure 6.6), using the same trick as above, we obtain

$$|[a - a_h](\tilde{u}_h, \bar{u}_h)| \leq C_I h^{-1} area(\Omega_h \setminus \Omega)\ \|\ \mathbf{grad}\ u\|_{0,\infty}\ \|\ \mathbf{grad}\ \bar{u}_h\|_{0,h}.$$

Notice that the area of the skin is bounded above by the $\delta_h length(\Gamma)$. Therefore, the third term can be estimated by

$$|[a - a_h](\tilde{u}_h, \bar{u}_h)| \leq C_{D3}h\ \|\ \mathbf{grad}\ u\|_{0,\infty}\ \|\ \mathbf{grad}\ \bar{u}_h\|_{0,h}.$$

Finally, the fourth term can be bounded by $|u - \tilde{u}_h|_{1,2}|\bar{u}_h|_{1,2}$ using the Cauchy–Schwarz inequality, which can be bounded in turn by $|u - \tilde{u}_h|_{1,2}\ \|\ \mathbf{grad}\ \bar{u}_h\|_{0,h}$ (please check!). It follows that this term can be estimated by the interpolation error in Ω measured in the (semi-) norm of $H^1(\Omega)$. We have already treated the term $|u - \tilde{u}_h|_{1,2}$ when Ω is convex. In case not, the novelty is that the interpolation error is to be considered only in the part $T^- := T \setminus \Delta_T$ (cf. Figure 6.6), since u is not defined in the skin. We cannot even extend u by zero to this set because the Hessian of the resulting function might not be defined in T. However, if we re-examine the calculations in Subsection 5.1.2 leading to the bound for the interpolation error $u - \tilde{u}_h$ in a triangle T, we conclude that it also applies to the subset T^-, in case $T^- \neq T$. Thus, the fourth and last term can be bounded above by $C_{D4} h \| H(u) \|_{0,2} \| \mathbf{grad} \, \bar{u}_h \|_{0,h}$. Summarizing, the following estimate holds:

$$\| \mathbf{grad} \, \bar{u}_h \|_{0,h} \leq h [C_{D1} h \| f \|_{0,\infty} + (C_{D2} + C_{D3}) \| \mathbf{grad} \, u \|_{0,\infty} + C_{D4} \| H(u) \|_{0,2}].$$

The final estimate of the error $|u - u_h|_{1,2}$ is obtained by the triangle inequality, that is, $|u - u_h|_{1,2} \leq |u - \tilde{u}_h|_{1,2} + |u_h - \tilde{u}_h|_{1,2}$ or yet by $|u - u_h|_{1,2} \leq |u - \tilde{u}_h|_{1,2} + \| \mathbf{grad} \, \bar{u}_h \|_{0,h}$.

Collecting the above results and those in equation [182] for Neumann boundary conditions, we can assert that in all cases, if a curved domain Ω is approximated by Ω_h, the $\mathcal{P}_1$ FE solution u_h satisfies for a suitable mesh-independent constant C_D,

$$|u - u_h|_{1,2} \leq C_D h [\| H(u) \|_{0,2} + \| f \|_{0,\infty} + \| \mathbf{grad} \, u \|_{0,\infty}] \tag{6.22}$$

Now let us see what happens if we use the $\mathcal{P}_2$ FEM to solve equation (4.3) posed in a curved domain Ω. Here again, the simplest thing to do is to approximate Ω by a polygon Ω_h. However, as already pointed out, if we proceed in this manner the expected second order of the approximation error will no longer hold. In order to explain why, we consider again a convex Ω. In this case, $\Omega_h \subset \Omega$, and we can extend by zero the functions in V_h defined like in the polygonal case, to the skin $\Omega \setminus \Omega_h$. We know that the consistency result (5.17) still holds and the solution $u_h \in V_h$ of $a(u_h, v) = F(v) \, \forall v \in V_h$ satisfies

$$|u - u_h|_{1,2} \leq 2 |u - \underline{u}_h|_{1,2}.$$

where $\underline{u}_h$ is the interpolate of u at all nodes but the mid-points of the edges of Γ_h. This is because by definition $\underline{u}_h$ belongs to V_h, and hence its values at such points are necessarily zero. From classical results on the FEM (see e.g. [45]), if all third-order partial derivatives of u belong to $L^2(\Omega)$, then the interpolation error $|u - \tilde{u}_h|_{1,2}$ is bounded above by an $O(h^2)$ term. Moreover, it can be asserted that in this case the gradient of u is bounded in Ω, according to the Sobolev embedding theorem [1]. We use this property to estimate the deviation $|\underline{u}_h - \tilde{u}_h|_{1,2}$, which will lead to an estimate of $|\underline{u}_h - u|_{1,2}$ and hence to the final estimate of $|u - u_h|_{1,2}$.

Let $\mathcal{S}_h$ be the subset of $\mathcal{T}_h$ consisting of the triangles that have an edge on Γ_h. Since $\underline{u}_h$ coincides with $\tilde{u}_h$ in all triangles that do not belong to $\mathcal{S}_h$, we clearly have

$$|\underline{u}_h - \tilde{u}_h|_{1,2}^2 = \sum_{T \in \mathcal{S}_h} \int_T |\mathbf{grad}(\underline{u}_h - \tilde{u}_h)|^2 \, dx dy.$$

If we examine more carefully the form of the terms in the above sum, we conclude that in every $T \in \mathcal{S}_h$, the difference $\underline{u}_h - \tilde{u}_h$ equals the shape function ϕ_{pq}, say with $p = 2$ and $q = 3$, associated with the mid-point lying on Γ_h, that is, $S_4^T \equiv S_{23}$, times $u(S_{23})$. The reader might determine as Exercise 6.6 a mesh-independent constant C_ϕ such that $|\mathbf{grad} \phi_{23}| \leq C_\phi h^{-1}$. Moreover, referring to Figure 6.7, an elementary Taylor expansion along the chord $\overline{P_T S_{23}}$, where P_T is the intersection with Γ of the perpendicular to Γ_h through S_{23}, yields $|u(S_4^T)| \leq \delta_h \| \mathbf{grad} \, u \|_{0,\infty}$. It follows that, for a suitable mesh-independent constant C_{D5},

$$|\underline{u}_h - \tilde{u}_h|_{1,2}^2 \leq C_{D5}^2 h^2 \| \mathbf{grad} \, u \|_{0,\infty}^2 \sum_{T \in \mathcal{S}_h} area(T).$$

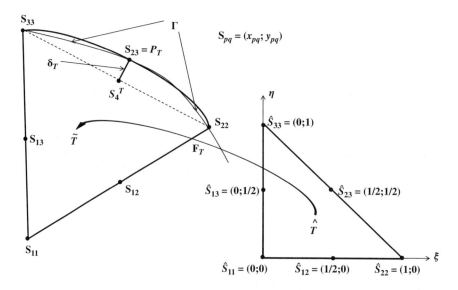

Figure 6.7   Mapping $\mathbf{F}_T$ from master triangle $\hat{T}$ onto curved triangle $\tilde{T}$

Since the above sum of areas cannot exceed one half of the length of Γ multiplied by h (please check!), collecting all the intermediate results, we finally obtain

$$|u - u_h|_{1,2} \leq \underline{C}_D(u)h^{3/2},$$

where the constant $\underline{C}_D(u)$ is expressed in terms of the third-order partial derivatives of u independently of h [45]. This estimate, corroborated by numerical experiments, confirms the erosion of the global approximation properties of the $\mathcal{P}_2$ FEM, if polygonal approximations of a curved domain Ω are employed.

From the above arguments, it seems intuitive that a simple way to recover second-order error estimates for the $\mathcal{P}_2$ FEM is to improve the boundary approximation. The natural idea is to shift the mid-point boundary nodes to Γ. However, this is only possible if Γ is approximated by curved portions instead of straight segments. In Figure 6.7, we illustrate the most employed boundary approximation in this case, namely, the one with arcs of parabola, each one of them passing through three points. These points are two vertices located on Γ of a boundary triangle T and the projection P_T of the boundary edge mid-point S_4^T onto Γ. The actual mesh triangle T becomes a curved triangle $\tilde{T}$ with two straight edges and one parabolic edge, as shown in Figure 6.7. However, if we are to fit this configuration into our construction of quadratic shape functions, we have to resort to isoparametric transformations of the master triangle $\hat{T}$ onto $\tilde{T}$. This means that the shape functions in the real domain $\tilde{T}$ are the images through a $\mathcal{P}_2$ mapping $\mathbf{F}_T$ from $\hat{T}$ onto $\tilde{T}$ of the six canonical quadratic shape functions $\hat{\phi}_{pq}$ defined in $\hat{T}$. Each one of these is related to a node $\hat{S}_{pq}$ of the master element, as illustrated in Figure 6.7. Denoting by $(x_{pq}; y_{pq})$ the coordinates of the six nodes S_{pq} of curved element $\tilde{T}$ (cf. Figure 6.7) for $1 \leq p \leq q \leq 3$, undoubtedly $\mathbf{F}_T = (F_{Tx}; F_{Ty})$ is uniquely defined by

$$F_{Tx}(\xi,\eta) = \sum_{p=1}^{3}\sum_{q=p}^{3} x_{pq}\hat{\phi}_{pq}(\xi,\eta) \quad \text{and} \quad F_{Ty}(\xi,\eta) = \sum_{p=1}^{3}\sum_{q=p}^{3} y_{pq}\hat{\phi}_{pq}(\xi,\eta),$$

where,

- $\hat{\phi}_{11} = (1-\xi-\eta)[2(1-\xi-\eta)-1]$; $\hat{\phi}_{22} = \xi(2\xi-1)$; $\hat{\phi}_{33} = \eta(2\eta-1)$;
- $\hat{\phi}_{12} = 4\xi(1-\xi-\eta)$; $\hat{\phi}_{13} = 4\eta(1-\xi-\eta)$; $\hat{\phi}_{23} = 4\xi\eta$.

It is important to note that in the interior of the domain, and in particular along mesh edges, a function v in the thus-constructed FE space V_h behaves like in the case of a polygonal domain. This means that the traces over those edges are quadratic functions, which implies that v is continuous. This function itself in any mesh triangle with three straight edges is an ordinary quadratic function; it is a rational function of x and y only in curved triangles $\tilde{T}$.

We refrain from elaborating any longer on practical calculations for the isoparametric formulation of the $\mathcal{P}_2$ FEM, such as the computation of element matrices and right-side vectors, since this was done in the previous subsection. For further details, we refer to classical books such as references [45] and [213].

As a conclusion, we stress that second-order error estimates in the norm of $H^1(\Omega)$ for a curved Ω do hold for the above-defined isoparametric $\mathcal{P}_2$ FEM, assuming that sufficiently accurate quadrature formulae are used to compute element matrices and right-side vectors (cf. [162]).

6.1.5 *Example 6.1: $\mathcal{P}_2$-FE Solution of the Equation $u - \Delta u = f$*

We demonstrated in the previous subsection that the expected order of convergence in the H^1-norm of the $\mathcal{P}_2$ FEM in the solution of the Poisson equation may undergo a half point downgrade, in case the equation domain Ω with a curved boundary is approximated by polygons Ω_h consisting of the union of mesh triangles with straight edges. The aim of this example is to show the numerical effect of such a domain approximation in the solution of second-order PDEs by the $\mathcal{P}_2$ FEM with different types of boundary conditions. First of all, we observe the method's convergence rates without boundary approximation effects, by solving a model problem in the unit square $(0,1) \times (0,1)$. Referring to reference [45], we note that the errors in the L^2-norm in the case of a polygonal domain are in $O(h^3)$, as long as all the third-order partial derivatives of u belong to $L^2(\Omega)$ (i.e. $u \in H^3(\Omega)$) [1]. In the sequel, we examine the case of a domain Ω with a curved boundary, successively approximated by Ω_h. In the latter framework, we consider two test problems in an ellipse in order to discard any super-convergence effect owing to symmetry. We consider ellipses centred at the origin defined by the equation $x^2 + (y/c)^2 = 1$, c being a strictly positive real number. In the computations, we took $c = 0.5$.

The variational formulations for each type of boundary conditions can be obtained in a straightforward manner. Their discrete counterparts are based on meshes with $4J^2$ triangles generated in the same way as in Example 5.3, both for the unit square and for the ellipse. In the latter case, we used the procedure depending on a single integer parameter J described in reference [166]. Notice that for each value of J, the mesh of the whole domain contains $8J^2$ triangles. However, the computational domain is restricted to the quarter of ellipse given by $x \geq 0$ and $y \geq 0$, for whose mesh with $2J^2$ triangles an illustration is supplied in Figure 6.8, taking $J = 32$ and $c = 0.5$. Homogeneous Neumann boundary conditions applied along the segments given by $\{(x; y)|y = 0; \ 0 < x < 1\}$ and $\{(x; y)|x = 0; \ 0 < y < c\}$ account for data and domain symmetry in this case. By watching Figure 6.8, the reader can figure out that the triangles in the sector given by $\pi/4 \leq \theta \leq \pi/2$ are a little smaller than those in the sector given by $0 \leq \theta \leq \pi/4$. However, the angles of all elements are not so different from each other. This confirms that the adopted procedure tends to generate triangles of similar shapes (cf. [166]).

In all cases, convergence rates were observed by taking $J = 2^m$ for $m = 2, 3, 4, 5, 6$. By definition, we set $h = 1/J$, which stands for a representative mesh step size.

Model problems in a unit square : We run simulations for two different problems called *model problem 1 and model problem 2*. Their manufactured exact solutions are $u(x, y) = (1 - x^2)(1 - y^2)$ and $u(x, y) = (1 - x^2)^2 + (1 - y^2)^2$, respectively, from which the right side f is determined. For model problem 1, we prescribe homogeneous mixed boundary conditions, namely, Neumann boundary conditions on the edges $x = 0$ and $y = 0$ and Dirichlet boundary conditions on the edges $x = 1$ and $y = 1$. For model problem 2, Neumann boundary conditions are applied everywhere. In Figure 6.9a and 6.9b, we show log-log plots of the errors in the H^1-(semi-)norm, in the L^2-norm and in the discrete L^∞-norm in terms of h for each model-problem. The slopes of the best-fitted straight lines in the mean square sense reflects the fact that the P_2 FEM is of the second order in $H^1(\Omega)$ and of the third order in both the L^2 and L^∞ senses. However, we must say that the right sides and the errors in mean square norms were not computed exactly, but rather with a numerical quadrature formula in triangles which is exact for polynomials of degree less than or equal to four. This could explain the small deviations from the order three for the L^2 errors observed in Figure 6.9, more particularly in the case of model problem 2.

Test problems in an ellipse : In a first battery of experiments named *test case 1*, homogeneous Dirichlet boundary conditions are considered. We take a manufactured exact solution given by $u(x, y) = (c^2x^2 + y^2 - c^2)(x^2 + c^2y^2 - c^2)$, and enforce $u = 0$ all over the boundary of the ellipse. The right-side function f is given by $f(x, y) = 4c^2(1 + c^2) - (2 + 12c^2 + 2c^4)(x^2 + y^2) + u(x, y)$. Figure 6.10a displays log-log plots of the errors in the H^1-(semi-)norm, in the L^2-norm and in the discrete L^∞-norm in terms of h. The slope of the best-fitted straight line in the mean-square sense confirms that

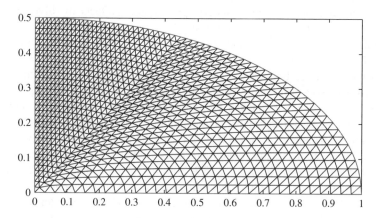

Figure 6.8   Mesh of a quarter of ellipse with $c = 0.5$ consisting of $2J^2$ triangles for $J = 32$

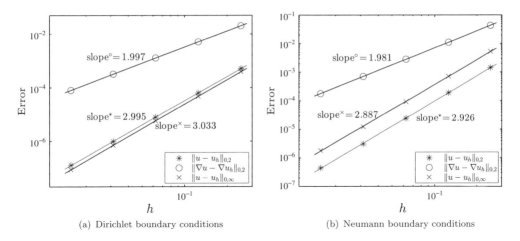

(a) Dirichlet boundary conditions   (b) Neumann boundary conditions

Figure 6.9 L^2 errors of u_h, **grad** u_h and maximum nodal errors for model problems 1 and 2

the errors are an $O(h^{3/2})$ in $H^1(\Omega_h)$. The errors in $L^2(\Omega_h)$ in turn are an $O(h^2)$, which is the best we can hope for. The reader might look for an explanation to the latter assertion, in the light of what we saw in Subsection 5.1.2 for the $\mathcal{P}_1$ FEM, even though the problem here is more delicate. As a matter of fact, the argument in that subsection leading to L^2-error estimates for the FE solution of the Poisson equation is held in check here, since the boundary term in the First Green's identity no longer vanishes when Ω is replaced by Ω_h. Finally, the discrete L^∞-errors appear to be close to an $O(h^2)$.

In a second test named *test case 2*, homogeneous Neumann boundary conditions are investigated. We take a manufactured exact solution given by $u(x,y) = (c^2x^2 + y^2 - c^2)^2$, and prescribe a zero normal derivative on the whole boundary of the computational domain. The right-side function f is given by $f(x,y) = 4c^2(c^2+1) - 12(c^4x^2 + y^2) - 4c^2(x^2+y^2) + u(x,y)$. We note that the problem at hand is uniquely solvable, in contrast to the Neumann problem for the Laplacian operator (cf. Example 5.1).

In Figure 6.10b, we show log-log plots of the errors in the (semi-)norm of $H^1(\Omega_h)$, in the norm of $L^2(\Omega_h)$ and in the discrete L^∞-norm, in terms of h. The slope of the best-fitted straight line in the mean-square sense indicates that the errors in $H^1(\Omega_h)$ behave roughly as an $O(h^2)$. This suggests that the method's optimal second order in this norm is essentially maintained, in case homogeneous Neumann boundary conditions are shifted from a curved boundary to the boundary of the approximating polygon. This result can be partially justified by the theory of interpolation in triangles (cf. [45]). Indeed, in contrast to the case of Dirichlet boundary conditions, we do not force the $\mathcal{P}_2$ interpolation by shifting to the boundary of Ω_h function values at points lying on the true boundary Γ. However, an additional source of error comes into play, owing to the fact that, instead of $\int_{\Omega_h} fv_h dxdy$, in the true problem the right side is $\int_\Omega f\tilde{v}_h dxdy$, $\tilde{v}_h$ being the prolongation of $v_h \in V_h$ to $\Omega \setminus \Omega_h$. Fortunately, this term admits an upper bound in $O(h^2)$, and hence in the case of homogeneous Neumann boundary conditions, the $\mathcal{P}_2$ FEM is globally a second-order method in $H^1(\Omega_h)$.

On the other hand, the errors in $L^2(\Omega_h)$ turn out to be no better than an $O(h^2)$. Even worse, they appear to be greater than the errors in the H^1-semi-norm, which can be viewed as a kind of bad surprise. This apparently strange behaviour relies on the fact that here optimal L^2-error

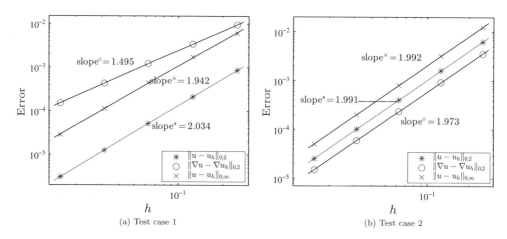

(a) Test case 1   (b) Test case 2

Figure 6.10 Errors in the H^1-(semi-)norm, in the L^2-norm and in the discrete L^∞-norm

estimates cannot be demonstrated by arguments similar to those employed in the analysis of the FE solution of other types of elliptic equations in curved domains with homogeneous Neumann boundary conditions (see e.g. [167]). Actually, the so-called **Aubin–Nitsche duality argument** employed so far to derive L^2 error estimates brings about no qualitative improvement with respect to the H^1 estimates. This shortcoming is essentially due to the integral of $f v_h$ in $\Omega \setminus \Omega_h$. Finally, we note that the errors measured in the discrete L^∞-norm are greater than the errors in $L^2(\Omega_h)$, and no better than an $O(h^2)$ either.

To conclude, we note that all suboptimal convergence rates pointed out in this example turn to optimal ones if the isoparametric technique is employed.

6.1.6   More about Implementation in Two-dimensional Space

In Chapter 4, we addressed the implementation of the $\mathcal{P}_1$ FEM in the framework of homogeneous mixed Dirichlet–Neumann boundary conditions. Here, we go a little further and consider implementation aspects in a wider context. Nonetheless, likewise Subsection 4.3.4, the reader should be aware of the fact that the procedures advocated in this subsection should be regarded as the author's suggestions only. There are actually several possibilities to write a FE code, among which the master element technique is very popular, but this is not what we do here. The reader is referred to reference [146] for further hints on FEM implementation.

Having in mind both objectiveness and sense of practical interest, we confine ourselves to the case of piecewise linear and piecewise quadratic approximations, applied to model problem (6.23) in a **polygonal domain** Ω. Still assuming that Γ consists of two disjoint complementary portions Γ_0 and Γ_1, we wish to solve

$$\begin{cases} -\Delta u = f \text{ in } \Omega \\ u = g_0 \text{ on } \Gamma_0 \\ \gamma u + \partial_\nu u = g_1 \text{ on } \Gamma_1 \end{cases} \qquad (6.23)$$

where g_0 and g_1 are given functions defined, respectively, on Γ_0 and Γ_1; $f \in L^2(\Omega)$ is given; and γ is a real coefficient $\gamma \geq 0$. For the moment, we just assume that $g_1 \in L^2(\Gamma_1)$, and that g_0 is continuous and defined everywhere on Γ_0. Notice that both assumptions are sufficient to guarantee that $u \in H^1(\Omega)$ (see e.g. [1]).

Problem (6.23) can be rewritten in the following equivalent Galerkin variational form:

$$\begin{cases} \text{Find} \quad u \in H^1(\Omega) \text{ with } \quad u = g_0 \text{ on } \Gamma_0 \text{ such that } \quad \forall v \in V \\ \int_\Omega (\mathbf{grad}\ \mathbf{u}|\mathbf{grad}\ \mathbf{v})dxdy + \gamma \int_{\Gamma_1} uv\ ds = \int_\Omega fv\ dxdy + \int_{\Gamma_1} g_1 v\ ds \end{cases} \tag{6.24}$$

where $V = \{v/v \in H^1(\Omega), v = 0 \text{ on } \Gamma_0\}$.

Problem (6.24) is equivalent to equation (6.23), as the reader may check by the same arguments we have used for variational formulations of boundary value problems so far. Moreover, both problems have a unique solution, even when $\gamma = 0$, provided $length(\Gamma_0) > 0$, or when $\gamma > 0$ and $length(\Gamma_0) = 0$ (see e.g. [31]). The uniqueness result can be better understood if, instead of u, we search for a function $\underline{u} \in V$ such that $u = \underline{u} + u_0$, where u_0 is any function of $H^1(\Omega)$ satisfying $u_0 = g_0$ on Γ_0. In doing so, it holds for $\underline{u}$:

$$\underline{u} \in V\ /\ a(\underline{u}, v) = F_0(v) \quad \forall v \in V \tag{6.25}$$

where

$$a(u, v) := \int_\Omega (\mathbf{grad}\ u|\mathbf{grad}\ v)\ dxdy + \gamma \int_{\Gamma_1} uv\ ds \tag{6.26}$$

$$F_0(v) := \int_\Omega fv\ dxdy + \int_{\Gamma_1} g_1 v\ ds - a(u_0, v). \tag{6.27}$$

The independence of u vis-à-vis the choice of u_0 can be established as Exercise 6.8. This implies the uniqueness of u, as the solution of equation (6.24).

Now, let $\mathcal{T}_h$ be a partition of Ω with M_h triangles satisfying the compatibility conditions specified in Chapter 4, and the further requirement that the transition points between Γ_0 and Γ_1 consist of a set of vertices of the partition, if applicable. In this subsection, we enlarge our implementation hints so as to include the P_2 FEM. For this reason, unlike Chapter 4, we denote the number of vertices of $\mathcal{T}_h$ by k_h. The number of vertices belonging to Γ_0 is denoted by l_h, whereas $n_h := k_h - l_h$. Furthermore, the expression **mesh nodes** refers to all the mesh points at which approximate solution values have to be determined in the equivalent SLAE, or are prescribed. Then, K_h represents the total number of mesh nodes, while L_h is the number of nodes belonging to Γ_0. It follows that the number of system unknowns is $N_h := K_h - L_h$. Of course, in the case of the P_1 FE, $K_h = k_h$, $L_h = l_h$ and $N_h = n_h$. For the P_2 FE, we have $K_h = k_h + j_h$ where j_h is the total number of edges of the triangulation, whereas $L_h = l_h + i_h$, i_h being the number of mesh edges contained in Γ_0. For coherence, we also set $I_h := i_h$ and $J_h = j_h$.

In order to apply the FEM to solve equations (6.25)~(6.27), we are first given a FE space W_h spanned by a set of continuous shape functions φ_i, associated with the nodes S_i of $\mathcal{T}_h$, $i = 1, 2, \cdots, K_h$, satisfying $\varphi_i(S_j) = \delta_{ij}, j = 1, 2, \cdots, K_h$. We assume that the K_h nodes are numbered in such a way that the last nodes are those located on Γ_0.

Naturally enough, we choose the approximation u_{0h} of u_0 to be the function in W_h given by

$$u_{0h} := \sum_{l=N_h+1}^{K_h} g_0(S_l)\varphi_l$$

Then, V_h being the subspace of W_h spanned by the functions $\varphi_l, l = 1, 2, \cdots, N_h$, the approximation $\underline{u}_h \in V_h$ of $\underline{u}$ is obtained by solving:

$$a(\underline{u}_h, v) = F_0(v) \quad \forall v \in V_h \tag{6.28}$$

As the reader knows, if we expand $\underline{u}_h$ as the sum $\sum_{j=1}^{N_h} u_j \varphi_j$, the problem to solve is equivalent to a SLAE $A_h \vec{u}_h = \vec{b}_h$ for the unknown vector $\vec{u}_h = [u_1, u_2, \cdots, u_{N_h}]^T$, where $A_h = \{a_{i,j}\}$ is the $N_h \times N_h$ matrix given by $a_{i,j} = a(\varphi_j, \varphi_i)$ and $\vec{b}_h \in \mathcal{R}^{N_h}$ is defined by $b_i = F_0(\varphi_i)$. From the basic property $\varphi_i(S_j) = \delta_{ij}$, we know that $u_j = \underline{u}_h(S_j), j = 1, 2, \cdots, N_h$.

In Subsection 4.3.4, we described how to assemble the system matrix and right-side vector for the $\mathcal{P}_1$ FEM by an element by element procedure. Here, we extend it in order to cover the $\mathcal{P}_k$ FEM for both $k = 1$ and $k = 2$, as applied to the solution of equation (6.23). We have

$$a_{i,j} = \sum_{T \in \mathcal{T}_h} \left[\int_T (\mathbf{grad} \ \varphi_j | \mathbf{grad} \ \varphi_i) \, dx dy + \gamma \int_{T \cap \Gamma_1} \varphi_j \varphi_i \, ds \right] \tag{6.29}$$

The terms in brackets are the contributions of a triangle T to the appropriate coefficients $a_{i,j}$. The number of possible nonzero values for each T equals J^2, where $J = 3$ for linear elements and $J = 6$ for quadratic elements. This is because each triangle contains only J nodes of the mesh. Notice that, owing to matrix symmetry, in practice only $J(J+1)/2$ different values must be determined. It is also noteworthy that the number of entries in the element matrix related to a triangle having a nonempty intersection with Γ_0 cannot be reduced, unless $g_0 = 0$. Indeed, in case g_0 is not identically zero, all these entries must be used in the computation of either A_h or $\vec{b}_h$, as seen below.

Akin to Subsection 4.3.4, we proceed in the following manner:

1. Set the initial values of all the $a_{i,j}$'s to zero;
2. Number the M_h triangles of the mesh as $T_1, T_2, \cdots, T_{M_h}$;
3. Number the nodes of the mesh not belonging to Γ_0, from one through $N_h = K_h - L_h$, and next those belonging to Γ_0 from $N_h + 1$ through K_h;
4. For each triangle T_m, introduce a local numbering of the nodes S_r^m, where r ranges between one and J; for $1 \le r \le 3$, S_r^m are the vertices of T_m numbered in the counterclockwise sense, but otherwise arbitrarily; S_{r+3}^m is the mid-point of the edge opposite to S_r^m. For convenience, we also use a double-subscript local numbering by setting $S_{rr} = S_r^m$, for $r = 1, 2, 3$, and setting $S_{23} = S_4^m$, $S_{13} = S_5^m$ and $S_{12} = S_6^m$;
5. Construct an **incidence matrix** for the mesh, that is, an $M_h \times J$ integer array $\mathcal{N} = \{n_{m,r}\}$, such that for $1 \le m \le M_h$ and $1 \le r \le J$, $n_{m,r}$ is the number in the mesh of the node corresponding to node S_r^m of triangle T_m;
6. Compute the $J \times J$ element matrix $E^m = \{e_{r,q}^m\}$ for triangle T_m by

$$e_{r,q}^m = \int_{T_m} (\mathbf{grad} \ \varphi_j | \mathbf{grad} \ \varphi_i) \, dx dy + \gamma \int_{T_m \cap \Gamma_1} \varphi_j \varphi_i \, ds, 1 \le r, q \le J, \tag{6.30}$$

with $i = n_{m,r}$ and $j = n_{m,q}$;
7. Successively for $m = 1, 2, \cdots, M_h$, add the coefficients $e_{r,q}^m$ to the value of $a_{i,j}$, with $i = n_{m,r}$ and $j = n_{m,q}$, for $r = 1, 2, \cdots, J$ and $q = 1, 2, \cdots, J$, provided $i, j \le N_h$. If $i > N_h$, $e_{r,q}^m$ is discarded $\forall j$. Otherwise, if $j > N_h$, use $e_{r,q}^m$ to update the right-side vector $\vec{b}_h$ as seen hereafter.

In addition to the above data structures, it is useful to make available pointers indicating whether the product $\varphi_i \varphi_j$ vanishes identically on Γ_1 or not. In the latter case, the second term on the right side of equation (6.29) has to be computed. Such pointers can be arranged in the form of an integer vector $\vec{\mu} = [\mu_1, \mu_2, \cdots, \mu_{K_h}]^T$, where $\mu_i = 1$ if S_i lies in the set $\bar{\Gamma}_1$ (i.e. the union of Γ_1 and the transition points between Γ_0 and Γ_1) and $\mu_i = 0$ if $S_i \notin \bar{\Gamma}_1$. In this way for a given triangle T_m, pairs of subscripts $(i; j)$ with $i = n_{m,r}$ and $j = n_{m,s}$ give rise to the computation of an integral over Γ_1 in equation (6.30) only if $\mu_i \mu_j \neq 0$.

Like in Chapter 4, an element-by-element assembling procedure is used to compute the right-side vector $\vec{b}_h$. Since we are incorporating several additional terms, we describe below procedures that extend to problem (6.28) the one given in Subsection 4.3.4. For the sake of clarity, the latter is repeated here.
We have

$$b_i = \sum_{m=1}^{M_h} \left\{ \int_{T_m} f\varphi_i \, dxdy - \left[\int_{T_m} (\mathbf{grad} \ u_{0h} | \mathbf{grad} \ \varphi_i) \, dxdy + \gamma \int_{T_m \cap \Gamma_1} u_{0h} \varphi_i \, ds \right] \right.$$

$$\left. + \int_{T_m \cap \Gamma_1} g_1 \varphi_i \, ds \right\}.$$

Each term in the above sum is a contribution to a component of the element vector $\vec{f}^m = \{f_r^m\}_{r=1}^J$ associated with triangle T_m. More precisely, for $i = n_{m,r}$, this component is

$$f_r^m = \int_{T_m} f\varphi_i \, dxdy - \left[\int_{T_m} (\mathbf{grad} \ u_{h0} | \mathbf{grad} \ \varphi_i) \, dxdy + \gamma \int_{T_m \cap \Gamma_1} u_{h0} \varphi_i \, ds \right]$$

$$+ \int_{T_m \cap \Gamma_1} g_1 \varphi_i \, ds, \text{ with } i = n_{m,r}.$$

Then, starting from zero values of b_i, $1 \leq i \leq N_h$ we sweep the mesh for $m = 1, 2, \cdots, M_h$ and compute element vectors $\vec{f}^m$ for each one of them, and add its component f_r^m to the current value of b_i, for $i = n_{m,r}$, provided $i \leq N_h$. If $i > N_h$, f_r^m is discarded.

At the procedure's term, the final values of the b_is will be available for $i = 1, 2, \cdots, N_h$. Notice, however, that some coefficients of the element matrices E^m interfere in the above calculations, since the terms of f_r^m involving u_{h0} may be expressed as follows:

$$\int_{T_m} -(\mathbf{grad} \ u_{h0} | \mathbf{grad} \ \varphi_i) \, dxdy - \gamma \int_{T_m \cap \Gamma_1} u_{h0} \varphi_i \, ds =$$

$$\sum_{j=N_h+1}^{K_h} -g_0(S_j) \left[\int_{T_m} (\mathbf{grad} \ \varphi_j | \mathbf{grad} \ \varphi_i) \, dxdy + \gamma \int_{T_m \cap \Gamma_1} \varphi_j \varphi_i \, ds \right]$$

The term in brackets is nothing but the coefficient of the element matrix E^m corresponding to a triangle T_m such that $T_m \cap \Gamma_0$ is not empty. Therefore, when dealing with such elements, every coefficient $e_{r,s}^m$ such that $j = n_{m,s} > N_h$ and $i = n_{m,r} \leq N_h$ must be pre-multiplied by $-g_0(S_j)$, and the resulting term added to the current value of b_i, together with $\int_{T_m} f\varphi_i \, dxdy + \int_{T_m \cap \Gamma_1} g_1 \varphi_i \, ds$.

Now, we turn our attention to details on the computation of the element matrices and element vectors themselves, for the two types of methods being considered. Henceforth, we assume that a table of coordinates $(x_i; y_i)$ of all the vertices S_i of $\mathcal{T}_h$ in a suitable orthonormal frame is available.

Linear elements: This case is a mere extension of the procedure indicated in Subsection 4.3.4, in order to incorporate the additional terms in equation (6.24). We recall that the area integral term of $e_{r,s}^m$ is just $(\mathbf{grad}\ \varphi_j | \mathbf{grad}\ \varphi_i) \times area(T_m)$, with $i = n_{m,r}$ and $j = n_{m,s}$. Moreover, in this case a non-vanishing product $\varphi_i \varphi_j$ is a quadratic function to be integrated over an edge e of T_m contained in $\bar{\Gamma}_1$, if any. Recalling the one-dimensional calculations of Chapter 1, this integral equals $2 length(e)/3$ if $i = j$ and $length(e)/3$ if $i \neq j$. Notice that if three vertices of T_m happen to belong to $\bar{\Gamma}_1$, the vertex belonging to both edges contained in $\bar{\Gamma}_1$, say S_i, gives rise to two such line integrals when computing $\gamma \int_{T_m \cap \Gamma_1} \varphi_j \varphi_i\, ds$ if $i = j$, and to just one otherwise.

In Subsection 4.3.4, we gave expressions for $\mathbf{grad}\ \varphi_i$ in an element having S_i as a vertex. In this respect, the reader is referred to equation (4.29) and the definitions given thereabout.

Finally, we turn our attention to the integrals $\int_{T_m} f \varphi_i\, dxdy$ and $\int_{T_m \cap \Gamma_1} g_1 \varphi_i\, ds$. In practical applications, both f and g_1 are often simple functions, such as piecewise polynomials of very low degree. In such cases, the above integrals can be calculated exactly. Nevertheless, in order to make this text as self-contained as possible, we assume that f and g_1 are arbitrary continuous functions in their respective definition domains. In this case, we may apply numerical integration to approximate these integrals in each triangle or on every edge of $\mathcal{T}_h$ contained in Γ_1, respectively.

For linear elements, the one-point Gaussian quadrature formula should do, without causing any harm to method's order of convergence.

Letting G_m denote the centroid of T_m, the one-point quadrature formula is recalled below:

$$\int_{T_m} g\, dxdy \simeq g(G_m) area(T_m), \forall g \in C^0(\bar{T}_m).$$

In our particular case, this leads to $\int_{T_m} f \varphi_i\ dx \simeq f(G_m) area(T_m)/3$.

Let now $T_m \cap \Gamma_1 \neq \emptyset$ and C_p^m denote the mid-point of an edge of T_m contained in Γ_1 (in practice, a certain S_{s+3}^m). Such edges are denoted by e_p^m, $1 \leq p \leq I(m)$, where $I(m)$ is either 1 or 2. Geometrically, e_p^m is identified by the vertex S_s^m of T_m it is opposite to. Then, the one-point quadrature writes

$$\int_{T_m \cap \Gamma_1} g\, ds \simeq \sum_{p=1}^{I(m)} g(C_p^m)\, length(e_p^m) \quad \forall g \in C^0(T_m \cap \Gamma_1)$$

The values at C_p^m of the functions φ_i concerned by these calculations may be either 0 or $1/2$. For instance, if T_m has two edges on Γ_1 and φ_i is associated with the vertex of T_m common to both edges, then

$$\int_{T_m \cap \Gamma_1} g_1 \varphi_i\, ds \simeq \sum_{p=1}^{2} g_1(C_p^m) length(e_p^m)/2$$

If φ_i is associated with a vertex other than the common one, say the one belonging to e_1^m, or T_m has only one edge on Γ_1, then

$$\int_{T_m \cap \Gamma_1} g_1 \varphi_i\, ds \simeq g_1(C_1^m) length(e_1^m)/2.$$

Quadratic elements: Although no major difficulty arises when handling the boundary integrals with a coefficient $\gamma > 0$, here we will consider only the case where $\gamma = 0$. Indeed, as seen above, the treatment of such integrals is a little technical. However, the same pointers μ_i indicating whether a node S_i is located on $\bar{\Gamma}_1$ or not can be used. Here higher order quadrature formulae are recommended to compute integrals along boundary edges if applicable. At least, the two-point Gaussian quadrature or Simpson's rule (see e.g [105]) should be employed, or even better, exact computation if possible. Notice that here this is perfectly possible, since the product $\varphi_i\varphi_j$ over an edge contained in Γ_1 is either a polynomial of degree four or vanishes identically. We leave to the reader as Exercise 6.9, the details on the necessary calculations, for the case $\gamma \neq 0$.

Under the assumption that $\gamma = 0$, the system matrix is given by

$$a_{i,j} = \sum_{m=1}^{M_h} \int_{T_m} (\mathbf{grad}\ \varphi_i | \mathbf{grad}\ \varphi_j)\, dxdy,$$

and the right side by

$$b_i = \sum_{m=1}^{M_h} \left[\int_{T_m} f\varphi_i\, dxdy - \int_{T_m} (\mathbf{grad}\ u_{h0} | \mathbf{grad}\ \varphi_i)\, dxdy + \int_{T_m \cap \Gamma_1} g_1\varphi_i\, ds \right],$$

for every i, j such that S_i and/or S_j are nodes of T_m, that is, $i = n_{m,r}$ and $j = n_{m,q}$ for some r or q, $1 \leq r, q \leq 6$.

Recalling the expressions of the local shape functions in a triangle for quadratic elements, the contribution $e_{r,q}^m$ of the element T_m to the matrix coefficient $a_{i,j}$ is seen to be the sum of integrals of the form

$$(\mathbf{grad}\ \lambda_p^m | \mathbf{grad}\ \lambda_s^m) \int_{T_m} d_{ps}\, dxdy$$

where d_{ps} is a polynomial of $\mathcal{P}_2(T_m)$ expressed in terms of the barycentric coordinates λ_p^m of T_m, $p = 1, 2, 3$. For example, taking r and q less than or equal to three, the associated local shape functions are given by $\phi_{rr} = \lambda_r^m(2\lambda_r^m - 1)$ and $\phi_{qq} = \lambda_q^m(2\lambda_q^m - 1)$. Thus, the above sum over p and s is reduced to a single term, namely, $e_{r,q}^m = (\mathbf{grad}\ \lambda_r^m | \mathbf{grad}\ \lambda_q^m) \int_{T_m} d_{rq}\, dxdy$, where $d_{rq} = (4\lambda_r^m - 1)(4\lambda_q^m - 1)$. We recall that the integral in a triangle T of the product of its rth and the qth barycentric coordinates is given by $area(T_m)(\delta_{rq} + 1)/12$. Therefore, after straightforward calculations, the coefficient $e_{r,q}^m$ of the element matrix is easily found to be $(4\delta_{rq} - 1)area(T_m)/3$ for $1 \leq r, q \leq 3$. In Exercise 6.10, the reader is asked to complete the calculations of the coefficients of E^m, for r greater than three, and all possible values of q (this will do owing to symmetry).

Now, we turn our attention to the computation of the integrals of $f\varphi_i$ in the triangles. Although the Three-point Gaussian quadrature formula using the edge mid-points integrates exactly polynomials of degree less than or equal to two (cf. [182]), its use here is not recommended. This is because the local shape functions ϕ_{pp} vanish at all those mid-points for $p = 1, 2, 3$ by definition, and therefore essential information will be lost in this approximation. Here, the use of the Seven-point quadrature formula is recommended (cf. [182]), which integrates exactly polynomials of degree less than or equal to three in a triangle. The seven points are S_r^m for $r = 1, 2, \cdots, 6$, plus the centroid G_m denoted by S_7^m for the purpose of this formula, which writes

$$\int_{T_m} g\, dxdy \simeq \sum_{r=1}^{7} \omega_r g(S_r^m) area(T_m),$$

where $\omega_r = 1/20$ for $r = 1, 2, 3$, $\omega_r = 2/15$ for $r = 4, 5, 6$ and $\omega_7 = 9/20$.

Recalling the expressions of the local shape functions ϕ_{pq}, for $1 \leq p, q \leq 3$, application of this formula yields

$$\int_{T_m} f\phi_{pq} \, dxdy \simeq area(T_m)[f(S_{pq}^m) - f(G_m)]/20, \text{ for } p = q \text{ and}$$

$$ \tag{6.31}$$

$$\int_{T_m} f\phi_{pq} \, dxdy \simeq area(T_m)[2f(S_{pq}^m) + 3f(G_m)]/15, \text{ for } p \neq q.$$

Checking the above values is proposed to the reader as Exercise 6.11.

Finally, Simpson's rule or the two-point Gaussian quadrature formula provides convenient approximations of the integral $\int_{T_m \cap \Gamma_1} g_1 \varphi_i \, ds$ over every edge e_j^m of $T_m \cap \Gamma_1$, in case the measure of this intersection happens to be strictly positive. If Simpson's rule is used to approximate the integral of a continuous function g in e_j^m, $1 \leq j \leq I(m)$, we have

$$\int_{e_j^m} g \, ds \simeq \frac{g(A_j^m) + 4g(C_j^m) + g(B_j^m)}{6} length(e_j^m),$$

where A_j^m and B_j^m are the end-points of e_j^m. In this case, the following values are obtained for a pertaining local shape function ϕ_{pq}, i.e., a non-vanishing one on $T_m \cap \Gamma_1$:

$$\int_{e_j^m} g_1 \phi_{pq} \, ds \simeq length(e_j^m)g_1(S_{pq}^m)/6, \text{ if } p = q \text{ and}$$

$$ \tag{6.32}$$

$$\int_{e_j^m} g_1 \phi_{pq} \, ds \simeq 2length(e_j^m)g_1(S_{pq}^m)/3, \text{ if } p \neq q,$$

as the reader may check as Exercise 6.12.

Once all the terms corresponding to the local shape function ϕ_{pq} are computed, their sum is added to the corresponding component b_i of $\vec{b}_h$ with $i = n_{m,r}$, where $r = p$ if $p = q$ and $r = 9 - (p + q)$ if $p \neq q$ (please check!).

6.2 Extensions to the Three-dimensional Case

To begin with, we make a general comment on discretisation methods for three-dimensional problems.

In the two-dimensional case, up to very recently, it was a common practice among FD users to construct **boundary fitted orthogonal curved grids**, in order to enable the method's application to problems posed in two-dimensional domains of arbitrary shape. The author refers to the review paper [191] for further explanations about this technique, among several other works on the subject, prior or subsequent to it. In the case of an irregular three-dimensional domain, in general it is very hard to adjust a FD grid in order to render the method computationally reliable or exploitable. This fact, together with rather cumbersome techniques to deal with Neumann boundary conditions, is certainly what explains the great success of the FEM and the FVM, and the gradual decline of the FDM, at least in the solution of PDEs posed in irregular spatial domains. It should be noted however that, even nowadays, the FDM is often preferred to solve some kinds of problems, owing to its simplicity. This is because it allows for an easy design of numerical schemes having suitable properties, aimed at mimicking the physics modelled by a given PDE. For instance, this can be the case of wave propagation, and more generally of hyperbolic equations posed in simple geometries.

In Chapter 1 and in Chapter 4, we studied the solution by the FDM, the $\mathcal{P}_1$ FEM and the FVM, of second-order differential equations defined in a segment and in a polygon, respectively. In this section, we give some highlights of these methods' three-dimensional analogs to solve the Poisson equation in a polyhedron Ω with boundary Γ. Emphasis is given on method description, and no reliability analyses are supplied. Nevertheless, we must say beforehand that results of this nature entirely analogous to those that hold for the two-dimensional counterparts can be expected of all the methods considered in this section.

As a model, we consider the case of mixed Dirichlet–Neumann boundary conditions, namely,

$$\begin{cases} -\Delta\, u = f \text{ in } \Omega \\ u = g_0 \text{ on } \Gamma_0, \\ \partial_\nu u = g_1 \text{ on } \Gamma_1. \end{cases} \qquad (6.33)$$

where, referred to a Cartesian coordinate system $(O; x, y, z)$, the Laplacian of u is given by $\Delta u := \partial_{xx} u + \partial_{yy} u + \partial_{zz} u$. Γ_0 and Γ_1 are two disjoint portions of Γ whose union is the whole Γ, Γ_0 being a closed set fulfilling $area(\Gamma_0) > 0$. $\vec{\nu} = (\nu_1; \nu_2; \nu_3)$ is the unit outer normal vector defined at every point of Γ, except along boundary edges or vertices $\partial_\nu u$ equals (**grad** $u|\vec{\nu}$). We assume that the intersection of Γ_0 and the closure $\bar{\Gamma}_1$ of Γ_1 is a spatial polygonal line contained in Γ_0 but not in Γ_1. This intersection is called the transition between Γ_0 and Γ_1.

6.2.1 Methods for Rectangular Domains

In case Ω is a rectangular parallelepiped (i.e. more vulgarly a rectangular box) say $(0, L_x) \times (0, L_y) \times (0, L_z)$, it may be simpler to solve elliptic equations like problem (6.33) by the three-dimensional analog of the Five-point FDM, by the $\mathcal{Q}_1$ FEM or by the Cell-centred FVM for rectangles, in connection with a rectangular mesh (cf. Subsections 4.2.1, 6.1.2 and 4.4.2). This is feasible, and not so difficult to implement, as long as the polygonal line separating Γ_0 and Γ_1 is the union of segments parallel to edges of Ω. Actually, in order to simplify the presentation, we assume that Γ_0 is the union of non-disjoint faces of Γ. An example is given in Figure 6.11, where Γ_0 consists of the faces given by $x = 0$, $x = L_x$, $y = 0$ and $y = L_y$.

For the sake of conciseness, we confine ourselves to a brief presentation of the Five-point FD scheme's analog and the $\mathcal{Q}_1$ FEM for rectangles. These are the **Seven-point FD scheme** and the **piecewise trilinear FEM** for rectangular domains, also called the $\mathcal{Q}_1$ FEM.

All these methods are based on the construction of a structured grid within Ω using planes orthogonal to the three coordinate axes $\vec{Ox}$, $\vec{Oy}$ and $\vec{Oz}$. For simplicity, the distance between the planes corresponding to each axis is assumed to be constant, equal to $h_x = L_x/n_x$, $h_y = L_y/n_y$ and $h_z = L_z/n_z$, respectively, where n_x, n_y and n_z are given integers greater than one.

Let us first consider the FDM, assuming that Γ_1 is empty. In this case, the grid points are the intersections of three such planes, that is, the points $M_{i,j,k}$ of coordinates (ih_x, jh_y, kh_z), with $0 \le i \le n_x$, $0 \le j \le n_y$ and $0 \le k \le n_z$.

Here again, we assume that both g_0 and f are uniquely defined at every point of Γ_0 and Ω, respectively. Then we may set $f_{i,j,k} := f(ih_x, jh_y, kh_z)$, together with

$$
\left.\begin{array}{l}
u_{0,j,k} = g_0(0, jh_y, kh_z) \text{ and} \\
u_{n_x,j,k} = g_0(L_x, jh_y, kh_z)
\end{array}\right\} \text{ for } 0 \le j \le n_y \text{ and } 0 \le k \le n_z,
$$

$$
\left.\begin{array}{l}
u_{i,0,k} = g_0(ih_x, 0, kh_z) \text{ and} \\
u_{i,n_y,k} = g_0(ih_x, L_y, kh_z)
\end{array}\right\} \text{ for } 0 \le i \le n_x \text{ and } 0 \le k \le n_z, \qquad (6.34)
$$

$$
\left.\begin{array}{l}
u_{i,j,0} = g_0(ih_x, jh_y, 0) \text{ and} \\
u_{i,j,n_z} = g_0(ih_x, jh_y, L_z)
\end{array}\right\} \text{ for } 0 \le i \le n_x \text{ and } 0 \le j \le n_y.
$$

Denoting by $u_{i,j,k}$ the approximation of $u(M_{i,j,k})$ for $1 \le i \le n_x - 1$, $1 \le j \le n_y - 1$ and $1 \le k \le n_z - 1$, here instead of **divided differences** for a non-uniform grid (cf. (4.5)), we define directly second-order difference operators involving seven approximate values of u about $M_{i,j,k}$ including $u_{i,j,k}$ itself. In Figure 6.12, we illustrate the underlying geometric arrangement, commonly called a **Seven-point FD stencil**. More precisely, setting,

$$
\left\{\begin{array}{l}
\delta_x^2(u_{i,j,k}) := \dfrac{2u_{i,j,k} - u_{i+1,j,k} - u_{i-1,j,k}}{h_x^2}, \\[2mm]
\delta_y^2(u_{i,j,k}) := \dfrac{2u_{i,j,k} - u_{i,j+1,k} - u_{i,j-1,k}}{h_y^2} \text{ and} \\[2mm]
\delta_z^2(u_{i,j,k}) := \dfrac{2u_{i,j,k} - u_{i,j,k+1} - u_{i,j,k-1}}{h_z^2}, \\[2mm]
\text{for } i = 1, \cdots, n_x - 1, j = 1, \cdots, n_y - 1 \text{ and } k = 1, \cdots, n_z - 1.
\end{array}\right. \qquad (6.35)
$$

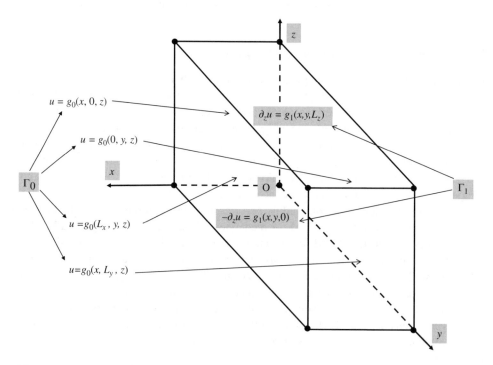

Figure 6.11   A rectangular Ω and an example of mixed boundary conditions for equation (6.33)

Quite naturally, and similarly to the two-dimensional case, the Seven-point FD scheme to solve equation (6.33) in a rectangular domain Ω with inhomogeneous Dirichlet boundary conditions writes as follows:

$$- \Delta_h(u_{i,j,k}) = f_{i,j,k} \text{ for } 1 \leq i \leq n_x - 1, 1 \leq j \leq n_y - 1 \text{ and } 1 \leq k \leq n_z - 1, \quad (6.36)$$

together with equation (6.34), where

$$\Delta_h(u_{i,j,k}) := \delta_x^2(u_{i,j,k}) + \delta_y^2(u_{i,j,k}) + \delta_z^2(u_{i,j,k}), \quad (6.37)$$

In the case of the FVM and the FEM, the above partition gives rise to a mesh $\mathcal{T}_h$ with $n_x \times n_y \times n_z$ rectangular cells or elements. More precisely, the mesh elements are the rectangles $([i-1]h_x, ih_x) \times ([j-1]h_y, jh_y) \times ([k-1]h_z, kh_z)$, for $i = 1, \cdots, n_x, j = 1, \cdots, n_y$ and $k = 1, \cdots, n_z$. We set $h = \max[h_x, h_y, h_z]$ and represent by Q a generic mesh element. In this way, we distinguish rectangular meshes from tetrahedral meshes studied in the sequel.

We presume that the reader will not find it difficult to extend to spatial rectangles the Cell-centred FVM for plane rectangles presented in Section 4.4.2 and studied in Part B. of Subsection 5.1.5. Akin to that method, we denote by $R_{i-1/2,j-1/2,k-1/2}$ the FV whose eight vertices are $M_{l,m,n}$ for $l = i-1$ or $l = i, m = j-1$ or $m = j$ and $n = k-1$ or $n = k$, where i, j and k range between one and n_x, n_y and n_z respectively.

It is rather intuitive that a balance equation for each one of those CVs can be achieved by simply computing the normal fluxes at the centres of the mesh faces, and then multiplying them by the corresponding areas. These can be either $h_x h_y$, $h_x h_z$ or $h_y h_z$. The right side in turn is given by $h_x h_y h_z f_{i,j,k}$ for each rectangular cell. Still assuming that Γ_1 is empty, the prescribed values g_0 are taken at the centre of the faces of boundary cells contained in Γ. The final form of the thus-constructed FV scheme should be worked out as Exercise 6.13.

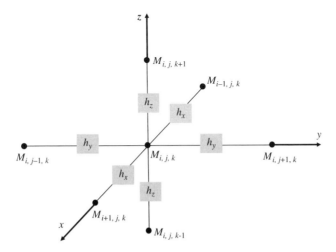

Figure 6.12 A Seven-point FD stencil for solving equation (6.33) in a parallelepiped

Next we focus on the $\mathcal{Q}_1$ FEM for rectangular elements, in connection with the mesh $\mathcal{T}_h$. Incidentally, for coherence with the notations previously used for the FEM, in the remainder of this subsection, the letter S refers to a mesh node rather than the letter M.

First of all, we must write equation (6.33) in (Galerkin) variational form. In this aim, we multiply both sides of equation (6.33) by a function v in $H^1(\Omega)$, that is, the set of functions which together with their first-order partial derivatives are square integrable in Ω. In doing so, we may integrate in Ω both sides of the resulting equation. Then, we apply First Green's identity for a sufficiently smooth bounded three-dimensional domain Ω, thereby obtaining,

$$-\int_\Omega \Delta\, uv\; dxdydz = -\oint_\Gamma \partial_\nu u\, v\; dS + \int_\Omega (\mathbf{grad}\; u|\mathbf{grad}\; v)\; dxdydz = \int_\Omega fv\; dxdydz.$$

Now we restrict the above equation to functions v that vanish on Γ_0, being otherwise arbitrary. Owing to the fact that $\partial_\nu u = g_1$ on Γ_1, we obtain the following variational formulation of equation (6.33):

$$\begin{cases} \text{Find}\quad u \in H^1(\Omega)\ \text{such that}\quad u = g_0\ \text{on}\ \Gamma_0\ \text{and} \\ \int_\Omega (\mathbf{grad}\; u|\mathbf{grad}\; v)\; dxdydz = \int_\Omega fv\; dxdydz + \oint_{\Gamma_1} g_1 v\; dS\ \forall v \in V, \end{cases} \tag{6.38}$$

where $V := \{v \in H^1(\Omega), v = 0\ \text{on}\ \Gamma_0\}$. We keep this notation even if $\Gamma_0 = \Gamma$.

Like in the two-dimensional case, we can assert that every solution of equation (6.38) is a solution of equation (6.33), and therefore the former has a unique solution, since Ω, f, g_0 and g_1 are assumed to be sufficiently smooth.

Now, we construct a FE analog of equation (6.38) using **trilinear elements**, for the case where Ω is a rectangular domain. As is often the case in practical situations, we assume that Γ_0 is the union of faces of Ω, for in this case we can make sure that a rectangular mesh fits the Dirichlet boundary conditions. If Γ_0 and Γ_1 are separated only by lines parallel to coordinate axes, which are not necessarily edges of Γ, the procedure we are going to describe can be applied with minor modifications. But in this case, in principle, non-uniform meshes will have to be used.

In the trilinear FEM, the function u is approximated by a continuous function u_h, whose restriction to every element of $\mathcal{T}_h$ is a linear function in each space variable separately. This means that u_h in an element of $\mathcal{T}_h$ is a linear combination of the monomials 1, x, y, z, xy, xz, yz and xyz. For a given $Q \in \mathcal{T}_h$, we define $\mathcal{Q}_1(Q)$ to be the set of trilinear functions in Q. Naturally enough, this algebraic structure of u_h extends to every function v belonging to the space of test functions V_h, which is a subset of the space W_h defined as follows:

$$W_h := \{w|\ w \in C^0(\bar{\Omega})\ \text{and}\ w_{|Q} \in \mathcal{Q}_1(Q)\ \forall Q \in \mathcal{T}_h\}.$$

The test function space V_h in turn is defined by

$$V_h := \{v|\ v \in W_h\ \text{and}\ v = 0\ \text{on}\ \Gamma_0\}.$$

We can assert that the above constructions of W_h and V_h are feasible under our assumptions on Γ_0, provided we require continuity of any function belonging to W_h at the mesh nodes. Indeed, the restriction of any such a function to an inner face F of the mesh common to two elements, orthogonal to a coordinate axis (say Oz), is a bilinear function of the two other variables (say, x and y), that is, a function of $\mathcal{Q}_1(F)$. Notice that by construction, the function values at the four vertices of F are the same for both elements sharing this face. Thus, recalling the uniqueness of the bilinear function that takes given values at the four vertices of a rectangle (cf. Subsection 6.1.2), the continuity of any function in W_h is ensured. A similar argument can be used to establish that a function $v \in W_h$ vanishes everywhere on Γ_0, provided it vanishes at all the mesh nodes belonging to Γ_0. Actually, this condition is enforced in order to construct functions $v \in V_h$, while the FE solution $u_h \in W_h$ at a mesh node $S \in \Gamma_0$ takes the prescribed value $g_0(S)$.

We skip further details on the construction of the space $\mathcal{Q}_1$ for the sake of conciseness, and refer in this respect to the vast literature on the FEM (see e.g [146]).

6.2.2   Tetrahedron-based Methods

In this subsection, we address the case where Ω is a polyhedron of arbitrary shape.

So far, we studied FEM and FVM to solve second-order boundary value problems constructed upon partitions of the equation's definition domain into disjoint intervals or triangles satisfying suitable compatibility conditions. In the three-dimensional case, the most used numerical methods of both types to solve the Poisson equation (6.33) are based on partitions of the domain into tetrahedra, which are the counterparts in $\Re^3$ of segments and triangles[1].

The FVM: Similarly to the two-dimensional case (cf. Section 4.3), the FVM is based on a partition of the equation domain into non-overlapping convex polyhedra, whose union is Ω. These CV should form an admissible mesh $\mathcal{T}_h$, which means that some conditions on its elements must be fulfilled. In particular, besides compatibility conditions of the same nature as those listed in Chapter 4 for triangular meshes (see also the $\mathcal{P}_1$ FEM hereafter), there should exist a representative point $\mathbf{x}_T$ in each CV T such that the line going through it and the representative point $\mathbf{x}_S$ of any other CV $S \in \mathcal{T}_h$ sharing a face with T is orthogonal to this face, and moreover $\mathbf{x}_T$ and $\mathbf{x}_S$ do not coincide.

In practice, the most used types of CV are either hexahedra with quadrilateral faces or tetrahedra. The former is usually employed in connection with structured meshes, owing to its obvious lack of geometric flexibility to fit into domains of arbitrary shape. Actually, we have basically addressed this case in Subsection 6.4.1, since a structured hexahedral mesh can be viewed as a mere deformation of a mesh consisting of parallelepipeds.

In this subsection we focus on tetrahedral meshes, and to begin with we must say that the generation of an admissible mesh of this type in the sense of the FVM is far from trivial. The search for **well-centred tetrahedrisations** that would lead to admissible tetrahedral meshes is still a subject of active research (see e.g. [198]). Addressing such a topic at beginner's level would be misleading and out of purpose. Furthermore, the reader should be aware of the fact that the ratio between the number of tetrahedra and the number of vertices in an arbitrary tetrahedrisation is quite high. In order to figure this out, it suffices to recall the rectangular mesh of a parallelepiped (cf. Subsection 6.4.1). A fine-shaped tetrahedrisation of such a domain is obtained by subdividing each rectangular cell $R_{i-1/2,j-1/2,k-1/2}$ into six tetrahedra. For example, this can be achieved by taking the diagonals of this cell joining vertices $M_{i-1,j-1,k-1}$ and $M_{i,j,k}$, and then the diagonals of the three faces of $R_{i-1/2,j-1/2,k-1/2}$ having one out of both vertices as an end-point. The resulting six tetrahedra have the chosen cell diagonal as

---

[1] The generalisation of these three geometric entities to $\Re^N$ is called an N-simplex.

a common edge and as opposite edge one out of the six cell edges that do not intersect this diagonal. The reader might exercise this construction, and check that it leads to a compatible tetrahedral mesh in the sense specified hereafter (cf. $\mathcal{P}_1$ FEM). In short, there are still $(n_x + 1)(n_y + 1)(n_z + 1)$ vertices in the resulting tetrahedrisation, for $6n_x n_y n_z$ CVs. Hence, the finer the mesh, the more the quotient between the number of vertices and the number of CVs approaches $1/6$.

In view of this, it would be certainly much less expensive to work with polyhedral CVs that have the vertices of a tetrahedrisation as representative points. Since the sizes of both meshes have comparable magnitudes, a priori replacing one mesh with the other would not bring about any significant loss in accuracy. This procedure is actually feasible if the original tetrahedral mesh is a **Delaunay triangulation** (cf. [108]) and a dual mesh is constructed upon its **Voronoi diagram** (see e.g. [68]). The resulting dual mesh consists of convex polyhedra with convex polygonal faces, whose representative points can be chosen to be the vertices of the original mesh. Notice that this is the three-dimensional analog of the dual polygonal mesh constructed about the vertices of a Delaunay triangulation of a plane domain, as described in Chapter 4. But in this case, we come up with the so-called FV–FE method. That is why we do not go any further with the FVM in the three-dimensional case, and switch to the tetrahedron-based $\mathcal{P}_1$ FEM.

The $\mathcal{P}_1$ FEM:

In Chapters 1 and 4, we studied the $\mathcal{P}_1$ FEM to solve second-order differential equations defined in a segment and in a polygon, respectively. Next, we consider the three-dimensional analog of this method applied to the case of an arbitrary polyhedron Ω, with mixed Dirichlet–Neumann boundary conditions. Before going into details on the $\mathcal{P}_1$ FEM for $N = 3$, we make a few considerations about geometric and algebraic structures related to a tetrahedron. In this context, the concept of barycentric coordinates plays a key role.

Let $S_1^T, S_2^T, S_3^T, S_4^T$ T be the vertices of a non-degenerated tetrahedron T, which means that T has a strictly positive volume. Otherwise stated, the four vertices of T are not contained in the same plane of $\Re^3$. We assume that the vertices are numbered according to the right-hand rule (see e.g. [207]).

Referring to Figure 6.13, where the case $r = 4$ is illustrated, we denote by F_r^T the face of T opposite to vertex S_r^T, and by h_r^T the length of the height of T orthogonal to F_r^T, $r = 1, 2, 3, 4$.

Let (x_r^T, y_r^T, z_r^T) be the coordinates of S_r^T, for $r = 1, 2, 3, 4$, in the orthonormal frame with origin O associated with the coordinates x, y, z. We recall that the volume of T denoted by $volume(T)$ is given by

$$volume(T) = \frac{1}{6} \begin{vmatrix} 1 & x_1^T & y_1^T & z_1^T \\ 1 & x_2^T & y_2^T & z_2^T \\ 1 & x_3^T & y_3^T & z_3^T \\ 1 & x_4^T & y_4^T & z_4^T \end{vmatrix} \tag{6.39}$$

By definition, the barycentric coordinate λ_r^T of T associated with vertex S_r^T is the unique linear function that satisfies $\lambda_r^T(S_q^T) = \delta_{rq}$. Denoting by $\det[\vec{a}_1, \vec{a}_2, \vec{a}_3, \vec{a}_4]$ the determinant of the 4×4 matrix, whose entries in the rth row are the components of the vector $\vec{a}_r \in \Re^4$, using equation (6.39), it is easy to check that the λ_r^T's are given by

$$\lambda_1^T(x, y, z) = \frac{\det[\vec{e}, \vec{e}_2^T, \vec{e}_3^T, \vec{e}_4^T]}{6volume(T)}$$

$$\lambda_2^T(x,y,z) = \frac{\det[\vec{e}_1^T, \vec{e}, \vec{e}_3^T, \vec{e}_4^T]}{6volume(T)}$$

(6.40)

$$\lambda_3^T(x,y,z) = \frac{\det[\vec{e}_1^T, \vec{e}_2^T, \vec{e}, \vec{e}_4^T]}{6volume(T)}$$

$$\lambda_4^T(x,y,z) = \frac{\det[\vec{e}_1^T, \vec{e}_2^T, \vec{e}_3^T, \vec{e}]}{6volume(T)},$$

where $\vec{e} = [1, x, y, z]^T$ and $\vec{e}_r^T = [1, x_r^T, y_r^T, z_r^T]^T, r = 1, 2, 3, 4$. Indeed, λ_r^T is a linear function of x, y and z that clearly satisfies $\lambda_r^T(S_q^T) = \delta_{rq}$.

Similarly to the two-dimensional case, the barycentric coordinates have several handy properties. The following ones can be very useful in FE computations:

1. λ_r^T vanishes identically on F_r^T;
2. The value of λ_r^T on the face parallel to F_r^T of the homothetic reduction of T with centre S_r^T and ratio $\rho, 0 < \rho \le 1$ is $1 - \rho$;
3. $\sum_{r=1}^4 \lambda_r^T \equiv 1$;
4. $(x; y; z) \equiv \sum_{r=1}^4 (x_r^T; y_r^T; z_r^T)\lambda_r^T(x, y, z)$;
5. The sum of the gradients of the four barycentric coordinates is the null vector at every point;
6. **grad** λ_r^T is the constant vector orthogonal to F_r^T, oriented from this face towards S_r^T, whose modulus is $[h_r^T]^{-1}$ (cf. Figure 6.13 and Property 1);
7. The values of all barycentric coordinates at the centroid of T is $1/4$.[2]

All the above properties can be demonstrated like in the two-dimensional case, except perhaps 7. However, starting from the analogous two-dimensional property, the result is easily

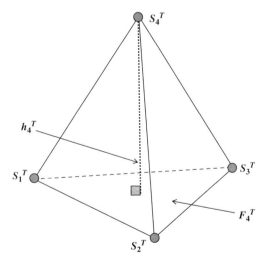

Figure 6.13   A tetrahedron T with pertaining vertices, face F_4^T and height h_4^T

---

[2] The centroid of a tetrahedron is the intersection of its four medians, which are the segments joining the centroid of each face to the opposite vertex.

seen to hold (cf. Exercise 6.14). Moreover, the fact that the set $\{\lambda_r^T\}_{r=1}^4$ forms a basis of the space $\mathcal{P}_1(T)$ of linear functions in T can be demonstrated in the same manner as in the two-dimensional case.

Likewise, the barycentric coordinates can be used to generate a basis for implementing the $\mathcal{P}_1$ FEM in three-dimensional space. This method is associated with a mesh $\mathcal{T}_h$ of Ω, consisting of the union of M_h tetrahedra, where h represents the maximum edge length of mesh tetrahedra.

The elements of $\mathcal{T}_h$ must satisfy the following compatibility conditions:

- All the tetrahedra in $\mathcal{T}_h$ are closed sets (i.e. they include their boundaries);
- $\bar{\Omega}$ is the union of all tetrahedra in $\mathcal{T}_h$;
- The intersection of two tetrahedra T_1 and T_2 in $\mathcal{T}_h$ is either empty, a common vertex, a common edge or a common face;
- Both $\bar{\Gamma}_0$ and $\bar{\Gamma}_1$ are the union of faces of tetrahedra belonging to $\mathcal{T}_h$.

Notice that the fourth condition means that the transition between Γ_0 and Γ_1 is necessarily a set of edges of tetrahedra in $\mathcal{T}_h$.

To mesh $\mathcal{T}_h$ corresponds a set of K_h vertices $\{S_i\}_{i=1}^{K_h}$. A shape function for the $\mathcal{P}_1$ FEM associated with vertex S_i is a function φ_i whose restriction to each tetrahedron of the mesh is a polynomial of degree less than or equal to one satisfying $\varphi_i(S_j) = \delta_{ij}$ $\forall j \in \{1, 2, \cdots, K_h\}$. Every vertex S_i belongs to a small set of elements of $\mathcal{T}_h$ denoted by Π_i. The restriction of φ_i to each T in Π_i is its barycentric coordinate related to vertex S_i, while $\varphi_i \equiv 0$ in every element of $\mathcal{T}_h$ not belonging to Π_i.

Remark 6.5 *In Remark 4.4, we explained why the piecewise linear shape functions φ_i form a basis of the space W_h of continuous piecewise linear functions in Ω, in connection with a triangular mesh. Clearly, enough analogous arguments apply to the three-dimensional case. The necessary adaptations relies upon a simple observation about functions defined by their values at tetrahedron vertices that are linear in each tetrahedron of the mesh: If two tetrahedra have a common face F, then the piecewise linear function is continuous on F, since the traces from both sides of this interface are linear functions in a triangle that have the same values at its three vertices.*∎

Similarly to the cases already treated, we denote by N_h the number of mesh nodes that do not belong to Γ_0. In order to simplify the presentation, we assume for the moment that the nodes belonging to Γ_0 are numbered from $N_h + 1$ through K_h. Notice that if we define a suitable integer pointer with K_h components in the solution procedure, the node numbering can be arbitrary. That is what we do in Example 6.2 hereafter.

Adopting however the above numbering systematic, let V_h be the subspace of W_h spanned by the functions φ_i for $i = 1, \cdots, N_h$. By construction, all such shape functions vanish identically on Γ_0 (please check!), and hence every function in V_h belongs to the space $V := \{v \in H^1(\Omega), v \equiv 0 \text{ on } \Gamma_0\}$. Here again, we set $u_{h0} := \sum_{i=N_h+1}^{K_h} g_0(S_k)\varphi_i$.

Now we are ready to apply the tools introduced here to solve the Poisson equation in three-dimensional space. For completeness, instead of the mixed problem (6.33) whose equivalent Galerkin variational form is equation (6.38), we consider the three-dimensional analog of equation (6.23). In order to simplify the presentation of corresponding implementation aspects in the next subsection, we assume right away that the given functions f, g_0 and g_1 are continuous in their respective definition domains. Under these assumptions, by a straightforward argument, the Galerkin variational form (equations (6.25)–(6.27)) in three-dimensional space is seen to be equivalent to equation (6.23). The $\mathcal{P}_1$ FEM applies to the discrete analog (6.28) of the continuous variational problem, and the FE approximation u_h is again the sum $\underline{u}_h + u_{0h}$. Like in all the other cases studied so far, this problem is equivalent to a SLAE with unknown vector $\vec{u}_h \in \Re^{N_h}$. The components of this vector are the values of $\underline{u}_h$ at the nodes that do not belong to Γ_0. These components are ordered according to the chosen node numbering. The system $N_h \times N_h$ matrix A_h and right-side vector $\vec{b}_h \in \Re^{N_h}$ can be computed by means of the element-by-element procedure described in the next subsection.

6.2.3 Implementation Aspects

The shape functions we introduced in Chapter 4 and in the previous subsection for tetrahedral meshes are by far the most used ones in FE codes. However clearly enough, three-dimensional computations are fairly more complex than two-dimensional ones. Still, the main ingredients needed to implement the FEM in this case are quite similar to those listed in Subsections 4.3.4 and 6.1.6. That is why, in this subsection, we confine ourselves to some specific issues of three-dimensional computations with linear FEs.

A first important aspect is the situation of the mesh vis-à-vis the interface separating Γ_0 and Γ_1. We recall the assumption that this interface is a spatial polygonal line, and that the mesh is constructed in such a way that no tetrahedron face is cut by it. In this way, both Γ_0 and Γ_1 can be viewed as the union of faces of tetrahedra, whose intersection with either one of them has a strictly positive measure in case $area(\Gamma_1) > 0$.

As usual, we assume that convenient numberings of the tetrahedra of the mesh $\mathcal{T}_h$ from one through M_h, and of the nodes from one through K_h, are available. The numbers of elements and nodes are linked together by an $M_h \times 4$ incidence matrix $\mathcal{N} = [n_{m,r}]$, for $1 \leq m \leq M_h$ and $1 \leq r \leq 4$, where $n_{m,r}$ is the number of the rth local node of the mth tetrahedron of $\mathcal{T}_h$. Given a tetrahedron $T \in \mathcal{T}_h$, the local numbering of the nodes S_r^T for r ranging from 1 to 4 can be performed as follows. First, an arbitrary vertex of T is selected to be S_1^T; then, the right-hand rule is applied to number the other three vertices in the order S_2^T, S_3^T, S_4^T (cf. Figure 6.13). We also assume that a table of coordinates $(x_i; y_i; z_i)$ of each vertex S_i of the mesh is available, for $i = 1, 2, \cdots, K_h$. Now, using the incidence matrix $\mathcal{N}$, the element matrices E^m, $m = 1, 2, \cdots, M_h$, are assembled in the very same manner as in the two-dimensional case, thereby generating the system matrix A_h. Likewise, element vectors $\vec{f}^m$ for $m = 1, 2, \cdots, M_h$ are assembled to form the right-side vector $\vec{b}_h$.

Next, we address the practical evaluation of the integrals appearing in equation (6.28). First, we note that, as long as $i = n_{m,r}$ and $j = n_{m,s}$ for certain r, s comprised between one and four, the computation of the element matrix coefficient $e_{r,s}^m$ requires the integration of $(\mathbf{grad}\varphi_i | \mathbf{grad}\varphi_j)$ in mesh tetrahedron T_m and the integration of $\varphi_i \varphi_j$ on any face of T_m contained in Γ_1. We know that φ_i restricted to a given T_m having S_i as a vertex equals a barycentric coordinate of this tetrahedron. Hence the components of $\mathbf{grad}\varphi_i$ are simply the first-order derivatives of the rth barycentric coordinate λ_r^m of T_m, provided $i = n_{m,r}$ for a certain r, $1 \leq r \leq 4$. These derivatives are obtained by replacing the rth row (i.e. the one filled in with the components of $\vec{e}$) in the numerators on the right side of equation (6.40) by $[0, 1, 0, 0]$ for the x-derivative, $[0,0,1,0]$ for the y-derivative and $[0,0,0,1]$ for the z-derivative. Indeed, the determinants in the numerators of equation (6.40) are linear functions of x, y and z, and these variables appear only in these rows.

Similarly to Subsection 4.3.4, the resulting numerators are denoted by Δ_{xr}^m, Δ_{yr}^m and Δ_{zr}^m, and with these definitions we have for $r = 1,2,3,4$,

$$\partial_x \lambda_r^m = \frac{\Delta_{xr}^m}{6volume(T_m)}; \ \partial_y \lambda_r^m = \frac{\Delta_{yr}^m}{6volume(T_m)}; \ \partial_z \lambda_r^m = \frac{\Delta_{zr}^m}{6volume(T_m)}$$

Notice that the numerators in the above expressions are the determinants of 3×3 matrices whose coefficients are those remaining after elimination of the rth row and the qth column of the original 4×4 matrix multiplied by $(-1)^{r+q+1}$, the value of q being 1 for the x derivative, 2 for the y derivative and 3 for the z derivative. As a result, if $\gamma = 0$, we have

$$e_{r,s}^m = \frac{1}{36volume(T_m)}[\Delta_{xr}^m \Delta_{xs}^m + \Delta_{yr}^m \Delta_{ys}^m + \Delta_{zr}^m \Delta_{zs}^m].$$

It is convenient to number the nodes according to a systematic aimed at optimising the solution procedure in a certain sense. In order to attain this goal, we suggest that, instead of the largest number rule for nodes at which the solution is prescribed, we use pointers arranged in the form of an integer vector $\vec{\mu} = [\mu_1, \mu_2, \cdots, \mu_{K_h}]$, such that $\mu_i = -2$ if $S_i \in \Gamma_1$, and otherwise $\mu_i = 0$ if the solution is not prescribed at S_i, $\mu_i = -1$ if the solution equals zero at S_i and $\mu_i = l > 0$ if the solution has a nonzero prescribed value d_l at S_i. The d_ls are arranged in a $(K_h - I_h)$-component array $\vec{d}$, where I_h is the number of nodes corresponding to unknown or zero values of the solution. A more delicate issue in the three-dimensional case is the treatment of inhomogeneous Neumann boundary conditions on Γ_1 (or a subset of it). When a face F_p^m of tetrahedron T_m opposite to its local vertex S_p^m is contained in Γ_1 for a certain $p \in \{1,2,3,4\}$, marking it somehow is mandatory. This is because integrals of $\varphi_i \varphi_j$ and $g_1 \varphi_i$ over F_p^m must be computed in an element-by-element matrix and right-side assembling procedure, in case S_i and S_j are vertices of F_p^m such that $\mu_i = -2$, and μ_j takes either the value -2 if u_j is a boundary unknown or a strictly positive value for a prescribed nonzero boundary datum $u_j = d_{\mu_j} = g_0(S_j)$ if $S_j \in \Gamma_0 \cap \overline{\Gamma}_1$ (please check!). A simple criterion we implemented in the FE code supplied in Example 6.2 is as follows. First, we note that $F_p^m \subset \Gamma_1$ only if its centroid G_p^m belongs to Γ_1. Checking whether this condition is fulfilled or not could be enough, but in order to rule out odd configurations, even unlikely ones, we check for the face's three vertices too. More specifically, in case all these four points belong to Γ_1, we assign to an associated pointer g_{mp}^a the value 'one' and the value 'zero' otherwise. Once we are sweeping the mesh tetrahedra in the matrix and right-side vector assembling routine, all we have to do is searching for 'one' values of pointer g_{mp}^a, while making sure that S_i and S_j are vertices of the corresponding face F_p^m, that is, $i = n_{m,r}$ and $j = n_{m,s}$ for certain $r,s \in \{1,2,3,4\}$, with $r \neq p$ and $s \neq p$. In this case, any vertex of F_p^m generically denoted by S_q^m for $q \in \{1,2,3,4\}$, $q \neq p$, can be easily identified by exclusion. In order to check whether a coefficient $e_{r,s}^m$ of element matrix E^m incorporates a boundary integral term or not, we suppose for convenience that an integer pointer γ_{prs}^m is available. By definition, $\gamma_{prs}^m = 1$ if both S_r^m and S_s^m are vertices of F_p^m, and $\gamma_{prs}^m = 0$ otherwise. We should clarify beforehand that, in the code of Example 6.2, such a pointer is not used. Instead, we search for the pertinent pairs $(r;s)$ directly for every face

F_p^m contained in Γ_1, element by element. Anyhow, such terms of $e_{r,s}^m$ are easily found to be

$$\gamma \int_{F_p^m} \lambda_r^m \lambda_s^m \; dS = \gamma \begin{cases} (1+\delta_{rs}) area(F_p^m)/12 & \text{if } \gamma_{prs}^m = 1 \\ 0 & \text{otherwise} \end{cases} \tag{6.41}$$

The area appearing in equation (6.41) can be determined by using the data structures already created at the element level to compute the volume integrals, according to the following considerations. F_p^m being the pth face of a tetrahedron T_m, by elementary geometry we know that $area(F_p^m) = 3volume(T_m)/h_p^m$, where h_p^m is the (length of the) height of T_m orthogonal to F_p^m. On the other hand, by Property 6 of the barycentric coordinates, $|\mathbf{grad}\,\lambda_p^m| = 1/h_p^m$. Since the values of $\mathbf{grad}\,\lambda_p^m$ are available, h_p^m can be readily determined and hence $area(F_p^m)$.

Now we switch to the computation of the right-side vector $\vec{b}_h$ whose ith component has the exact value $b_i = -a(u_{0h}, \varphi_i) + \int_\Omega f\varphi_i dxdydz + \int_{\Gamma_1} g_1\varphi_i dS$.

The terms whose sum over the elements is $-a(u_h^0, \varphi_i)$ are handled as the element matrices E^m are generated. Their values are $-d_l e_{r,s}^m$, where $l = \mu_j > 0$ for $j = n_{ms}$ and $i = n_{m,r}$ only for i such that $\mu_i = 0$ or $\mu_i = -2$. We refer to the two-dimensional case for more details.

On the other hand, like in the two-dimensional case, the integral $\int_{T_m} f\varphi_i dxdydz$ with $i = n_{m,r}$ for a given r, $1 \le r \le 4$ can be conveniently approximated by the one-point Gaussian quadrature formula. More precisely, $\int_{T_m} f\varphi_i \simeq f(G_m)volume(T_m)/4$, where G_m is the centroid of T_m, that is, G_m is the point of T_m whose barycentric coordinates are all equal to $1/4$. Better accuracy is attained by applying a Gaussian quadrature formula using four points, which is exact for quadratic functions. The barycentric coordinates of the four points are $(\alpha, \beta, \beta, \beta)$, $(\beta, \alpha, \beta, \beta)$, $(\beta, \beta, \alpha, \beta)$ and $(\beta, \beta, \beta, \alpha)$, with $\beta \simeq .1381966$ and $\alpha = 1 - 3\beta$ (cf. [184]). The latter formula was implemented in the code of Example 6.2.

Finally, the integrals on Γ_1 on the right side can be approximated by the one-point Gaussian quadrature formula on each triangular face of the mesh contained in Γ_1. In this aim, we may proceed as follows. First of all, the pointer μ_i indicates whether the (local) node S_r^m of a given tetrahedron T_m is located on Γ_1 or not, where $i = n_{m,r}$, $1 \le r \le 4$. In case $\mu_i = -2$, the answer is yes and hence the contributions to b_i of $\int_{F_p^m} g_1\varphi_i dS$ corresponding to the faces F_p^m of T_m contained in Γ_1, for a certain p, $1 \le p \le 4$, $p \ne r$, have to be computed if S_i is one of its vertices. In this case, the corresponding contribution can be approximated by $area(F_p^m)g_1(G_p^m)/3$, where G_p^m is the centroid of F_p^m, that is, the point of T_m whose barycentric coordinates equal $1/3$, except λ_p^m whose value is zero. More accurate results are obtained with the Three-point Gaussian quadrature formula in a triangle, using the mid-points of the edges of F_p^m. In this case, the integral over this face of $g_1\varphi_i$ is approximated by the sum $g_1(C_{pr1}^m) + g(C_{pr2}^m)$ multiplied by $area(F_p^m)/6$, where C_{pr1}^m and C_{pr2}^m are the mid-points of the edges of F_p^m intersecting at $S_i = S_r^m$ with $i = n_{m,r}$, as the reader can easily check.

Remark 6.6 *Lagrange $\mathcal{P}_k$ FEM and $\mathcal{Q}_k$ FEM in two space variables have three-dimensional versions, isoparametric or not, for every $k \ge 1$. The former are based on tetrahedra while the latter are designed for convex hexahedra with quadrilateral faces. For descriptions of such methods, we refer to the classical books on the FEM cited throughout the text.* ∎

6.2.4 Example 6.2: A MATLAB Code for Three-dimensional FE Computations

In this example, computational tools are set at the reader's disposal for solving by the $\mathcal{P}_1$ FEM the Poisson equation in a bounded three-dimensional domain. We consider that only Dirichlet and Neumann boundary conditions are prescribed, which means that $\gamma = 0$. More precisely, we supply a code for this purpose programmed in the MATLAB environment, using the formulae given in Subsections 6.2.2 and 6.2.3, along with the recommended procedures considered in Subsection 6.2.3. We recall that both Dirichlet and Neumann boundary conditions are identified by the pointer $\vec{\mu}$ represented by the variable *ibound* in the main program. Therefore, $ibound(i)$ equals 0 or -2 in case the solution is unknown at mesh node S_i, the value -2 indicating that, in addition to this, an inhomogeneous Neumann boundary condition applies to S_i. The numbering of the unknowns $u(j)$, where j varies from one up to their total number represented by the variable *newnpt* in the main program, is obtained from the node numbering with the help of a correcting array *icorr*. The latter is built in such a way that if $ibound(i) = 0$ or $ibound(i) = -2$, the unknown at S_i is the jth unknown where $j = i - icorr(i)$. The value of $ibound(i)$ equals -1 if homogeneous Dirichlet boundary conditions are prescribed at S_i. In case inhomogeneous Dirichlet boundary conditions apply to S_i, then $ibound(i) = l > 0$, where l is such that the lth value of $\vec{d}$, represented in the code by the array *dbc* of prescribed nonzero boundary values, equals the solution at S_i.

The code is rather long for obvious reasons, but the version given below is richly commented, so that the reader should not find it difficult to understand. Three parts surrounded by boxes should be completed by external sources, namely, the FE mesh supplied in file *file_mesh_data*, SLAE solving functions named *crout* and *cholesky* for banded matrices, and function *bc* generating boundary condition data (see also Exercises 6.16 and 6.17).

```
function main
% NUMERICAL PARAMETERS OF THE CODE
exact = 1;
% SOLVER = 1 -> CROUT
% SOLVER = 0 -> CHOLESKY
SOLVER = 1;
% MAXIMUM NUMBER OF ELEMENTS
NELEM    = 6000;
% MAXIMUM NUMBER OF NODES
NPT      = 1331;
% MAXIMUM NUMBER OF INHOMOGENEOUS DIRICHET DATA
NDIRICH = 700;
% MAXIMUM  MATRIX BAND WIDTH = AT LEAST 3 * BW (TRUE HALF-BAND WIDTH)  - 2
BWMAX    = 150;
% ARRAY ALLOCATION
FACE     = zeros(NELEM,4); NNO    = zeros(NELEM,4);    H      = zeros(NELEM,4);
X        = zeros(NELEM,4); Y      = zeros(NELEM,4);    Z      = zeros(NELEM,4);
DX       = zeros(NELEM,4); DY     = zeros(NELEM,4);    DZ     = zeros(NELEM,4);
XG       = zeros(NELEM,4); YG     = zeros(NELEM,4);    ZG     = zeros(NELEM,4);
VOLT     = zeros(NELEM,1); XYZ    = zeros(3,NPT);      DBC    = zeros(NDIRICH,1);
IPIVO    = zeros(NPT,1);   ICORR  = zeros(NPT,1);      IBOUND = zeros(NPT,1);
U        = zeros(NPT,1);   AMAT   = zeros(NPT,BWMAX);  BF     = zeros(NPT,1);
% INPUT/OUTPUT TEXT FILES
file_resul  = fopen('RESUL.TXT','w');
% BARYCENTRIC COORDINATES OF A 4-POINT NUMERICAL INTEGRATION FORMULA
ALPHA = 0.58541020;
BETA  = 0.13819660;
```

```
%xxxxxxxxxxxxxxxxxxxxxxxxxxxxxxxxxxxxxxxxxxxxxxxxxxxxxxxxxxxxxxxxxxxxxxxxxxxxxxxxxxx
%MESH INPUT BY FUNCTION MESH: ELEMENT VERTEX COORDINATES X,Y,Z AND INCIDENCE MATRIX NNO%
file_mesh_data = fopen('meshdata.txt');                                              %
NELEM = fscanf(file_mesh_data,'%i',[1]);                                             %
NPT   = fscanf(file_mesh_data,'%i',[1]);                                             %
for i=1:NELEM                                                                        %
  [ X(i,1) X(i,2) X(i,3) X(i,4) ] = fscanf(file_mesh_data,'%f %f %f %f ',[1 1 1 1]); %
  [ Y(i,1) Y(i,2) Y(i,3) Y(i,4) ] = fscanf(file_mesh_data,'%f %f %f %f ',[1 1 1 1]); %
  [ Z(i,1) Z(i,2) Z(i,3) Z(i,4) ] = fscanf(file_mesh_data,'%f %f %f %f ',[1 1 1 1]); %
  [ NNO(i,1) NNO(i,2) NNO(i,3) NNO(i,4) ] = fscanf(file_mesh_data,'%i %i %i %i ',[1 1 1 1]); %
end                                                                                  %
%xxxxxxxxxxxxxxxxxxxxxxxxxxxxxxxxxxxxxxxxxxxxxxxxxxxxxxxxxxxxxxxxxxxxxxxxxxxxxxxxxxx
%   CREATING ARRAY OF NODAL MESH COORDINATES
for i = 1:NELEM
  for j = 1:4
    XYZ(1,NNO(i,j)) = X(i,j); XYZ(2,NNO(i,j)) = Y(i,j); XYZ(3,NNO(i,j)) = Z(i,j);
  end
end
% CALCULATING MATRIX HALF-BAND WIDTH
BW = 0;
for k=1:NELEM
  for i=1:4
    l = NNO(k,i);
    for J=1:4
      m = NNO(k,J);
      if ( (l-m) > BW )
        BW = l-m;
      end
    end
  end
end
BW = BW + 1;
% DETERMINING GEOMETRIC DATA RELATED TO THE TETRAHEDRA
% VOLT(K) = VOLUME OF THE K-TH TETRAHEDRON
% (DX(K,I),DY(K,I),DZ(K,I)) = GRAD[I-TH BARYCENTRIC COORDINATE OF THE K-TH TETRAHEDRON]
% H(K,I) = HEIGHT OF THE K-TH TETRAHEDRON PERPENDICULAR TO ITS I-TH FACE
[ H,VOLT,DX,DY,DZ,XG,YG,ZG,FACE ] = geometry (X,Y,Z,NNO,NELEM);
%xxxxxxxxxxxxxxxxxxxxxxxxxxxxxxxxxxxxxxxxxxxxxxxxxxxxxxxxxxxxxxxxxxxxxxxxxxxxxxxxxxx
% BOUNDARY CONDITIONS DATA CREATED BY FUNCTION bc, NAMELY,                           %
% JDBC = NUMBER OF NODES AT WHICH INHOMOGENEOUS DIRICHLET BOUNDARY CONDITIONS APPLY. %
% AT THE J-TH MESH NODE SJ:                                                          %
% IBOUND(J) =-1 -> SOLUTION = 0 ;                                                    %
% IBOUND(J) > 0 -> NON ZERO PRESCRIBED VALUE = DBC(IBOUND(J)) ;                      %
% IBOUND(J) =-2 -> INHOMOGENEOUS NEUMANN BOUNDARY CONDITION (IHNBC);                 %
% IBOUND(J) = 0 -> SOLUTION UNKNOWN (WITHOUT AN IHNBC) ;                             %
% DBC(L) = L-TH INHOMOGENEOUS DIRICHLET BOUNDARY CONDITION (L=IBOUND(J)) ;           %
% F BEING THE FACE OPPOSITE TO THE I-TH VERTEX OF THE K-TH TETRAHEDRON:              %
% FACE(K,I) = 1 IF AN INHOMOGENEOUS NEUMANN BOUNDARY CONDITION APPLIES TO F;         %
% FACE(K,I) = 0 OTHERWISE;                                                           %
% XG(K,I),YG(K,I),ZG(K,I) ARE THE COORDINATES OF THE CENTROID OF F.                  %
[JDBC,IBOUND,DBC,FACE] = bc(NPT,NELEM,NDIRICH,XYZ,NNO,XG,YG,ZG);                      %
%xxxxxxxxxxxxxxxxxxxxxxxxxxxxxxxxxxxxxxxxxxxxxxxxxxxxxxxxxxxxxxxxxxxxxxxxxxxxxxxxxxx
%    COMPUTING SYSTEM  MATRIX AND RIGHT SIDE VECTOR
[AMAT,BF,NEWPTP] = ab(ICORR,X,Y,Z,H,NELEM,NPT,BW,BWMAX,NDIRICH,DBC,NNO,DX,DY,DZ, ...
                   VOLT,IBOUND,ALPHA,BETA,FACE,XG,YG,ZG,SOLVER);
%xxxxxxxxxxxxxxxxxxxxxxxxxxxxxxxxxxxxxxxxxxxxxxxxxxxxxxxxxxxxxxxxxxxxxxxxxxxxxxxxxxx
% SOLVING THE LINEAR SYSTEM OF ALGE-
BRAIC EQUATIONS                                                                      %
if ( SOLVER == 0 )                                                                   %
  [AMAT,BF] = cholesky(AMAT,BF,NEWPTP,BW,0);                                         %
else                                                                                 %
```

```
   [AMAT,BF,IPIVO] = crout(AMAT,BF,IPIVO,NEWPTP,BW,0);                                    %
end                                                                                       %
%xxxxxxxxxxxxxxxxxxxxxxxxxxxxxxxxxxxxxxxxxxxxxxxxxxxxxxxxxxxxxxxxxxxxxxxxxxxxxxxxxxxxxxxxxxx
for I=1:NEWPTP
  U(I) = BF(I);
end
% INSERTING PRESCRIBED BOUNDARY VALUES IN THE SOLUTION VECTOR
[U] = insert(U,NPT,NEWPTP,IBOUND,DBC,NDIRICH)
% PRINTING APPROXIMATE SOLUTION
for I=1:NPT
 fprintf(file_resul,'%i, X=%f, Y=%f, Z=%f, U=%f  \r\n',I,XYZ(1,I),XYZ(2,I),XYZ(3,I),U(I));
end
% PRINTING ERRORS IF THE EXACT SOLUTION IS KNOWN
if ( exact == 1 )
  impress( X,Y,Z,U,VOLT,XYZ,NNO,NPT,NELEM );
end
fclose(file_resul);
end
% END OF MAIN PROGRAM
%++++++++++++++++++++++++++++++++++++++++++++++++++++++++++++++++++++++++++++++++++++++++++++
% ab - FUNCTION FOR DETERMINING ELEMENT MATRICES AND RIGHT SIDE VECTORS
%        AND ASSEMBLING GLOBAL MATRIX AND RIGHT SIDE VECTOR
function [AMAT,BF,NEWPTP] = ab(ICORR,X,Y,Z,H,NELEM,NPT,BW,BWMAX,NDIRICH,DBC,NNO,DX,DY,DZ,
                            ...VOLT,IBOUND,ALPHA,BETA,FACE,XG,YG,ZG,SOLVER)
AMAT  = zeros(NPT,BWMAX);  BF     = zeros(NPT,1);
XI  = zeros(4);    YI  = zeros(4);    ZI  = zeros(4);
BL  = zeros(4,4);  AE  = zeros(4,4);  IND = zeros(3);
XV  = zeros(3);    YV  = zeros(3);    ZV  = zeros(3);
% INITIALIZING SYSTEM BANDED MATRIX AMAT AND RIGHT SIDE VECTOR BF
for I=1:NPT
  if ( SOLVER == 0 )
    for J=1,BW
      AMAT(I,J)=0.0;
    end
  else
    for J=1:3*BW-2
      AMAT(I,J)=0.0;
    end
  end
  BF(I)=0.0
end
% DETERMINING INDEX CORRECTOR ICORR OF UNKNOWN NUMBERING
[ICORR] = corrind ( IBOUND, NPT );
% DETERMINING THE NUMBER OF UNKNOWNS
[NEWPTP] = count ( IBOUND, NPT );
% COMPUTING THE SYSTEM MATRIX AMAT AND RIGHT SIDE VECTOR BF
% BARYCENTRIC COORDINATES OF THE INTEGRATION POINTS
for I=1:4
  for J=1:4
    BL(I,J) = BETA/4.0;
  end
  BL(I,I) = ALPHA/4.0;
end
% COORDINATES XI,YI,ZI OF THE INTEGRATION POINTS
GAMMA = ALPHA - BETA;
for K=1:NELEM
  SUMX = 0; SUMY = 0; SUMZ = 0;
  for I=1:4
    SUMX = SUMX + X(K,I); SUMY = SUMY + Y(K,I); SUMZ = SUMZ + Z(K,I);
  end
  for I=1:4
```

```
   XI(I)  = SUMX*BETA + X(K,I)*GAMMA;
   YI(I)  = SUMY*BETA + Y(K,I)*GAMMA;
   ZI(I)  = SUMZ*BETA + Z(K,I)*GAMMA;
end
% 4 x 4 ELEMENT MATRIX AE
for I=1:4
  for J=I:4
    AE(I,J) = ( DX(K,I)*DX(K,J) + DY(K,I)*DY(K,J) + DZ(K,I)*DZ(K,J) )*VOLT(K);
    AE(J,I) = AE(I,J);
  end
end
% ASSEMBLING SYSTEM MATRIX AMAT AND RIGHT SIDE BF
for I=1:4
  L = NNO(K,I);
  if ( IBOUND(L) > 0 ) | (IBOUND(L) == -1)
    continue
  end
  LL = L - ICORR(L);
  for J=1:4
    M = NNO(K,J);
    if ( IBOUND(M) == -1 )
      continue
    end
    if ( IBOUND(M) <= 0 )
      M = M - ICORR(M);
      if ( SOLVER == 0 )
        if (M < LL)
          continue
        end
        M = M - LL + 1;
      else
        M = M - LL + BW;
      end;
      AMAT(LL,M) = AE(I,J) + AMAT(LL,M);
    else
      % TAKING INTO ACCOUNT INHOMOGENEOUS DIRICHLET BOUNDARY CONDITIONS
      BF(LL) = BF(LL) - AE(I,J)*DBC(IBOUND(M));
    end
  end
  % VOLUME INTEGRAL TERM OF BF
  for N=1:4
    BF(LL) = BF(LL) + VOLT(K)*F(XI(N),YI(N),ZI(N))*BL(N,I);
  end
  % TAKING INTO ACCOUNT INHOMOGENEOUS NEUMANN BOUNDARY CONDITIONS
  % WITH A 3-POINT INTEGRATION OF G1*I-TH SHAPE FUNCTION ON A BOUNDARY FACE
  if ( IBOUND(L) ~= -2 )
    continue
  end
  for N = 1:4
    if ( N == I )
      continue
    end
    if ( FACE(K,N) == 1 )
      IC = 0;
      for J=1:4
        if (J==I) | (J==N)
          continue
        end
        IC      = IC + 1;
        XI(IC) = ( X(K,I) + X(K,J) )/2.0;
        YI(IC) = ( Y(K,I) + Y(K,J) )/2.0;
```

```
            ZI(IC) = ( Z(K,I) + Z(K,J) )/2.0;
        end
        AREAF = 3.0*VOLT(K)/H(K,N);
        for J=1:2
            BF(LL) = BF(LL) + G1(XI(J),YI(J),ZI(J))*AREAF/6.0;
        end
      end
    end
  end
end
end
%++++++++++++++++++++++++++++++++++++++++++++++++++++++++++++++++++++++++++++++++++
% geometry - PROCEDURE FOR CALCULATING GEOMETRIC QUANTITIES OF THE TETRAHEDRA
function [H,VOLT,DX,DY,DZ,XG,YG,ZG,FACE] = geometry (X,Y,Z,NNO,NELEM)
XT    = zeros(4,1);  YT    = zeros(4,1); ZT    = zeros(4,1);
XP    = zeros(3,1);  YP    = zeros(3,1); ZP    = zeros(3,1);
CK    = zeros(4,3);
H     = zeros(NELEM,4); DX   = zeros(NELEM,4); DY    = zeros(NELEM,4);
DZ    = zeros(NELEM,4); VOLT = zeros(NELEM,1); XG    = zeros(NELEM,4);
YG    = zeros(NELEM,4); ZG   = zeros(NELEM,4);
FACE  = zeros(NELEM,4);
% SWEEPING THE MESH
for K=1:NELEM
  % COORDINATES OF THE i-th VERTEX S(I) OF THE K-TH TETRAHEDRON
  for I=1:4
    XT(I) = X(K,I); YT(I) = Y(K,I); ZT(I) = Z(K,I);
  end
  % CALCULATING THE COMPONENTS OF VECTORS V12,V13,V14 WITH V1I=-S(I)-S(1)
  for I=1:3
    XP(I) = XT(I+1) - XT(1); YP(I) = YT(I+1) - YT(1); ZP(I) = ZT(I+1) - ZT(1);
  end
  % CALCULATING THE VOLUME OF THE K-TH TETRAHEDRON
  VOLTR = abs( XP(3)*(YP(1)*ZP(2)-YP(2)*ZP(1))     ...
             + YP(3)*(XP(2)*ZP(1)-XP(1)*ZP(2))        ...
             + ZP(3)*(XP(1)*YP(2)-XP(2)*YP(1)) )/6.0;
  % CHECKING WHETHER LOCAL VERTEX NUMBERING CONFORMS TO THE RIGHT HAND RULE
  VOL   = -(XT(1)*YT(2)*ZT(3)+XT(3)*YT(1)*ZT(2)+XT(2)*YT(3)*ZT(1)    ...
           -XT(3)*YT(2)*ZT(1)-XT(2)*YT(1)*ZT(3)-XT(1)*YT(3)*ZT(2))   ...
          +XT(4)*(YT(2)*ZT(3)+YT(1)*ZT(2)+YT(3)*ZT(1)                ...
           -YT(2)*ZT(1)-YT(1)*ZT(3)-YT(3)*ZT(2))                     ...
          -YT(4)*(XT(2)*ZT(3)+XT(1)*ZT(2)+XT(3)*ZT(1)                ...
           -XT(2)*ZT(1)-XT(1)*ZT(3)-XT(3)*ZT(2))                     ...
          +ZT(4)*(XT(2)*YT(3)+XT(1)*YT(2)+XT(3)*YT(1)                ...
           -XT(2)*YT(1)-XT(1)*YT(3)-XT(3)*YT(2));
  if ( abs(VOL) <= 10^(-4) )
    if ( VOL < 0 )
      % CORRECTION OF THE LOCAL VERTEX NUMBERING IF NECESSARY
      NTEMP   = NNO(K,2); NNO(K,2) = NNO(K,3); NNO(K,3) = NTEMP;
      XTEMP   = X(K,2); X(K,2)   = X(K,3); X(K,3)   = XTEMP;
      YTEMP   = Y(K,2); Y(K,2)   = Y(K,3); Y(K,3)   = YTEMP;
      ZTEMP   = Z(K,2); Z(K,2)   = Z(K,3); Z(K,3)   = ZTEMP;
    end
  end
  % DETERMINING LINEAR SHAPE FUNCTION COEFFICIENTS CK
  for I=1:4
    L = 1;
    for J=1:3
      if ( I == J )
        L = L + 1;
      end
      XP(J) = XT(L); YP(J) = YT(L); ZP(J) = ZT(L);
```

```matlab
    L      = L + 1;
  end
  % CALCULATING COEFFICIENTS CK(I,J) OF THE LINEAR SHAPE FUNCTION LAMBDA(I)
  % I.E. LAMBDA(I)= CK(I,1)*X+CK(I,2)*Y+CK(I,3)*Z+CK(I,4)
  CK(I,1) = ((-1.)^I)   * (YP(2)*ZP(3)+YP(1)*ZP(2)+YP(3)*ZP(1)      ...
                           -YP(2)*ZP(1)-YP(1)*ZP(3)-YP(3)*ZP(2))/VOL;
  CK(I,2) = ((-1.)^(I+1)) * (XP(2)*ZP(3)+XP(1)*ZP(2)+XP(3)*ZP(1) ...
                           -XP(2)*ZP(1)-XP(1)*ZP(3)-XP(3)*ZP(2))/VOL;
  CK(I,3) = ((-1.)^I)   * (XP(2)*YP(3)+XP(1)*YP(2)+XP(3)*YP(1)      ...
                           -XP(2)*YP(1)-XP(1)*YP(3)-XP(3)*YP(2))/VOL;
  end
  %   VOLUME OF THE K-TH TETRAHEDRON
  VOLT(K) = VOLTR;
  % DERIVATIVES OF THE i-TH BARYCENTRIC COORDINATE lambda(K,i)
  for i=1:4
    DX(K,i) = CK(i,1); DY(K,i) = CK(i,2); DZ(K,i) = CK(i,3);
  end
  % DETERMINING FOR THE K-TH TETRAHEDRON
  % H(K,I) = HEIGHT RELATED TO THE I-TH VERTEX S(I);
  % XG(K,I),YG(K,I),ZG(K,I) COORDINATES OF FACE CENTROID OPPOSITE TO S(I)
  for I = 1:4
    GRADLAMB = sqrt(DX(K,I)^2+DY(K,I)^2+DZ(K,I)^2);
    H(K,I)   = 1.0/GRADLAMB;
    SUMX = 0.0;   SUMY = 0.0; SUMZ = 0.0;
    for J=1:4
      if ( J ~= I )
        SUMX = SUMX+X(K,J);   SUMY = SUMY+Y(K,J);   SUMZ = SUMZ+Z(K,J);
      end
    end
    XG(K,I) = SUMX/3.0;  YG(K,I) = SUMY/3.0;  ZG(K,I) = SUMZ/3.0;
  end
end
end
%+++++++++++++++++++++++++++++++++++++++++++++++++++++++++++++++++++++++++++++++++++++
% count - PROCEDURE FOR UNKNOWN COUNTING
function [K] = count (IFRON, NNOE)
K = 0;
for I = 1:NNOE
  if ( IFRON(I) == 0 ) | ( IFRON(I) <= -2 )
    K = K + 1;
  end
end
end
%+++++++++++++++++++++++++++++++++++++++++++++++++++++++++++++++++++++++++++++++++++++
% corrind - PROCEDURE FOR THE ADJUSTMENT OF UNKNOWN NUMBERING AFTER REMOVING
%           INDICES CORRESPONDING TO PRESCRIBED VALUES FROM THE SOLUTION VECTOR
function [IC] = corrind ( IP, NPT )
IC = zeros(NPT,1);
IC(1) = 0;
for i = 2:NPT
  IC(i) = IC(i-1);
  if ( IP(i-1) ~= 0 ) & ( IP(i-1) > -2 )
    IC(i) = IC(i) + 1;
  end
end
end
%+++++++++++++++++++++++++++++++++++++++++++++++++++++++++++++++++++++++++++++++++++++
% insert - PROCEDURE FOR INSERTING PRESCRIBED VALUES IN THE SOLUTION VECTOR
function [B] = insert(B,NNOSU,NNOSN,IBOUND,DBC,N3)
K = NNOSN + 1;
for I=NNOSU:-1:1
```

```
L = IBOUND(I);
if ( L == 0 ) | ( L <= -2 )
   K     = K - 1;
   B(I) = B(K);
else
   if ( L < 0 )
      B(I) = 0.0;
   end
   if ( L > 0 )
      B(I) = DBC(L);
   end
end
end
end
```

6.3 Exercises

6.1 Extend the error estimates in the norm $|\cdot|_{1,2}$ established in Subsection 5.1.2 for
equation $-\Delta u = f \in L^2(\Omega)$, to the case of inhomogeneous Dirichlet boundary
conditions $u = g_0$ on Γ. Consider only the case of exact integration. Next, extend
the second-order error estimate (5.46) in the norm of $L^2(\Omega)$ to the case where Ω is a
rectangle $(0, L_x) \times (0, L_y)$. In this aim, assume that the second-order derivative of g_0
with respect to the abscissa along every edge e of Γ is well-defined and belongs to
$L^2(e)$, and prove the following **trace property** of $\partial_\nu v$ when $v \in H^2(\Omega) \cap H_0^1(\Omega)$: for
a suitable constant C_ν $[\oint_\Gamma |\partial_\nu v|^2 ds]^{1/2} \le C_\nu \parallel H(v) \parallel_{0,2}^2$. Check that an analogous
result is sufficient for estimate (5.46) to hold in the inhomogeneous case, when Ω is an
arbitrary convex polygon.

6.2 Extend the error estimates in $H^1(\Omega)$ for the Dirichlet–Poisson problem established
in Subsection 5.1.2 to equation $u - \Delta u = f \in L^2(\Omega)$ with inhomogeneous Neumann
boundary conditions $\partial_\nu u = g_1$ on Γ, assuming that u belongs to $H^2(\Omega)$. Consider the
case where f is simple enough to allow exact integration of the underlying term on the
right side, but g_1 is a continuous function to be replaced in the approximate problem by
its linear interpolate g_{1h} at the mesh vertices located on the boundary. First, assuming
that Ω is a rectangle, derive a trace property of a function $v \in H^1(\Omega)$ to check that the
$\mathcal{P}_1$ FEM approximation is of the first order in the norm $\parallel \cdot \parallel_{1,2}$ of $H^1(\Omega)$, as long as
$dg_1/ds \in L^2(e)$ for every straight portion e of Γ, s being the abscissa along e. Then,
switch to the case where Ω is an arbitrary polygon and take for granted the same **trace
property** that holds for a rectangular domain, namely: Any function $v \in H^1(\Omega)$ has
a prolongation to a square integrable function defined on Γ still denoted by v, satisfy-
ing $[\oint_\Gamma |v|^2 ds]^{1/2} \le C_\Gamma \parallel v \parallel_{1,2} \forall v \in H^1(\Omega)$, where C_Γ is a constant. Conclude that the
same first-order error estimate for a rectangle applies to a polygonal Ω.

6.3 Check that equations (6.4), (6.5) and (6.6) are equivalent to ODEs (6.1), (6.2) and (6.3),
respectively. Prove that under the indicated assumptions, a solution to any of these three
problems is necessarily unique.

6.4 Consider the variants of ODEs (6.1), (6.3) and (6.2) with inhomogeneous boundary conditions. Set them in equivalent variational forms, and check the issue of solution uniqueness.

6.5 Check that the space Q_1 for an arbitrary quadrilateral contains all complete polynomials of degree less than or equal to one.

6.6 Let λ_p and λ_q be two distinct barycentric coordinates of an arbitrary triangle T of a mesh with maximum edge length h belonging to a quasi-uniform family of triangulations. Set $\phi_{pq} := 4\lambda_p\lambda_q$. Find the expression of the constant C_ϕ independent of h in terms of the constant c in equation (5.43), such that $\max\limits_{\mathbf{x}\in T} |[\mathbf{grad}\ \phi_{23}](\mathbf{x})| \leq C_\phi h^{-1}$.

6.7 Let u_h be the nonisoparametric P_2-FE solution with a regular family of triangulations, of the Poisson equation $-\Delta u = f$ in a smooth convex curved domain Ω of $\Re^2$ with homogeneous Dirichlet boundary conditions everywhere on the boundary Γ. We assume that $f \in H^1(\Omega)$, which implies that $u \in H^3(\Omega)$. Find an explanation for the fact that the norm of $u - u_h$ in $L^2(\Omega_h)$ is no better than an $O(h^2)$, Ω_h being the polygon with boundary Γ_h, consisting of the union of mesh triangles. In this aim, the reader can admit the validity of the following **trace property:** for every function $v \in H^2(\Omega) \cap H_0^1(\Omega)$ the outer normal derivative $\partial_{\nu_h} v$ of v along Γ_h is well-defined, and there exists a constant $\tilde{C}_\nu$ independent of h such that $[\oint_{\Gamma_h} |\partial_{\nu_h} v|^2 ds]^{1/2} \leq \tilde{C}_\nu \| H(v) \|_{0,2}$. Moreover, she or he can use the following **regularity result:** $\forall g \in L^2(\Omega)$ the solution of the Poisson equation $-\Delta v = g$ in Ω with homogeneous Dirichlet boundary conditions belongs to $H^2(\Omega)$, and there exists a constant $C(\Omega)$ such that $\| H(v) \|_{0,2} \leq C(\Omega) \| g \|_{0,2}$.

6.8 Show that $u := u + u_0$ is invariant with respect to the choice of u_0, where u is the unique solution to equations (6.25), (6.26) and (6.27).

6.9 Let φ_i be the shape function associated with node S_i for the P_2 FE approximation of problems (6.25), (6.26) and (6.27), taking $\gamma > 0$. Specify the possible nonzero contributions to the underlying linear system matrix of the term having γ as a coefficient, in terms of the positions of pairs of nodes $(S_i; S_j)$ distinct or not, vis-à-vis the mesh and Γ_1. Calculate all nonzero integrals of the form $\int_{\Gamma_1} \varphi_i\varphi_j ds$.

6.10 Let T be a triangle with barycentric coordinates λ_i for $i = 1, 2, 3$. For r greater than three and all possible values of q, determine the coefficients $e_{r,q}$ of the element matrix related to T for the P_2 FE approximation of problems (6.25), (6.26) and (6.27) with $\gamma = 0$, as a summation of terms of the form $(\mathbf{grad}\lambda_p|\mathbf{grad}\lambda_s)d_{ps}area(T)$ where d_{ps} are scalars independent of T. Identify the relevant values of p and s for each pair $(r; q)$, after simplifying the expressions as much as possible using the identity $\sum_{i=1}^3 \mathbf{grad}\ \lambda_i \equiv \vec{0}$. Check that $\sum_{q=1}^6 e_{r,q} = 0$ for $r = 4, 5, 6$, and explain why.

6.11 Check the values on the right side of equation (6.31) by application of the Seven-point quadrature formula to approximate the integral of $f\phi_{pq}$ in a mesh triangle T_m.

6.12 Check the values on the right side of equation (6.32) by application of Simpson's rule to approximate the integral of $g_1\phi_{pq}$ along an edge e_j^m opposite to a vertex S_j^m of a mesh triangle T_m having a non-empty intersection with Γ_1. Take into account the various possible pairs $(p; q)$.

6.13 Write down the equations for a rectangular Cell-centred FV scheme with variable cell sizes, to solve the Poisson equation in a rectangular three-dimensional domain $\Omega = (0, L_x) \times (0, L_y) \times (0, L_z)$ with inhomogeneous Dirichlet boundary conditions. Use notations analogous to those of Subsection 4.4.2 for the two-dimensional case. Approximate the integral on the right side by the one-point quadrature formula using the cell centre. Indicate how the scheme has to be modified in order to incorporate homogeneous Neumann boundary conditions on face $z = L_z$ instead of Dirichlet boundary conditions.

6.14 Demonstrate Property 7 of the barycentric coordinates of a tetrahedron.

6.15 Let T be a tetrahedron with barycentric coordinates λ_i, $i = 1, 2, 3, 4$ and $volume(T) > 0$. Show that $\int_T \lambda_i \lambda_j dx dy dz = (1 + \delta_{ij}) volume(T)/20$ (hint: use the master tetrahedron $\hat{T}$ whose vertex coordinates are $(0; 0; 0)$, $(1; 0; 0)$, $(0; 1; 0)$ and $(0; 0; 1)$).

6.16 Study the MATLAB code of Example 6.2 in detail, and identify in particular the implementation of the formulae and procedures recommended in Subsections 6.2.2 and 6.2.3.

6.17 The purpose of this exercise is to complete the MATLAB code of Example 6.2, and test it in the solution of the following equation: $-\Delta u = 2(y + z)$ in $(0, 1) \times (0, 1) \times (0, 1)$, with $u(0, y, z) = u(1, y, z) = 0$ for $0 \le y, z \le 1$, $u(x, 0, z) = z(x - x^2)$ for $0 \le x, z \le 1$, $u(x, y, 0) = y(x - x^2)$ for $0 \le x, y \le 1$ and $[\partial_y u](x, 1, z) = [\partial_z u](x, y, 1) = x - x^2$ for $0 < x, y, z < 1$, whose exact solution is $u(x, y, z) = (y + z)(x - x^2)$. Use a MATLAB package to generate tetrahedral meshes starting from a uniform $n \times n \times n$ mesh with n^3 equal cubes for variable n. If preferred, generate yourself a mesh with $6n^3$ tetrahedra by subdividing each cube into six tetrahedra having its diagonal parallel to the line $x = y = z$ as a common edge. In this case, use sequential numbering of the nodes face by face orthogonal to the x-axis, so that the matrix band width is minimised. Next, program function bc to create the boundary condition arrays necessary to run the code, together with the functions $g0$, $g1$, f and fu for the proposed test case. Program also function $impress$ in order to compute the errors in the L^2 and L^∞ norms, and check convergence in these norms by taking $n = 2^m$ for $m = 2, 3, \cdots$, up to the largest possible value of m. Compute the errors in the L^2 norm using the same four-point integration formula in a tetrahedron used in the code.

7

Miscellaneous Complements

> When it is not in our power to
> follow what is true, we ought
> to follow what is most
> probable.
>
> René Descartes

In this chapter, we address some scattered selected topics on the the FDM, FEM and FVM that do not really fit into the aim and framework of the previous chapters. The guiding criteria of our topic selection are their relevance and the light shed on some complementary aspects of the numerical methods for PDEs, overlooked in the studies already carried out, owing to a lack of pertinence. Of course, the general approach adopted throughout the book is roughly maintained in the material that follows. However, this chapter's four sections are basically disconnected, as far as the corresponding subjects are concerned.

Chapter outline: In Section 7.1, we study fourth-order PDEs, whose treatment involves several new considerations. Section 7.2 is devoted to the advection–diffusion equations, which play a fundamental role in numerical simulations of countless physical phenomena. Section 7.3 addresses the error control through mesh refinement via a posteriori error estimates. In contrast to the material presented so far, we do not attempt to treat this topic in a rigorous manner, but rather draw the reader's attention to this subject as one of the main pillars of contemporary numerical simulations. Finally, we address in Section 7.4 essential aspects of the numerical solution of non-linear PDEs. Two numerical examples are given in this section, incorporating a detailed description of some techniques to handle problem non-linearities.

7.1 Numerical Solution of Biharmonic Equations in Rectangles

A biharmonic equation in two-space variables is a linearised form of fourth-order PDEs that model some fundamental problems in continuum mechanics. Just to give two examples, we could mention the **thin plate-bending** problem (cf. [118]) and plane **viscous incompressible**

Numerical Methods for Partial Differential Equations:
An Introduction Finite Differences, Finite Elements and Finite Volumes, First Edition. Vitoriano Ruas.
© 2016 John Wiley & Sons, Ltd. Published 2016 by John Wiley & Sons, Ltd.
Companion Website: www.wiley.com/go/ruas/numericalmethodsforpartial

flow in terms of the **stream function** (see e.g. [157]). In these equations, the dominant differential operator is the square of the Laplacian, that is, the operator $\Delta^2(\cdot) = \Delta[\Delta](\cdot)$. This means that $\Delta^2(\cdot) = \partial_{xxxx}(\cdot) + 2\partial_{xxyy}(\cdot) + \partial_{yyyy}(\cdot)$.

7.1.1 Model Fourth-order Elliptic PDEs

The model problem studied in this section is the linear biharmonic equation, which is posed as follows: Given a function f defined in a bounded two-dimensional domain Ω assumed to have a sufficiently smooth boundary Γ, the problem consists of finding a function u that satisfies

$$\Delta^2 u = f \text{ in } \Omega. \tag{7.1}$$

In the case of thin plate bending, for instance, Ω represents the plate's mid-plane and $u(x, y)$ its deflexion orthogonal to Ω at a point with coordinates $(x; y)$, under the action of an area density of forces proportional to f, pointing in the same direction (see e.g. [118]). Of course, equation (7.1) must be supplemented with suitable boundary conditions, and a minimum regularity of u and f has to be required, otherwise this equation will not have a unique solution. In this section, we will search for u in the space $H^2(\Omega)$, by assuming that $f \in L^2(\Omega)$. Although the latter assumption implies that we can expect more regularity of u, for the purpose of setting up equivalent variational formulations, $H^2(\Omega)$ will be the natural working space as seen hereafter.

In this chapter, we consider only the following essential boundary conditions for equation (7.1):

$$u = g_0; \ \partial_\nu u = g_1 \text{ on } \Gamma, \tag{7.2}$$

where g_0 and g_1 are functions having suitable continuity and differentiability properties on Γ. In the case of the plate-bending problem, most frequently $g_0 = g_1 = 0$ (i.e. the boundary conditions are homogeneous). This corresponds to a **clamped plate**, which means that the plate undergoes no displacement and no rotation along its boundary. In the case of incompressible flow, usually g_0 or g_1 are not zero, but in contrast f is most frequently equal to zero. In order to simplify the presentation, we will focus on homogeneous boundary conditions, but the reader will certainly be able to identify the necessary modifications in the methodology to be studied, in order to accommodate inhomogeneous boundary conditions. In view of this, we will definitively study numerical methods to solve the following:

Model biharmonic equation

$$\begin{array}{|l}
\Delta^2 u = f \text{ in } \Omega, \\
u = \partial_\nu u = 0 \text{ on } \Gamma,
\end{array} \tag{7.3}$$

where $f = g/K$, g being the area density of forces applied to the plate and K a coefficient depending on physical characteristics of the plate. More specifically, $Ed^3/12(1\sigma^2)$, where E is Young's modulus and σ is Poisson's ratio of the material the plate is made of. d in turn is the plate's width. Moreover, the study will be limited to the case of a rectangular Ω.

In Remark 2.4, we explained the difference between essential and natural boundary conditions taking the bar problem (P_1) as a model. In solid mechanics, the solution of an equilibrium problem is usually a kinematic entity (e.g. a deflexion function or a displacement field) that minimises a certain energy functional defined for **kinematically admissible** quantities of the kind. This means that all of them must satisfy certain boundary conditions, which are thus called essential. In contrast to the case of the bar, all the admissible deflexions of a clamped plate must satisfy both deflexion and deflexion normal derivative zero boundary conditions. Hence, in this case, both conditions are essential. The solution itself can eventually satisfy specific boundary conditions carrying a physical meaning, other than essential ones. In this case, the former are natural boundary conditions. Just to give an example, we may consider a rectangular thin plate, whose boundary is simply supported. This means that the deflexion u still vanishes on Γ but $\partial_\nu u$ is not prescribed thereupon (i.e. rotations are allowed along the boundary). But in this case, besides the essential boundary condition satisfied by all the admissible deflexions, the solution will necessarily satisfy the zero momentum boundary condition $\Delta u = 0$, which is thus a natural boundary condition. In this case, the rectangular plate-bending problem reduces to two Poisson equations with homogeneous Dirichlet boundary conditions, that is,

$$\begin{cases} -\Delta w = f \text{ in } \Omega \text{ and } w = 0 \text{ on } \Gamma \\ \text{followed by} \\ -\Delta u = w \text{ in } \Omega \text{ and } u = 0 \text{ on } \Gamma. \end{cases}$$

Of course, it is no point addressing again the solution of Poisson problems in this chapter, and hence we will concentrate on the clamped plate problem.

Remark 7.1 *Whenever a simply supported plate has a curved boundary, the problem does not reduce to the solution of a sequence of two Poisson equations. Indeed, in this case the zero momentum boundary condition is $\sigma \Delta u + (1 - \sigma)\partial_{nn}u = 0$, where σ is Poisson's ratio of the material the plate is made of, and $\partial_{nn}u$ denotes the second-order outer normal derivative of u along the curved boundary. Notice that in principle, $\partial_{nn}u$ differs from the derivative in the direction of the outer normal $\vec{\nu}$ of the derivative of u in the same direction (i.e., $\partial_\nu u = (\mathbf{grad}\, u | \vec{\nu})$). This is because the continuous variation of $\vec{\nu}$ along Γ must be taken into account to determine $\partial_{nn}u$.∎*

7.1.2   The 13-point FD Scheme

Like in the case of second-order PDEs, a simple way to solve equation (7.3) is the FDM. For the sake of simplicity, we consider only the case of uniform grids. Using divided differences, there is no essential difficulty to treat grids with variable spacings, but in the case of the biharmonic equation the description of the scheme becomes a little clumsy.

Let the plate occupy the domain $\Omega = (0, L_x) \times (0, L_y)$ in the plane equipped with a Cartesian coordinate system $(O; x, y)$. The uniform grid has spacings $h_x = L_x/n_x$ and $h_y = L_y/n_y$ for given integers $n_x \geq 3$ and $n_y \geq 3$. Like in Chapter 4, the grid points with coordinates $(ih_x; jh_y)$ are denoted by $M_{i,j}$, for $i = 0, 1, \ldots, n_x$ and $j = 0, 1, \ldots, n_y$.

Let $w = \Delta u$. Denoting by $u_{i,j}$ the approximate values of u at $M_{i,j}$ for $i = 1, \ldots, n_x - 1$ and $j = 1, \ldots, n_y - 1$, we know from Subsection 5.1.1 that second-order approximations of w at the same points are given by

$$w(ih_x, jh_y) \simeq [u_{i,j-1} + u_{i,j+1} - 2u_{i,j}]/h_x^2 + [u_{i-1,j} + u_{i+1,j} - 2u_{i,j}]/h_y^2,$$

where $u_{0,j} = u_{n_x,j} = 0$ for $j = 0, 1, \ldots, n_y$ and $u_{i,0} = u_{i,n_y} = 0$ for $i = 0, 1, \ldots, n_x$. Now, in order to extend the approximation of w to the boundary points $M_{0,j}$, $M_{n_x,j}$ for $j = 1, \ldots, n_y - 1$ and $M_{i,0}$, M_{i,n_y} for $i = 1, \ldots, n_x - 1$, we proceed as follows:

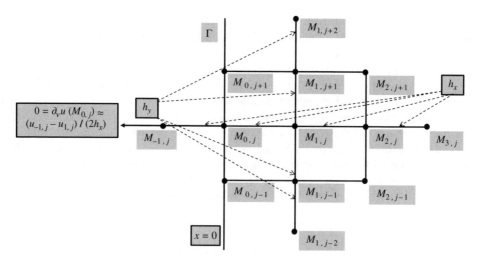

Figure 7.1 The 13-point FD stencil centred at point $M_{1,j}$, for $1 \leq j \leq n_y - 1$

Referring to Figure 7.1, similarly to Chapter 1, we use the fictitious point technique to take into account the boundary conditions $[\partial_x u](M_{0,j}) = [\partial_x u](M_{n_x,j}) = 0$ for $j = 1, \ldots, n_y - 1$ and $[\partial_y u](M_{i,0}) = [\partial_y u](M_{i,n_y}) = 0$ for $i = 1, \ldots, n_x - 1$. This naturally leads to the definitions $u_{-1,j} := u_{1,j}$, $u_{n_x+1,j} := u_{n_x-1,j}$ for $j = 1, \ldots, n_y - 1$ and $u_{i,-1} := u_{i,1}$, $u_{i,n_y+1} := u_{i,n_y-1}$ for $i = 1, \ldots, n_x - 1$. In doing so w is definitively approximated at $M_{i,j}$ by $\Delta_h(u_{i,j})$ applying the same recipe as in Section 4.2, that is,

$$
\begin{cases}
\text{For } i = 1, \ldots, n_x - 1 \text{ and } j = 0, 1, \ldots, n_y, \\
\text{and for } i = 0 \text{ or } i = n_x \text{ and } j = 1, \ldots, n_y - 1, \\
\Delta_h(u_{i,j}) := \dfrac{u_{i-1,j} + u_{i+1,j} - 2u_{i,j}}{h_x^2} + \dfrac{u_{i,j-1} + u_{i,j+1} - 2u_{i,j}}{h_y^2}.
\end{cases}
\tag{7.4}
$$

Naturally enough, Δw at an inner grid point $M_{i,j}$ can be handled by the following:

Approximation of $\Delta^2 u$ at $M_{i,j}$

$$
\boxed{
\begin{array}{l}
\text{For } i = 1, \ldots, n_x - 1 \text{ and } j = 1, \ldots, n_y - 1, \\[4pt]
\Delta_h^2(u_{i,j}) := \dfrac{\Delta_h(u_{i+1,j}) + \Delta_h(u_{i-1,j}) - 2\Delta_h(u_{i,j})}{h_x^2} \\[10pt]
+ \dfrac{\Delta_h(u_{i,j+1}) + \Delta_h(u_{i,j-1}) - 2\Delta_h(u_{i,j})}{h_y^2}
\end{array}
}
\tag{7.5}
$$

If we replace the values of $\Delta_h(u_{k,l})$ involved in the expression of $\Delta_h^2(u_{i,j})$ (i.e., values of k (resp. l) equal to either i, $i - 1$ or $i + 1$ (resp. j, $j - 1$ or $j + 1$)) by using equation (7.4), we figure out that there are exactly 13 points $M_{k,l}$ involved in the resulting expression. The combination of equations (7.4) and (7.5) yields rather lengthy expressions, which are left to the reader as a part of Exercise 7.1. Nevertheless, we give below the resulting scheme for the case where Ω is a square and a squared grid is employed, i.e. $n_x = n_y = n$. Setting $h := h_x = h_y$ we have

The 13-point FD scheme with a squared grid for the

biharmonic equation in a square

$$
\begin{aligned}
&\text{For } i = 1, \ldots, n-1 \text{ and } j = 1, \ldots, n-1, \\
&20u_{i,j} + 2(u_{i-1,j-1} + u_{i+1,j-1} + u_{i-1,j+1} + u_{i+1,j+1}) \\
&-8(u_{i-1,j} + u_{i+1,j} + u_{i,j-1} + u_{i,j+1}) + (u_{i-2,j} + u_{i+2,j} + u_{i,j-2} + u_{i,j+2}) \\
&= h^4 f(ih, jh).
\end{aligned}
\tag{7.6}
$$

This scheme is stable in a sense specified in [94], and its local truncation error for a solution sufficiently smooth is an $O(h^2)$ (cf. Exercise 7.1). This yields error estimates in the same sense of order 3/2 (cf. [94]). Further studies on the 13-point FD scheme are proposed to the reader in Exercises 7.1 and 7.5. More details can be found in several classical books dealing with the FDM, such as references [52] or [71] and [94].

7.1.3   Hermite FEM in Intervals and Rectangles

In the two-dimensional case, there are not as many FEM for solving fourth-order boundary value problems as for second-order ones. Indeed, for reasons to be clarified hereafter, Lagrange FEs are not suited to the solution of biharmonic equations or problems alike. In the case of fourth-order boundary value problems, we must resort to **Hermite FEs**. This means that the solution degrees of freedom (i.e. unknowns), involve derivative values besides function values. In this way, continuity of both can be enforced at inter-element boundaries. On the other hand, constructing FEs belonging to this class is much more difficult. As a matter of fact, very few options are available, and we cannot really talk about families of elements like the $\mathcal{P}_k$ and the $\mathcal{Q}_k$ FEs. The only exception is the one-dimensional case, in which a family of Hermite FEs defined upon disjoint intervals, starting from degree three, can be defined. We will work this out in more details in the sequel. For the moment, let us just write equation (7.3) in standard variational form, in order to examine a little better the restrictions pointed out above. Assuming only that Ω has no boundary singularities, first of all we multiply both sides of equation (7.1) with a function v and integrate in Ω. We make the assumption that $v \in H^2(\Omega)$ because we next apply First Green's indentity twice to obtain successively:

$$
\int_\Omega \Delta\Delta u \, v \, dxdy = \oint_\Gamma \partial_\nu \Delta u \, v \, ds - \int_\Omega (\mathbf{grad}\Delta u | \mathbf{grad}\, v) \, dxdy = \int_\Omega f \, v \, dxdy,
$$

$$
\int_\Omega \Delta\Delta u \, v \, dxdy = \oint_\Gamma \partial_\nu \Delta u \, v \, ds - \oint_\Gamma \Delta u \, \partial_\nu v \, ds + \int_\Omega \Delta u \Delta v \, dxdy = \int_\Omega f \, v \, dxdy.
$$

Now, analogously to the case of (P$_3$), we choose v to satisfy $v = 0$ and $\partial_\nu v = 0$ on Γ. In this case, we simply have $\int_\Omega \Delta u \Delta v \, dxdy = \int_\Omega f v \, dxdy$. Denoting by $H_0^2(\Omega)$ the subset of $H^2(\Omega)$

consisting of functions satisfying both boundary conditions[1], we thus have the

Standard Galerkin variational formulation of equation (7.3)

$$\boxed{\begin{array}{l} \text{Find } u \in H_0^2(\Omega) \text{ such that} \\ \int_\Omega \Delta u \Delta v \; dxdy = \int_\Omega fv \; dxdy \; \forall v \in H_0^2(\Omega). \end{array}} \tag{7.7}$$

The other way around, the reader can easily check as Exercise 7.2 that equation (7.7) implies (7.3), assuming a suitable regularity of u. Incidentally, equation (7.7) has a unique solution, which is a consequence of the Lax–Milgram theorem (see e.g. [45]). Moreover, provided Ω is sufficiently smooth (and polygons are included in this category), the following stability property holds for problem (7.7):

$$\|H(u)\|_{0,2} \leq C_P^2 \|f\|_{0,2} \tag{7.8}$$

where C_P is the constant of the Friedrichs–Poincaré inequality. Indeed, first we note that $\|H(w)\|_{0,2} = \|\Delta w\|_{0,2}$ for every $w \in H_0^2(\Omega)$. Inspiring herself or himself in the argument of Subsection 5.1.2, the reader can check this equality as Exercise 7.3. Therefore, taking $v = u$, by the Cauchy–Schwarz inequality followed by the Friedrichs–Poincaré inequality, we have, successively,

$$\|H(u)\|_{0,2}^2 \leq \|f\|_{0,2} \|u\|_{0,2},$$

$$\|H(u)\|_{0,2}^2 \leq C_P \|f\|_{0,2} \|\mathbf{grad}\; u\|_{0,2}. \tag{7.9}$$

Now, we note that by First Green's identity $\|\mathbf{grad}\; u\|_{0,2}^2 = -\int_\Omega \Delta u \; u \; dxdy$. Hence, applying the Cauchy–Schwarz inequality on the right side of this identity followed by the Friedrichs–Poincaré inequality, we easily derive (please check!)

$$\|\mathbf{grad}\; u\|_{0,2} \leq C_P \|H(u)\|_{0,2}. \tag{7.10}$$

Plugging equation (7.10) into (7.9), we obtain (7.8).

The next step is to introduce a FE analog of equation (7.7), which will lead to an approximation u_h of u associated with a given mesh. Why is Hermite interpolation necessary? The answer to this question is to be found in the next two paragraphs.

Two approaches are commonly used to solve a boundary value problem in variational form by the FEM. The **internal approximation** in which the FE space spanned by given shape functions is a subspace of the working space. This gives rise to a **conforming FEM**. Notice that in the case of equation (7.7), the working space is $H_0^2(\Omega)$. The other approach is the **external approximation** in which the underlying FE space is not a subspace of the working space. In this case, the method is called a **non-conforming FEM**. This allows some flexibility and hence simpler FE constructions. For example, a condition like $\partial_\nu u_h = 0$ everywhere on Γ can be relaxed, by enforcing it only at a finite number of boundary points. But in this case, obviously $u_h \notin H_0^2(\Omega)$, and there is a price to pay. First of all, the variational formulation for a non-conforming FEM does not inherit the nice stability property equation (7.8) of the continuous problem, and hence an *ad hoc* stability analysis must be carried out in each case. Furthermore, the easy-to-prove consistency result using the FE interpolate of u cannot be applied to a non-conforming approximation. In this connection, it is generally considered that

---

[1] Both conditions are perfectly compatible with functions in $H^2(\Omega)$ (see e.g. [126]).

Iron's **patch-test** (see e.g. [45]) implies the consistency of a non-conforming FEM, but we decline to further elaborate on this issue, because definitively we will only consider a conforming method in this chapter. For further information on non-conforming FEM for the biharmonic equation, the author refers to reference [120], and for conforming ones to reference [45].

So far, only a partial answer to our question on Hermite interpolation has been given: somehow the normal derivative must be a method's degree of freedom, otherwise the vanishing rotation condition for a clamped plate will not be enforced at all. But there is another condition even more stringent: if we are to ensure that a piecewise polynomial function lies in $H^2(\Omega)$, then it must be continuously differentiable at inter-element boundaries (i.e. of the C^1-class). Indeed, if the function is only continuous at such interfaces we can compute its first-order derivatives like in the case of the $\mathcal{P}_1$ FEM. However, the gradient of such a function will be discontinuous at the interfaces, and hence neither $H(u)$ nor Δu will be defined thereupon. Notice that somehow a non-conforming FEM must mimic this C^1 property, by satisfying for instance the continuous differentiability property at least at some points of the inter-element boundaries. But this is only possible if normal derivatives are degrees of freedom in the FE space, and hence in all cases the use of Hermite FE is a must to solve biharmonic problems.

Remark 7.2 *The above assertion on Hermite FEs applies to the standard Galerkin formulation (equation). If some other techniques are used, such as **discontinuous Galerkin methods** (see e.g. [48]), then this continuous differentiability condition, and also a mere continuity requirement, can be disregarded even for second-order problems. A little more information about this approach is provided in the Appendix.*∎

Before pursuing the description of the FE solution of equation (7.7), it is advisable to introduce the FEM based on Hermite interpolation in the framework of the beam-bending model problem posed in an interval $(0, L)$. More precisely, given $f \in L^2(0, L)$ proportional to a force field, find the deflexion u of the beam satisfying

The clamped beam-bending equation

$$
\boxed{\begin{aligned}
&u^{(iv)} = f \text{ in } (0, L) \\
&u(0) = u(L) = u'(0) = u'(L) = 0.
\end{aligned}} \tag{7.11}
$$

Of course, unless f is too complex, equation (7.11) can be solved by hand, but this is not the point. The reader certainly understood that we are using this equation just to present some concepts inherent to Hermite FEs. In this aim, let us first set it in variational form. Notice that, as a one-dimensional counterpart of equation (7.3) all the boundary conditions in equation (7.11) are essential, and hence they must be incorporated into the working space. Hence, this space is $H_0^2(0, L)$, that is, the subspace of $H^2(0, L)$ consisting of functions v satisfying $v(0) = v(L) = v'(0) = v'(L) = 0$.

Using elementary integration by parts, it is easy to see that the equivalent Galerkin variational form of equation (7.11) is the symmetric problem

$$
\begin{cases}
\text{Find } u \in H_0^2(0, L) \text{ such that} \\
\int_0^L u'' v'' \, dx = \int_0^L fv \, dx \ \forall v \in H_0^2(0, L).
\end{cases} \tag{7.12}
$$

We next consider a mesh $\mathcal{T}_h$ of $(0, L)$ consisting of the closure T_i of n disjoint intervals (x_{i-1}, x_i) for $i = 1, n$ with $0 = x_0 < x_1 < \ldots < x_{n-1} < x_n = L$. We use again the notation h_i for $x_i - x_{i-1}$, together with $h := \max_{1 \leq i \leq n} h_i$. For the same reasons already pointed out in the case of equation (7.7), a conforming FEM associated with this mesh must be based on continuously differentiable functions at all the x_i's for $i = 1, 2, \ldots, n - 1$. Moreover, we should be able to enforce the essential boundary conditions exactly. Then a natural solution method would be based on spaces V_h associated with $\mathcal{T}_h$ defined as follows for some integer $k > 0$:

$$V_h := \{v \mid v \in C^1[0, L], v \in \mathcal{P}_k(T_i), i = 1, \ldots, n, v(0) = v(L) = v'(0) = v'(L) = 0\}.$$

In the FEM, polynomials of a certain type are defined in the elements independently from each other using certain degrees of freedom. Then they are assembled together by enforcing coincidence of degrees of freedom at inter-element boundaries, so as to satisfy required continuity or differentiability properties. For example, in the case of the $\mathcal{P}_1$ FEM the degrees of freedom are only function values at two points in each interval. These are chosen to be the interval end-points in order to ensure the continuity of the assembled piecewise linear function by requiring coincidence of the function values at these points from each side. In the case of V_h, we must enforce the coincidence not only of function values, but also of first-order derivatives at all interval end-points. For this reason, four degrees of freedom pertaining to each element are necessary. Therefore, the minimum possible value of k is three. Indeed, a cubic function in an interval is defined by a linear combination of the monomials $1, x, x^2, x^3$, and hence there is hope for constructing a cubic function $\varphi \in \mathcal{P}_3(T_i)$ satisfying four conditions, namely,

$$\varphi(x_{i-1}) = d_1; \ \varphi(x_i) = d_2, \varphi'(x_{i-1}) = d_3; \ \varphi'(x_i) = d_4,$$

for any set of given real numbers d_1, d_2, d_3, d_4. Actually, using the auxiliary functions

$$\phi_0(x) := 1 - 3x^2 + 2x^3 \text{ and } \phi_1(x) := x - 2x^2 + x^3,$$

we construct four shape functions, namely,

- $\varphi_1^i(x) = \phi_0([x - x_{i-1}]/h_i)$;
- $\varphi_2^i(x) = \phi_0([x_i - x]/h_i)$;
- $\varphi_3^i(x) = h_i \phi_1([x - x_{i-1}]/h_i)$;
- $\varphi_4^i(x) = h_i \phi_1([x_i - x]/h_i)$.

so that $\varphi := \sum_{l=1}^4 d_l \varphi_l^i$ fulfils the required conditions, as the reader can easily check. The above list indicates that at a global level, there are two types of Hermite shape functions associated with an inner vertex of coordinate $x = x_i$, that is, for $i = 1, \ldots, n - 1$, spanning the corresponding FE space V_h: 'odd' shape functions φ_{2i-1} satisfying $\varphi_{2i-1}(x_j) = \delta_{ij}$ and $\varphi'_{2i-1}(x_j) = 0$ for $j = 0, 1, \ldots, n$, and 'even' shape functions φ_{2i} satisfying $\varphi_{2i}(x_j) = 0$ and $\varphi'_{2i}(x_j) = \delta_{ij}$, for $j = 0, 1, \ldots, n$. Both are illustrated in Figure 7.2. In short, in case $k = 3$, a function in V_h is a linear combination of the shape functions in the set $\{\varphi_k\}_{k=1}^{2(n-1)}$. In particular, the solution u_h of the FE counterpart of equation (7.12) obtained by replacing $H_0^2(0, L)$ with V_h is given by $\sum_{k=1}^{2(n-1)} u_k \varphi_k$, where $u_{2i-1} = u_h(x_i)$ and $u_{2i} = u_h'(x_i)$ for $i = 1, 2, \ldots, n - 1$. Checking this assertion is a simple task left to the reader.

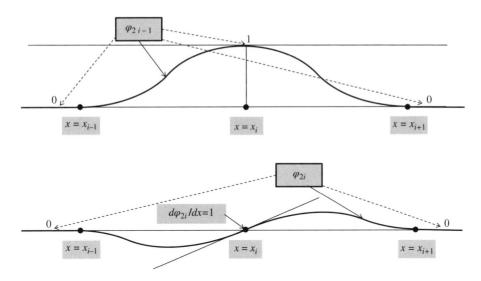

Figure 7.2   The Hermite piecewise cubic shape functions φ_{2i-1} and φ_{2i}, $1 \leq i \leq n-1$

We do not insist on the equivalent SLAE to determine these degrees of freedom of u_h, for it is similar to those we have dealt with so far. We do not elaborate either on FE spaces V_h with $k > 3$, although it is possible to define them in a similar manner whatever k, by enforcing the continuity of shape functions and their derivatives at interval end-points. This is because such constructions are seldom useful in practice.

Remark 7.3 *Using quadratic* **splines**, *it is possible to construct a function of the C^1-class, whose restriction to T_i belongs to $\mathcal{P}_2(T_i)$ for every i. However, in this case the functions in the approximation space cannot be defined in the intervals independently from each other. As a consequence, a given degree of freedom will lose its local character, since it will influence far away portions of the domain. This situation would be in conflict with the original concept of the FEM. For more information about splines, the reader can consult reference [2].*∎

After this brief introduction to Hermite FEs in one-dimensional space, we go back to equation (7.3). Still assuming that Ω is the rectangular domain $(0, L_x) \times (0, L_y)$, let $\mathcal{R}_h$ be a non-uniform partition of it into rectangles, satisfying the usual compatibility conditions. Similarly to the non-uniform FD grid introduced in Section 4.2, we denote the x coordinates of the mesh vertices by $0 = x_0 < x_1 < \ldots < x_{n_x-1} < x_{n_x} = L_x$ and its y coordinates by $0 = y_0 < y_1 < \ldots < y_{n_y-1} < y_{n_y} = L_y$. We recall that the mesh spacings are $h_i = x_i - x_{i-1}$ for $i = 1, \ldots, n_x$ and $k_j = y_j - y_{j-1}$ for $j = 1, \ldots, n_y$. We also set $h = \max\{ \max_{1 \leq i \leq n_x} h_i, \max_{1 \leq j \leq n_y} k_j\}$.

We will next describe the **Bogner–Fox–Schmit** FEM [31], which is a conforming method based on $\mathcal{Q}_3$ functions in a rectangle. More precisely, we define a space W_h as follows:

$$W_h = \{v| \; v \in C^1(\bar{\Omega}), v_{|R} \in \mathcal{Q}_3(R) \; \forall R \in \mathcal{R}_h\}.$$

We admit for the moment that the set W_h is not empty, and endeavour to construct a subset V_h of W_h with the following characteristics.

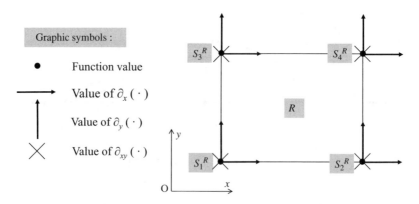

Figure 7.3   The Bogner–Fox–Schmit rectangular element

Referring to Figure 7.3, where a rectangle $R \in \mathcal{R}_h$ is represented, we first define the local degrees of freedom associated with a function $v \in V_h$ restricted to R to be its values, together with those of its partial derivatives $\partial_x v$, $\partial_y v$ and $\partial_{xy} v$ at the four vertices S_i^R of R, $i = 1, 2, 3, 4$. Let us number these degrees of freedom for a given function in $\mathcal{Q}_3(R)$ as $\mathcal{F}_{ij}$ for $i = 1, 2, 3, 4$ and $j = 1, 2, 3, 4$, where subscript i stands for the vertex number and j for the type of degree of freedom. More precisely, $j = 1$ refers to a function value, $j = 2$ to an x-derivative, $j = 3$ to a y-derivative and $j = 4$ to a cross second-order derivative. Noticing that a function in $\mathcal{Q}_3$ is a linear combination of the 16 monomials $1, x, y, x^2, xy, y^2, x^3, x^2y, xy^2, y^3, x^3y, x^2y^2, xy^3, x^3y^2, x^2y^3$ and x^3y^3, the following property is to be expected, although it must be confirmed: Given a set of 16 real values d_{ij}, $1 \le i, j \le 4$ of the degrees of freedom specified above, we can define a unique function $\varphi \in \mathcal{Q}_3$ such that, for $1 \le i, j \le 4$,

$$\varphi(S_i^R) = d_{i1}; \quad [\partial_x \varphi](S_i^R) = d_{i2}; \quad [\partial_y \varphi](S_i^R) = d_{i3}; \quad [\partial_{xy} \varphi] = d_{i4}.$$

This is possible indeed, since we can exhibit 16 canonical basis functions $\varphi_{ij}^R \in \mathcal{Q}_3(R)$, for $i, j = 1, 2, 3, 4$, such that

- $\varphi_{i1}^R(S_j^R) = \delta_{ij}; \quad [\partial_x \varphi_{i1}^R](S_j^R) = [\partial_y \varphi_{i1}^R](S_j^R) = [\partial_{xy} \varphi_{i1}^R](S_j^R) = 0;$
- $[\partial_x \varphi_{i2}^R](S_j^R) = \delta_{ij}; \quad \varphi_{i2}^R(S_j^R) = [\partial_y \varphi_{i2}^R](S_j^R) = [\partial_{xy} \varphi_{i2}^R](S_j^R) = 0;$
- $[\partial_y \varphi_{i3}^R](S_j^R) = \delta_{ij}; \quad \varphi_{i3}^R(S_j^R) = [\partial_x \varphi_{i3}^R](S_j^R) = [\partial_{xy} \varphi_{i3}^R](S_j^R) = 0;$
- $[\partial_{xy} \varphi_{i4}^R](S_j^R) = \delta_{ij}; \quad \varphi_{i4}^R(S_j^R) = [\partial_x \varphi_{i4}^R](S_j^R) = [\partial_y \varphi_{i4}^R](S_j^R) = 0.$

It follows that the function $\varphi := \sum_{i=1}^{4} \sum_{j=1}^{4} d_{ij} \varphi_{ij}^R$ satisfies the required conditions.

The expressions of the basis functions φ_{ij}^R can be found in many books or articles, and in this respect we refer for instance to reference [86].

Now, naturally enough, our FE space V_h consists of functions in W_h that are continuous, as much as their gradients and cross second-order derivatives, at all the inner mesh vertices, and whose values together with those of their gradients and cross second-order derivatives vanish at all the boundary vertices. If these conditions allow for the construction of functions of the C^1-class, then we will have validated a posteriori the initial assumption that W_h is not empty. Let us see how this works.

First of all, we note that the trace of a function in V_h on a mesh inner edge e_x parallel to the Ox axis is a cubic function in terms of x. By construction, this function together with the x derivatives at the edge end-points have the same values in both rectangles sharing e_x. Taking into account what we showed for the one-dimensional Hermite interpolation with cubics, it follows that the cubic traces from both sides of e_x must be the same. The same argument obviously applies to inner edges parallel to the Oy axis. Next, we examine the situation of normal derivative traces along an inner edge e_x. The normal derivative along e_x is a partial derivative with respect to y, and as the reader may check (cf. Exercise 7.4), such traces are also cubic functions in terms of x. Therefore, by the same argument as above, continuity of the normal derivative along e_x will be ensured if this cubic function together with its x partial derivatives at both ends of this edge coincide for both elements sharing it. By construction again, this condition is fulfilled since the y derivatives and their x-derivatives, that is, the cross second-order derivatives, are continuous at all the mesh vertices.

Finally, the reader can check as a part of Exercise 7.4 that, as long as the four degrees of freedom at all the vertices located on Γ vanish, every function in $v \in V_h$ satisfies the boundary conditions $v = \partial_\nu v = 0$ everywhere on Γ. This establishes that V_h is indeed a subspace of $H_0^2(\Omega)$, which means that the following problem is an internal approximation of the biharmonic equation with homogeneous essential boundary conditions, namely,

The Bogner–Fox–Schmit FE counterpart of equation (7.7)

$$\boxed{\begin{array}{l} \text{Find } u_h \in V_h \text{ such that} \\ \int_\Omega \Delta u_h \Delta v \; dxdy = \int_\Omega f v \; dxdy \; \forall v \in V_h. \end{array}} \tag{7.13}$$

The total number of degrees of freedom defining a function in V_h is $N_h = 4(n_x - 1)(n_y - 1)$. Since we are dealing with a structured mesh, the unknown numbering may follow the line-by-line systematic already described in similar situations. Here, the only peculiarity is that the numbers of the four degrees of freedom attached to each vertex should be preferably consecutive: for instance, $4i - 3$ for the function value, $4i - 2$ for the x derivative, $4i - 1$ for the y derivative and $4i$ for the cross second-order derivative, where i sweeps the mesh from one through N_h.

The shape functions underlying the $N_h \times N_h$ matrix and the right-side vector for the SLAE corresponding to equation (7.13) are constructed by putting together basis functions for elements forming a patch of elements surrounding a node. This allows us to proceed in a standard manner. Then both are obtained by assembling 16×16 element matrices and 16 component element vectors. The solution of the resulting SLAE yields the degrees of freedom of u_h, namely, approximations of u, **grad** u and $\partial_{xy} u$ at the inner mesh vertices. By interpolation using the local shape functions, we can determine u_h everywhere in Ω.

We refrain from elaborating on error estimates for u_h at this introductory level. We just quote known results stating that, as long as all the fourth-order partial derivatives of u belong to $L^2(\Omega)$ and the family of meshes in use is quasi-uniform, then for a constant $C(u)$ independent of h that vanishes if all the fourth-order partial derivatives of u equal zero, it holds that (cf. [45])

$$\| u - u_h \|_{1,2} + \| H(u - u_h) \|_{0,2} \leq C(u) h^2. \tag{7.14}$$

The above error estimate is logical for two reasons. First of all, it means that $u = u_h$ if u happens to belong to $\mathcal{P}_3$ but not to $\mathcal{P}_k$ for $k > 3$. Indeed, it is so because $\mathcal{Q}_3$ contains $\mathcal{P}_3$ but not $\mathcal{P}_4$. On the other hand, second-order convergence is the best we can hope for, since the norm of $H^2(\Omega)$ came into play in the biharmonic case. This implies a one-point downgrade in a method's order, as compared to estimates for second-order equations, whose working norm is $\| \cdot \|_{1,2}$.

Finally, it would be interesting to compare the 13-point FD scheme with the Bogner–Fox–Schmit FE, in the light of their costs. In Exercise 7.5, we propose to the reader the determination of their respective number of unknowns and bandwidths in comparable situations.

Remark 7.4 *A result qualitatively equivalent to equation (7.14) holds for the FE approxima-tion of equation (7.11) using the Hermite cubics studied in this subsection.*■

7.2 The Advection–Diffusion Equation

The advection–diffusion equation, also called the **convection–diffusion equation** depending on context, is the linear paradigm of several non-linear PDEs that model important physical events. Outstanding examples are the incompressible and compressible Navier–Stokes equations governing viscous flow (cf. [15]). The advection–diffusion equation itself has countless applications in engineering and science. The equation carries this name because it models two coupled processes: diffusion and advection of an unknown quantity, which can be as diverse as a pollutant concentration in air or water, or heat distribution in fluids. Once again, it would be out of purpose going into details on phenomenon modelling in this book. Nevertheless, such an aspect of the equation to be studied is briefly addressed hereafter just to keep ideas clear.

7.2.1 A Model One-Dimensional Equation

Let us consider the following problem. We wish to determine the temperature u of a fluid in motion in a straight cylindrical tube whose total length equals L, with mean cross section velocity $\vec{v}(x) = v(x)\vec{e}_x$, where $\vec{e}_x$ is the unit vector parallel to the tube's axis of symmetry, and x is the abscissa along this axis, with $x = 0$ at the tube's 'left end'. Assuming that heat conduction is also taking place, and that the medium's conductivity is constant equal to $p > 0$, the heat flux accross a section at abscissa x in the sense of increasing x is given by $-pu'(x)$. Further assuming that the velocity v is constant, for any subdomain of the tube, say, $\omega = (a_\omega, b_\omega)$, the incoming and outgoing heat fluxes are given by $pu'(a_\omega) - vu(a_\omega)$ and $-pu'(b_\omega) + vu(b_\omega)$. Moreover, if a distribution g of heat sources is acting along the tube while the heat transfer goes on, the balance of (thermal) energy requires that

$$pu'(a_\omega) - vu(a_\omega) - pu'(b_\omega) + vu(b_\omega) = \int_{a_\omega}^{b_\omega} g\, dx.$$

This readily implies

$$\int_{a_\omega}^{b_\omega} [-pu'' + vu' - g]dx = 0 \; \forall \omega \subset (0, L).$$

Finally, assuming that the temperature is kept fixed at both ends of the tube, say $u(0) = u(L) = 0$, recalling the arguments developed in Section 1.1, we immediately derive the advection–diffusion equation in its simplest expression, namely,

The one-dimensional advection–diffusion equation

$$\boxed{\begin{array}{l} -pu'' + vu' = g \text{ in } (0, L) \\ u(0) = u(L) = 0 \end{array}} \tag{7.15}$$

It is easy to see that this equation has a unique solution, and clearly enough, provided g is not too wild, it can be solved by hand. But once again, this is not the point; (7.15) is just the basic model that will allow us to introduce some popular numerical techniques aimed at overcoming delicate issues often encountered in the solution of advection–diffusion equations. Let us see right away the main source of problems.

First, we set $P := L|v|/p$ and $\epsilon = P^{-1}$. P is a dimensionless parameter called the **Péclet number**. This parameter points in particular to two important situations. The advection-dominant case corresponds to a high value of P. Conversely, whenever P is not so large, the diffusion will influence the phenomenon being modelled practically everywhere in the domain, and convection will not be dominant.

Switching to the dimensionless coordinate $\bar{x} = x/L$ and defining $\bar{u}(\bar{x}) := u(x)$ for all $x \in [0, L]$, a simple manipulation of the ODE in equation (7.15) yields the following equation for $\bar{u}$:

$$-\epsilon \bar{u}'' \pm \bar{u}' = \bar{g} \text{ where } \bar{g}(\bar{x}) = g(x)L/|v|.$$

supplemented with the boundary conditions $\bar{u}(0) = \bar{u}(1) = 0$. Let us drop the bars above the variables x, u and g and definitively work with the **dimensionless equation** (cf. [5]):

$$-\epsilon u'' \pm u' = g \text{ with } u(0) = u(1) = 0. \tag{7.16}$$

Taking $g \equiv 1$, the analytical solution of the above ODE is easily seen to be

$$u(x) = x - \frac{e^{\pm x\epsilon^{-1}} - 1}{e^{\pm \epsilon^{-1}} - 1}. \tag{7.17}$$

In Figure 7.4, we sketch a plot of this function u assuming that $\epsilon \ll 1$: the solution behaves like $y(x) = x$ everywhere for $x \geq 0$, except in a narrow **boundary layer** of width $\simeq \epsilon$ close to $x = 1$, where it abruptly changes of pattern in order to satisfy the boundary condition $u(1) = 0$.

Now suppose that, just for fun, we attempt to solve equation (7.16) by one of the numerical methods studied in this book. We guess it is not necessary to describe such procedures, to convince the reader that an accurate numerical solution would require very small grid or mesh sizes in the interior of the boundary layer. Notice that, in practice, the Péclet number can attain very large values, in case convection strongly dominates diffusion. For example, values of P as high as one million or more are perfectly compatible with realistic physical situations. Then the question is: Would it be reasonable to try to tackle a boundary layer behaviour by means of such fine discretisations? The answer seems to be no, but even if it were yes, there are other numerical issues typical of this class of equations that must be handled with much care. That is what we endeavour to show in the remainder of this section. However, before starting, we should make a few comments.

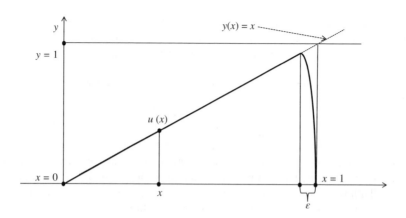

Figure 7.4   Solution of the advection–diffusion equation (7.16) with a boundary layer

With a few exceptions, the numerical methods for equation (7.15) to be studied are well-known, and at this writing can be considered as classical techniques. Incidentally, there is a rather vast literature on this subject owing to its undeniable importance. However, the presentations in this section are partially homemade, in the sense that several relevant aspects are not to be found elsewhere, to the best of the author's knowledge. There is another reason for this approach. Numerical methods for the advection–diffusion equation can be viewed as an advanced topic. Going into its rigorous treatment would require mathematical resources more sophisticated than those employed elsewhere in this book.

Most advection–diffusion phenomena of practical interest require a two- or a three-dimensional modelling, specially because in general analytical solutions are not available in this case. However, the main obstacles to overcome in order to solve the advection–diffusion equation numerically show up even in the one-dimensional case. Therefore, for the sake of brevity, we shall restrict to the latter framework the presentation of specific methodology to treat this problem. Nevertheless, we will address in Subsection 7.2.4 the time-dependent counterpart of equation (7.15). Extensions to higher dimensions will be the object of a few comments only.

For comprehensive studies on the topic, the author could cite references [116] and [103], among many others.

7.2.2   Overcoming the Main Difficulties with the FDM

From now on, instead of the dimensionless form (equation (7.16)), we rewrite the ODE in equation (7.15) as the

Working form of the advection–diffusion equation

$$-\epsilon u'' + w u' = f \text{ with } u(0) = u(L) = 0, \tag{7.18}$$

where $w = v/|v|$, $\epsilon = p/|v|$ and f stands for $g/|v|$. This means that we still work with the original variable x sweeping $(0, L)$. This will allow in particular to consider a bounded

Table 7.1 u_{50} versu $u(1/2)$ for $\epsilon = 10^{-m}$ and $h = 10^{-2}$, u solves equation (7.18) with $w = f = L = 1$.

$\epsilon \rightarrow$	10^{-1}	10^{-2}	10^{-3}	10^{-4}	10^{-5}
$u(1/2) \rightarrow$	0.493307149	0.500000000	0.500000000	0.500000000	0.500000000
$u_{50} \rightarrow$	0.493334839	0.500000000	0.499999998	0.380825080	0.049834063

continuously varying v in $(0, L)$ a little later. In this case, $\|v\|_{0,\infty}$ will replace $|v|$ in the above definitions of w, f and ϵ. Notice that in this case too, equation (7.18) still has a unique solution, according to the well-known theory of linear ODEs.

Since we are dealing with a second-order two-point boundary value ODE, all the methods presented in Chapter 1 can be applied to solve it. However, now the term vu' replaces qu, and this makes a big difference. In this subsection, we will consider a solution commonly adopted in the framework of the FDM. In order to simplify the presentation, we restrict ourselves to uniform grids, leaving the non-uniform case to the next subsections, where the FEM and the FVM will be addressed.

Recalling the uniform grid and pertaining notations of Section 1.2, a natural FDM to solve equation (7.18), is based on centred FD providing a second-order local truncation error, that is,

$$\begin{cases} \text{Setting } u_0 = u_n = 0, \text{ find } u_i \text{ for } i = 1, 2, \ldots, n-1 \text{ such that} \\ \epsilon(2u_i - u_{i-1} - u_{i+1})/h^2 + w(u_{i+1} - u_{i-1})/(2h) = f_i \text{ for } i = 1, 2, \ldots, n-1. \end{cases} \quad (7.19)$$

Unfortunately, this scheme is subject to stringent stability conditions. Let us examine this issue a little closer. We have

$$u_i = h^2 f_i/(2\epsilon) + (1/2 - \alpha)u_{i+1} + (1/2 + \alpha)u_{i-1} \text{ where } \alpha = wh/(4\epsilon).$$

A numerical example is not a proof, but it turns out to be a valid argument to supply convincing indications about the behaviour of the above scheme: we take $L = 1$, $f \equiv 1$ and $\epsilon = 10^{-m}$ for m ranging from 1 to 5. Keeping $h = 10^{-2}$, we display in Table 7.1 $u(1/2)$ together with the corresponding approximate values u_{50} rounded to eight decimals for the five values of ϵ. The sharp increase of the errors so far from the boundary layer as ϵ diminishes does not happen by chance. Such a bad behaviour is inherent to the **Centred FD scheme**. However, it can be remedied, as seen hereafter. As a matter of fact, recalling the stability analyses carried out in Subsection 2.2.1, we may legitimately conjecture that we must have $|\alpha| \leq 1/2$, if we wish to enjoy nice stability properties in the maximum norm for this Centred scheme. But this requires $h \leq 2\epsilon$, which is prohibitive if $\epsilon \ll 1$. Furthermore, our computations strongly suggest that this scheme is suspect. As a conclusion, it is advisable to discard it, at least if ϵ is small.

Inspired by the Upwind scheme studied in Chapter 3, let us try the following alternative: Letting $w^+ = \max[w, 0]$ and $w^- = \max[-w, 0]$, we set up

The Upwind FD scheme for equation (7.18)

Setting $u_0 = u_n = 0$, find u_i such that for $i = 1, 2, \ldots, n-1$ $\epsilon(2u_i - u_{i-1} - u_{i+1})/h^2 + w^+(u_i - u_{i-1})/h - w^-(u_{i+1} - u_i)/h = f_i.$	(7.20)

Let us see what has changed with respect to the Centred FD scheme we have just discarded. Now u_i satisfies

$$(2\epsilon + hw^+ + hw^-)u_i = h^2 f_i + (\epsilon + hw^+)u_{i-1} + (\epsilon + hw^-)u_{i+1}.$$

Since the coefficients of u_{i-1} and u_{i+1} are both strictly positive, and moreover their sum equals the coefficient of u_i on the left side, there is no way for scheme (7.20) to be unstable. This property is interesting because it holds even if ϵ is very small. An interpretation of this conclusion relies upon the following argument:

Let us add and subtract to the left side of equation (7.20) the term $(w^+ + w^-)(u_{i+1} + u_{i-1})/(2h)$. After straightforward manipulations, taking into account that $w^+ + w^- = |w| = 1$ and $w^+ - w^- = w$, we come up with a different but equivalent form of equation (7.20), namely,

$$\begin{cases} \text{Setting } u_0 = u_n = 0, \text{ find } u_i \text{ such that for } i = 1, 2, \ldots, n-1 \\ (\epsilon + h/2)(2u_i - u_{i-1} - u_{i+1})/h^2 + w(u_{i+1} - u_{i-1})/(2h) = f_i. \end{cases} \tag{7.21}$$

We observe that in fact the Upwind scheme corresponds to the initial centred scheme, if we replace the diffusion coefficient ϵ with $\epsilon_h := \epsilon + h|w|/2$. Otherwise stated, equation (7.20) adds **numerical diffusion** to the equation, sometimes called **artificial diffusion**. This is what makes the difference in terms of stability. Indeed, replacing α by $\alpha_h = wh/(4\epsilon_h)$, the counterpart of condition $|\alpha| \leq 1/2$ is seen to be $h/(4\epsilon + 2h) \leq 1/2$, which holds true for every h. The problem with numerical diffusion is that, if too much of it is added, then big accuracy losses might be observed locally, in particular close to boundary layers. As pointed out at the beginning of this section, it is difficult to simulate accurately in the pointwise sense narrow boundary layers. Hence, in advection–diffusion computations at high Péclet numbers, the practitioner could be satisfied with a good accuracy in the mean-square sense, by giving up very precise representations of the solution in certain narrow regions. In the numerical example using a $\mathcal{P}_1$ FE scheme given in Subsection 7.2.4, we will show this in more detail.

Incidentally, thanks to the fact that the coefficients of the Upwind FD scheme form a partition of unit, it satisfies a DMP of the same kind as the one that holds for the Three-point FDM applied to (P_1). Owing to this similarity, we address this issue rather briefly here, and recommend the reader to refresh her or his understanding of the DMP by reporting to Section 2.2.1 if necessary.

Assume that we are solving the problem $-\epsilon u'' + wu' = f$ in $(0, L)$ with $u(0) = a$ and $u(L) = b$ with the Upwind scheme (7.20) by modifying the homogeneous boundary conditions into $u_0 = a$ and $u_n = b$. Let u_M be the maximum of all the u_i's for $i = 0, 1, \ldots, n$, and assume that $M \neq 0$ and $M \neq n$. Now, if $f_i \leq 0$ for every i, from the property of the scheme's coefficients, it is easily seen that $u_M \leq \max[u_{M-1}, u_{M+1}]$. Therefore, necessarily $u_{M-1} = u_{M+1} = u_M$ and, like in the case of the Three-point FD scheme, we establish that all the u_i's must be equal. This contradicts the assertion that $M \neq 0$ or $M \neq n$ (it is even absurd if $a \neq b$!). It follows that $\max_{1 \leq i \leq n-1} u_i \leq \max[a, b]$ if $f_i \leq 0 \; \forall i$. This establishes the DMP for equation (7.20).

Again and again, similarly to the case of the Three-point FD scheme, we set $F := \max_{x \in [0,L]} |f(x)|$. Then, the above DMP allows us to derive a

Pointwise stability inequality for scheme (7.20)

$$\boxed{\max_{1 \leq i \leq n-1} |u_i| \leq C_{upw} F.} \tag{7.22}$$

The proof of equation (7.22) is left to the reader as Exercise 7.6. As a matter of fact, using the technique of analysis of Subsection 2.2.1, she or he is supposed to find $C_{upw} = L^2/(2\epsilon)$.

Strictly speaking, this inequality shows that there cannot be a solution sudden explosion whatever h, but everyone will certainly agree that this bound is not satisfactory if ϵ is very small.

Remark 7.5 *A first-order convergence result in the pointwise sense can be established on the basis of equation (7.22). It suffices to derive the local truncation error, assuming that a solution's third-order derivative is bounded in $[0, L]$. This proof is left to the reader as Exercise 7.7. However, as pointed out in this chapter, such a result is rather of academic use only, in case ϵ happens to be very small.* ∎

One can obtain more realistic stability inequalites for practical purposes using mean square norms instead of the maximum norm. It turns out, however, that the Upwind scheme (7.20) can be viewed as a particular case of a Petrov–Galerkin formulation of equation (7.15) combined with a piecewise linear FE discretization, for which a stability result in this sense will be derived in Subsection 7.2.4.

7.2.3   Example 7.1: Numerical Study of the Upwind FD Scheme

In this example, we illustrate the stabilising effect of upwinding. We attempt to approximate the solution of equation (7.16) for $g \equiv 1$ taking the plus sign and $\epsilon = 10^{-4}$. Both the exact solution and the numerical results are plotted in Figures 7.5a and 7.5b at the grid points using two different FDM, namely, the Centred FD scheme (7.19) and scheme (7.20). We computed with a uniform grid containing 101 grid points. As the reader can observe, the centred scheme is blatantly unstable, while the Upwind scheme reproduces the exact solution almost to machine precision at all grid points. Actually, such a high accuracy holds true, except in the region very close to the abscissa $x = 1$, where the FD solution drops to zero within an interval of length $h = 0.01$, instead of the much smaller $O(\epsilon)$ width for the exact solution. This observation just expresses the fact that a very narrow boundary layer can barely be reproduced by a single-scale

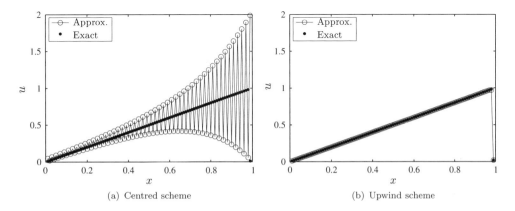

(a) Centred scheme   (b) Upwind scheme

Figure 7.5   FD solution and exact solution for $\epsilon = 10^{-4}$ and $h = 0.01$

numerical method[2]. Notice, however, that the main focus of this example is the stability issue, in which respect Figure 7.5 provides clear illustrations.

7.2.4 The SUPG Formulation

The SUPG is a technique designed to be used in the FE context. The acronym stands for **streamline upwind Petrov–Galerkin**, which corresponds to a formulation of the advection–diffusion equation aimed at obtaining stable and possibly accurate solutions even in high Péclet number simulations. The creators of this technique are acknowledged as Brooks and Hughes in a paper dating back to 1982 [38].

Originally, the Petrov–Galerkin method was used to obtain approximate solutions of PDEs whose highest order term is of odd order. In the variational formulation of this kind of problem, the space in which the solution is searched for (i.e. the space of **trial functions**) and the space of **test functions** cannot be the same. Clearly enough, this restriction needs not to apply to a corresponding FE approximate problem, but in any case the underlying discrete problem will not be symmetric. For example, this would be the case of a basic transport equation – that is, a pure **advection equation**, namely, $ou' = f$ in $(0, L)$, with the boundary condition $u(0) = 0$, where o is a non-vanishing function. If f is given in $L^2(0, L)$, then we have $u \in H^1(0, L)$ and the natural variational formulation of this equation would be

$$\text{Find } u \in U \text{ such that } \int_0^L ou'v \, dx = \int_0^L fv \, dx \; \forall v \in V,$$

where $U := \{u|\, u \in H^1(0, L), u(0) = 0\}$ and $V = L^2(0, L)$. This is a Petrov–Galerkin formulation. Notice that even in the case of the pure advection equation, we can obtain a symmetric formulation by 'testing' both sides of it with the function ov' instead of v, where v now belongs to U. This gives rise to a symmetrised Petrov–Galerkin formulation, that is,

$$\text{Find } u \in U \text{ such that } \int_0^L o^2 u'v' \, dx = \int_0^L ofv' \, dx \; \forall v \in U.$$

It is very easy to check that both formulations are equivalent to the original equation. Another possibility is a weighting of both formulations, leading to the asymmetric formulation

$$\text{Find } u \in U \text{ such that } \int_0^L [ou'v + wo^2 u'v'] dx = \int_0^L f(v + wov') dx \; \forall v \in U.$$

where the weight w is a strictly positive real number. Its role is to balance the importance of the symmetric and asymmetric Petrov–Galerkin formulations, but we note that still in this case the trial and test function spaces remain the same. This weighted formulation lies on the basis of the SUPG technique. Incidentally, we note that nothing prevents the weight w from varying with x.

Strictly speaking, the Petrov–Galerkin formulation is not well-suited to the advection–diffusion equation (7.15) since it is a second-order boundary value problem. However, if convection strongly dominates diffusion, to a large extent u behaves like

---

[2] A **multiscale** approach could yield better results, using for instance **subscales** (see e.g. [49]). These techniques are playing an increasingly important role in modern numerical simulation of certain phenomena, but their presentation has to be left to higher level studies.

the solution of a pure advection equation, as we saw in the introductory part of this section. That is where the SUPG formulation comes into play. However, in contrast to the model one-dimensional pure advection equation, the SUPG formulation is designed to act directly on the FE approximate problem. This is because the weighting function w will decrease with the mesh sizes, as seen below. This means that in some sense in the limit as the mesh size goes to zero, only the original asymmetric formulation will remain. Let us see how this works.

We are given a uniform mesh $\mathcal{T}_h$ with mesh size $h = L/n$ for a given integer $n > 2$, and a $\mathcal{P}_1$ FE discretisation of equation (7.15). The corresponding FE space is

$$V_h := \{v|\ v \in C^0[0, L], v_{|T} \in \mathcal{P}_1(T)\ \forall T \in \mathcal{T}_h, v(0) = v(L) = 0\}.$$

The reader should pay attention to the fact that, here, V_h is a subspace of the space denoted in the same manner in Chapters 1, 2 and 3.

Now, we choose the weight w to be Ch for a suitable constant $C > 0$. Then, we set the

SUPG formulation of equation (7.18)

Find $u_h \in V_h$ such that $\forall v \in V_h$
$$\int_0^L [\epsilon u_h' v' + w u_h' v]dx + Ch \int_0^L w^2 u_h' v' dx = \int_0^L fv\,dx + Ch \int_0^L wfv' dx. \tag{7.23}$$

Using the $\mathcal{P}_1$ shape functions φ_i and writing as usual $u_h = \sum_{j=1}^{n-1} u_j\varphi_j$, and then setting $v = \varphi_i$ for $i = 1, \ldots, n-1$, the reader can check as Exercise 7.8 that in case $w = \pm 1$, equation (7.23) is equivalent to the following:

SUPG $\mathcal{P}_1$ FE scheme to solve equation (7.18)

Find u_i satisfying for $i = 1, \ldots, n-1$,
$$(\epsilon + Ch)\frac{2u_i - u_{i-1} - u_{i+1}}{h} + w\frac{u_{i+1} - u_{i-1}}{2} =$$
$$\int_{(i-1)h}^{ih} f(x)\left[\frac{x - ih + h}{h} + Cw\right]dx + \int_{ih}^{(i+1)h} f(x)\left[\frac{x - ih - h}{h} - Cw\right]dx. \tag{7.24}$$

Now, assuming that f is continuous, we set $f_i = f(ih)$. Then, if we use the trapezoidal rule to compute the integrals on the right side of equation (7.24), we obtain

$$(\epsilon + Ch)\frac{2u_i - u_{i-1} - u_{i+1}}{h} + w\frac{u_{i+1} - u_{i-1}}{2} = hf_i + hCw\frac{f_{i-1} - f_{i+1}}{2}.$$

Then, we note that if $C = 1/2$ and we divide both sides of it by h, except for the second term on the right side, scheme (7.24) reproduces the Upwind FD scheme (7.20). Notice that, as long as f is continuously differentiable, the factor $(f_{i-1} - f_{i+1})/2$ is an $O(h)$, and therefore, for not so large C, the SUPG term $Cw[f_{i-1} - f_{i+1}]/2$ on the right side represents just a small perturbation of f_i. We conclude that the SUPG technique is a sort of FE counterpart of the Upwind FD scheme (7.20). Of course, in the case of equation (7.23) we go much further, since this scheme can be applied to more general problems by taking different velocities w. Actually, similarly to Chapter 1, if suitable numerical quadrature formulae are employed to approximate integrals involving this function, we can get upwind FD analogs of equation (7.15), from the SUPG $\mathcal{P}_1$ formulation for variable w. We do not further elaborate on this issue for it has been developed in detail in Section 1.4.

Anyhow, owing to this equivalence, the SUPG FE scheme enjoys the same stability properties as its FD counterpart with $w = \pm 1$. We do not insist very much on this property either, as
far as the maximum norm is concerned. On the other hand, it is possible to obtain an exploitable
stability inequality in the mean-square sense. Indeed, taking $v = u_h$ in equation (7.23), we get

$$\int_0^L (\epsilon + Ch) u_h'{}^{|2} dx + \int_0^L w u_h' u_h dx = \int_0^L f u_h \, dx + Ch \int_0^L f w u_h' dx. \qquad (7.25)$$

Now using integration by parts, since $u_h(0) = u_h(L) = 0$ and w is constant, we note that
$\int_0^L w u_h' u_h dx = - \int_0^L w u_h' u_h dx = 0$. Hence, using the Cauchy–Schwarz inequality together
with the Friedrichs–Poincaré inequality, after some trivial manipulations on the right side of
equation (7.25) that we suggest the reader to carry out as Exercise 7.9, we obtain the

Stability inequalities for scheme (7.23) in the mean-square sense

$$\boxed{\begin{aligned} (\epsilon + Ch) \, \| \, u_h' \|_{0,2} &\leq \| \, f \, \|_{0,2}(C_P + hC) \\ (\epsilon + Ch) \, \| \, u_h \|_{0,2} &\leq \| \, f \, \|_{0,2} C_P (C_P + hC) \end{aligned}} \qquad (7.26)$$

Notice that we can distinguish two important cases, namely, $\epsilon << Ch$ and $h \leq \epsilon$, in order to
make sure that the solution norms remain under control. The stability inequality (7.26) could
sound strange, for it is ineffective if $h \leq \epsilon$ and ϵ is small. However, this situation is utopian
in practice, while in contrast, computations with $h >> \epsilon/C$ turn out to be quite realistic.

Now, as far as consistency is concerned, it is worth examining a little closer the situation of
the SUPG scheme, for after all it is based on a variational approach. We do this for an arbitrary
w with $\| \, w \|_{0,\infty} = 1$, assuming that $u'' \in L^2(\Omega)$ and that the integrals are computed exactly,
for simplicity.

Recalling Subsection 2.3.2, we replace u_h by the $\mathcal{P}_1$ interpolate $\tilde{u}_h$ of u defined in Chapter
2, to obtain two residual functions $R_h(u)$ and $S_h(u)$, satisfying $\forall v \in V_h$,

$$\epsilon(\tilde{u}_h'|v')_0 + (w\tilde{u}_h'|v)_0 + Ch(w^2\tilde{u}_h'|v')_0 - (f|v + Chwv')_0 = (R_h(u)|v)_0 + (S_h(u)|v')_0. \qquad (7.27)$$

Using the fact that u is the solution of equation (7.15), these residual functions are found to be
$R_h(u) = w[\tilde{u}_h - u]'$ and $S_h(u) = \epsilon Chwu'' + (\epsilon + Chw^2)[\tilde{u}_h - u]'$.

Since $\| \, w \|_{0,\infty}$ is bounded, estimates in terms of an $O(h)$ can be easily derived for both
residual functions in the L^2-norm, respectively. This, combined with equation (7.26), yields
first-order error estimates for $\| \, u - u_h \|_{1,2}$. The calculations leading to this result from
equation (7.27) on should be carried out in detail as Exercise 7.10, until the final estimate is
specified. Notice that, once again, the error estimates to be found are rather academic if ϵ is
very small.

In complement to the material presented in this subsection and in the preceding one, we
should add a few words about artificial diffusion.

We saw that the Upwind FD scheme adds artificial diffusion to the equation. Due to the fact
that this method is practically identical to the SUPG FE scheme for $C = 1/2$, at least in this
case the latter adds the same amount of artificial diffusion. Then some questions immediately
arise: What happens for different values of C? Is it possible to reduce artificial diffusion to an
optimum without penalising stability? The answer to the first question can be given as follows.
A value of C in the interval $[0, 1/2)$ corresponds to a weighted (convex) combination of the
Upwind FD scheme and the Centred FD scheme, in the sense that the more we approach zero,

the less artificial diffusion is added, but at the price of downgrading stability. The answer to the second question is yes, but it requires a more careful analysis. Actually, optimality is attained if the constant C in the added Petrov–Galerkin terms is replaced by a certain function $C_{opt}(h)$ (see e.g. [78]). More specifically, C_{opt} is the

Optimal coefficient of the SUPG added terms

$$C_{opt} = \frac{1}{2}\left[coth(P_h) - \frac{1}{P_h}\right] \quad \text{with } P_h = \frac{h}{2\epsilon}. \tag{7.28}$$

We remind the reader that the function $coth(x)$ is the hyperbolic cotangent given by $\dfrac{e^{2x}+1}{e^{2x}-1}$. The coefficient P_h in turn is called the **element Péclet number**.

As shown in reference [78], if C_{opt} replaces C in equation (7.23), then the nodal values supposed to approximate the test solution of Figure 7.4 are exact. This is the usual argument for concluding the optimality of this choice.

Following similar principles, we finally note that the best results for the FDM are obtained with a suitable convex combination of the centred and the upwind approximations of the term wu'. This means tuning added artificial diffusion to the optimal level.

Remark 7.6 *The SUPG formulation is of the* **Galerkin least-squares** *type, in the sense that it is a linear combination of the standard Galerkin formulation and the* **least-squares formulation** *of the advection–diffusion equation. The acronym* **GLS** *is commonly used when referring to it. A pure least-squares formulation of the latter is a symmetric formulation, which would write as follows for a suitable space* V_h:

$$\sum_{i=1}^{n}\int_{x_{i-1}}^{x_i}[\epsilon u_h'' + wu_h'][\epsilon v'' + wv']dx = \sum_{i=1}^{n}\int_{x_{i-1}}^{x_i}f[\epsilon v'' + wv']dx \;\; \forall v \in V_h.$$

Notice that in the above formulation, the equation is tested with an expression of the same type in terms of the test functions v. *For more explanations about least-squares FEM of PDEs, the author refers the reader to reference [30].* ■

7.2.5   *Example 7.2: Numerics of the SUPG Formulation for the* $\mathcal{P}_1$ *FEM*

The aim of this example is to show the performance of the SUPG formulation for different values of the constant C. Two test problems were solved with both a uniform and a non-uniform mesh of $(0,1)$ consisting of 1024 intervals. For the uniform mesh, $h \simeq 0.00098$ and $C_{opt} = 0.489760$. The elements in the non-uniform mesh are sequentially numbered from the leftmost to the rightmost. Their lengths decrease from left to right by groups of 256 or 128 intervals with equal lengths. The groups together with the corresponding interval lengths are specified in Table 7.2. In the same table, we supply the value of C_{opt} for each interval length.

Test case 1: We first tested the SUPG method in the solution of equation (7.15) under the same conditions as in Example 7.1. We recall that in this case the exact solution is given by equation (7.17), and that the SUPG method reproduces the exact solution at the nodes,

Table 7.2   Values of C_{opt} for different sizes of the variable FE mesh.

Element number →	1–256	257–384	385–512	513–640	641–768	769–896	897–1024
Mesh size ≃ →	0.00213	0.00142	0.00071	0.00057	0.00043	0.00028	0.00014
C_{opt} →	0.495307	0.492960	0.485920	0.482400	0.476533	0.464800	0.429601

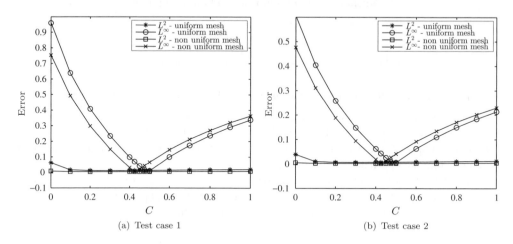

(a) Test case 1   (b) Test case 2

Figure 7.6   Absolute errors in L^2 and L^∞ for $\epsilon = 10^{-5}$ and $n = 1024$ in terms of C

in case the constant C equals C_{opt} for all the intervals. In Figure 7.6a, we display the error measured in both the L^2-norm and the discrete L^∞-norm in terms of the constant C taken in the computations for all the intervals. As one can see, the L^2 errors are very small for every C. As for the L^∞-errors, a minimum is attained for $C = C_{opt}$ in the case of the uniform mesh, while in the case of the non-uniform mesh the best results are obtained with C close to the value of C_{opt} corresponding to the smallest mesh size.

Test case 2: We also tested the SUPG method in the solution of equation (7.15) for $f(x) = -e^{-x}$ and $w \equiv 1 - \epsilon$. In this case, the exact solution is given by

$$u(x) = e^{-x} + \frac{e^{-\frac{1}{\epsilon}} - 1 + (e-1)e^{\frac{(1-\epsilon)x-1}{\epsilon}}}{1 - e^{1-\frac{1}{\epsilon}}}.$$

Here again, we computed with constant values of C for all mesh intervals. Figure 7.6b reflects the very same behaviour as in test case 1, even though the numerical values are no longer exact at the mesh nodes.

Summarizing, we emphasised in this example the importance of doing simulations of advection–diffusion phenomena using the SUPG technique with the formulation's constant C close to C_{opt} everywhere in the domain.

7.2.6   An Upwind FV Scheme

Following the ideas presented in Subsection 7.2.2, we next study a (Cell-centred) FVM to solve equation (7.18). In this aim, we use a non-uniform mesh and pertaining notations for this kind of method, introduced in Subsection 1.4.2. For the sake of simplicity, instead of using the values w^+ and w^-, we assume that $w = 1$. Of course, the case $w = -1$ can be treated in a similar manner, by taking corresponding approximations of u' in the opposite upwind direction. In this way, we study in this subsection as a simplified model, an Upwind scheme very close to the one considered in reference [68], but not quite the same.

Setting $u_{-1/2} = 0$, $u_{n+1/2} = 0$ and $f_{j-1/2} := \int_{x_{j-1}}^{x_j} f(x)\,dx/h_j$, we set the

Upwind FV scheme for equation (7.18)

Find $u_{j-1/2}$ such that for $j = 1, 2, \ldots, n$,

$$\epsilon\left[\frac{u_{j-1/2} - u_{j-3/2}}{h_{j-1/2}} - \frac{u_{j+1/2} - u_{j-1/2}}{h_{j+1/2}}\right] + u_{j-1/2} - u_{j-3/2} = h_j f_{j-1/2} \qquad (7.29)$$

As everyone now certainly knows, the first step in the reliability analysis of a numerical scheme is to establish a stability inequality for it. Like in the case of Cell-centred FV schemes proposed in the preceding chapters, the above FV scheme is not consistent in the usual FD sense. More precisely, except for special meshes, the local truncation errors are an $O(1)$ in the case of equation (7.29), as the reader can check (cf. Exercise 7.11). Hence, we shall resort to the concept of stability in the mean-square sense, by applying the scheme to a problem more general than equation (7.29), namely,

$$\begin{cases} \text{Find } u_{j-1/2} \text{ such that, for } j = 1, 2, \ldots, n, \\ \epsilon\left[\frac{u_{j-1/2} - u_{j-3/2}}{h_{j-1/2}} - \frac{u_{j+1/2} - u_{j-1/2}}{h_{j+1/2}}\right] + u_{j-1/2} - u_{j-3/2} \\ = h_j f_{j-1/2} + \mathcal{F}^+_{j-1/2} + \mathcal{F}^-_{j-1/2} \end{cases} \qquad (7.30)$$

where $\mathcal{F}^+_{j-1/2}$ and $\mathcal{F}^-_{j-1/2}$ stand for given incoming and outgoing fluxes with respect to CV $T_j = (x_{j-1}, x_j)$. These fluxes are required to be bounded independently of the mesh and conservative in the sense that $\mathcal{F}^+_{j-1/2} = -\mathcal{F}^-_{j+1/2}$ for $j = 1, 2, \ldots, n-1$, and are completed by $\mathcal{F}^+_{-1/2} := -\mathcal{F}^-_{1/2}$ and $\mathcal{F}^-_{n+1/2} := -\mathcal{F}^+_{n-1/2}$.

Our goal is to derive a stability inequality in the mean-square sense. We should stress beforehand that this is not only because the discretisation is not uniform. The reason is inherent to FV schemes, for in general they are not consistent in the FD sense.

Here again, we resort to a functional framework close to the one of Part B of Subsection 5.1.5. In this aim, we need some definitions and preliminary results.

First of all, we define a piecewise constant function u_h by

$$u_h = \sum_{j=1}^{n} u_{j-1/2}\chi_{T_j},$$

where χ_{T_j} is the **characteristic function** of T_j. Next, we define a discrete derivative u'^h of u_h, namely, a constant function in every interval $(x_{j-3/2}, x_{j-1/2}]$ with value equal to $(u_{j-1/2} - u_{j-3/2})/h_{j-1/2}$, for $j = 1, 2, \ldots, n+1$. We recall that the L^2-norm of a function v_h whose value is v_j in every element of a partition of $(0, L)$ into m disjoint intervals I_j with length k_j for $j = 1, \ldots, m$ is given by $\|v_h\|_{0,2} = \left[\sum_{j=1}^{m} k_j v_j^2\right]^{1/2}$. Taking down this expression, the reader is invited to establish as Exercise 7.12, the validity of the following discrete Friedrichs–Poincaré inequality:

$$\|u_h\|_{0,2} \le L \, \|u_h'^h\|_{0,2}. \tag{7.31}$$

Now, let us multiply both sides of equation (7.30) with $u_{j-1/2}$ and add up the resulting relations from $j = 1$ through $j = n$. Let $\mathcal{F}_j$ be the outgoing flux $\mathcal{F}_{j-1/2}^+$ for $j = 1, \ldots, n-1$, $\mathcal{F}_0 := \mathcal{F}_{1/2}^-$ and $\mathcal{F}_n := \mathcal{F}_{n-1/2}^+$. Rearranging the summations, similarly to Part B of Subsection 5.1.5, after some straightforward manipulations to be checked by the reader, we obtain

$$\left\{ \begin{aligned} &\sum_{j=2}^{n} \left[\epsilon \frac{(u_{j-1/2} - u_{j-3/2})^2}{h_{j-1/2}} + (u_{j-1/2}^2 - u_{j-1/2}u_{j-3/2}) \right] + \frac{u_{1/2}^2(\epsilon + h_{1/2})}{h_{1/2}} + \\ &\frac{\epsilon u_{n-1/2}^2}{h_{n+1/2}} = \sum_{j=1}^{n} h_j f_{j-1/2} u_{j-1/2} - \sum_{j=1}^{n-1} \mathcal{F}_j(u_{j+1/2} - u_{j-1/2}) + \mathcal{F}_0 u_{1/2} + \mathcal{F}_n u_{n-1/2}. \end{aligned} \right. \tag{7.32}$$

Next, we apply the same trick as in Part B of Subsection 5.1.5. to the flux terms on the right side of equation (7.32). More precisely, we apply Young's inequality to the term $-u_{j-3/2}u_{j-1/2}$ on the left side with $\delta = 1$. After straightforward calculations,

$$\left\{ \begin{aligned} &\epsilon \sum_{j=1}^{n+1} \frac{(u_{j-1/2} - u_{j-3/2})^2}{h_{j-1/2}} + \frac{u_{n-1/2}^2 + u_{1/2}^2}{2} \\ &\le \sum_{j=1}^{n} h_j f_{j-1/2} u_{j-1/2} + \mathcal{F}_{\max} \sum_{j=0}^{n} \sqrt{h_{j+1/2}} \frac{|u_{j+1/2} - u_{j-1/2}|}{\sqrt{h_{j+1/2}}}. \end{aligned} \right. \tag{7.33}$$

where $\mathcal{F}_{\max} := \max_{0 \le j \le n} |\mathcal{F}_j|$. Notice that $\sum_{j=0}^{n} h_{j+1/2} = L$. Thus, setting $f_h := \sum_{j=1}^{n} f_{j-1/2}\chi_{T_j}$ and using the definition of u_h together with the Cauchy–Schwarz inequality on the right side of equation (7.33), we further obtain

$$\left\{ \begin{aligned} &\epsilon \sum_{j=1}^{n+1} h_{j-1/2} \frac{(u_{j-1/2} - u_{j-3/2})^2}{h_{j-1/2}^2} \le \|f_h\|_{0,2} \, \|u_h\|_{0,2} \\ &+ \sqrt{L}\mathcal{F}_{\max} \left[\sum_{j=1}^{n+1} h_{j-1/2} \frac{(u_{j-1/2} - u_{j-3/2})^2}{h_{j-1/2}^2} \right]^{1/2}. \end{aligned} \right. \tag{7.34}$$

Finally, recalling the expression of $u_h'^h$ and using equation (7.31), equation (7.34) yields

$$\epsilon \, \|u_h'^h\|_{0,2}^2 \le \|f_h\|_{0,2} \, \|u_h\|_{0,2} + \sqrt{L}\mathcal{F}_{\max} \|u_h'^h\|_{0,2}. \tag{7.35}$$

and therefore two

Mean-square stability inequalities for the Upwind FV scheme

$$\boxed{\begin{aligned} \|u_h'^h\|_{0,2} &\le [L \, \|f_h\|_{0,2} + \sqrt{L}\mathcal{F}_{\max}]/\epsilon, \\ \|u_h\|_{0,2} &\le [L^2 \, \|f_h\|_{0,2} + \sqrt{L^3}\mathcal{F}_{\max}]/\epsilon. \end{aligned}} \tag{7.36}$$

Next, we examine the consistency of scheme (7.29). Similarly to Part B of Subsection 5.1.5, all we have to do is to establish the consistency of the fluxes across the boundaries of the CVs, when $u_{j-1/2}$ is replaced in the scheme by $u(x_{j-1/2})$ for pertaining values of j. For CV $T_j, j = 1, 2, \ldots, n$, the exact and resulting approximate fluxes are, respectively,

- Incoming flux: $\tilde{F}_{j-1/2}^- = \epsilon u'(x_{j-1}) - u(x_{j-1}); \; F_{j-1/2}^- = \epsilon \frac{u(x_{j-1/2}) - u(x_{j-3/2})}{h_{j-1/2}} - u(x_{j-3/2})$,
- Outgoing flux: $\tilde{F}_{j-1/2}^+ = -\epsilon u'(x_j) + u(x_j); \; F_{j-1/2}^+ = \epsilon \frac{u(x_{j-1/2}) - u(x_{j+1/2})}{h_{j+1/2}} + u(x_{j-1/2})$,

completed by $\tilde{F}_{-1/2}^+ = -\tilde{F}_{1/2}^-; F_{-1/2}^+ = -F_{1/2}^-$, and $\tilde{F}_{n+1/2}^- = -\tilde{F}_{n-1/2}^+; F_{n+1/2}^- = -F_{n-1/2}^+$. It is clear that the property $F_{j-1/2}^+ = -F_{j+1/2}^-$ (resp. $\tilde{F}_{j-1/2}^+ = -\tilde{F}_{j+1/2}^-$) holds for $j = 1, \ldots, n-1$.

Assuming that u is twice continuously differentiable, let us evaluate the numerical flux residuals $R_{j-1/2}^+(u) := \tilde{F}_{j-1/2}^+ - F_{j-1/2}^+ = -R_{j+1/2}^-(u)$ at the end-points x_j for $n > j > 0$, that is, from the right and from the left of x_j, respectively. By elementary Taylor expansions, we have

$$u(x_{j-1/2}) = u(x_j) - h_j u'(x_j)/2 + h_j^2 u''(\xi_j^-)/8$$
$$u(x_{j+1/2}) = u(x_j) + h_{j+1} u'(x_j)/2 + h_{j+1}^2 u''(\xi_j^+)/8,$$

where $x_{j-1/2} \le \xi_j^- \le x_j$ and $x_j \le \xi_j^+ \le x_{j+1/2}$. Recalling that $h_{j+1/2} = (h_j + h_{j+1})/2$, after simple manipulations, it follows that

$$R_{j-1/2}^+(u) = \epsilon [h_{j+1}^2 u''(\xi_j^+) - h_j^2 u''(\xi_j^-)]/(8h_{j+1/2}) + h_j u'(x_j)/2 - h_j^2 u''(\xi_j^-)/8. \quad (7.37)$$

These relations are completed with the residuals at $x = 0$ and $x = L$, namely, $R_{-1/2}^+(u) = \epsilon[-u'(0) + u(x_{1/2})/h_{1/2}]$ and $R_{n-1/2}^+(u) = \epsilon[-u'(L) - u(x_{n-1/2})/h_{n+1/2}] - u(x_{n-1/2})$, which yield for $0 \le \xi_0^+ \le h_{1/2}$ and $L - h_{n-1/2} \le \xi_n^- \le L$,

$$\begin{cases} R_{-1/2}^+(u) = \epsilon h_{1/2} u''(\xi_0^+)/2, \text{ and} \\ R_{n-1/2}^+(u) = -\epsilon h_{n+1/2} u''(\xi_n^-)/2 + h_{n+1/2} u'(L) - h_{n+1/2}^2 u''(\xi_n^-)/2. \end{cases} \quad (7.38)$$

Now we introduce the function $\bar{u}_h := \sum_{j=1}^{n} \bar{u}_{j-1/2} \chi_{T_j}$, with $\bar{u}_{j-1/2} := u(x_{j-1/2}) - u_{j-1/2}$. By straightforward calculations, $\bar{u}_h$ is seen to solve equation (7.30) when we set $f_{j-1/2} = 0$ for all j, and take $\mathcal{F}_{j-1/2}^- = R_{j-1/2}^+$ and $\mathcal{F}_{j-1/2}^- = R_{j-1/2}^-$. Noticing that $h_{j+1/2} \ge \max[h_j, h_{j+1}]/2$ for $j = 1, \ldots, n-1$, $h_{1/2} = h_1/2$ and $h_{n+1/2} = h_n/2$ and setting $h := \max_{1 \le j \le n} h_j$, on the basis of equations (7.37) and (7.38), we can assert that the corresponding value of $\mathcal{F}_{\max}$ is bounded above by $[\epsilon h/2 + h^2/8] \|u''\|_{0,\infty} + h \|u'\|_{0,\infty}/2$ (please check!). Recalling equation (7.36), the function $\bar{u}_h$ is readily seen to satisfy

$$\begin{cases} \|\bar{u}_h'\|_{0,2} \le \sqrt{L} \left[\left(\frac{h}{2} + \frac{h^2}{8\epsilon} \right) \|u''\|_{0,\infty} + \frac{h}{2\epsilon} \|u'\|_{0,\infty} \right], \\ \|\bar{u}_h\|_{0,2} \le \sqrt{L^3} \left[\left(\frac{h}{2} + \frac{h^2}{8\epsilon} \right) \|u''\|_{0,\infty} + \frac{h}{2\epsilon} \|u'\|_{0,\infty} \right]. \end{cases} \quad (7.39)$$

Now, we further define $\tilde{u}_h := \sum_{j=1}^{n} u(x_{j-1/2})\chi_{T_j}$. Using elementary interpolation theory, the reader should not find it difficult to prove as Exercise 7.13 that

$$\|\tilde{u}_h - u\|_{0,2} \leq \sqrt{L}h \, \|u'\|_{0,\infty}/2. \tag{7.40}$$

Finally, combining equation (7.39) and (7.40), by the triangle inequality we derive the

Mean-square error estimate for the Upwind FV scheme

$$\|u_h - u\|_{0,2} \leq \sqrt{L^3} \left[\left(\frac{h}{2} + \frac{h^2}{8\epsilon} \right) \|u''\|_{0,\infty} + \frac{(L+\epsilon)h}{2\epsilon L} \|u'\|_{0,\infty} \right]. \tag{7.41}$$

As a conclusion, the Upwind FV scheme is first-order convergent in the L^2-norm for a non-uniform mesh. Notice that, here again, this is a qualitative result of little quantitative use, in case ϵ is very small.

7.2.7 A FE Scheme for the Time-dependent Problem

In the last part of this section, we deal with a FE scheme to solve the time-dependent advection–diffusion equation that can be easily applied to the case of multiple space variables. Like in the previous subsections, however, we will focus on the one-dimensional case for the sake of conciseness. Before starting, let us make a brief review of different strategies to treat advection in a variational framework.

One of the first FE techniques employed to model advection is the Lesaint–Raviart method (cf. [124]). As for convection–diffusion, the earliest known contribution is due to Heinrich et al. [97]. A little later, the Japanese school gave relevant contributions to the subject, as it is well reported in [103]. In this respect, the pioneering work of Tabata (cf. [187]) together with [115] among others should be quoted. Since the mid-1980s a widespread manner to deal with dominant advection has been the use of stabilizing procedures based on the space mesh parameter, among which the SUPG technique described in Subsection 7.2.3 (see also [38]) appears to be the most popular. Among many others, an interesting approach based on this technique was proposed in reference [82]. An interesting reference on the FEM to deal with advection-diffusion problems and related topics is the book of Knabner and Angermann [116].

As far as time-dependent problems are concerned, it turns out that the time step plays a better stabilizing role, provided a formulation well suited to the equations to be solved is employed. A good illustration of this assertion in the case of the time-dependent Navier–Stokes equations can be found in reference [50]. The author himself and co-workers gave a contribution in this direction, in the case of the advection–diffusion equations with dominant advection, discretised in space with piecewise linear FEs using a classical Galerkin approach, combined with a non-standard Forward Euler scheme for the time integration. Actually, this subsection is devoted to this method, which follows similar principles to the one long exploited by Kawahara et al. for simulating convection dominated phenomena (cf. [114], among several other earlier or later papers by this author and co-workers). Actually, an introduction to this technique was supplied in Section 3.2.4 restricted to the one-dimensional transport equation, which is a pure advection equation.

An outline of this subsection is as follows. We first specify the problem to solve and make some assumptions on the data. Next, we describe the type of discretisation corresponding to the FE scheme, and especially the weighted manner to deal with the **mass matrices**[3] on both sides of the discrete equations. In the sequel, we give stability results for this scheme in the sense of the space and time maximum norm, and quote the error estimates that hold for this scheme, derived in reference [172]. Finally, we illustrate the scheme's performance by means of some numerical tests in one space dimension.

The time-dependent advection–diffusion equation consists of finding a scalar valued function $u(\boldsymbol{x}, t)$ defined in $\bar{\Omega} \times [0, \Theta]$, Ω being a bounded open subset of $\mathfrak{R}^N$ with boundary Γ, for $N = 1, 2$ or 3, such that

$$\begin{cases} \partial_t u + (\vec{v} \mid \mathbf{grad}\ u) - p\Delta u = f & \text{in } \Omega \times (0, \Theta] \\ u = g \text{ on } \Gamma \times (0, \Theta] \\ u = u_0 \text{ in } \Omega \text{ for } t = 0. \end{cases} \tag{7.42}$$

where p is a (positive) diffusivity constant and $\vec{v}$ is a given (advective) convective velocity at every time t, assumed to be uniformly bounded in $\bar{\Omega} \times [0, \Theta]$. The data f and g are, respectively, a given forcing function belonging to $L^\infty[\Omega \times (0, \Theta)]$, and a prescribed boundary value $\forall t$.

Henceforth, we confine ourselves to the case where $\Omega = (0, L)$, in which g reduces to a pair of functions $[a(t); b(t)]$ such that $u(0, t) = a(t)$ and $u(L, t) = b(t)\ \forall t$. We further assume that $u_0 \in C[0, L]$ and that $a(t)$ and $b(t)$ are bounded in $[0, \Theta]$. Moreover, we consider only normalised dimensionless lengths, time and velocity $w = v/\mathrm{V}$ where $\mathrm{V} := \max_{x \in [0,L];\ t \in [0,\Theta]} |v(x, t)|$. In doing so, p is replaced by ϵ in equation (7.42) and v is replaced by w. Then like before, the inverse of ϵ is the Péclet number $P := \mathrm{V}L/p$. Finally, without loss of essential aspects, we assume that $L = 1$.

Taking all these simplifications into account, equation (7.42) reduces to

$$\begin{cases} \partial_t u + w \partial_x u - \epsilon \partial_{xx} u = f & \text{in } (0, 1) \times (0, \Theta] \\ u(0, t) = a(t) \text{ and } u(1, t) = b(t)\ \forall t \in (0, \Theta] \\ u(x, 0) = u_0(x)\ \forall x \in (0, 1). \end{cases} \tag{7.43}$$

We will work with the equivalent standard Galerkin formulation of equation (7.43), namely,

$$\begin{cases} \text{For every } t \in (0, \Theta], \text{ find } u(\cdot, t) \in H^1(0, 1) \text{ with } \partial_t u(\cdot, t) \in L^2(0, 1), \\ u(0, t) = a(t), u(1, t) = b(t)\ \forall t \in (0, \Theta] \text{ and } u(x, 0) = u_0(x)\ \forall x \in (0, 1) \\ \text{such that, } \int_0^1 [\partial_t u + w \partial_x u] v\ dx + \epsilon \int_0^1 \partial_x u v'\ dx = \int_0^1 f v\ dx\ \forall v \in H_0^1(0, 1). \end{cases}$$

Next we consider a partition $\mathcal{T}_h$ of $(0, 1)$ into intervals T_j with maximum edge length equal to h. We also need a second mesh parameter ρ, namely, the minimum length of all the elements in $\mathcal{T}_h$. We assume that $\mathcal{T}_h$ belongs to a quasi-uniform family of meshes, which means that there exists a constant $\tilde{c} > 0$ independent of h such that $h/\rho \leq \tilde{c}$ for all meshes in such a family.

---

[3] The expression mass matrix is commonly used in the engineering community. It stems from the fact that this matrix applies to discretised time derivatives, which multiplied by a mass density accounts for the inertia of the medium under study. In fluid mechanics, this derivative is the first-order time derivative, while in solid mechanics it is usually a second-order time derivative. On the other hand, in solid mechanics, the matrix related to body's deformation, that is, to terms involving second- or fourth-order spatial partial derivatives depending on the model, is called the **stiffness matrix**.

Recalling the notation $\mathcal{P}_1(T)$ introduced in Section 1.3, for $T \in \mathcal{T}_h$, we also revisit the following spaces associated with $\mathcal{T}_h$:

$$W_h := \{v \mid v \in C^0[0,1] \text{ and } v|_T \in \mathcal{P}_1(T), \forall T \in \mathcal{T}_h\},$$
$$V_h := W_h \cap H_0^1(0,1).$$

We further introduce for any pair $g(t) := (a(t); b(t))$ of real functions a and b bounded in $[0, \Theta]$, the following manifold of W_h:

$$V_h^g(t) := \{v \in V_h \mid v(0) = a(t), v(1) = b(t)\}.$$

Now, let u_h^0 be the field of $V_h^g(0)$ that interpolates u_0 at the mesh nodes, and $\tau > 0$ be a time step given by $\tau = \Theta/l$ where l is a non-negative integer. Defining $g^k := [a(k\tau); b(k\tau)]$, together with f^k by $f^k(\cdot) = f(\cdot, k\tau)$ and $w^k(\cdot) := w(\cdot, k\tau)$ in $(0,1)$, for $k = 1, 2, \ldots, l$, ideally we wish to determine approximations $u_h^k(\cdot)$ of $u(\cdot, k\tau)$ for $k = 1, 2, \ldots$, by solving the following FE discrete set of equations, corresponding to the Forward Euler scheme:

$$\begin{cases} \text{For } k = 1, 2, \ldots, l, \text{ find } u_h^k \in V_h^g(k\tau) \text{ satisfying } \forall v \in V_h, \\ \int_0^1 u_h^k v \, dx = \int_0^1 u_h^{k-1} v \, dx \\ + \tau \left[\int_0^1 f^{k-1} v \, dx - \int_0^1 w^k [u_h^{k-1}]' v \, dx - \epsilon \int_0^1 [u_h^{k-1}]' v' \, dx \right]. \end{cases} \quad (7.44)$$

Now, we expand u_h^k into a sum of the form, $u_h^k = \sum_{j=0}^n u_j^k \varphi_j$, where φ_j is the shape function of V_h associated with the jth node of $\mathcal{T}_h$, say S_j, $u_j^k \in \Re$ is the value of u_h^k at S_j, the dimension of V_h being $n - 1$. We assume that the nodes S_j are numbered in such a manner that the corresponding abscissae are x_j satisfying $0 = x_0 < x_1 < x_2 < \ldots < x_{n-1} < x_n = 1$. Now, we choose v successively equal to φ_i, for $i = 1, 2, \ldots, n-1$, and we approximate for every k, $\int_0^1 w^k [\varphi_j]' \varphi_i dx$ by $\int_0^1 w_i^k [\varphi_j]' \varphi_i dx$ and $\int_0^1 f^k \varphi_i dx$ by $\int_0^1 f_i^k \varphi_i dx$, where $w_i^k := w^k(S_i)$, $f_i^k := f^k(S_i)$.

Still denoting the resulting values of $u_h^k(S_j)$ by u_j^k for $j = 1, \ldots, n-1$ and setting $u_0^k = a(k\tau)$ and $u_n^k = b(k\tau)$, the unknown coefficients u_j^k for $j = 1, 2, \ldots, n-1$ and $k = 1, 2, \ldots, l$, are recursively determined by solving the following SLAE:

$$\sum_{j=1}^{n-1} m_{i,j}^C u_j^k = \sum_{j=0}^n [m_{i,j}^C - \tau a_{i,j}^{k-1}] u_j^{k-1} + \tau b_i^k, \text{ for } i = 1, \ldots, n-1, \quad (7.45)$$

where the coefficients $m_{i,j}^C$, $a_{i,j}^k$ and b_i^k are given by

$$m_{i,j}^C = \int_0^1 \varphi_j \varphi_i dx; \quad a_{i,j}^k = \int_0^1 [w_i^k \varphi_j' \varphi_i \, dx + \epsilon \varphi_j' \varphi_i'] dx; \quad b_i^k = \int_0^1 f_i^{k-1} \varphi_i dx. \quad (7.46)$$

Actually, since for every $1 \le i \le n-1$, φ_i vanishes at $x = 0$ and $x = 1$ and w_i^k is constant, by integration by parts we easily derive $\int_0^1 w_i^k \cdot [\varphi_i]' \varphi_i = 0$. Hence, we may rewrite a_{ij}^k in equation (7.46) as

$$a_{i,j}^k = \int_0^1 [(1 - \delta_{ij})w_i^k \varphi_j' \varphi_i + \epsilon \varphi_j' \varphi_i'] dx. \tag{7.47}$$

The pointwise stability of scheme (7.44) is not ensured in general. Therefore, stabilizing techniques have been introduced such as upwinding (cf. [187] and [10]), in which the integral corresponding to the advection term is computed only in the element(s) situated upwind to the node S_i, with respect to w_i^k. The strategy adopted here is based on the use of different quadrature formulae, according to the side of equation (7.44), to approximate the **consistent mass matrix**, that is, the matrix whose coefficients are the $m_{i,j}^C$'s, where for convenience we let j vary from zero through n, while $i = 1, 2, \ldots, n-1$. In the particular choice made in this work, on the left side of equation (7.45), the integral in the expression of $m_{i,j}^C$ is approximated by the trapezoidal rule, and on the right side the approximate value of $m_{i,j}^C$ denoted by $m_{i,j}^W$ is obtained by an asymmetric quadrature formula (at least for non-uniform meshes) specified below. We recall that the trapezoidal rule consists of approximating the integral of a continuous function ψ in $(0, 1)$ by

$$\mathcal{J}_h(\psi) = \sum_{T \in \mathcal{T}_h} \frac{length(T)}{2} \sum_{m=1}^2 \psi(S_m^T),$$

where the S_m^T's denote the end-points of T, with $m = 1, 2$.

We recall that the **support** of φ_i is the closure of the set in which this shape function does not vanish identically (i.e. the interval $J_i := [x_{i-1}, x_{i+1}]$ for $i = 1, \ldots, n-1$). Then, setting $\Pi_i := x_{i+1} - x_{i-1}$, the approximation of the consistent mass matrix on the left side of equation (7.45) is nothing but the well-known **lumped mass matrix**, $\{m_{i,j}^L\}$, whose coefficients are $m_{i,j}^L = \Pi_i \delta_{ij}/2$, $0 \le j \le n$ and $1 \le i \le n-1$.

In doing so and scaling the linear equations by the factor $\Pi_i/2$, the unknown nodal values of u_h^k still denoted by u_j^k are determined by

The weighted mass explicit scheme to solve equation (7.43)

Compute recursively for $k = 1, 2, \ldots, l$,

$$u_i^k = \sum_{j=0}^n [\tilde{m}_{i,j} - \tau \tilde{a}_{i,j}^{k-1}]u_j^{k-1} + \tau \tilde{b}_i^k, \text{ for } i = 1, \ldots, n-1, \tag{7.48}$$

where $\tilde{m}_{i,j} = 2m_{i,j}^W/\Pi_i, \tilde{a}_{i,j}^k = 2a_{i,j}^k/\Pi_i$ and $\tilde{b}_i^k = f_i^{k-1}$

---

as one easily concludes for $\tilde{b}_i^k$. The coefficients $m_{i,j}^W$ on the right side are determined as follows: An inner node S_i has two neighbouring nodes S_{i-1} and S_{i+1}. Let $\Pi_{i\pm1}^i$ be the measure fractions associated with $S_{i\pm1}$ given by

$$\Pi_{i-1}^i = \frac{h_i}{2} \text{ and } \Pi_{i+1}^i = \frac{h_{i+1}}{2} \tag{7.49}$$

where $h_i = x_i - x_{i-1}$, $i = 1, 2, \ldots, n$. Notice that

$$\Pi_{i-1}^i + \Pi_{i+1}^i = \frac{\Pi_i}{2}. \tag{7.50}$$

Let $\omega_{i\pm1}^i$ be strictly positive weights satisfying

$$\omega_{i-1}^i \Pi_{i-1}^i + \omega_{i+1}^i \Pi_{i+1}^i = \frac{\Pi_i}{6} \tag{7.51}$$

Now, we define the $(n-1) \times (n+1)$ **weighted mass matrix** $M^W := \{m_{i,j}^W\}$ by

The weighted mass matrix M^W

$$m_{i,i\pm1}^W = \omega_{i\pm1}^i \Pi_{i\pm1}^i; \quad m_{i,i}^W = m_{i,i}^C = \frac{\Pi_i}{3}; \quad m_{i,j}^W = 0 \text{ for } |i-j| > 1. \tag{7.52}$$

It is interesting to note that equation (7.52) implies that

$$m_{i,i-1}^W + m_{i,i}^W + m_{i,i+1}^W = m_{i,i}^L, \forall i \in \{1, 2, \ldots, n-1\}. \tag{7.53}$$

We observe as well that the choice of the weights is not unique. For instance, if we take $\omega_{i\pm1}^i = \frac{1}{3}$ for every node S_i, the coefficients $m_{i,j}^W$ will be nothing but the $m_{i,j}^C$'s. However, except for the case of uniform meshes, this is not the right choice if one wishes to have a consistent scheme, as seen hereafter.

Next, we show how τ must be related to the spatial mesh parameter, in such a way that scheme (7.48) is stable in the sense of L^∞. First, we have to define the following quantities:

- $W = \max_{t \in [0,\Theta]} \max_{1 \le i \le n-1} |w(x_i, t)|$; (here $W = 1$, but we leave W as such for more general problems);
- $\omega = \min_{1 \le i \le n-1} \min[\omega_{i-1}^i, \omega_{i+1}^i]$.

It is interesting to note that equations (7.51) and (7.50) imply that $\omega \le \frac{1}{3}$.

Now we claim that if τ fulfil

$$\tau \le \frac{\omega \rho^2}{W\rho + \epsilon}, \tag{7.54}$$

then the $(n-1) \times (n+1)$ matrix $C^k = \{c_{i,j}^k\}$ given by $c_{i,j}^k = \tilde{m}_{i,j} - \tau \tilde{a}_{i,j}^{k-1}$, is a non-negative matrix having a unit row-norm $\|\cdot\|_1$:

$$\begin{cases} c_{i,j}^k \ge 0 & \forall i \in \{1, 2, \ldots, n-1\} \text{ and } \forall j \in \{0, 1, \ldots, n\} \\ \sum_{j=0}^{n} c_{i,j}^k = 1 & \forall i \in \{1, 2, \ldots, n-1\} \end{cases} \tag{7.55}$$

Let us justify this property. First, we treat the coefficients $c_{i,i}^k$, which are given by

$$c_{i,i}^k = \frac{2}{\Pi_i}(m_{i,i}^W - \tau a_{i,i}^{k-1}), \tag{7.56}$$

where $m_{i,i}^W$ is defined by equation (7.52) and $a_{i,i}^k$ is given by $a_{i,i}^k = \int_{x_{i-1}}^{x_{i+1}} \epsilon|\varphi_i'|^2 dx \ \forall k$. Then, straightforward calculations lead to

$$a_{i,i}^{k-1} \le \epsilon \rho^{-2} \Pi_i. \tag{7.57}$$

It follows from equation (7.52) and (7.57) that $c_{i,i}^k \ge 0 \ \forall i \in \{1, 2, \ldots, n-1\}$ if $\tau \le \frac{\rho^2}{3\epsilon}$. Since $\omega \le 1/3$, this condition is satisfied if equation (7.54) holds.

Next, we switch to the coefficients $c_{i,j}^k$ for $i \ne j$. Noticing that $c_{i,j}^k = 0$ if S_j does not belong to J_i, for $j = i \pm 1$, we have

$$c_{i,i\pm1}^k = \frac{2}{\Pi_i}(m_{i,i\pm1}^W - \tau a_{i,i\pm1}^{k-1}), \tag{7.58}$$

where $m_{i,i\pm1}^W$ is defined by equation (7.52) and $a_{i,i\pm1}^k$ is given by

$$a_{i,i\pm1}^k = \int_0^1 [w_i^k \varphi_{i\pm1}' \varphi_i + \epsilon \varphi_{i\pm1}' \varphi_i'] dx. \tag{7.59}$$

From equation (7.52), equation (7.49) and the definition of w, we trivially have $m_{i,i\pm1}^W \geq \frac{\omega}{2}|x_i - x_{i\pm1}|$. Moreover, we note that $\epsilon \int_0^1 \varphi_{i\pm1}' \varphi_i' dx < 0, \forall i$. Hence, referring to Exercise 3.4, after straightforward calculations, we conclude that for all k, $c_{i,i\pm1}^k \geq 0$ provided $\tau \leq \frac{\omega \rho}{W}$. This condition in turn is fulfilled if equation (7.54) holds.

To complete the argument, we note that, according to well-known properties of the shape functions φ_j, we have $\sum_{j=0}^n \varphi_j = 1$ everywhere in $[0,1]$. Hence, by linearity, we easily derive

$$\sum_{j=0}^n a_{i,j}^k = 0, \forall i \in \{1, 2, \ldots, n-1\} \text{ and } \forall k = 1, 2, \ldots, l. \tag{7.60}$$

Moreover, from equation (7.53) we derive $\sum_{j=0}^n m_{i,j}^W = m_{i,i}^W + m_{i,i-1}^W + m_{i,i+1}^W = m_{i,i}^L$. Thus,

$$\sum_{j=0}^n m_{i,j}^W = \frac{\Pi_i}{2} \ \forall i \in \{1, 2, \ldots, n-1\}. \tag{7.61}$$

Then, using the definitions $\tilde{a}_{i,j}^k = 2\frac{a_{i,j}^k}{\Pi_i}$ and $\tilde{m}_{i,j} = 2\frac{m_{i,j}^W}{\Pi_i}$, together with equations (7.58), (7.60) and (7.61), we readily conclude that for every $i \in \{1, 2, \ldots, n-1\}$, $\sum_{j=0}^n c_{i,j}^k = 1$.

As a consequence of the property we just established, the FE solution sequence $\{u_h^k\}_k$ generated by equation (7.48), given by $u_h^k = \sum_{j=0}^n u_j^k \varphi_j$, satisfies the

Maximum norm stability inequality for schemes (7.48)–(7.52)

Under the condition (7.54), we have for $1 \leq k \leq l$,

$$\| u_h^k \|_{0,\infty} \leq \max \left\{ \max_{1 \leq m \leq k} \max[|a(m\tau)|, |b(m\tau)|], \| u^0 \|_{0,\infty} \right\} + \tau \sum_{m=1}^k \| f^{m-1} \|_{0,\infty} \tag{7.62}$$

This result is proven in reference [172] in a wider context. The reader can use arguments very close to those in Section 3.2 for the Forward Euler scheme, to establish the validity of equation (7.62) as Exercise 7.14.

It is interesting to observe that condition (7.54) reflects the particular nature of equation (7.43). If ϵ is very small as compared to ρ, the time step τ should be roughly an $O(\rho)$, like in the case of explicit schemes for hyperbolic equations. On the other hand, in the diffusion dominant case, or if ρ is small enough to be close to ϵ, then stability requires a time step proportional to the square of the mesh size. Notice that this is the case of explicit schemes for parabolic equations.

Finally, we specify a condition on the weights $w^i_{\pm 1}$ under which a variant of schemes (7.48)–(7.52) proposed in reference [172] to solve equation (7.43) is consistent. In this variant, the weighted mass matrix coefficient $m^W_{i,j}$ is replaced by a convex combination of it with the coefficient $m^L_{i,j}$ of the lumped mass matrix, more precisely by $\beta m^L_{i,j} + (1 - \beta) m^W_{i,j}$ with $\beta = \epsilon/(\epsilon + \rho)$. Consistency is achieved by enforcing $w^i_{i-1} \Pi^i_{i-1} = w^i_{i+1} \Pi^i_{i+1}$, which combined with equation (7.51) leads to a

Consistency condition on the weights

$$
w^i_{i+1} = \frac{h_i}{3h_{i+1}}; \quad w^i_{i-1} = \frac{h_{i+1}}{3h_i} \tag{7.63}
$$

Now we make the following assumptions on f, w and u:

- f, w and $\partial_x w$ are bounded in $L^\infty[(0, 1) \times (0, \Theta)]$;
- $\partial_{xx} u$, $\partial_{xt} u$, $\partial_{xxt} u$ and $\partial_{tt} u$ are bounded in $L^\infty[(0, 1) \times (0, \Theta)]$.

Under these assumptions together with equation (7.63), we are lead to the following:

Error estimate for the weighted mass scheme in reference [172]

satisfying equation (7.63)

$$
\begin{aligned}
&\text{If } \tau \leq \kappa \frac{\rho^2}{3} \text{ with } \kappa = \min\left[\frac{1}{\varepsilon}, \frac{1}{W(\varepsilon+\rho)}\right], \text{ for } C > 0 \text{ independent of } u, h, \tau: \\
&\max_{1 \leq k \leq l} \| u^k - u^k_h \|_{0,\infty} \leq Ch \max_{0 \leq s \leq \Theta} \{\| \partial_{xx} u(\cdot, s) \|_{0,\infty} + \| \partial_{xt} u(\cdot, s) \|_{0,\infty} \\
&+ h \| \partial_{xxt} u(\cdot, s) \|_{0,\infty} + h \| \partial_{tt} u(\cdot, s) \|_{0,\infty} + \| \partial_x w(\cdot, s) \|_{0,\infty} (\| \partial_x u(\cdot, s) \|_{0,\infty} \\
&+ h \| \partial_{xx} u(\cdot, s) \|_{0,\infty}) + \| w(\cdot, s) \|_{0,\infty} \| \partial_{xx} u(\cdot, s) \|_{0,\infty} + \| \partial_x f(\cdot, s) \|_{0,\infty}\}.
\end{aligned} \tag{7.64}
$$

As a matter of fact, equation (7.64) is a particular case of a more general result proven in reference [172]. The reader can figure out how complex issues are inherent to the error analysis of scheme (7.48)–(7.52)–(7.63) by solving Exercise 7.15.

7.2.8 Example 7.3: Numerical Study of the Weighted Mass FE Scheme

To close this section, we show some numerical results for a one-dimensional problem with a sharp boundary layer.

Two schemes are experimented and compared: scheme (7.48)–(7.52)–(7.63) and the one resulting from a combination of the lumped mass matrix on the left side and the consistent mass matrix on the right side. This corresponds to a first-order consistent counterpart of the second-order consistent method considered in reference [114] for uniform meshes. Hereafter, we call this scheme the **basic Kawahara scheme**.

More specifically, problem (7.43) was solved with $a(t) = b(t) = 0 \; \forall t \in (0, \Theta]$, $u_0(x) = 0 \; \forall x \in (0, 1)$ and $w = 1$. Setting $u_\epsilon(x) := x - \dfrac{1 - e^{x/\epsilon}}{1 - e^{1/\epsilon}}$, which is nothing but the function depicted in Figure 7.4, the manufactured exact solution is given by $u(x, t) = (1 - e^{-t})u_\epsilon(x)$. Akin to u_ϵ, u presents a boundary layer of width $O(\epsilon)$ close to the point $x = 1$ for every $t > 0$. The corresponding right-side datum f is given by $f(x, t) := 1 + e^{-t}[u_\epsilon(x) - 1]$. The results supplied below are restricted to a given time, namely, $\Theta = 1$, as they are sufficiently representative of the behaviour of the numerical methods being experimented. This is due to the solution's decaying exponential term.

In order to figure out the influence of the schemes in the numerical results, double precision was used combined with non-uniform spatial meshes with $n + 1$ nodes, n being an even number. The corresponding step sizes are h_i for $i = 1, 2, \ldots, n$ where $h_{2k} = h/R$ and $h_{2k-1} = h$, for $k = 1, 2, \ldots, n/2$, R being a real number greater than one. It follows that $\rho = h/R = 2/[n(1 + R)]$. We determine $\tau = \Theta/l$ for each n, as the largest possible value that satisfies equation (7.54). Notice that the pair of weights for the weighted mass scheme are either $(R; R^{-1})/3$ or $(R^{-1}; R)/3$, according to the parity of the node subscript. As for the basic Kawahara scheme, both weights are equal to $1/3$ for every node. In the tables that follow, the weighted mass scheme is referred to as **WMS** and the basic Kawahara scheme as **BKS**.

First, we take $\epsilon = 10^{-2}$ and $R = 5$. We display in Tables 7.3 and 7.4 the relative errors for the indicated values of n, in $L^\infty(0, 1)$ and in $L^2(0, 1)$, respectively. It is noteworthy observing that the maximum errors are not attained at the leftmost inner mesh node (i.e. the closest inner node to the abscissa $x = 1$, as one might expect), but rather at a point not so far from it. Both points are different for each scheme. Anyway, the error at such points is rather large, since they lie inside the boundary layer. Next, we compute approximate solutions for $\epsilon = 10^{-5}$ taking again $R = 5$. We display in Table 7.5 the absolute errors of the approximations of $u(0.5, 1)$ computed with both schemes being experimented for increasing n. The corresponding relative errors measured in the norm of $L^2(0, 1)$ at time $t = \Theta$ in terms of n are shown in Table 7.6.

In the case of a moderately dominant convection (i.e. for $\epsilon = 10^{-2}$, we can assert that WMS performs globally better than BKS, as could be expected. Indeed, the errors for the former are significantly smaller than for the latter. Moreover, convergence is observed for both schemes

Table 7.3 Relative errors in $L^\infty(0, 1)$ at time $\Theta = 1$, for $\epsilon = 10^{-2}$ and $R = 5$.

$n \rightarrow$	64	128	256	512	1024
$WMS \rightarrow$	77.812%	72.447%	57.208%	40.572%	26.687%
$BKS \rightarrow$	81.755%	71.490%	66.125%	58.275%	49.888%

Table 7.4 Relative errors in $L^2(0, 1)$ at time $\Theta = 1$ for $\epsilon = 10^{-2}$ and $R = 5$.

$n \rightarrow$	64	128	256	512	1024
$WMS \rightarrow$	36.111%	24.704%	15.898%	9.566%	5.407%
$BKS \rightarrow$	38.547%	29.670%	22.840%	17.275%	12.438%

Table 7.5 Absolute errors of approximations of $u(0.5, 1)$ for $\epsilon = 10^{-5}$ and $R = 5$.

$n \rightarrow$	256	512	1024	2048	4096
$WMS \rightarrow$	0.105×10^{-4}	0.522×10^{-5}	0.260×10^{-5}	0.130×10^{-5}	0.650×10^{-6}
$BKS \rightarrow$	0.652×10^{-3}	0.322×10^{-3}	0.159×10^{-3}	0.791×10^{-4}	0.394×10^{-4}

Table 7.6   Relative errors in $L^2(0,1)$ at time $\Theta = 1$ for $\epsilon = 10^{-5}$ and $R = 5$

n $\rightarrow$	256	512	1024	2048	4096
WMS $\rightarrow$	20.093%	14.268%	10.113%	7.163%	5.074%
BKS $\rightarrow$	18.248%	12.956%	9.193%	6.530%	4.654%

in the L^2- and L^∞-norms, but at better rates for WMS. As for the convection largely dominant case with $\epsilon = 10^{-5}$, it is not possible to detect convergence in the maximum norm as n
increases, for either schemes. As a matter of fact, we observed that the maximum errors occur
at the grid point next to $x = 1$, and therefore this effect is not surprising at all. After all, it
is well-known that the mesh has to be extremely refined locally, in order to reduce numerical
errors in the interior of such a narrow boundary layer, which was not done here. We refer to
similar numerical examples given in reference [172] for more details on this issue. Nevertheless, as one moves away from the boundary layer, WMS is found to be very accurate in the
pointwise sense; BKS in turn is also accurate but much less than WMS, as long as $R \neq 1$. The
results given in Table 7.5 are particularly representative of this behaviour of the schemes being
experimented. On the other hand, from Table 7.6, we infer that for this value of ϵ both schemes
seem to converge in the sense of L^2, with a slight advantage of BKS over WMS in terms of
error magnitudes, at least up to the degree of refinement we have attained.

An interesting conclusion about the experiments reported here is that schemes of the type
extensively exploited by Kawahara et al. (see e.g. [114] and references therein) are inexpensive numerical tools and acceptably accurate in all cases, as much as their weighted mass
modification [172]. The observation that the smaller the value of ϵ, the better the performance
of BKS in the mean square sense is particularly impressive. In spite of this fact, BKS does
not seem to be optimally convergent in the strict mathematical sense for non-uniform meshes.
WMS in turn is an asserted reliable alternative from this point of view, at an equivalent cost.

7.3   Basics of a Posteriori Error Estimates and Adaptivity

Several error estimates were established for the methods we studied throughout the seven
chapters of this book. All these results are **a priori error estimates** in the sense that they
predict the behaviour of the numerical solution, in terms of both stability and order of convergence. However, unless empirical techniques are employed such as Richardson's extrapolation
like in Example 4.2, these estimates provide only a qualitative answer on what can be expected
of a numerical method. Incidentally, we recall that Richardson's extrapolation is a strategy to
come very close to a limiting unknown exact solution, on the basis of a bunch of numerical
results obtained for the same problem with different discretisation levels (see e.g. [35]). As a
matter of fact, in principle, an a priori error bound depends on the exact solution of the differential equation, which, except for very particular cases, is not known. Therefore, since long
authors endeavoured to find useful **a posteriori error estimates** for a numerical solution. This
means error bounds computed only from problem data and the numerical solution itself. The
FE school was very active in this field, and several important works emerged in the late 1970s
and the 1980s. Among early contributions in this direction, we could quote references [12],
[117], [18], [20] and [211]. A posteriori error estimation is still a subject of active research for
both the FEM and the FVM. They are usually applied in the framework of technical problems
far more complex than most of the linear differential models we studied in this book. In

addition to this, even for linear problems the rigorous study of these estimations requires some knowledge most readers might not master. From the beginning, it was not our intention to address mathematical problems that fall into this category. However, owing to its great importance in contemporary numerical solution of PDEs, this topic was included in this book's final chapter.

As a complement, we address as briefly an important corollary of a posteriori error estimation, namely, **mesh adaptivity**. As we should clarify, most **a posteriori error estimators** are local in the sense that they evaluate the numerical errors at the level of elements or to the most in their immediate vicinity. In general, the error distribution is not uniform. For instance, the numerical solution may have steep variations in certain regions, thereby representing an expected physical reality. In this case, it seems reasonable to try to improve thus revealed less smooth behaviour of the solution, by increasing the number of mesh nodes in such regions. That is where the concept of mesh adaptivity comes into play: If the number of nodes increase in certain regions, eventually to the detriment of others, in case the total number of nodes is to remain constant, then there must be **remeshing**. As a consequence, if the problem is linear for example, a new system matrix and right-side vector must be computed, and the underlying SLAE solved once again for the newly located unknowns. Notice that this procedure can be repeated several times until acceptable error bounds based on the a posteriori error estimation are obtained.

The principle of mesh adaptivity is explained in rather simple terms in reference [111]. In contrast, as pointed out above, even by simplifying things with all one's might, a rigorous and complete treatment of a posteriori error estimates requires concepts not well suited to this book's level. Therefore, in the remainder of this subsection, we give just a quick overview of both connected subjects.

7.3.1   A Posteriori Error Estimates

Let us consider the solution by the $\mathcal{P}_1$ FEM of the Poisson equation in a polygon Ω, with mixed homogeneous Dirichlet–inhomogeneous Neumann boundary conditions. More precisely, $u = 0$ on Γ_0 and $\partial_\nu u = g$ on Γ_1 where $\Gamma_0 \cap \Gamma_1 = \emptyset, \Gamma_0 \cup \Gamma_1 = \Gamma, length(\Gamma_0) > 0$. Like before, we assume that Γ_0 contains its transition points with Γ_1.

V being the space defined in Section 4.3, we recall that the variational form of this problem writes $a(u, v) = F(v) \; \forall v \in V$, where

$$a(u, v) := \int_\Omega (\mathbf{grad}\; u | \mathbf{grad}\; v) dx dy \text{ and } F(v) := \int_\Omega fv \; dx dy + \oint_{\Gamma_1} gv \; ds.$$

V_h is the $\mathcal{P}_1$ FE subspace of V associated with a family of triangulations $\mathcal{T}_h$ of Ω satisfying the compatibility conditions specified in Section 4.3. We assume the $\mathcal{T}_h$ belongs to a quasi-uniform family of triangulations of Ω. Let $T \in \mathcal{T}_h$ and h_T be the maximum edge length of T. We recall that $h = \max_{T \in \mathcal{T}_h} h_T$.

The corresponding approximate problem consists of finding $u_h \in V_h$ such that $a(u_h, v) = F(v) \; \forall v \in V_h$. We would like to determine **computable upper and lower bounds** for the error function $u - u_h$ in the norm of V, that is, $\|u - u_h\|_{1,2}$. Following reference [201], there are at least six different techniques to achieve this. Here, we will focus on two of them, namely, the **explicit residual based method** and the **averaging method** (also called the **gradient recovery method**).

The residual based method:

First, we note that $a(u - u_h, v) = F(v) - a(u_h, v) \; \forall v \in V$. The right side is the **residual** in the exact problem when u is replaced with u_h tested with $v \in V$. We denote it by $r(u_h, v)$, that is, r is a bilinear form with the first argument in V_h and second argument in V. We denote the boundary of T by ∂T, and the unit normal vector along any edge e of the triangulation by $\vec{\nu}_e$ taken in a sole direction for elements intersecting at e, except for a boundary edge where it is necessarily oriented outwards the domain. The set of all the edges of the triangles in $\mathcal{T}_h$ are assigned to three disjoint sets, namely, $\mathcal{E}_h$ consisting only of inner edges, $\mathcal{E}_{h1}$ consisting of edges contained in Γ_1 and the complementary set $\mathcal{E}_{h0}$ which will play no role. Let $e \in \mathcal{E}_h$, $h_e := length(e)$ and T and T' be two triangles of $\mathcal{T}_h$ whose intersection is e. Assuming that $\vec{\nu}_e$ is oriented from T onto T', we denote by $[\, |\partial_{\nu_e} u_h| \,]_e$ the (constant) jump of the normal derivative of u_h across $e \in \mathcal{E}_h$, that is, $([\mathbf{grad}\, u_h]_{|T} - [\mathbf{grad}\, u_h]_{|T'} | \vec{\nu}_e)$. Applying First Green's identity in each element T, since $\Delta u_h = 0$ in every T, after straightforward calculations, we obtain

$$r(u_h, v) = \int_\Omega fv \, dxdy + \sum_{e \in \mathcal{E}_{h1}} \int_e [g - \partial_{\nu_e} u_h] v \, ds + \sum_{e \in \mathcal{E}_h} \int_e [\, |\partial_{\nu_e} u_h| \,]_e v ds. \qquad (7.65)$$

Notice that the two summation terms in the residual express the fact that, contrary to the expected behaviour of exact solution normal derivatives, those of the approximate solution have jumps across inter-element boundaries. Therefore, it makes sense to evaluate those jumps as a measure of how much a numerical solution differs from the exact solution at a local level. In order to do this, we further define a constant function f_T in every $T \in \mathcal{T}_h$ given by $f_T = \dfrac{\int_T f \, dxdy}{area(T)}$ together with another constant function g_e in every $e \in \mathcal{E}_{1h}$ by $g_e = \dfrac{\int_e g \, ds}{length(e)}$. We denote by $\|\cdot\|_{0,T}$ and by $\|\cdot\|_{0,e}$ the norms in $L^2(T)$ and $L^2(e)$, respectively. Finally, for $T \in \mathcal{T}_h$, let $\mathcal{E}_T$ be the set of edges of T belonging to $\mathcal{E}_h$ and $\mathcal{E}_{T1}$ be the set of edges of T belonging to $\mathcal{E}_{1h}$. In doing so, we define a quantity η_T, namely, the

Local explicit residual a posteriori error estimator

$$\eta_T = \left[h_T^2 \, \|f_T\|_{0,T}^2 + \frac{1}{2} \sum_{e \in \mathcal{E}_T} h_e \, \|[|\partial_{\nu_e} u_h|]_e\|_{0,e}^2 + \sum_{e \in \mathcal{E}_{T1}} h_e \, \|g_e - \partial_{\nu_e} u_h\|_{0,e}^2 \right]^{\frac{1}{2}}. \qquad (7.66)$$

η_T is intended to represent an upper bound for the residual in the sense that $r(u_h, v) \leq C_T \eta_T$ $\|\mathbf{grad}\, v\|_{0,T}$, if we take $v \in V$ vanishing everywhere in the domain but in a small subdomain surrounding T, C_T being a constant independent of h_T. In this respect, it is a good error indicator to assess the quality of the numerical solution locally, and thus be used for mesh refinement.

Now, if we define the

Global residual a posteriori error estimator

$$\eta := \left[\sum_{T \in \mathcal{T}_h} (\eta_T^2 + h_T^2 \, \|f - f_T\|_{0,T}^2) + \sum_{e \in \mathcal{E}_{T1}} h_e \, \|g - g_e\|_{0,e}^2 \right]^{1/2}, \qquad (7.67)$$

it can be asserted that there exists a mesh-independent constant c_T such that

$$\|u - u_h\|_{1,2} \leq c_T \eta.$$

This is the final a posteriori error estimate using the explicit residual-based method. The constant c_T can be estimated in order to obtain more or less sharp estimates, depending on the case (cf [200]). Moreover, η_T also provides local lower bounds for the error $u - u_h$ in a patch of elements surrounding T, measured in a suitable natural norm. For further details, we refer to reference [200].

The averaging method:
In this method, the same fundamentals of the residual-based a posteriori error estimation for the Poisson equation are exploited, but in a more heuristic manner. As pointed out above, in general the gradient of the approximate solution is discontinuous at inter-element boundaries. Here, instead of setting forth estimations in terms of normal derivative jumps, we attempt to smoothen the gradient of u_h into a continuous field $\mathbf{G}(u_h)$ determined by local averaging. Actually, for those well acquainted with concepts and ordinary operations in Hilbert spaces (see e.g. [52]), in principle $\mathbf{G}(u_h)$ is the **orthogonal projection** of **grad** u_h onto the space of continuous fields $W_h \times W_h$, where W_h is the $\mathcal{P}_1$ FE space incorporating no boundary conditions. As a matter of fact, irrespective of mastering the concept of orthogonal projection or not, the reader should take note that $\mathbf{G}(u_h)$ fulfils,

$$\mathbf{G}(u_h) \in W_h \times W_h \text{ and } (\mathbf{G}(u_h)|\mathbf{w})_0 = (\mathbf{grad}\ u_h|\mathbf{w})_0\ \forall \mathbf{w} \in W_h \times W_h. \qquad (7.68)$$

Notice that we have $\mathbf{G}(u_h) = [G_x(u_h), G_y(u_h)]^T$. Thus, we must solve two SLAEs to determine $G_x(u_h)$ and $G_y(u_h)$, each one of them being at least as costly as the system we must solve to compute u_h itself. This is because there are two unknowns per mesh node instead of one per mesh node not belonging to Γ_0 in the case of u_h. Therefore, the procedure as such is not so reasonable, if we are to apply it recursively in order to successive refine the mesh. However, recalling the lumped mass technique, we can diagonalise the matrix underlying equation (7.68). In this case, $\mathbf{G}(u_h)$ can be determined in a straightforward manner, that is, without solving SLAEs. Let us see how this works.

First of all, we recall the two-dimensional counterpart of the trapezoidal rule, yielding an approximation $\mathcal{J}_h(\psi)$ of the integral of a continuous function ψ in Ω, namely,

$$\mathcal{J}_h(\psi) = \sum_{T \in \mathcal{T}_h} \frac{area(T)}{3} \sum_{m=1}^{3} \psi(S_m^T), \qquad (7.69)$$

where S_m^T, $m = 1, 2, 3$ are the vertices of a triangle T. Let us apply this formula to compute the left side of equation (7.68), taking $\mathbf{w}$ equal to either $[\varphi_i, 0]^T$ or $[0, \varphi_i]^T$, φ_i being the shape function associated with mesh node S_i. Notice that in the former case for instance, this integral is nothing but the integral in Ω of the product $G_x(u_h)\varphi_i$, whereas in the latter case it equals the integral of $G_y(u_h)\varphi_i$ in Ω. Let Π_i be the union of mesh triangles having S_i as a vertex. Appying equation (7.69), from the fundamental properties of the shape functions it easily follows that

$$\begin{cases} G_x(S_i) \sum_{T \subset \Pi_i} area(T)/3 = (\partial_x u_h|\varphi_i)_0, \\ G_y(S_i) \sum_{T \subset \Pi_i} area(T)/3 = (\partial_y u_h|\varphi_i)_0. \end{cases} \qquad (7.70)$$

Now, using the notation $[u_{h,x}^T, u_{h,y}^T]^T$ for $[\mathbf{grad}\ u_h]_{|T}$, we observe that in each triangle $T \subset \Pi_i$ this field is constant. Hence, taking into account that $\int_T \varphi_i dx dy = area(T)/3$ if $T \subset \Pi_i$, the right sides of equation (7.70) are given by

$$\begin{cases} (\partial_x u_h | \varphi_i)_0 = \sum_{T \subset \Pi_i} u_{h,x}^T area(T)/3, \\ (\partial_y u_h | \varphi_i)_0 = \sum_{T \subset \Pi_i} u_{h,y}^T area(T)/3. \end{cases} \tag{7.71}$$

Combining equations (7.70) and (7.71), we immediately conclude that

$$\begin{cases} G_x(S_i) = \sum_{T \subset \Pi_i} u_{h,x}^T area(T)/area(\Pi_i), \\ G_y(S_i) = \sum_{T \subset \Pi_i} u_{h,y}^T area(T)/area(\Pi_i). \end{cases} \tag{7.72}$$

Equation (7.72) can be interpreted as follows:

$$[\mathbf{G}(u_h)](S_i) = \sum_{T \subset \Pi_i} [\mathbf{grad}\ u_h]_{|T} area(T)/area(\Pi_i), \text{ for } i = 1, 2, \ldots, K_h, \tag{7.73}$$

where K_h is the number of mesh nodes. This means that the value of the orthogonal projection $\mathbf{G}(u_h)$ of $\mathbf{grad}\ u_h$ at mesh node S_i is obtained by the averaging of $\mathbf{grad}\ u_h$ in the domain Π_i about vertex S_i specified on the right side of equation (7.73).

It turns out that the quantity $\gamma_T := \|[\mathbf{G}(u_h) - \mathbf{grad}\ u_h]_{|T}\|_{0,T}$ related to an element $T \in \mathcal{T}_h$, that is,

$$\gamma_T = \left[\int_T \left\{ |[G_x(u_h)]_{|T} - u_{h,x}^T|^2 + |[G_y(u_h)]_{|T} - u_{h,y}^T|^2 \right\} dx dy \right]^{1/2} \tag{7.74}$$

is an excellent easy-to-compute local error estimator [3].

7.3.2 Mesh Adaptivity: h, p and h–p Methods

Skipping details for the sake of brevity, we see below some highlights on how γ_T or η_T can be used to improve FEM's accuracy through **mesh adaptivity**. FE and FV users have seen the importance of this technique steadily increase since the 1980s. In many branches of activity, it became an inseparable companion. For example, this is true of **inverse problems** governed by a PDE, whose solution is aimed at reconstructing field data. For applications of mesh adaptivity in this framework, we refer to the work of Beilina *et al.* [24].

Let us be given a small tolerance $\epsilon > 0$ for the quantities γ_T. First, we assume that only one mesh triangle, say $T_0 \in \mathcal{T}_h$, fulfils $\gamma_{T_0} \geq \epsilon$. Referring to Figure 7.7, we can subdivide T_0 into four triangles using the edge mid-points of T_0. Of course, if M_h is the total number of triangles in $\mathcal{T}_h$, the new triangulation with $M_h + 3$ triangles will not be a compatible mesh in the FE (or the FV) sense. Then, assuming that T_0 does not have any edge contained in Γ_1 for simplicity, we may adjust the three triangles T_1, T_2 and T_3 sharing an edge with T_0 in such a way that each one of them is split into two triangles by the segment joining the edge mid-point of T_0 to the opposite vertex of the neighbouring triangle (cf. Figure 7.7). Then, we come up with a

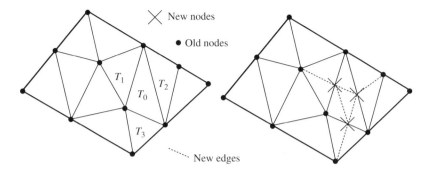

Figure 7.7   Subdivision of T_0 into four triangles and of T_1, T_2 and T_3 into two triangles

final compatible mesh with $M_h + 6$ triangles, which is then used for the computation of a new approximate solution, say $\underline{u}_h$. Quantities $\underline{\gamma}_T$ related to $\underline{u}_h$ are computed in the same manner as γ_T for the original mesh, and the tolerance criterion is applied again. In case there are still triangles T in the locally refined mesh such that $\underline{\gamma}_T \geq \epsilon$, we repeat the remeshing procedure, until the tolerance criterion is fulfilled.

The above procedure was described in a deliberately simplified context, just to illustrate the principles that mesh adaptivity is based upon. Some remarks on everyday mesh refinement are in order to complete this brief introduction to the subject.

First of all, we note that the procedure we have just described is the so-called **h-version** of mesh adaptivity. Provided the exact solution is sufficiently smooth in the mesh portion to be refined, though with steep gradients, instead of subdividing the triangles not fulfilling the tolerance criterion into four or two triangles, we may change the approximate solution representation in each one of them using Lagrange quadratic interpolation having as additional nodes the edge mid-points (cf. Subsection 6.1.3). However, we cannot do the same for the triangles ensuring the transition from quadratic interpolation to linear interpolation, corresponding to those subdivided into two triangles in the refinement h-version. But this is not really a problem. It suffices to assume that the quadratic function in those transition triangles is such that their restriction to edges common to another triangle where only linear interpolation is performed is itself a linear function. This reduces to considering that the values of the quadratic function at such edge mid-points is the mean value of those at the respective end-points. In this approach, the computational refinement can go on by increasing the degree p of the interpolating function in triangles that do not fulfil the tolerance criterion. This is the so-called **p-version** of adaptivity. It is also possible to combine both techniques, thereby giving rise to the **h-p-version** of adaptivity. We do not further elaborate on procedures based on Lagrange interpolation of increasing order, since we did not present FE spaces of degree higher than two in two dimensions. We refer to reference [45] for a description of Lagrange interpolation of arbitrary order in triangles and to reference [13] for a thorough study on the h, p and $h - p$ methods. It is also noteworthy that, in principle, every step of a mesh refinement procedure affects a large number of mesh elements. For this reason, a sensible choice of remeshing techniques is recommended. There are plenty of them in the literature, and it would be difficult to be exhaustive in this respect. A remarkable example of efficient mesh refinement algorithms can be found in reference [164] and the references therein.

7.4   A Word about Non-linear PDEs

As mathematical models of countless problems of interest in real-life applications, PDEs are mostly non-linear. For this reason, in contemporary numerical simulations of underlying physical events, practitioners have to deal with the solution of non-linear systems of algebraic equations of increasing complexity. In most cases, it is not reasonable to attempt to examine all details pertaining to an adopted numerical procedure, and users are satisfied with the a posteriori validation of their approach based on the coherence of the results, often as per their own judgment. This means that emphasis is given to the solution procedure in the aim of reducing costs, rather than on formal reliability studies. That is roughly what we intend to show in this book's final section, even though this does not mean that we approve such an attitude. Whatever a problem's complexity, there should be concern about accuracy control. However, we have to overlook such aspects here as a matter of choice, owing to obvious limitations to treat such a vast subject in a single section.

Before starting, a few comments on the relationship between linear and non-linear, as far as numerical methods for PDEs are concerned, are of paramount importance.

This book was intended to give insight on the most used numerical methods for solving PDEs. However as a beginner-oriented text, the presentation of these techniques was confined to the linear paradigms of the three types of PDEs in both one and two space variables, with a short extension to the three-dimensional case. In many cases, these linear PDEs are the **linearised form** of the true non-linear equation. For this reason too, the adopted approach in this book is mandatory in some sense. Indeed, methods that fail to work for a linearised equation will certainly work even worse if one attempts to apply it to the full non-linear form of the equation. But then the question is: to which extent can the reliability of a numerical method for a certain type of linear PDE predict its performance as applied to a complete non-linear PDE of a similar nature? The study of non-linear PDEs opens the gate to a much more complex universe, and for this reason unfortunately the answer to this question cannot be given in simple terms. For instance, even if the existence of a solution to a non-linear PDE is guaranteed, is general its uniqueness is not. There are phenomena known as bifurcation or turning points depending on the variation of a real parameter defining in some sense the importance of non-linear terms. Generally speaking, if this parameter is close to zero the non-linear equation has a behaviour similar to the underlying linearised model. In particular, this means an equation with a unique solution. On the other hand, as this parameter goes beyond critical values the solution will no longer be unique. Some non-linear PDEs also behave in a way rather difficult to deal with. Even smooth data and domains can give rise to discontinuous solutions. A particularly representative example is the apparently gentle **Bürgers equation** of gas dynamics, which is a sort of non-linear transport equation, whose transport velocity at every $(x; t)$ is the solution at this point itself. In an infinite one-dimensional domain, this equation writes,

$$\partial_t u + u \partial_x u = 0 \text{ for } t > 0 \text{ with given } u_0(x) \text{ such that } u(x, 0) = u_0(x) \ \forall x.$$

If we take $u_0(x) = x$, then a possible solution is given by $u(x, t) = x/(t + 1)$ for $t > 0$, as one can easily check. Now let us take $u_0(x) = -x$. If we change $t + 1$ into $t - 1$ in the above expression of $u(x, t)$, in principle the new function satisfies the Bürgers equation, together with the initial condition. However, what happens when $t = 1$? The gas particles located at the abscissae x at time $t = 0$, and transported with initial velocity equal to $-x$ will all collide at time $t = 1$! From this time on, the unknown function u will continue to evolve, even though not according to the above analytic expression. This example was rather academic, but in practice a phenomenon known as *shock wave* occurs for flows modelled by the Bürgers equation. In mathematical terms, this consists of a curve $t = g(x)$ in the $x - t$ plane, along which the solution is discontinuous. We refer for instance to references [111] and [67] for more explanations about the Bürgers equation and an introduction to numerical methods to deal with it, even in the presence of shock waves.

As a rule, whenever the dominant differential operator in terms of order of differentiation is linear, a non-linear PDE behaves somehow like its linear counterpart, provided the data are sufficiently small. Just to give an example, let us consider the following equation in a bounded two-dimensional domain Ω with boundary Γ:

$$-\Delta u + u^3 = f \text{ in } \Omega \text{ with } u = 0 \text{ on } \Gamma.$$

Multiplying both sides of this equation by u and integrating in Ω, and further using the Cauchy–Schwarz and the Friedrichs–Poincaré inequalities, followed by Young's inequality with $\delta = 1$, we obtain

$$\|\mathbf{grad}\, u\,\|_{0,2}^2 + 2 \,\|u^2\|_{0,2}^2 \le C_P^2 \,\|f\|_{0,2}^2 .$$

The solution gradient is seen to be bounded by the norm of f times a constant, but better than this, the norm of the squared solution in $L^2(\Omega)$ is also bounded by the same quantity. Then, provided the latter is not so large (i.e. f is sufficiently small), we can study the numerical solution of the above equation following recipes in all similar to those we exploited throughout the text, based on the Lax–Richtmyer equivalence theorem. Of course, this is not as direct and simple as for the linearised form $-\Delta u + cu = f$ for a certain non-negative constant c. However, the simple fact that $u^3 = u^2 \times u$, combined with the bound for the norm of u^2 which also holds for the numerical counterpart u_h, allows for this. We observe, however, that even so the reliability analysis becomes fairly more complicated than in the linear case, and for this reason we do not further elaborate on them.

A final comment to conclude this brief introduction to the non-linear world is as follows. For the general case, a certain number of mathematical tools not considered in this text are available, to handle non-linear PDEs together with their numerical solutions, such as **Brower's fixed point theorem**. We refer to reference [47] for more details on this matter.

7.4.1   Example 7.4: Solving Non-linear Two-point Boundary Value Problems

Along the same line of thought as in the introduction of this section, we next endeavour to solve a very simple non-linear ODE as a model. However, it is the author's expectation that the reader will be able to extrapolate and figure out what could be done to solve any other non-linear boundary value PDE.

More precisely, we consider the numerical solution by the FDM of the following non-linear boundary value ODE:

Given $p, r, f \in C^0[0, L]$, find u such that

$$- (pu')' + ru - (1+r)u^2 + u^3 = f \text{ in } (0, L), u(0) = 0, u'(L) = 0. \tag{7.75}$$

This equation models a special case of combustion, in which a reaction–diffusion process takes place along a straight cylindrical channel with length L, occupied by a certain medium, without transversal effects [209]. In this case, p and r are the (local) diffusion and reaction coefficients, respectively, both being strictly positive. Notice that for suitably small data, u will be necessarily small enough for the non-linear terms to be negligible. In such a case, the model reduces to the linear ODE (P_1) with $q = r$.

Taking $p = r = 1$, let us be given a uniform FD grid with size $h = L/n$ for a certain integer $n > 1$. Denoting $f(ih)$ by f_i and the approximations of $u(ih)$ by u_i for $i = 1, 2, \ldots, n$, similarly to Section 1.2, the latter satisfy

$$\frac{2u_i - u_{i-1} - u_{i+1}}{h^2} + u_i - 2u_i^2 + u_i^3 = f_i, \text{ for } i = 1, 2, \ldots, n \text{ with } u_{n+1} = u_{n-1}, u_0 = 0.$$

These equations correspond to a non-linear system of n algebraic equations with n unknowns. Leaving aside deeper analyses, we assume that it has a solution $\vec{u}_h = [u_1, u_2, \ldots, u_n]^T$ close to an initial guess $\vec{u}_h^0 = [u_1^0, u_2^0, \ldots, u_n^0]^T$, more precisely satisfying $|\vec{u}_h^0 - \vec{u}_h| \leq \rho$ for ρ sufficiently small. There are many methods to solve this non-linear system, and in this respect we refer to classical books on the subject such as reference [150]. Here, we select Newton's method as one of the most effective. For a given n-component vector $\vec{v} = [v_1, v_2, \ldots, v_n]^T$, we set

$$F_i(\vec{v}) := \frac{2v_i - v_{i-1} - v_{i+1}}{h^2} + v_i - 2v_i^2 + v_i^3 - f_i, \text{ for } i = 1, 2, \ldots, n-1,$$

together with

$$F_n(\vec{v}) := \frac{2(v_n - v_{n-1})}{h^2} + v_n - 2v_n^2 + v_n^3 - f_n.$$

Notice that the system to solve is $F_i(\vec{u}_h) = 0$ for $i = 1, 2, \ldots, n$. Now we define the gradient of $F_i(\vec{v})$ to be the n-component vector whose jth component is $G_{i,j}(\vec{v}) := \partial F_i(\vec{v})/\partial v_j$. This definition gives rise to an $n \times n$ **Jacobian matrix** $\mathcal{G}(\vec{v}) = \{G_{i,j}(\vec{v})\}$ associated with $\vec{v}$, which is nothing but the gradient of the n-component field $\mathcal{F}(\vec{v}) = [F_1(\vec{v}), \ldots, F_n(\vec{v})]^T$. Notice that $G_{i,j}(\vec{v}) = 0$ if $|j - i| > 1$. Therefore, $\mathcal{G}(\vec{v})$ is a tridiagonal matrix whose entries are $G_{i,i}(\vec{v}) = 2/h^2 + 1 - 4v_i + 3v_i^2$, for $i = 1, 2, \ldots, n$, $G_{i,i-1}(\vec{v}) = -1/h^2$ for $i = 2, \ldots, n-1$ and $G_{n,n-1}(\vec{v}) = -2/h^2$, $G_{i,i+1}(\vec{v}) = -1/h^2$ for $i = 1, 2, \ldots, n-1$. Then the first iteration of Newton's method yields a vector $\vec{u}_h^1 = [u_1^1, \ldots, u_n^1]^T$ satisfying,

$$\mathcal{G}(\vec{u}_h^0)[\vec{u}_h^1 - \vec{u}_h^0] = -\mathcal{F}(\vec{u}_h^0).$$

The same procedure is recursively applied to determine approximations $\vec{u}_h^{k+1}$ of $\vec{u}_h$ for $k = 1, 2, \ldots$, by **Newton's iterations**, that is,

$$\mathcal{G}(\vec{u}_h^k)[\vec{u}_h^{k+1} - \vec{u}_h^k] = -\mathcal{F}(\vec{u}_h^k),$$

until the maximum norm of $\vec{u}_h^k - \vec{u}_h^{k+1}$ becomes smaller than a given tolerance δ, or, for better scaling, the product of δ with the maximum norm of $\vec{u}_h^{k+1}$. This means that in the latter case, the stop criterion is based on the relative increment between two successive iterations, and on the absolute increment in the former case. Notice that a sufficiently large maximum number of iterations must be stipulated, in order to provide for the eventuality of iteration divergence. Anyway, as long as the initial guess is not too far from an expected solution of

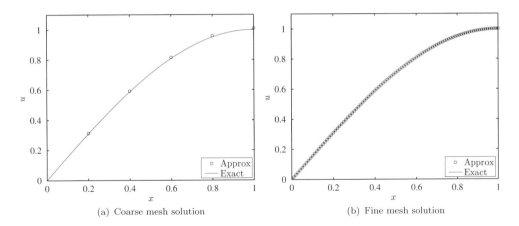

(a) Coarse mesh solution   (b) Fine mesh solution

Figure 7.8   Exact and FD solution of a non-linear ODE

the non-linear system, as a rule convergence of this procedure to this solution does occur at an asymptotic rate better than linear, except for especial cases, and to the most quadratic (see e.g. [46]). More precisely, $\lim_{k\to\infty} \frac{\|\vec{u}_h^{k+1} - \vec{u}_h\|_\infty}{\|\vec{u}_h^k - \vec{u}_h\|_\infty^\alpha}$ equals a constant for some α, $1 < \alpha \leq 2$. The well-posedness issue of the SLAE to solve at the kth iteration deserves careful attention. In case the matrix $\mathcal{G}(\vec{u}_h^k)$ is singular or very ill-conditioned (cf. [46]), a possible way out is to shift the components of $\vec{u}_h^k$ by a small quantity and recompute the matrix.

In order to check the performance of the combination of Newton's method and the FD discretisation, we next give some numerical results obtained for a **manufactured solution**. We take $u(x) = \sin(\pi x/2)$, which satisfies the prescribed boundary conditions at $x = 0$ and $x = 1$. Then, we calculate $f = -u'' + u - 2u^2 + u^3$ to obtain $f(x) = \sin(\pi x/2)\{(\pi/2)^2 + 1 - \sin(\pi x/2)[2 - [\sin(\pi x/2)]]\}$. We ran a MATLAB code supplied hereafter for solving this ODE by Newton's iterations, starting from a zero initial guess and taking a tolerance for the absolute increment between two successive iterations equal to 10^{-5}. Two different meshes were tested, namely, a very coarse one with $n = 5$ and a finer one with $n = 100$. In both cases, the stop criterion was satisfied after only five iterations, which is undoubtedly a remarkable performance.

Figures 7.8a and 7.8b displays the exact solution and the approximate solution at the grid points for $n = 5$ and $n = 100$, respectively. As one can infer from both figures, the discretisation errors are very small, even for the coarse mesh.

The following MATLAB code was used to generate these numerical results. It is rather self-explanatory, but the reader is advised to read it through in order to consolidate her or his understanding of Newton's iterations to solve non-linear systems of algebraic equations. Such a task, among others, is the object of Exercise 7.16.

```
function main
MaxNumNewtonIter = 100; delta = 1e-5;  n = 5;
a                = 0;   b     = 1;      h = (b-a)/n; u0 = 0;
x  = zeros(n,1); uold = zeros(n,1);   unew  = zeros(n,1);
DF = zeros(n,n); F    = zeros(n,1);
for k = 1:MaxNumNewtonIter
    x(1)    = h;
```

```
DF(1,1)  = 2/h^2 + 1 - 4*uold(1) + 3*uold(1)^2;
DF(1,2)  = -1/h^2;
F(1)     = -(- u0/h^2 + ( 2*uold(1)/h^2 + uold(1) - 2*uold(1)^2
           + uold(1)^3 ) ... - uold(2)/h^2 - f(x(1)) ) ;
for i = 2:n-1
  x(i) = i*h;
  DF(i,i-1) = -1/h^2;
  DF(i,i  ) = 2/h^2 + 1 - 4*uold(i) + 3*uold(i)^2;
  DF(i,i+1) = -1/h^2;
  F(i)      = -(- uold(i-1)/h^2 + ( 2*uold(i)/h^2 + uold(i)
              -2*uold(i)^2 + ... uold(i)^3 ) - uold(i+1)/h^2 - f(x(i)) ) ;
end
x(n)      = b;
DF(n,n-1) = -1/h^2;
DF(n,n  ) = ( 2/h^2 + 1 - 4*uold(n) + 3*uold(n)^2 )/2;
F(n)      = -( - uold(n-1)/h^2 + ( 2*uold(n)/h^2 + uold(n)
            - 2*uold(n)^2 + ... uold(n)^3 ) - uold(n-1)/h^2 - f(x(n)) )/2 ;
unew = linsolve( DF, F );
max=0;
for i=1:n
  if ( abs(unew(i)) > max )
    max = abs(unew(i));
  end
end
unew = unew + uold;
if ( max < delta )
  break
end
uold = unew;
end
figure(1);
z = linspace( 0, 1, 100 );
hfig = plot( x, unew, 'ko', z, sin(pi*z/2), '-k' );
xlabel('$x$','FontSize', 18,'interpreter','latex');
ylabel('$u$','FontSize', 18,'interpreter','latex');
legend('Approx','Exact','Location','southeast');
end
function y = f( x )
   y = sin(pi*x/2)*( pi^2/4 + 1 - 2*sin(pi*x/2) + (sin(pi*x/2))^2 );
end
```

To conclude, it is important to point out that non-linear differential equations discretised by the FVM and the FEM can be solved in a similar manner. For example, in the case of the above model ODE, if the FVM is used, the non-linear terms will be replaced with their integrals in the CVs. A field $\mathcal{F}$ and corresponding Jacobian matrix $\mathcal{G}$ very similar to those we determine above for the FDM will result from such a discretisation. In the case of the $\mathcal{P}_1$ FEM, however, the non-linear algebraic equations will not quite have the same form, since they will result from the following variational form: $a(u, v) = F(v) \; \forall v \in V$, where

$$a(u, v) := \int_0^L [pu' v' + (u - 2u^2 + u^3)v]dx; \quad F(v) := \int_0^L fv \, dx,$$

V being the space defined in Section 1.3, together with its FE counterpart V_h. Actually, if V is replaced by V_h and u by $u_h \in V_h$ in the above equations, a system of n non-linear algebraic equations $F_i(\vec{u}_h) = 0$ will have to be solved, each F_i corresponding to the choice $v = \varphi_i$, $i = 1, 2, \ldots, n$ (i.e. the $\mathcal{P}_1$ FE shape functions). However, now the off-diagonal coefficients of the Jacobian matrix will no longer be constant because second and third powers of $u_h = \sum_{j=1}^n u_j \varphi_j$ will couple products of powers of u_i and u_j if $|j - i| \le 1$. The reader is invited to write down the non-linear systems to solve at each Newton's iteration for both the Vertex-centred FVM and the $\mathcal{P}_1$ FEM for the same mesh with n equally spaced CVs or elements, as Exercise 7.17.

7.4.2 Example 7.5: A Quasi-explicit Method for the Navier–Stokes Equations

Newton's method is a a sort of master key to be employed in the efficient solution of discretised non-linear PDEs, with very few exceptions. However, the underlying procedure couples all the unknowns in a sequence of varying non-linear systems, whose solution can be very expensive in terms of computational effort, depending on the space dimension and mesh or grid size. Therefore, practitioners are often searching for less costly though reliable solution algorithms, well adapted to a particular type of non-linear PDE they have to deal with. Once more, we do not intend to go into details, but to conclude this section it seems important to minimally address what solution algorithms for non-linear PDEs other than Newton's iterations are about. This will be done in the framework of a *homemade* method to solve the **Navier–Stokes equations** [171], which model stationary or time-dependent flows of an incompressible Newtonian fluid (see e.g. [15]) in a bounded N-dimensional domain Ω, $N = 2$ or $N = 3$, with boundary Γ. Here, only numerical results based on the $\mathcal{P}_1$ FEM matter, which illustrate an inexpensive solution strategy as compared to Newton's method.

Skipping details for the sake of conciseness, we consider only stationary flows which are governed by

The stationary incompressible Navier–Stokes equations

Find a velocity $\vec{u} = [u_1, \ldots, u_N]^T$ and a pressure p such that
$\left. \begin{array}{l} -\mu \Delta \vec{u} + [\mathbf{grad} \; \vec{u}]\vec{u} + \mathbf{grad} \; p = \vec{f}, \\ div \; \vec{u} = 0, \end{array} \right\}$ in Ω;   (7.76)
$\vec{u} = \vec{g}$ on Γ

where $\vec{f}$ is a given density of forces, and $\vec{g}$ is a prescribed boundary velocity satisfying the global conservation property $\oint_\Gamma (\vec{g}|\vec{v})dS = 0$ (because $div \; \vec{u} = 0$). Like in the linear elasticity system, $\mathbf{grad} \; \vec{u}$ is a second-order $N \times N$ tensor. For $N = 3$, its $i - j$ component is $\partial u_i / \partial x_j$ where $x_1 = x$, $x_2 = y$ and $x_3 = z$. The unknowns $\vec{u}$, p and domain are assumed to be in normalised dimensionless form. In this case, the coefficient μ is the inverse of the **Reynolds number** Re (cf. [15]), which is given by $Re = \rho V L/\eta$ where ρ is fluid's density, V and L are characteristic velocity and length, and η is fluid's viscosity. Notice that, similarly to the advection–diffusion equation, μ measures the importance of the terms containing second-order derivatives with respect to those containing first-order derivatives. More precisely, a large μ indicates dominant viscous effects, while a small μ points to dominance of mass transfer by inertia. Notice that the pressure is determined up to an additive constant, which can be fixed by prescribing for instance $p = 0$ at a given point of Ω.

The method employed to obtain the computational results shown hereafter incorporates the **deviator Cauchy stress tensor** σ as an additional unknown field. For a Newtonian viscous fluid, this tensor is expressed by $\sigma = 2\mu D(\vec{u})$, $D(\vec{u})$ being the symmetric gradient of $\vec{u}$ given by $D(\vec{u}) := [\mathbf{grad}\ \vec{u} + (\mathbf{grad}\ \vec{u})^T]/2$, known as the **strain rate tensor**. In doing so, the stationary incompressible Navier–Stokes equations can be rewritten as follows:

$$\left\{\begin{array}{l} \text{Find } \vec{u}, p \text{ and } \sigma \text{ such that} \\ -\mathbf{Div}\sigma + [\mathbf{grad}\ \vec{u}]\vec{u} + \mathbf{grad}\ p = \vec{f}, \\ div\ \vec{u} = 0 \\ \sigma = 2\mu D(\vec{u}), \\ \vec{u} = \vec{g} \text{ on } \Gamma \text{ and } p(M) = 0. \end{array}\right\} \text{ in } \Omega; \qquad (7.77)$$

where M is a given point of Ω. A popular technique to linearise time-independent non-linear PDEs like equation (7.76) is based on a pseudo time integration. This consists of applying a FD discretisation to a fictitious first-order time derivative added to the left side of the equation, with (pseudo) time step τ. In the case of equation (7.77), among other possibilities, this can be achieved by adding $\partial_t \vec{u}$ to the first equation and $\lambda\partial_t\sigma$ to the third equation, where $\lambda > 0$ is a constant numerical parameter. Then these terms are approximated by $[\vec{u}^k - \vec{u}^{k-1}]/\tau$ and $\lambda[\sigma^k - \sigma^{k-1}]/\tau$, for $k = 1, 2, \ldots$, starting from a suitably chosen $\vec{u}^0$ satisfying $\vec{u}^0 = \vec{g}$ on Γ, and setting $\sigma^0 := 2\mu D(\vec{u}^0)$.

Schematically, assuming that $\vec{u}^{k-1}$ and σ^{k-1} are known, we solve at the kth iteration,

$$\left\{\begin{array}{l} \text{Find } \vec{u}^k, p^k \text{ and } \sigma^k \text{ such that} \\ \vec{u}^k = \vec{u}^{k-1} + \tau\{\mathbf{Div}\sigma^{k-1} - [\mathbf{grad}\ \vec{u}^{k-1}]\vec{u}^{k-1} - \mathbf{grad}\ p^k + \vec{f}\}, \\ div\ \vec{u}^k = 0 \\ \sigma^k = \frac{1}{\tau+\lambda}[\lambda\sigma^{k-1} + 2\mu\tau D(\vec{u}^k)], \\ \vec{u}^k = \vec{g} \text{ on } \Gamma \text{ and } p^k(M) = 0. \end{array}\right\} \text{ in } \Omega; \qquad (7.78)$$

for $k \geq 1$. Solution algorithm (7.78) is explicit but for the coupling of $\vec{u}^k$ and p^k in the first two equations. Both unknowns can be decoupled, taking the divergence of all the terms in the first equation except $\vec{u}^k$. In doing so, we come up with a new second equation to replace $div\ \vec{u} = 0$. Recalling that $div\ \mathbf{grad}\ p = \Delta p$, interchanging the resulting second equation and the first equation, after straightforward manipulations, instead of equation (7.78) we solve

$$\left\{\begin{array}{l} \text{Find } \vec{u}^k, p^k \text{ and } \sigma^k \text{ such that} \\ -\Delta p^k = -div\ \vec{u}^{k-1}/\tau - div(\mathbf{Div}\sigma^{k-1} - [\mathbf{grad}\ \vec{u}^{k-1}]\vec{u}^{k-1} - \vec{f}), \\ \vec{u}^k = \vec{u}^{k-1} + \tau(\mathbf{Div}\sigma^{k-1} - [\mathbf{grad}\ \vec{u}^{k-1}]\vec{u}^{k-1} - \mathbf{grad}\ p^k + \vec{f}), \\ \sigma^k = \frac{1}{\tau+\lambda}[\lambda\sigma^{k-1} + 2\tau\mu D(\vec{u}^k)]. \\ \vec{u}^k = \vec{g} \text{ on } \Gamma \text{ and } p^k(M) = 0. \end{array}\right\} \text{ in } \Omega; \qquad (7.79)$$

Notice that, taking the divergence of both sides of the second equation of (7.79) and adding up both sides of the resulting relation to the first equation multiplied by τ, we derive $div\ \vec{u}^k = 0$. Equation (7.79) is said to be quasi-explicit since only the pressure p^k is determined by solving a PDE. More specifically, p^k is the solution of a Poisson equation with inhomogeneous Neumann boundary conditions, which are formally determined by multiplying both sides of the second equation by $\vec{v}$ (cf. [171] and references therein). However, since we are dealing with a variational (weak) formulation, we do not need to bother about these boundary conditions, for they will be automatically fulfilled, though only in a weak sense.

Convergence of $(\vec{u}^k; p^k; \sigma^k)$ to $(\vec{u}; p; \sigma)$ in an appropriate norm is ensured, provided μ is sufficiently large or τ is sufficiently small (cf. [171]).

Equation (7.79) is set in a suitable GLS variational form, and then a classical discretisation of $\vec{u}^k$, p^k and σ^k with P_1 FEs is performed, thereby yielding corresponding approximations $\vec{u}_h^k$, p_h^k and σ_h^k. We refer to reference [171] for the adaptation to this formulation of the quasi-explicit solution algorithm (7.79), among other pertaining details. In particular, the reader should be aware that the solution method for this system of equations in Galerkin formulation with linear FE representations of the three fields turns out to be unstable. This is due to the violation of two Babuška–Brezzi (inf-sup) conditions (see e.g. [169]).

It is important to stress the fact that the mass lumping technique is used to determine $\vec{u}_h^k$ and σ_h^k from the FE analog of the second and third equations of (7.79). This means that the nodal values of both fields are determined by sweeping the mesh in a completely explicit manner (i.e. node by node and component by component). p_h^k in turn is computed by solving the SLAE underlying the discretised first equation of (7.79). Storage requirements are reduced to a minimum, if this SLAE is solved by an iterative method. Actually, in these computations we used the **pre-conditioned conjugate gradient method** (see e.g. [63]) with a pre-conditioning matrix obtained by incomplete Cholesky factorisation. This means that the latter has the same sparsity structure as the original matrix. Notice that in this case, only the nonzero coefficients of fixed matrices have to be stored along the iterations, in the total amount of circa l times the number of nodes multiplied by two to take into account the pre-conditioning matrix. l accounts for the half number of neighbours of a typical node, since the matrix is symmetric. The value of l is about 5 (resp. 14) in two (resp. three) space dimensions. Moreover, since for every k we take p_h^{k-1} as an initial guess, convergence of this linear system solver occurs after one or two iterations, except for the very first values of k.

Next we supply some numerical results illustrating the performance of this algorithm, taking as test problem the **circular Couette flow** or **Taylor–Couette flow** (see e.g. [44]). In this particular problem, a viscous incompressible fluid is confined between two concentric cylinders with radii r_i and $r_e > r_i$. The outer cylinder stands still, while the inner cylinder is rotating with angular velocity ω. If the magnitude of a dimensionless parameter Ta called the Taylor number does not exceed a critical value Ta_c, the flow will be laminar (i.e. not turbulent) and invariant with respect to the axial coordinate z. Ta_c in turn is attained for a critical value of Re depending on r_i and r_e (see e.g. [6]). If $Ta \leq Ta_c$, it is possible to determine exact analytic expressions for $\vec{u}$ and p, and consequently for σ, depending only on the radial coordinate of the polar coordinate system $(O; r, \theta)$ of the concentric cylinders' cross-section plane, whose origin O lies on their axis of symmetry. Moreover, we may consider that the Navier–Stokes equations hold for $\vec{u}$ and p, in an annulus Ω with inner radius r_i and outer radius r_e. Γ is the union of the concentric circles with radii r_i and r_e. The velocity field components in the polar coordinate basis $[\vec{e}_r; \vec{e}_\theta]$ are u_r and u_θ, respectively, for which the following Dirichlet boundary conditions apply: $u_r = 0$ on Γ, $u_\theta = \omega r_i$ for $r = r_i$ and $u_\theta = 0$ for $r = r_e$. Field $\vec{f}$ vanishes identically, and the exact solution is given by

$$u_r \equiv 0; \; u_\theta = A(r_e^2/r - r); \; p = A^2(r^2 - 4r_e^2 \log r - r_e^4/r^2 + 4r_e^2 \log r_e)/2,$$

with $A = \omega r_i^2/(r_e^2 - r_i^2)$. Notice that the pressure vanishes for $r = r_e$ and this is the condition we use to fix the additive constant up to which p is defined.

In a battery of experiments with $Ta < Ta_c$, we took $r_i = 1/2$, $r_e = 1$, $\omega = 1/2$ and $Re = 18.75$, We worked in Cartesian coordinates $(O; x, y)$ of the plane of Ω. This means that we search for two velocity components u_x and u_y in the Cartesian frame in terms of x and y. This is because a priori a computer code ignores symmetries if they are not prescribed. It is noteworthy that in some cases, solving a non-linear problem by forcing symmetries can be harder than without doing it.

In order to observe the convergence of the three fields to the exact solution in the L^2-norm, we solved the problem with meshes containing an equal number of elements in each radial level corresponding to n_r equal subdivisions of annulus' radius, and the subdivision of the domain in the azimuthal sense into n_θ equal sectors, for increasing values of n_r and n_θ. The triangles are obtained by subdividing the thus-generated $n_r \times n_\theta$ trapezoids by means of one of their two diagonals, in such a way that a given diagonal never shares its end-points with the other ones both in the same radial level and in the same sector (see Figure 7.9). In doing so, the mesh contains m triangles with $m = 2n_r \times n_\theta$ but has symmetries with respect to neither x nor y. The mesh in a quarter annulus is illustrated in Figure 7.9 for $n_r = 2$ and $n_\theta = 16$.

Notice that the union of the mesh triangles is not contained in Ω, since Γ consists of both convex and concave portions. The value of τ decreases with the mesh size in the way indicated in Table 7.7, where we display relative errors in the L^2-norm of $\vec{u}$, p and σ, for $n_r = 2, 4, 8, 16$ and $n_\theta = 16, 32, 64, 128$. A tolerance of 0.2×10^{-6} for the velocity maximum increment between two successive iterations was used, and we took $\lambda = .5$. The total number of iterations k_f to attain the tolerance is also shown, rounded to 1000. In order to speed up the convergence in those experiments, we used the combination of mass lumping on the left side with the consistent mass on the right side in the velocity equation (cf. Subsection 7.2.7), in the middle of the simulations. As one can infer from Table 7.8, τ must diminish roughly

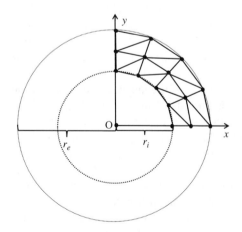

Figure 7.9   Mesh of a quarter annulus for the simulation of circular Couette flow

Table 7.7   Relative errors of $\vec{u}$, p and σ in the L^2-norm.

$m \rightarrow$	64	256	1024	4096
$\tau \rightarrow$	0.00050	0.00040	0.00002	0.00001
$k_f \times 10^{-3} \rightarrow$	8	15	47	112
$\vec{u} \rightarrow$	0.05159502	0.04407774	0.02531385	0.00823048
$p \rightarrow$	0.04668763	0.07072221	0.02191661	0.00143939
$\sigma \rightarrow$	0.26958132	0.13792677	0.06432909	0.02446557

linearly as the mesh is refined, while the total number of iterations to satisfy the tolerance criterion sharply increases. The main conclusion of the above results is the fact that the approximations of the three unknown fields generated by the combination of a $\mathcal{P}_1$ FE – GLS formulation and the algorithm (7.79) can be qualified as reasonable. This is because convergence can be detected as the number of elements increase, though at an unclear rate. On the other hand, the method's main limitation in the solution of this non-linear problem is the rather large number of iterations necessary to attain convergence. However, this is often the case of numerical methods to solve non-linear PDEs designed to reduce costs. Notice that if Newton's iterations were used to solve the same problem by coupling all the three fields, much less iterations would be necessary to attain convergence. But in this case, an $m \times m$ matrix would have to be computed at every iteration, where m equals 6 (resp. 10) times the number of nodes in two (resp. three) space dimensions, in principle in the form of a banded matrix. Notice that this matrix would have to be factorised, in case a direct method or even some iterative method was employed, and this can be really very costly. In short, the solution of large systems resulting from the discretisation of non-linear PDEs requires good compromises between cost and accuracy. That is what we intended to show by means of this test case, among a multitude of other examples to be found in the literature.

For a comprehensive study on the numerical solution of the incompressible Navier–Stokes equations by the FEM, the reader can consult Glowinski's book [90].

7.5   Exercises

7.1   Show that, as long as the solution of the biharmonic equation (7.3) is sufficiently smooth, the local truncation error of FD scheme (7.6) is an $O(h^2)$ with $h = h_x = h_y$. Then develop (7.4)–(7.5) to obtain the 13-point FD scheme analogous to equation (7.6), for solving the biharmonic equation in a rectangle with $h_x \neq h_y$. What happens to the local truncation error in this case?

7.2   Check that equation (7.7) implies equation (7.3), assuming a suitable regularity of u. To be rigorous, the reader should further admit the validity of the following **density property**: For every function $g \in L^2(\Omega)$, given a real number $\epsilon > 0$ arbitrarily small, there exists a function $v \in H_0^2(\Omega)$ such that $\| g - v \|_{0,2} \leq \epsilon$ (see e.g. Adams, 1975).

7.3   Show that $\| H(w) \|_{0,2} = \| \Delta w \|_{0,2}$ for every $w \in H_0^2(\Omega)$ (although the result is perfectly true, in the calculations the reader may assume a little more regularity of w).

7.4   Check that the traces of the normal derivative of a function v in the space V_h associated with the Bogner–Fox–Schmit FE along a mesh edge parallel to the x-axis (resp. y-axis) are a cubic function in terms of x (resp. y). Conclude that, as long as the four values $v(P)$, $[\partial_x v](P)$, $[\partial_y v](P)$ and $[\partial_{xy} v](P)$ at all the mesh vertices P located on the boundary Γ vanish, every function in $v \in V_h$ satisfies the boundary conditions $v = \partial_\nu v = 0$ everywhere on Γ.

7.5 Determine the number of unknowns and matrix bandwidth for the 13-point FD scheme with a $2n_x \times 2n_y$ uniform grid, and the Bogner–Fox–Schmit (BFS) FE to solve equation (7.3) with the corresponding $n_x \times n_y$ mesh. Assume an unknown numbering minimising the maximum bandwidths, and take $n_y \leq n_x$. Compare also matrix sparsity for both methods in the same situation.

7.6 Prove the stability inequality (7.22) for scheme (7.20) taking $L = 1$ (hint: use a technique inspired by the stability analysis for the Three-point FD scheme leading to equation (2.3)).

7.7 Assuming that the solution to the advection–diffusion equation has a continuous third-order derivative in $[0, L]$, prove the first-order convergence in the pointwise sense of FD scheme (7.20) starting from equation (7.22).

7.8 Assuming that $w = \pm 1$, check that equation (7.23) gives rise to equation (7.24).

7.9 Starting from equation (7.25), prove the stability inequalities (7.26) for the SUPG FE scheme with $w = \pm 1$.

7.10 Derive an estimate for $|u - u_h|_{1,2}$, assuming that w is a differentiable function everywhere in $[0, L]$, having a non-positive first-order derivative in $(0, L)$. Specify the underlying constant based on the fact that necessarily $w \in L^\infty(0, L)$.

7.11 Check that the local truncation error of the FV scheme (7.29) in the FD sense is an $O(1)$. Consider the case of a non-uniform mesh.

7.12 Establish the validity of the discrete Friedrichs–Poincaré inequality (7.31).

7.13 Prove that estimate (7.40) holds.

7.14 Establish the validity of equation (7.62) (hint: use arguments similar to those exploited in Section 3.2 in the stability analysis of the Forward Euler scheme).

7.15 Check that the local truncation error of schemes (7.48)–(7.52) is a term of the form $\tau O(1)$, even under suitable solution regularity conditions and assuming that equation (7.63) holds.

7.16 Study the MATLAB code supplied in Example 7.4, until all its steps and instructions are fully understood. Modify the code in order to treat inhomogeneous Dirichlet and Neumann boundary conditions at the right and left ends, respectively, and recover FD approximations of another manufactured solution satisfying them.

7.17 Write down the systems of non-linear algebraic equations to solve at each Newton's iteration, for both a Vertex-centred FV and a $\mathcal{P}_1$ FE discretisation applied to equation (7.75) for $r = 0$ and a constant p. Consider a mesh consisting of n equally spaced intervals.

The lesson is over, the time is up.

Appendix

> We can only see a short
> distance ahead, but we can
> see plenty there that needs to
> be done.
>
> Alan Turing

Other Numerical Techniques for PDEs: A Brief Survey

In this Appendix, a short account is given of effective numerical methods for solving PDEs and relevant related problems, lying outside this book's coverage. Nine methods were selected, out of which six are based exclusively on Galerkin variational formulations. The presentation of these methods is restricted to time-independent equations, but the reader will certainly know how to handle time-dependent problems with a spatial discretisation performed by means of any of these approaches. The boundary element method is treated in more detail, since it is based on principles not really akin to those we have dealt with along the book's seven chapters. The presentation of seven discretisation methods is permeated by brief descriptions of domain decomposition and multigrid methods. The latter are popular techniques that can be combined with most of the former, in order to reduce both implementation costs and processing errors. On the other hand, the author decided to exclude from this Appendix some approaches for the numerical solution of PDEs of current use. This is because either they are not as universal as the methods addressed in this book, or their proper description cannot be undertaken in sufficiently simple and short terms, as the author is committed to in this Appendix. **Monte Carlo methods** (cf. [96]), **wavelet methods** (see e.g. [51]) and **lattice Boltzmann methods** (see e.g. [185]) and [153]) fall into this category.

Numerical Methods for Partial Differential Equations:
An Introduction Finite Differences, Finite Elements and Finite Volumes, First Edition. Vitoriano Ruas.
© 2016 John Wiley & Sons, Ltd. Published 2016 by John Wiley & Sons, Ltd.
Companion Website: www.wiley.com/go/ruas/numericalmethodsforpartial

A.1 The Boundary Element Method

The **boundary element method (BEM)** is designed for certain linear PDEs posed in unbounded domains, but can also be used in the case of bounded domains. In order to shorten this brief description of the method, let us consider an example right away. Many authors describe the BEM taking as a model the Laplace equation (see e.g [111]). Here, we consider the **Helmholz equation** $\Delta u + cu = 0$ in an unbounded domain Ω_e of $\Re^3$, where c is a nonzero constant. If $c > 0$, the Helmholz equation plays an important role in wave propagation applications, in which case one usually sets $c = k^2$, $k \in \Re$, and enters the domain of complex numbers. Ω_e is assumed to be the complementary of a bounded simply connected domain Ω_i with boundary Γ sufficiently smooth. Once specific conditions are prescribed both on Γ and at infinity, the Helmholz equation has a unique solution in Ω_e (cf. [192] and [58]).

Let P be a fixed point of $\Re^3$ not belonging to Γ, and M a varying point in the space. Let also S_R be the ball with centre at the origin and radius $R > 0$ containing Ω_i, and ∂S_R be its boundary. We denote by $\Omega_{e,R}$ the domain comprised between Γ and ∂S_R and by $\vec{\nu}$ the unit outer normal vector to its boundary denoted by $\partial\Omega_{e,R}$. I being the identity operator, let $G(P, M)$ be the function of M satisfying the equation $\int_{\Omega_{e,R}}[\Delta + k^2 I]G(P,M)v(M)dM = v(P)$ for every $v \in C^0(\overline{\Omega_{e,R}})$ with $P \in \Omega_{e,R}$. G is called the **Green's function** of the Helmholz operator $\Delta + k^2 I$. It is defined everywhere but for $M = P$, and its analytic expression is $G(M,P) = e^{-ik|\overrightarrow{PM}|}/(4\pi|\overrightarrow{PM}|)$.

We may formally apply First Green's identity twice in the domain $\Omega_{e,R}$, assuming that the singularities of G in this domain for $M = P$ do not invalidate the calculations (as they effectively don't). In this case, we obtain for all $P \in \Omega_{e,R}$:

$$0 = \int_{\Omega_{e,R}} G(P, M)[\Delta u + k^2 u](M)dM =$$

$$\oint_{\Gamma \cup \partial S_R} G(P, M)\frac{\partial u}{\partial \nu}dS(M) - \oint_{\Gamma \cup \partial S_R} u(M)\frac{\partial G(P, M)}{\partial \nu}dS(M) + u(P).$$

Assuming a certain behaviour of u at infinity[1], the integrals on ∂S_R tend to zero as $R \to \infty$. Then, the solution u at every point $P \in \Omega_e$ can be determined by the formula

$$u(P) = -\oint_\Gamma G(P, M)\frac{\partial u}{\partial \nu}dS(M) + \oint_\Gamma u(M)\frac{\partial G(P, M)}{\partial \nu}dS(M).$$

Notice that there is no singularity in the above expression since M belongs to Γ but not P. However, a singular behaviour is to be expected as P approaches Γ, and the study of this limiting process is the main point to be addressed in order to formulate the integral equation the BEM is based upon. If, for instance, Dirichlet boundary conditions are prescribed, say $u = g$ on Γ, the quantity necessary to determine u is thus $w := \partial u/\partial \nu$ on Γ, which becomes the only problem unknown. Actually, the true unknown in the BEM is a companion scalar function γ defined only on Γ that can be either a single, or a double-layer potential or a combination of both, depending on the type of boundary conditions on Γ one decides to let coincide for the underlying exterior and interior equation (cf. [67]). In the case of the Helmholz equation, with

---

[1] More precisely, that u tends to zero as $1/R$ and Sommerfeld's radiation condition is satisfied [178].

C being an arbitrary nonzero real constant, the solution can be expressed by

$$u(P) = \oint_{\Gamma} \gamma(M) \left[\frac{\partial G(P, M)}{\partial \nu} + C i G(M, P) \right] dS(M), \forall P \in \Omega_e.$$

Anyway, the problem dimension decreases from 3 to 2, and moreover we have to operate only in the bounded domain Γ. This is more than a substantial gain!

The problem to solve in order to determine γ is a **Fredholm integral equation**, which can be set in a variational form based on integrals on Γ. Its description is not so simple, for it requires the manipulation of singular kernels. For this reason, we skip details and refer to reference [33], [67] or [111] for a precise description of this technique.

Here we would just like to emphasise that at the discrete level, the BEM consists of partitioning Γ into a finite number M_h of disjoint (boundary) elements, most commonly segments in two dimensions and triangles in three dimensions. In each element, a number q of approximate values of γ (at least one in the method's simplest version) are computed satisfying a discrete analog of the equations to determine this boundary unknown. In two dimensions, $q - 1$ is the degree of the polynomial approximating γ in every boundary element, whose choice determines the method's order. Here again, the unknown values satisfy a SLAE whose $(qM_h) \times (qM_h)$ matrix is fully populated in general. Although the problem dimension diminishes, in contrast to the FDM, FEM and FVM, the matrix for the BEM is neither symmetric nor positive-definite. Moreover, the fact that it is no longer sparse has bad consequences in terms of storage for problems posed in bounded domains, especially in the three-dimensional case: depending on the method employed to solve the SLAE for the FEM, for example, we may have less matrix entries to store than for the BEM in its equivalent version in terms of order. In view of this, the computational effort to solve SLAEs for the BEM can be discouraging.

In conclusion, the BEM is undoubtedly an efficient technique to solve linear PDEs posed in exterior (unbounded) domains, as long as Green's functions are available for the underlying operator. Its advantages over methods like the FEM or the FVM is to be questioned when the PDE is defined in a bounded domain, especially in the three-dimensional case.

As for rigorous mathematical studies on **boundary integral equations** and the BEM, several authors can be cited. In this respect, the author refers for instance to references [99] and [143].

A.2   Methods Based on Galerkin Variational Formulations

Since the FEM became a widespread technique to solve practically all types of boundary value differential equations, a series of alternatives based on variational formulations of such problems arose. Generally speaking, the main motivation is to attain enhanced accuracy and efficiency at a time, or yet to be better adapted to characteristics of specific classes of PDEs. We have already mentioned two examples of such techniques in Chapters 4 and 7, namely, the XFEM and the h, p and $h - p$ versions of the FEM, respectively. We briefly describe in this section five other methods of the kind that have already become or are likely to become the preferred tools within some communities of practitioners, according to context.

A.2.1   The Discontinuous Galerkin Method

Earlier versions of the **discontinuous Galerkin method** (DGM), appeared in the 1970s (see e.g. [17]). This method can be viewed as a sort of nonconforming FEM, but in a sense different from the one we described in Section 7.1 in general terms. First of all, the DGM is based on meshes that may not be compatible in the FE or FV sense. In the case of triangulations, for example, this means that the intersection of two neighbouring triangles can be just a proper subset of an edge partially shared by them. But this is not the main point. The shape functions in the DGM are completely discontinuous in the sense that they are nonzero only in the interior of a given element. For instance, in the case of linear elements and triangular meshes, there are three shape functions per triangle that can be its barycentric coordinates. However, a neighbouring triangle sharing an edge or part of it with a given triangle will have its own shape functions. Since in the Galerkin formulation each shape functions correspond to one unknown, there are $3M_h$ unknowns in the underlying SLAE. In principle, this increases sharply the number of nonzero entries to store as compared to the standard $\mathcal{P}_1$ FEM, which has the same order. Indeed, in the latter case two triangles account roughly for one mesh node, and hence the total number of unknowns of the DGM as compared to the standard FEM is multiplied by a factor of about six for the same mesh. In the three-dimensional case, the number of shape functions for the same mesh increases by a factor greater than 20! In addition to this, the standard Galerkin variational form cannot be applied, for it is restricted to continuous shape functions or at most to shape functions having minimum continuity properties at inter-element boundaries. Jump terms involving trial and test function discontinuities but also normal derivatives on the boundaries of the elements multiplied by suitable numerical coefficients, known as **interior penalty terms**, must be added to the terms already appearing in the standard Galerkin formulation. We refer to reference [9] for more details on different possible variational formulations for the DGM. Notice that the addition of jump terms renders implementation of DGMs significantly more complex than for standard Galerkin methods.

   In view of all this, one might ask oneself why the DGM has been a relatively successful numerical approach since it was developed some decades ago. The answer is at least twofold. First of all, some category of PDEs have discontinuous solutions or very sharp gradients. This is what happens for instance to non-linear hyperbolic equations such as Bürgers equation, or yet to advection–diffusion equations at high Péclet numbers. In this case, numerical methods with intrinsic discontinuities can produce better results than those forcing solution continuity such as the FEM. Another interesting feature of the DGM is the fact that whatever the degree of the local polynomials in use, the mass matrix in a time-dependent problem is a block diagonal matrix. Each block is associated with nodes pertaining to only one element. For this reason, each block can be easily inverted before the solution process starts. It follows that explicit solution methods can be easily implemented, in contrast to the case of the FEM with piecewise polynomials of degree greater than one, for which a banded mass matrix must necessarily be factorised, before the time marching procedure starts.

   Generally speaking, the DGM is a technique well suited to use in association with certain type of solution algorithms, such as the one described above. In some cases, this can make up for the a priori demerit in terms of degree of freedom count. For further information on the DGM method, we refer to reference [9] or [48].

A.2.2 Spectral Methods

Classical **spectral methods** are linked to Ritz–Galerkin formulations, but there are methods of this type based on **collocation**, sometimes called pseudo-spectral methods (see e.g. [53])[2]. In earlier versions of spectral methods, approximating polynomials were defined in the whole domain, in contrast to the FEM or the DGM. In this case, method's use was restricted to rectangular shaped domains. Later versions associated with **domain decomposition algorithms** are well adapted to subdivisions of the domain into smaller portions. An illustration to the case of subdivisions into triangles can be found in reference [160]. Whatever the case, the goal is to obtain enhanced approximation properties by increasing the degree of the polynomial interpolation in the domain, as long as the exact solution is smooth. The choice of particular sets of points at which the latter are defined allow for method's efficient implementation. These points play a role equivalent to the nodes in the FEM. The use of **Chebyshev polyomials** with corresponding sets of points in each Cartesian direction were popularised from the beginning. The orthogonality of the polynomial basis functions in the L^2-inner product leads to diagonal mass matrices in the case of time-dependent problems, irrespective of their degree. However, a priori the stiffness-type matrix associated with the equation's differential operator is full for each element, which can be a drawback as compared to other methods such as the FEM, in case polynomials of a very high degree are employed.

For an introductory description of spectral methods, the author refers to reference [27].

A.2.3 Meshless Methods

As the name suggests, **meshless methods** are numerical techniques for solving PDEs based on a set of M points (i.e. nodes, that are not necessarily connected through a mesh or grid structure). One of the main motivations for setting up such a technique was avoiding mesh generation, which can be a burden in the three-dimensional case [148], or yet the successive generation of new meshes (a process known as **remeshing**), to conform to new shapes of an evolving structure [26]. There are many different versions of meshless methods, some of them being called **the finite point method**. For a survey on this topic, we refer for instance to reference [129] or [203]. Just to give a significant example, we could quote the one based on **radial basis functions**. In this version, the approximation space is formed by linear combinations of functions of the form $\phi(|\mathbf{x} - \mathbf{x}_i|)$, where $\mathbf{x}_i$ is the location of a generic node in the meshless point set. The function ϕ is monotonic, non-negative and vanishes identically outside a ball centred at $\mathbf{x}_i$ with a certain strictly positive radius. The approximate solution u_h of the underlying variational problem is given by an expression of the form $\sum_{i=1}^{M} u_i w_i(\mathbf{x})$ where w_i is multiple of ϕ. However, in contrast to the FEM, in principle the coefficients u_i of this linear combination are different from $u_h(\mathbf{x}_i)$, since the corresponding basis functions are not supposed to have the property $w_i(\mathbf{x}_j) = \delta_{ij}$. Since meshless methods emerged into current use in the 1990s, they do not seem to have sufficiently convinced practitioners about

---

[2] Collocation is an old numerical method to solve an ODE of the mth order, in which an expansion of the solution as a linear combination of M given functions is assumed. These functions must be at least m times continuously differentiable so that the approximate solution is required to satisfy the ODE at a set of M collocation points belonging to the equation's definitions interval. The collocation technique can be applied to boundary value PDEs, but in this case handling boundary conditions properly may not be so easy.

their universality, as compared to the three methods studied in detail in this book. Among other shortcomings, we could mention the fact that even Dirichlet boundary conditions cannot be enforced as simply and directly as for those three methods. On the other hand, several authors contributed to validate from the rigorous mathematical point of view, the reliability of meshless methods observed in practice (see e.g. [65] and [14]).

A.2.4   Hybrid FEM

Hybrid methods were developed almost in parallel to earlier studies on Galerkin FEM, such as those studied in Chapters 4 and 6. They can be viewed as variants of the standard variational forms in terms of a single unknown called the **primal field** or the underlying mixed formulation set up by introducing an additional unknown called the **dual field**. One of the main contributions in this direction is due to Fraeijs de Veubeke [212]. A basic characteristics of the hybrid formulation is the fact that continuity requirements along inter-element boundaries on piecewise polynomial approximations of unknown functions or fields are relaxed. Instead, Lagrange multipliers defined only on the boundary of the elements are introduced in order to ensure the continuity of the solution only, in a suitable weak sense. In this manner, similarly to the discontinuous Galerkin method in some cases, the resulting SLAE can be simplified by elimination of solution degrees of freedom, which are defined solely at element level. Taking as a model the Poisson equation $-\Delta u = f$ in a polygon Ω with Dirichlet or Neumann boundary conditions, two types of hybrid FEM are generally considered: the **primal hybrid method** and the **dual hybrid method**. The former is derived from the formulation $\int_{\Omega}(\mathbf{grad}\ u|\mathbf{grad}\ v)dxdy = \int_{\Omega}fv\ dxdy$ for every v in a space V, which is approximated by means of a standard Lagrange FE space V_h. In this case, both the FE solution u_h and the test functions $v \in V_h$ are defined at the element level only, and their continuity is not enforced. In order to make up for such discontinuities, terms consisting of the sum of integrals along the element boundaries of u_h or v multiplied by a polynomial function of a certain type defined only thereupon are added to the left side. This procedure gives rise to a sort of mixed formulation, whose unknowns are the a priori discontinuous u_h and a Lagrange multiplier defined only on the boundary of the elements. In the dual hybrid FEM, a similar principle applies to the dual unknown $\vec{p}$ in the standard mixed Galerkin formulation (cf. Subsection 4.3.2). More precisely, the continuity of normal components of the FE counterpart of $\vec{p}$ along the element edges is relaxed, and additional terms involving integrals along the edges are introduced. On the other hand, the primal unknown can be eliminated from the mixed system in an element-by-element basis. For further details on hybrid FEM, we refer to references [37] and [165]. An outstanding application of this technique can be found in reference [159], among many other works.

A.2.5   Isogeometric Analysis

The **isogeometric analysis** (**IGA**) is a numerical approach in common engineering practice similar to a classical FE analysis (**FEA**). The main difference relies on the fact that instead of the usual Lagrange interpolation the solution and test functions are linear combinations of splines commonly employed in CAD (computer-aided design), CAM (computer-aided manufacture) and CAE (computer-aided engineering) to represent objects in the space (cf. [73]). More specifically, IGA operates with **NURBS**, which is an acronym for **non-uniform**

rational B-splines. NURBS are well-known by specialists in computer graphics, in particular owing to the high accuracy that can be attained in the representation of complex surfaces. One of the reasons for this is the fact that NURBS can be many times differentiable functions. Incidentally, in their simplest C^0 form, NURBS are nothing but classical $\mathcal{P}_1$ functions. IGA's basic idea is to speed up design development in different fields of applied sciences where structures to be studied as diverse as aircraft fuselages, human bones, ship hulls, automobile parts and objects alike are duly described by means of this kind of representation. In this manner, the same data structure can be employed to represent unknown fields defined on the objects under study, thereby bypassing a costly conversion step from one representation to the other. Although the first papers on the B-spline FE approach appeared over 15 years ago [113], its fast dissemination among engineers and researchers is relatively recent. For pioneering contributions grounded on similar ideas, we can quote the work by Oñate and Flores of 2005 (see e.g. [154] and [149]). However from the popularizing stand-point, one can say that IGA can be credited to T. J. R. Hughes and collaborators from the Texas University at Austin. The book [54] co-authored by him, Austin Cottrell and Bazilevs is the main reference on this method for the time being. By consulting this work, the reader interested in integration of CAD and FEA will certainly acquire more insight on the principles IGA is based upon, and to some extent about the present state-of-the-art.

A.2.6   The Virtual Element Method

The **virtual element method (VEM)** is the most recent technique based on variational formulations, among those briefly described in this Appendix. It stems from the **mimetic FDM**, which is a technique created in an attempt to mimic FD principles, as applied to distorted grids of arbitrary polygonal or polyhedral form. Among the most active researchers on this method are members of the Italian school at Pavia University, including F. Brezzi and L. D. Marini [25]. At this writing, we can say that the VEM is still incipient and its application range is relatively limited. However, in the author's opinion, this method is bound to become a powerful tool for the numerical simulation of complex events governed by boundary value PDEs. Above all, this evaluation is based on the method's geometric flexibility, for the elements can be polygons or polyhedra of arbitrary shape, even non-convex. Another interesting feature of the VEM is the fact that the variational unknowns are attached only to nodes located at element interfaces. In view of this, geometric restrictions such as those inherent to the FVM can be disregarded. Furthermore, a rigorous theoretical background for the method does exist, at least for some classes of linear problems that have already been studied.

A.3   Domain Decomposition Techniques

A **domain decomposition method (DDM)** consists of splitting a boundary value PDE defined in a bounded domain Ω into smaller problems posed in subdomains of Ω. The underlying sub-problems are solved separately at each step of an iterative procedure, whose aim is to couple them through data exchange between adjacent subdomains on their interfaces. Besides reducing the size of matrix SLAEs, DDMs were designed to favor the use of **parallel computing**, since the problems in the subdomains are solved independently at each step. Furthermore, these methods are also used as pre-conditioners for Krylov space iterative methods in the whole domain, such as the conjugate gradient method or GMRES.

There are basically two kinds of DDM: non-overlapping and overlapping methods. As these terms suggest, in the first type of method only unknowns attached to the interface of two neighboring subdomains domain can be common to them; in contrast, in an overlapping method, the common unknowns pertain to narrow strips contained in adjacent subdomains. The continuity of the solution across subdomain interfaces can be enforced by representing the value of the solution on the interface of neighboring subdomains by the same unknown. This is called a primal DDM. In dual methods, the continuity of the solution across subdomain interfaces is enforced by Lagrange multipliers. In the framework of a numerical solution based on Galerkin formulations of the sub-problems, **mortar methods** are an important class of dual non over-lapping DDM. Similarly to the DGM, in a mortar method the meshes of adjacent subdomains do not match on the interfaces. More details about this kind of technique can be found in reference [89]. A relatively recent text book on DDM is reference [194].

A.4   Multi-grid Methods

The first work on **multigrid methods** is generally credited to Fedorenko [70]. As the name suggests, a multigrid method is based on the use of nested grids or meshes to resolve the same problem. The basic idea is to solve alternately a SLAE coupling the problem unknowns attached to a coarse grid or mesh, and then solve only local problems for unknowns attached to at least one finer grid or mesh by means of a few iterations of an iterative method such as Gauss–Seidel's. There are at least three reasons to justify the use of multigrid methods. The first one is the fact that the use of a direct method in connection with a coarse grid reduces significantly factorisation costs, and thereby harmful effects of round-off errors. The second reason is that the banded matrix related to the finest grid is neither stored nor handled by a direct method. This certainly represents a big advantage in terms of RAM usage. Finally, multigrid methods are fast converging iterative procedures, owing to the following argument. Assume that the solution is expanded into a Fourier series. The solution modes with large amplitudes can be well resolved on the coarsest grid or mesh. On the other hand, the short amplitude modes will be efficiently approximated by an iterative method, but this only couples the nodal unknowns in the immediate vicinity of a given node on the finer grids or meshes [32]. For the very last time, we illustrate the procedure in its simplest expression, that is, in the framework of the one-dimensional problem (P$_1$) solved by the Three-point FDM with uniform grids. Moreover, we do this for the so-called *V-cycle*, the most basic multigrid method.

We are given two grids: a fine (and final one) with size h and an even number of grid points $n > 2$, and a coarse grid with size $2h$. The points of the fine grid are $x_i = ih$ for $i = 0, 1, \cdots, n$ and those of the coarse grid are x_{2i} for $i = 0, 1, \cdots, n/2$. First, we take a vector $[u_1^0, u_2^0, \cdots, u_n^0]^T$ as an intial guess and compute a vector of approximations of u, say $\vec{u}_h^1 := [u_1^1, u_2^1, \cdots, u_n^1]^T$ at the fine grid points by an iterative method, such as the Gauss–Seidel method; only a few iterations should be enough at this stage. In this case, setting $u_{n+1}^0 = u_{n-1}^0$ and assuming that only one Gauss–Seidel iteration has been run, the new approximation vector $\vec{U}_h^1$ is given by

$$U_1^1 = \frac{h^2 f_1 + u_2^0}{2}; \ U_i^1 = \frac{h^2 f_i + U_{i-1}^1 + u_{i+1}^0}{2} \text{ for } i = 2, \cdots, n-1; \ U_n^1 = \frac{h^2 f_n}{2} + U_{n-1}^1.$$

In case more than one GaussSeidel iterations are used, formulae of the same kind yield successive approximation vectors, the last one being $\vec{U}_h^1$. Now, let $\vec{r}_h^1$ be the residual vector $\vec{r}_h^1 :=$ $A_h \vec{U}_h^1 - \vec{b}_h$. The cycle goes on by restricting $\vec{r}_h^1$ to the coarse grid, by simply discarding its components corresponding to the odd grid points, or by means of some averaging process. This step, called residual restriction, generates an $n/2$-component vector $\vec{R}_{2h}^1$. Then, an $n/2$-component correction vector $\vec{D}_{2h}^1 := [D_1^1, \cdots, D_{n/2}^1]^T$ is determined as the solution of $A_{2h} \vec{D}_{2h}^1 = \vec{R}_{2h}^1$, where A_{2h} is the FD matrix associated with the coarse grid. A direct or an iterative method can be used at this level. The next step in the cycle consists of prolongating $\vec{D}_{2h}^1$ into an n-component vector $\vec{d}_h^1 := [d_1^1, d_2^1, \cdots, d_n^1]$, such that $d_{2i}^1 = D_i^1$ and $d_{2i-1}^1 = [D_{i-1}^1 + D_i^1]/2$, for $i = 1, 2, \cdots, n/2$. Then, $\vec{U}_h^1$ is updated to $\vec{u}_h^1 := \vec{U}_h^1 - \vec{d}_h^1$. The cycle is closed by taking $\vec{u}_h^1$ as initial guess of a few Gauss–Seidel iterations on the fine grid, if necessary. This last step is called solution relaxation and smoothing. Eventually, the cycle can be repeated a certain number of times until satisfactory convergence to the vector of unknowns at the fine grid points is attained, but running only one cycle can often be amazingly efficient.

For more complete information about different types of multigrid cycles as applied to the FDM and the FEM, together with related techniques, the author refers the reader for instance to reference [95].

References

[1] ADAMS, R.A., **Sobolev Spaces**, Academic Press, New York, 1975.

[2] AHLBERG, J.H., NILSON, E.N. and WALSH, J.L., **The Theory of Splines and Their Applications**, Academic Press, New York, 1967.

[3] AINSWORTH, M. and ODEN, J.T., **A Posteriori Error Estimation in Finite Element Analysis, Computer Methods in Applied Mechanics and Engineering**, John Wiley & Sons, Hoboken, NJ, 2000.

[4] AKIN, J.E., **Finite Elements for Analysis and Design**, Computational Mathematics and Applications, Academic Press, 2000.

[5] ALLEN III,, M.B. and ISAACSON, E.L., **Numerical Analysis for Applied Science, Pure and Applied Mathematics**, a Wiley-Interscience Series of Texts, Monographs and Tracts, New York, 1998.

[6] ANDERECK, C.D., LUI, S.S. and SWINNEY, H.L., Flow regimes in a circular Couette system with independently rotating cylinders, J. Fluid Mech. 164 (1986), 155–183.

[7] ARANTES E OLIVEIRA, E.R., Theoretical foundations of the finite element method, Int. J. Solids and Struct., 4 (1968), 929–952.

[8] ARGYRIS, J.H., **Energy Theorems and Structural Analysis**, Butterworth, London, 1960.

[9] ARNOLD, D.N., BREZZI, F., COCKBURN, B. and MARINI, L.D., Unified analysis of discontinuous Galerkin methods for elliptic problems, SIAM J. Numer. Anal. 39-5 (2002), 1749–1779.

[10] BABA, M. and TABATA, M., On a conservative upwind finite element scheme for convective-diffusion equations, RAIRO-Analyse Numérique 15-1 (1981), 3–25.

[11] BABUŠKA, I., The finite element method with Lagrangian multipliers, Numerische Mathematik, 20 (1973), 179–192.

[12] BABUŠKA, I. and RHEINBOLDT, W.C., A posteriori error estimates for the finite element method, Int. J. Numer. Meth. Engrg. 12 (1978), 1597–1615.

[13] BABUŠKA, I. and GUO, B.Q., The h, p and h-p version of the finite element method: basis theory and applications, Advances Engin. Software, 15-3-4 (1992), 159–174.

[14] BABUŠKA, I. and MELENK, J.M., The partition of unit method, International Journal for Numerical Methods in Engineering, 40-4 (1997), 581–772

[15] BACHELOR, G.K., **An Introduction to Fluid Dynamics**, Cambridge University Press, Cambridge, 2000.

Numerical Methods for Partial Differential Equations:
An Introduction Finite Differences, Finite Elements and Finite Volumes, First Edition. Vitoriano Ruas.
© 2016 John Wiley & Sons, Ltd. Published 2016 by John Wiley & Sons, Ltd.
Companion Website: www.wiley.com/go/ruas/numericalmethodsforpartial

[16] BAI, M. and ELSWORTH, D., **Coupled Processes in Subsurface Deformation, Flow and Transport**, American Society of Civil Engineers Press, Reston, VA, 2000.

[17] BAKER, G.A., Finite element methods for elliptic equations using non-conforming elements, Math. Comput. 31–137 (1977), 45–59.

[18] BANK, R.E. and WEISER, A., Some a posteriori error estimators for elliptic partial differential equations. Math. Comput. 44 (1985), 283–301.

[19] BARANGER, J., MAÎTRE J.-F. and OUDIN, F., Connection between finite volume and mixed finite element methods, ESAIM: Mathematical Modelling and Numerical Analysis, 30-4 (1996), 445–465.

[20] BARANGER, J. and ELAMRI, H., Estimateurs a posteriori d'erreur pour le calcul adaptatif d'écoulements quasi-newtoniens, RAIRO-Anal. Numér., 25 (1991), 31–48.

[21] BARBEIRO, S., Supraconvergent cell-centered scheme for two-dimensional elliptic problems, Applied Numerical Mathematics, 59 (2009), 56–72

[22] BARBEIRO, S., **Personal communication, Universidade de Coimbra**, Portugal, 2015.

[23] BARROS, F.B., PROENÇA, S.P.B. and DE BARCELLOS, C.S., On error estimator and p-adaptivity in the generalized finite element method, Internat. J. Numer. Methods Engrg., 60 (2004), 2373–2398.

[24] BEILINA, L. (ed.), **Inverse Problems and Applications**, Springer Proceedings in Mathematics and Statistics, Springer, Berlin, 2015.

[25] BEIRÃO DA VEIGA, L, BREZZI, F., CANGIANI, A., MANZINI, G., MARINI, L.D. and RUSSO, A., Basic principles of virtual element methods, Math. Models Methods Appl. Sci., 23-1 (2013), 199, DOI:10.1142/S0218202512500492.

[26] BELYTSCHKO, T., KRONGAUZ, I., ORGAN, D. and FLEMING, M., Meshless Method: an overview and recent developments, Comp. Meth. Appl. Mech. Engin., 139 (1996),3–47.

[27] BERNARDI, C. and MADAY, Y., **Spectral Methods**, in Handbook of Numerical Analysis, P.G. Ciarlet and J.L. Lions eds., Springer, Berlin, 1997.

[28] BERTRANDIAS. J.-P., **Mathématiques pour l'Informatique, Tome 1: Analyse Fonctionnelle, Collection U**, Armand Collin, Paris, 1970.

[29] BLUM, E.K., **Numerical Analysis and Computation: Theory and Practice**, Addison-Wesley, Reading MA, 1972.

[30] BOCHEV, P. and GUNZBURGER, M.D., **Least Squares Finite Element Methods**, Applied Mathematical Sciences 166, Springer, Berlin, 2009.

[31] BRAESS, D., **Finite Elements Theory, Fast Solvers, and Applications in Solid Mechanics**, Cambridge University Press, Cambridge, 1997.

[32] BRANDT, A., Multi-level adaptive solutions to boundary-value problems, Math. Comput., 31 (1977), 333–390.

[33] BREBBIA, C.A. and DOMINGUEZ, J., **Boundary Elements: An Introductory Course**, WIT Press, Computational Mechanics Publications, Ashurst, UK, 1996.

[34] BRENNER, S.C. and SCOTT, L.R., **The Mathematical Theory of Finite Element Methods**, Texts in Applied Mathematics 15, Springer, Berlin, 2008.

[35] BREZINSKI, C. and REDIVO-ZAGLIA, M., **Extrapolation Methods. Theory and Practice**, North-Holland, Amsterdam, 1991.

[36] BREZZI, F., On the existence, uniqueness and approximation of saddle-point problems arising from Lagrange multipliers, RAIRO Anal. Numér., 8-2 (1974), 129–151.

[37] BREZZI, F. and FORTIN, M. (eds.), Mixed and Hybrid Finite Element Methods, Springer Series in Computational Mathematics, Volume 15, Springer, Berlin, 1991.

[38] BROOKS, A.N. and HUGHES, T.J.R., Streamline upwind/Petrov Galerkin formulations for convection dominated flows with particular emphasis on the incompressible Navier-Stokes equations, Computer Methods in Applied Mechanics and Engineering, 32 (1982), 199–259.

[39] CAI, Z., On the finite volume element method, Numerische Mathematik, 58-1 (1990/91), 713–735.

[40] CÉA, J., Approximation variationnelle des problèmes aux limites, Annales de l'Institut Fourier, 14-2 (1964), 345–444.

[41] CHARNEY, J.G., FJØRTOFT, R. and VON NEUMANN, J., Numerical integration of the barotropic vorticity equation, Tellus 2 (1950), 237–254.

[42] CHATZIPANTELIDIS, P., LAZAROV, R.D. and THOMÉE, V., **Error Estimates for a Finite Volume Element Method for Parabolic Equations in Convex Polygonal Domain**, Wiley Inter-Science (www.interscience.wiley.com), Hoboken, NJ, 2004, DOI: 10.1002/num.20006

[43] CHEN, L., Finite Volume Methods, www.math.uci.edu/chenlong/226/FVM.pdf

[44] CHOSSAT, P. and IOOSS, G., **The Couette-Taylor Problem**, Applied Mathematical Sciences 102, Springer, Berlin, 1994, DOI: 10.1007/978-1-4612-4300-7

[45] CIARLET, P.G., **The Finite Element Method for Elliptic Problems**, North Holland, Amsterdam, 1978.

[46] CIARLET, P.G., **Introduction to Numerical Linear Algebra and Optimization**, Cambridge University Press, Cambridge, 1989.

[47] CIARLET, P.G., **Linear and Nonlinear Functional Analysis with Applications**, SIAM, Philadelphia, 2013.

[48] COCKBURN, B., KARNIADAKIS, G.E., SHU, C.-W. (eds.), Discontinuous Galerkin Methods. **Theory, Computation and Applications**, Lecture Notes in Computational Science and Engineering, Springer, Berlin 2000.

[49] CODINA, R., Stabilization of incompressibility and convection through orthogonal sub-scales in finite element methods, Computer Methods in Applied Mechanics and Engineering, 190 (2000), 1579–1599.

[50] CODINA, R. and ZIENKIEWICZ, O.C., CBS versus GLS stabilization of the incompressible Navier-Stokes equations and the role of the time step as a stabilization parameter, Comm. Numer. Meth. Engin., 18, (2002), 99–112.

[51] COHEN, A., **Numerical Analysis of Wavelet Methods**, Elsevier, Amsteram, 2003.

[52] COLLATZ, L., **The Numerical Treatment of Differential Equations**, Grundlehren der mathematischen Wissenschaften 60, Springer Verlag, Berlin, 1960.

[53] COSTA, B., Spectral methods for partial differential equations, CUBO Math. 6-4 (2004),1–32.

[54] COTTRELL, J. A., HUGHES, T.J.R., BAZILEVS, Y., **Isogeometric Analysis: Toward Integration of CAD and FEA**, John Wiley & Sons, Hoboken, NJ, 2009.

[55] COURANT, R., FRIEDRICHS, K. and LEWY, H., Über die partiellen Differenzengleichungen der mathematischen Physik, Math. Annal. 100-1 (1928) 32–74.

[56] COURANT, R., Variational methods for the solution of problems of equilibrium and vibrations, Bull. Amer. Math. Soc., 49 (1943), 1–23.

[57] COURANT, R. and HILBERT, D., **Methods of Mathematical Physics**, John Wiley, New York, 1962.

[58] COURANT, R. and HILBERT, D. **Methods of Mathematical Physics. Partial Differential Equations**, 2, Wiley-Interscience, New York, 1965.

[59] CUMINATO, J.A. and RUAS, V., Comput. Appl. Math., 34 (2015), 1009–1033.

[60] DAUTRAY, R., and LIONS, J.-L. **Mathematical Analysis and Numerical Methods for Science and Technology, Vol. 2: Functional and Variational Methods**, Springer-Verlag, Berlin, 1988.

[61] DE VAHL DAVIS, **G. Numerical Methods in Engineering and Science**, Allen & Unwin Ltd., London, 1986.

[62] DISKIN, B. and THOMAS, J.L., Notes on accuracy of finite-volume discretization schemes on irregular grids, Journal Applied Numerical Mathematics, 60-3 (2010), 224–226

[63] DONGARRA, J.J., DUFF, I.S., SORENSEN, D.C. and VAN DER VORST, H.A., **Solving Linear Systems on Vector and Shared Memory Computers**, SIAM, Philadelphia, 1991.

[64] DOUGLAS Jr.,, J. and DUPONT, T., Galerkin methods for parabolic equations, SIAM J. Numer. Anal., 7-4 (1970), 575–626.

[65] DUARTE, C.A. and ODEN, J.T., H-p clouds an h-p meshless method, Numer. Meth. Part. Diff. Equa., 12-6 (1996), 643–766.

[66] ERN, A. and GUERMOND, J.-L., **Theory and Practice of Finite Elements, Applied Mathematical Sciences 159**, Springer, New York, 2004.

[67] EUVRARD, D., **Résolution numérique des équations aux dérivées partielles de la physique, de la mécanique et des sciences de l'ingénieur. Différences finies, éléments finis, méthode des singularités**, Masson, Paris, 1990.

[68] EYMARD, R., GALLOUËT, T.R. and HERBIN, R., The finite volume method, In: **Handbook of Numerical Analysis, Vol. VII**, Ciarlet, P.G. and Lions, J.-L. eds, North Holland, Amsterdam, 2000.

[69] FALK, R.S., Lecture notes on numerical solution of PDEs (online), **Chapter 1, Finite Difference Methods for Elliptic Equations**, Department of Mathematics, Rutgers, The State University of New Jersey, 2012.

[70] FEDORENKO, R.P., A relaxation method for solving elliptic difference equations, USSR Comput. Math. Math. Phys. 1–5 (1961) 1092–1096.

[71] FERREIRA, J.A. and GRIGORIEFF, R.D., On the supraconvergence of elliptic finite difference methods, App. Numer. Math., 28 (1998), 275–292.

[72] FERZIGER, J.H. and PERIČ, M., **Computational Methods for Fluid Dynamics**, third rev. ed., Springer, Berlin, 2002.

[73] FORSEY, D.R. and BARTELS, R.H.. Hierachical b-spline refinement, Comput. Graph., 22 (1988), 205–212.

[74] FORSYTH, P.A. and SAMMON, P.H., Quadratic convergence for cell-centered grids, Appl. Num. Math., 4 (1988), 377–394.

[75] FRAEIJS DE VEUBEKE, B., Displacement and equilibrium models in the finite element method, in: **Stress Analysis**, O.C. Zienkiewicz and G.S. Holister eds., John Wiley & Sons, 1965, pp. 145–197.

[76] FREY, P.J. and GEORGE, P.-L., **Mesh Generation**, John Wiley & Sons (ISTE), Hoboken, NJ, 2008.

[77] FRIEDMAN, A., **Partial Differential Equations of Parabolic Type**, Dover Books on Mathematics, New York, 2008.

[78] FRIES, T.-P. and MATTHIES, H.G., **A Review of Petrov-Galerkin Stabilization Approaches and an Extension to Meshfree Methods**, Informatikbericht Nr.: 2004-01, Institute of Scientific Computing, Technische Universität Braunschweig, 2004.

[79] FUBINI, G., Sugli integrali multipli, Opere Scelte. 2 Cremonese (1958), 243–249.

[80] FUJII, H., Some remarks on finite element analysis of time-dependent field problems. In: **Theory and Practice in Finite Element Structural Analysis**, University of Tokyo Press, Tokyo, 1973, pp. 91–106.

[81] FUJITA, H. and SUZUKI, T., Evolution problems, in: P.G. Ciarlet and J.-L. Lions eds., **Handbook of Numerical Analysis II**, North-Holland, Amsterdam, 1991, pp. 789–928.

[82] GALEÃO, A.C. and DUTRA DO CARMO, E.G., A consistent upwind Petrov-Galerkin method for convection-dominated problems, Comput. Meth. Appl. Mech. Engin., 68 (1988), 81–95.

[83] GANTMACHER, F.R. **The Theory of Matrices**, Chelsea Publishing Company, New York, 1959.

[84] GEAR, C.W., **Numerical Initial Value Problems in Ordinary Differential Equations**, Prentice-Hall, Englewood Cliffs, NJ, 1971.

[85] GERSCHGORIN, S., Fehlerabschätzung für das Differenzenverfahren zur Lösung partieller Differentialgleichungen, Z. Angew. Math. Mech., 10 (1930), 373–382.

[86] GILEVA, L., SHAYDUROV, V. and DOBRONETS, B., The triangular Hermite finite element complementing the Bogner-Fox-Schmit rectangle, Appl. Math., 4 (2013), 50–56.

[87] GIRAULT, V. and RAVIART, P.-A., **Finite Element Methods for the Navier-Stokes Equations**, Springer, New York, 1986.

[88] GLOWINSKI, R., **Numerical Methods for Nonlinear Variational Problems**, Springer-Verlag, New York, 1984.

[89] GLOWINSKI, R., PAN, J. and PERIAUX, J., **Fictitious Domain / Domain Decomposition Methods for Partial Differential Equations, Chapter 11 of Domain-Based Parallelism and Problem Decomposition Methods in Computational Science and Engineering**, Keyes, D.E., Saad, Y. and Truhlar, D.G. eds., SIAM, Philadelphia, 1995, pp. 177–192.

[90] GLOWINSKI, R., Finite element methods for incompressible viscous flow, In: **Handbook of Numerical Analysis**, Vol. 9, P.G. Ciarlet and J.L. Lions, eds., North-Holland, Amsterdam, 2003, p. 3–1176.

[91] GODUNOV, S. K., A Difference scheme for numerical solution of discontinuous solution of hydrodynamic equations, Math. Sbornik, 47 (1959), 271–306 (translated US Joint Publ. Res. Service, JPRS 7226, 1969).

[92] GRISVARD, P., **Elliptic Problems in Nonsmooth Domains**, Pitman, 1985.

[93] GURTIN, G.E., The Linear Theory of Elasticity, **Handbuch der Physik**, vol. VI, a/2, Springer Verlag, New York, 1972.

[94] HACKBUSCH, W., **Elliptic Partial Differential Equations**, Springer Series in Computational Mathematics 18, Springer Verlag, Berlin-Heidelberg, 1992.

[95] HACKBUSCH, W., **Multi-Grid Methods and Applications**, Springer, Berlin, 2003.

[96] HAMMERSLEY, J.M. and HANDSCOMB, D.C., Monte Carlo Methods, **Monographs on Applied Probability and Statistics**, Chapman and Hall Ltd, London, 1979.

[97] HEINRICH, J., HUYAKORN, P.S., ZIENKIEWICZ, O.C. and MITCHELL, A.R., An "Upwind" finite element scheme for two-dimensional convective-transport equations, Intl. J. Numer. Meth. Engin., 11 (1977), 131–143.

[98] HIRSCH, C., Numerical Computation of Internal and External Flows. **The Fundamentals of Computational Fluid Dynamics**, 2nd ed., Elsevier, Amsterdam, 2007.

[99] HSIAO, G. C. and WENDLAND, W. L. (eds.), **Boundary Integral Equations**, Applied Mathematical Sciences 164, Springer, New York: 2008.

[100] HUGHES, T.J.R., **The Finite Element Method. Linear Static and Dynamic Finite Element Analysis**, Prentice Hall, Englewood Cliffs, NJ, 1987.

[101] HUNTER, J.K., **Lecture notes on partial differential equations (online)**, Chapter 6, Parabolic Equations, Department of Mathematics, University of California at Davis, 2012.

[102] IDELSOHN, S. and OÑATE, E., Finite volumes and finite elements: two 'good friends', Num. Meth. Engin., 37 - 19 (1994), 3323–3341.

[103] IKEDA, T., **Maximum Principle in Finite Element Models for Convection-Diffusion Phenomena**, North-Holland Mathematics Studies 76, Lecture Notes in Numerical and Applied Analysis vol. 4, North Holland, Amsterdam, 1983.

[104] IRONS, B.M., A frontal solution program for finite element analysis, Intl. J. Numer. Meth. Engin., 2 (1970), 5–32.

[105] ISAACSON, E. and KELLER, H.B., **Analysis of Numerical Methods**, Dover, Mineola, NY, 1994.

[106] JAFFRÉ, J., ROBERTS, J.E. and SAADA, A., Cell-centered discretization methods for 2nd order elliptic problems, Lecture Notes, ENIT, Tunisia, 2008, http://www.tn.refer.org/unesco/semestre1/Cours-sem1/Jaffre.pdf/

[107] JAMET, P., Estimations d'erreur pour des éléments finis droits presque dégénérés, RAIRO Anal. Numér. 10 (1976), 43–61.

[108] JOE, B., Construction of three-dimensional Delaunay triangulations using local transformations, Comp. Aided Geo. Des., 8 (1991), 123–142.

[109] JOHN, F., **Partial Differential Equations**, Applied Mathematical Sciences, nr. 1, Springer, New York, 1982.

[110] JOHNSON, C., **Numerical Solution of Partial Differential Equations by the Finite Element Method**, Cambridge University Press, Cambridge, 1987.

[111] JONES, F., **Lebesgue Integration on Euclidean Space**, Jones and Barlett Publishers, Boston, 1993.

[112] JOST, J., **Partial Differential Equations, Graduate Texts in Mathematics**, Springer, Berlin, 1998.

[113] KAGAN, P., FISCHER, A. and BAR-YOSEPH, P.Z., New b-spline finite element approach for geometrical design and mechanical analysis, Intl. J. Numer. Meth. Engin., 41 (1998), 425–458.

[114] KAWAHARA, M., TAKEUCHI, N. and YOSHIDA, Y., Two step finite element methods for tsunami wave propagation analysis, Intl. J. Numer. Meth. Engin., 12 (1978), 331–351.

[115] KIKUCHI, F. and USHIJIMA, T., On finite element methods for convection dominated phenomena, Math. Meth. Appl. Sci., 4 (1982), 98–122.

[116] KNABNER, P. and ANGERMANN, L., **Numerical Methods for Elliptic and Parabolic Partial Differential Equations**, Springer, New York, 2003.

[117] LADEVÈZE, P. and LEGUILLON, D. Error estimate procedure in the finite element method and appli- cations. SIAM J. Numer. Anal., 20 (1983), 485–509.

[118] LANDAU, L.D. and LIFSHITZ, E.M., **Theory of Elasticity**, Elsevier, Amsterdam, 1986.

[119] LARSSON, S. and THOMEE, V., **Partial Differential Equations with Numerical Methods**, Texts in Applied Mathematics 45, Springer Verlag, Berlin-Heidelberg, 2003.

[120] LASCAUX, P. and LEASINT, P, Some nonconforming finite elements for the plate bending problem, ESAIM, 9-R1 (1975), 9–53

[121] LASCAUX, P. and THEODOR, R., **Analyse numérique matricielle appliquée à l'art de l'ingénieur**, Masson, Paris, 2004.

[122] LAX, P. and MILGRAM, A. N., Parabolic equations. Contributions to the theory of partial differential equations, Ann. Math. Stud., 33 (1954), 167–190.

[123] LAX, P. and RICHTMYER, R.D., Survey of the stability of linear finite difference equations, Comm. Pure Appl. Math. 9 (1956), 267–293.

[124] LESAINT, P. and RAVIART, P.-A., On a finite element method for solving the neutron transport equation, in: **Mathematical Aspects of Finite Elements in Partial Differential Equations**, C. de Boor ed., Academic Press, 1974, pp. 89–123.

[125] LEVEQUE, R.J., **Finite Volume Methods for Hyperbolic Problems**, Cambridge University Press, Cambridge, 2002.

[126] LIONS, J.L., **Problèmes aux limites dans les équations aux dérivées partielles**, Presses de l'Université de Montréal, Montreal, 1965.

[127] LIONS, J.L. and MARCHUK, G. I. eds., **Méthodes Mathématiques de l'Informatique 4, sur les méthodes numériques en sciences physiques et économiques**, Dunod, Paris, 1974.

[128] LIONS, J.-L. and MAGENES, E., **Problèmes aux limites non homogènes et applications**, Dunod, Paris, 1968.

[129] LIU, G.R., **Meshfree Methods: Moving Beyond the Finite Element Method**, 2nd ed., CRC Press, Boca Raton, FL, 2010.

[130] LONG, C., Finite Volume Methods, www.math.uci.edu/chenlong/226/FVM.pdf

[131] LUBLINER, J. and PAPADOPOULOS, P., **Introduction to Solid Mechanics: An Integrated Approach**, Springer, New York: 2014.

[132] LUENBERGER, D. G., **Optimization by Vector Space Methods**, John Wiley & Sons, New York, 1969.

[133] MARCHUK, G.I. and SHAIDUROV, V.V., **Difference Methods and Their Extrapolations**, Series of Applications of Mathematics 19, Springer, New York 1983.

[134] MEYERS, M.A. and CHAWLA, K.K., **Mechanical Behavior of Materials**, Prentice Hall, Englewood Cliffs, NJ, 1999.

[135] MIAN, M. A., **Petroleum Engineering Handbook for the Practicing Engineer (Vol. 1)**, Penn Well Publishing Company, Tulsa, OK, 1992.

[136] MIKHLIN, S.G., **Variational Methods in Mathematical Physics**, MIR, Moscow, 1957.

[137] MINKOWYCZ, W.J., SPARROW, E.M. and , MURTHY, J.Y., **Handbook of Numerical Heat Transfer**, Wiley, Chichester, 2006.

[138] MITCHELL, A.R., **Computational Methods in Partial Differential Equations**, John Wiley & Sons, London, 1969.

[139] MITCHELL, A.R. and GRIFFITHS, D.F., **The Finite Difference Method in Partial Differential Equations**, John Wiley & Sons, London, 1980.

[140] MIYAUCHI. M., KITAGIRI, M. and KAMITANI, A., Automatic numerical element generation by boundary-fitted curvilinear coordinate system, Period. Polytech. Elec. Eng., 38-1 (1994), 29–43.

[141] MOËS, N., DOLBOW, J. and BELYTSCHKO, T., A finite element method for crack growth without remeshing, Intl. J. Numer. Meth. Engin., 46-1 (1999), 131–150.

[142] MORTON, K.W. and MAYERS, D.F., **Numerical Solution of Partial Differential Equations**, Cambridge University Press, Cambridge, 2005.

[143] NEDELEC, J.-C., **Approximation des équations intégrales en mécanique et en physique**, Lecture Notes, Centre de Mathématiques Appliquées, Ecole Polytechnique, Palaiseau, France, 1977.

[144] NITSCHE, J. A., L^{∞}-convergence of finite element approximations, **Proceedings of the C.N.R. Conference on Mathematical Aspects of Finite Element Methods**, Rome, 1975 (also in: Lecture Notes in Mathematics, **606**, Springer, Berlin, pp. 261–274, 1977).

[145] NITSCHE, J. A., L^{∞}-convergence of finite element Galerkin approximations for parabolic problems, RAIRO Numer. Anal. 13 (1979), 31–54.

[146] ODEN, J.T., BECKER, E.B. and CAREY, G.F., **Finite Elements: Computational Aspects**, Vol. 3, Prentice Hall, Englewood Cliffs, NJ, 1981.

[147] OMNES, P., On the second-order convergence of finite volume methods for the Laplace equation on Delaunay-Voronoi meshes, ESAIM, EDP Sciences, 45-4 (2011), 627–650.

[148] OÑATE, E., IDELSOHN, S., ZIENKIEWICZ, O.C. and TAYLOR, R.L., A finte point method in computational mechanics. Applications to convective transport and fluid flow, Int. J. Num. Meths. Engin., 39 (1996), 3839–3866.

[149] OÑATE, E. and FLORES F.G., Advances in the formulation of the rotation-free basic shell triangle, Comput. Methods Appl. Mech. Eng., 194 (2005) 2406–2443.

[150] ORTEGA, J.M. and RHEIMBOLDT, W.C., **Iterative Solution of Nonlinear Equations in Several Variables**, Academic Press, New York, 1970.

[151] PATANKAR, S.V. and SPALDING, D.B., A calculation procedure for heat, mass and momentum transfer in three-dimensional parabolic flows, Intl. J. Heat Mass Flow, 15-10 (1972), 1787–1806.

[152] PATANKAR, S. V., **Numerical Heat Transfer and Fluid Flow**, Taylor-Francis, Philadelphia, 1980.

[153] PAVARINO, L. F. and TOSELLI, A. (eds.), **Recent Developments in Domain Decomposition Methods**, Lecture Notes in Computational Science and Engineering 23, Springer, New York, 2012.

[154] PIMENTA, P.M. and WRIGGERS, P. eds., **New Trends in Thin Structures: Formulation, Optimization and Coupled Problems**, CISM Courses and Lectures 519, Springer, New York, 2011.

[155] PREPARATA, F. R. and SHAMOS, M. I., **Computational Geometry: An Introduction**, Springer-Verlag, New York, 1985.

[156] PROTTER, M.H. and WEINBERGER, H.F., **Maximum Principle in Differential Equations**, Springer, Heidelberg, 1984.

[157] QUARTAPELLE, L., **Numerical Solution of the Incompressible Navier-Stokes Equations**, Springer Basel AG, Basel, 1993.

[158] QUARTERONI, A., SACCO, R. and SALERI, F., **Numerical Mathematics**, Springer, Berlin, 2000.

[159] RADU, F.A., BAUSE, M., PRECHTEL, A. and ATTINGER, S., A mixed hybrid finite element discretization scheme for reactive transport in porous media, Numer. Math. Adv. Appl., (2008), 513–520– (also in **Proceedings of ENUMATH 2007**, the 7th European Conference on Numerical

Mathematics and Advanced Applications, Graz, Austria, September 2007, Karl Kunisch, Günther Of, Olaf Steinbach eds.).

[160] RAPPETI, F. and PASQUETTI, R., Spectral element methods on unstructured meshes: which interpolation points? Numer. Algor., 2010, DOI: 10.1007/s11075-010-9390-0

[161] RAVIART, P.-A. and THOMAS, J.-M., **Mixed finite element methods for second-order elliptic problems**, Lecture Notes in Mathematics 606, Springer Verlag, Berlin-Heidelberg, 1977, pp. 292–315.

[162] RAVIART, P.-A. and THOMAS, J.-M., **Introduction à l'Analyse Numérique des Equations aux Dérivees Partielles**, Masson, Paris, 1983.

[163] RICHTMYER, R.D. and MORTON, K.W., **Difference Methods for Initial Value Problems**, John Wiley & Sons, London, 1967.

[164] RIVARA, M.C., Local modification of meshes for adaptive and/or multigrid finite-element methods, J. Comput. Appl. Math., 36-1 (1991), 79–89.

[165] ROBERTS, J. and THOMAS, J.-M., Mixed and Hybrid Finite Element Methods, Research Report RR-0737, HAL Id: inria-00075815, 1987, on-line access https://hal.inria.fr/inria-00075815

[166] RUAS, V., Automatic generation of triangular finite element meshes, Comp. Math. Appl., 5-2 (1979), 125–140.

[167] RUAS, V., Méthodes d'éléments finis pour l'élasticité non linéaire incompressible et diverses applications à l'approximation des problèmes aux limites, Thèse de Doctorat d'Etat, Université Pierre et Marie Curie, Paris, 1982.

[168] RUAS, V., On the strong maximum principle for some piecewise linear finite element approximations of non positive type, J. Facul. Sci. Univ. of Tokyo, Section I-A, 29 (1982), 473–491.

[169] RUAS, V., Finite element methods for the three-field Stokes system in $\Re^3$: Galerkin methods, Math. Model. Numer. Anal., 30-4 (1996), 489–525.

[170] RUAS, V., An Introduction to the Mathematical Foundations of the Finite Element Method. Lecture notes, Fachbereich Mathematik, Universität Hamburg, 2001, www.icmc.usp.br/gustavo .buscaglia/cursos/mef/mef.pdf

[171] RUAS, V. and BRASIL Jr,, A.P., Explicit solution of the incompressible Navier-Stokes equations with linear finite elements, Appl. Maths. Letters, 20-9 (2007), 1005–1010.

[172] RUAS, V., BRASIL J., A.P. and TRALES, P.R., An explicit method for convection-diffusion equations, Japan J. Indus. Appl. Math., 26 (2009), 65–91.

[173] RUAS, V., Hermite finite elements for second order boundary value problems with sharp gradient discontinuities, J. Comput. Appl. Maths., 246 (2013) 234–242.

[174] RUAS, V., KISCHINHEVSKY, M. and LEAL TOLEDO, R., Elementos finitos em formulação mista de mínimos quadrados para a simulação da convecção-difusão em regime transiente, Revista Internacional de Métodos Numéricos para Cálculo y Diseño en Ingeniería, 29 (2013), 21–28.

[175] RUAS, V., BRANDÃO, D. N. and KISCHINHEVSKY, M., Hermite finite elements for diffusion phenomena, J. Comput. Phys., 235 (2013), 542–564.

[176] SAAD, Y., **Iterative Methods for Sparse Linear Systems**, 2nd ed.. SIAM, Philadelphia, 2003.

[177] SAMARSKII, A. A., **Introduction to the Theory of Difference Schemes**, Nauka, Moscow (in Russian), 1971.

[178] SCHOT, S.H., Eighty years of Sommerfeld's radiation condition, Hist. Math., 19-4 (1992), 385–401.

[179] SCHWARTZ, L., **Théorie des distributions**, Hermann et Cie. éditions, Paris, 1950.

[180] SCOTT, R., Optimal L^∞ estimates for the finite element method on irregular meshes, Math. Comp. 30 (1976), 681–697.

[181] SOUTHWELL, R.V., **An Introduction to the Theory of Elasticity**, Dover, New York, 1969.

[182] STRANG, G. and FIX, G.J., **An Analysis of the Finite Element Method**, Prentice Hall, Englewood Cliffs, NJ, 1973.

[183] STRANG, G., **Linear Algebra and Its Applications**, 4th ed., Thomson Brooks/Cole, New York, 2006.

[184] STROUD, A.H., **Approximate Calculation of Multiple Integrals**, Prentice-Hall, Englewood Cliffs, NJ, 1971.

[185] SUKOP, M.C. and THORNE, D.T., **Lattice Boltzmann Modeling: An Introduction for Geoscientists and Engineers**, Springer, New York, 2007.

[186] SZABÓ, B. and BABUŠKA, I., **Introduction to Finite Element Analysis. Formulation, Verification and Validation**, Wiley, Chichester, 2011.

[187] TABATA, M., Uniform convergence of the upwind finite element approximation for semilinear parabolic problems, J. Math. Kyoto Univ., 18-2 (1978), 327–351.

[188] TEMAM, R., **Navier-Stokes Equations**, North Holland, Amsterdam, 1977.

[189] THOMAS, J.-M., Sur l'analyse numérique des méthodes d'éléments finis hybrides et mixtes, Thèse de Doctorat d'Etat, Université Pierre et Marie Curie, Paris, 1977.

[190] THOMEE, V., Finite Difference Methods for Linear Parabolic Equations, In: **Handbook of Numerical Analysis, Finite Difference Methods, Part 1**, Ciarlet, P.G. and Lions, J.-L. eds, North Holland, Amsterdam, 1989.

[191] THOMPSON, J.F., WARSI, Z.U.A. and MASTIN, C.W., Boundary-fitted coordinate systems for numerical solution of partial differential equations: –A review, J. Comp. Phys., 47-1 (1982), 1–108.

[192] TIKHONOV, A.N. and SAMARSKII, A.A., **Differentialgleichungen der mathematischen Physik**, Deutsch. Verlag Wissenschaft, 1959 (trans. Russian).

[193] TIKHONOV, A.N. and SAMARSKII, A.A., Homogeneous difference schemes on non uniform nets, Zh. Vychisl. Mat. i. Mat. Fiz., 2 (1962), 812–832 (in Russian).

[194] TORO, E.F., **Riemann Solvers and Numerical Methods for Fluid Dynamics**, Springer-Verlag, Berlin, 1999.

[195] TOSELLI, A. and WIDLUND, O., **Domain Decomposition Methods: Algorithms and Theory**, Springer Series in Computational Mathematics 34, Berlin, New York, 2004.

[196] TURNER, M.J., CLOUGH, R.W., MARTIN, H.C. and TOPP, L.J., Stiffness and deflexion analysis of complex structures, J. Aeronautical Sci., 23 (1956), 805–823.

[197] VAN DER VORST, H.A., Bi-CGSTAB: A fast and smoothly converging variant of Bi-CG for the solution of non symmetric linear systems, SIAM J. Sci. and Stat. Comput., 13-2 (1992), 631–644.

[198] VANDERZEE, E., ZHARNOTZKY, V., HIRANI, A.N., GUOY, D. and RAMOS, E.A., Geometric and combinatorial properties of well-centered triangulations in three and higher dimensions, Comp. Geom. Theor. Appl., 46-6 (2013), 700–724.

[199] VARGA, R.S., **Matrix Iterative Analysis**, Springer Series on Computational Mathematics, 27, Springer, Berlin, 2009.

[200] VERFÜRTH, R., **A Review of A Posteriori Error Estimation and Adaptive Mesh Refinement Techniques**, Wiley, Chichester and B.G. Teubner, Stuttgart, 1996.

[201] VERFÜRTH, R., **Adaptive Finite Element Methods**, Lecture Notes Winter Term 2013/14, Fakultät für Mathematik, Ruhr-Universität Bochum, Germany, 2013.

[202] VIALLON, M.-C., **Personal communication, Département de Mathématiques**, **Université de Saint-Etienne**, France, 2006.

[203] VIANA, S.A., RODGER, D. and LAI, H., Overview of meshless methods, ICS Newsletter, **14** (2007), 4 pp.

[204] WAHLBIN, L., **Superconvergence in Galerkin Finite Element Methods**, Series Lecture Notes in Mathematics 1605, Springer, Berlin, 1995.

[205] WATKINS, T., **An Introduction to Tensor Analysis**, San José State University, applet-magic.com, Silicon Valley & Tornado Alley, CA, 2015.

[206] WEIR, A.J., **Lebesgue Integration and Measure**, Cambridge University Press, Cambridge, 1974.

[207] WEISSTEIN, E. W., Right-hand rule, MathWorld–A Wolfram Web Resource, http://mathworld.wolfram.com/Right-HandRule.html

[208] YOUNG, D.M. and GREGORY, R.T., **A Survey of Numerical Mathematics**, Dover, Mineola, NY, 1988.

[209] ZELDOVICH, Y.B. and FRANK-KAMENETSKY, D.A., Acta Physicochim. USSR 9 (1938), 341.

[210] ZIENKIEWICZ, O.C., **The Finite Element Method in Engineering Science**, McGraw-Hill, Maidenhead, 1971.

[211] ZIENKIEWICZ, O.C. and ZHU, J.Z., A simple error estimator and adaptive procedure for practical engineering analysis. Int. J. Numer. Meth. Engin., 24 (1987), 337–357.

[212] ZIENKIEWICZ, O.C., Displacement and equilibrium models in the finite element method by B. Fraeijs de Veubeke, In: **Stress Analysis**, eds. O. C. Zienkiewicz and G. S. Holister, John Wiley & Sons, 1965, pp. 145–197.

[213] ZIENKIEWICZ, O. C., TAYLOR, R. L. and ZHU, J. Z., **The Finite Element Method: Its Basis and Fundamentals**, Elsevier, Amsterdam, 2013.

Index

a posteriori error estimates(tor), 295, 296
  global residual, 297
  local explicit residual, 297
a priori error estimates, 226, 295
adjoint problem, 52
admissible
  deflexion, 264
  displacement, 42, 58, 263
  kinematically, 263
  mesh, 121, 122, 246
    FV, 182, (restricted), 185, 186
    tetrahedral, 246
  partition, *see* mesh
  physically, 37
  triangulation, 122
advection, 273, 287, 290
  dominant, 287
  equation, 279
advection-diffusion
  equation, 273–274, 279, 307, 315
  phenomena, 275, 283
  time-dependent, 287–288
advective, *see* convective
almost everywhere, 33
analytical solution, xix, xx, 108, 132
angle, 114, 127, 158, 172
  bound, 162, 164, 165
  minimum, 160, 162, 165
    criterion, 167, 168

obtuse, 122–123, 126–128, 205
  right, 22, 123–124, 191
approximation, xxii, xxxi, 8, 14, 19, 21
  derivative, 10
  external and internal, 267
  successive, xxix, 26
artificial diffusion, *see* numerical diffusion
Aubin-Nitsche
  duality argument, 217, 223, 235
  trick, 163
averaging method, 296

B-spline, 318
Babuška-Brezzi condition, 189, 308
Backward Euler scheme, 65, 193
  error estimate, 83
  local truncation error, 82, 195
  matrix form, 193
  stability, 78, 194
balance
  energy, 273
  equation, 19, 20, 127, 182, 186, 244
  relation, *see* balance equation
banded matrix, xxxiii, 26, 98, 138
bar deformation (elongational), 2
barycentric coordinates
  segment, 110, 214
  tetrahedron, 247–250
  triangle, 108, 113, 156, 170, 224

Numerical Methods for Partial Differential Equations:
An Introduction Finite Differences, Finite Elements and Finite Volumes, First Edition. Vitoriano Ruas.
© 2016 John Wiley & Sons, Ltd. Published 2016 by John Wiley & Sons, Ltd.
Companion Website: www.wiley.com/go/ruas/numericalmethodsforpartial

beam bending (clamped), 268
BEM, *see* boundary element method
BES, *see* Backward Euler scheme
biharmonic equation, 263, 271
    conforming FEM, 270
    Thirteen-point FD scheme, 266
bilinear
    form, 103, 216, 222, 226, 297
    function, 218, 220–221
    mapping, 218, 223
BKS, *see* Kawarara scheme
Bogner-Fox-Schmit, 270, 271
boundary condition, 3–5, 8–9, 62–67, 125
    Dirichlet, 24, 35, 62, 94, 103, 226, 244
    Dirichlet-Neumann (mixed), 95, 103,
        218, 233, 235, 242, 247, 296
    essential, 18, 42, 263, 269, 271
    homogeneous, 24, 163, 177, 212, 271
    inhomogeneous, 24, 39, 78, 151, 217
    natural, 42, 263, 264
    Neumann, 25, 44, 98–99, 165, 296
    Poisson equation, 94–96
    Robin, 25, 212, 217
    transport equation, 64
boundary element method, 313
boundary integral, 240, 251
    equations, 314
boundary layer, 274–277, 293
boundary value PDE, *see* boundary value
    problem
boundary value problem, xix, 2, 24, 318
    fourth-order, 263, 266
    model, 24
    non-linear, 302
    ODE, 276
    second-order, 93, 212, 246, 279
    two-point, 2, 214, 276, 302
bounded below, *see* lower bound
Bürgers equation, 301, 315

canonical basis, 132, 225, 270
Cartesian mesh, 177, 184–185, 205
Cauchy stress tensor, 119
    deviator, 307
Cauchy-Schwarz inequality, 33, 51, 173
Cell-centred FV scheme, 20, 27, 54, 74

    Poisson equation, 126, 135, 177, 244
Centred FD, 70, 88
    scheme, 176, 276, 278, 282
CFD, *see* computational fluid dynamics
CFL condition, 81, 86, 88, 90
characteristic
    function, 176, 189, 284
    length, 12, 307
    physical
        bar, 3
        plate, 263
    polynomial, xxvii
    velocity, 307
Chebyshev polyomials, 316
Cholesky
    decomposition, xxxiii, 138–139
        existence, 140
        incomplete, 308
    factorisation, *see* Cholesky
        decomposition
    method, xxxiii, 26, 197
closure, 94, 242, 269, 290
CNS, *see* Crank-Nicolson scheme
collocation, 316
computational effort, 66, 165, 306, 314
computational fluid dynamics, 186
condition number, xxviii
conforming FEM, 267, 269–270
conjugate gradient method, xxxi, 26, 140
    pre-conditioned (ning), 140, 308, 319
conservation, 306
    laws, 63
    principle, 18, 306 (property)
conservative,
    fluxes, 22, 131,182, 284
    form, 1, 3–6
consistency, xxiii, 42, 50, 82, 227, 230
    condition (weights), 293
    Euler schemes, 195
    fluxes (numerical), 185, 286
    modified Three-point FD scheme, 44
    order, 42–43, 49, 152
    $\mathcal{P}_1$ FEM, 46, 160
    SUPG scheme, 281

consistent, 31, 42, 47, 84, 201
    FD sense, 48, 284
continuous, xx
    piecewise linear, 111, 187, 249
control volume, *see* CV
convection-diffusion, *see*
       advection-diffusion
convective velocity, 288
convergence, xxiii, xxxi, 31, 48
    FDM, 151
    FV scheme, 52, 169, 175, 177, 186
    iterative method, xxix
    order, 49, 54–56, 58, 160, 167
    $\mathcal{P}_1$ FEM, 154, 163, 170
       Crank-Nicolson, 198
    rate, *see* order of convergence
    suboptimal, 235
convex, 103–104, 163, 199, 225–230
    combination, 281–282, 293
    polygon (gonal), 121, 155, 186, 226
    polyhedron 246–247
    quadrilateral, 183, 190, 218–220
Couette flow (circular), 308–309
Courant-Friedrichs-Levy, *see* CFL
Crank-Nicolson method, *see* CNS
Crank-Nicolson scheme, 66
    consistent, 201
    error estimate, 202, 204
    stability, 79, 199, 204
Crout decomposition, 139–140
Crout's method, xxxii, 26, 93
CV, 17, 73, 121–124, 170
    boundary, 127
    cell-centred, 21
    fictitious, 128, 182
    polygonal, 122
    rectangular, 177–178
    representative point, 19, 27, 54, 184, 247
    upwind, 73–74
    vertex-centred, 18, 123–124

Darcy pressure, 131
DDM, *see* domain decomposition method
deformation
    bar, 2, 6, 24, 58
    membrane, 92

plane, 117
degree of freedom (FEM), 120–121, 212,
    266–271, 315–317
Delaunay triangulation, 122, 126, 205, 247
density
    fluid, 307
    force, 2, 6, 36, 42, 118, 263, 306
    mass, 62, 288
DGM, *see* discontinuous Galerkin method
differential equations, xix
dimensionless, 118, 274, 288, 307–308
Dirichlet, 24, 62, 95, 103, 122, 129, 217
Dirichlet-Neumann, 95, 103, 212, 217, 242
discontinuous Galerkin method, 268, 315
discrete, xx, 23, 54, 78, 165, 198, 279, 289
    analog, 20, 25, 98, 120, 204, 250, 314
    balance equation, 18–20
    counterpart, *see* discrete analog
    derivative, 22, 24, 56, 171, 284
    flux, 127
    Friedrichs-Poincaré inequality, 179, 285
    gradient, 171, 174, 178, 187
    variational problem, 23–24
discrete maximum principle, *see* DMP
discrete minimum principle, 36
discretisation
    FD, 61, 65, (total) 70, 193, 307
    FE, 280
    FV, 178
    lattice, xxiv, 1, 7, 12, 30
    level, 35, 49, 295
    method, xx, xxiv, 35, 48
    parameter, 31, 35, 42, 48, 66, 79
    space, 61, 64–65, 70, 75, 88, 196–199
    space-time, 64, 75, 88, 192
    time, 61, 65–66, 75, 88, 195, 198
displacement, 2, 36–37, 41
    admissible, 42, 58, 263
    degrees of freedom, 120–121
    field, 117–121, 263
    minimising, 42
    prescribed, 24
distribution of forces, *see* density of forces

divergence, 102, 137, 308
  operator, 95
  tensor, 119
  theorem, 102, 119, 124, 135, 165
divided difference, 11, 243, 264
  operator (rectangular grid), 96
DMP, 36, 52, 78, 152, 277
domain decomposition, 316, 318, 320
dual field (variable), 189, 317
dual hybrid FEM, 317
dual mesh, 124, 128, (semi) 178–180, 247
  non overlapping method, 319
  partition, *see* dual mesh
duality argument, *see* Aubin-Nitsche

elastic solid
  homogeneous isotropic, 117
  incompressible, 120
elasticity (linear) equations, 118
electric field, 92
element (mesh), 12, 16–22, 75, 105, 160
  area, 108, 161, 170, 178, 218
  matrix, 111–115, 121
  maximum edge length, 161
  minimum length, 288
  numbering, 111–112, 127, 135, 138
  vector (right-side), 112, 115, 121
element-by-element, 111–112, 170
elliptic equations, xx, 94, 100, 205, 263
  curved domain, 230, 235
  parallelepiped, 242
elliptic PDEs, *see* elliptic equations
energy, 41, 263
  balance, 42, 273
  bar, 41, 42, 58
  conservation, 63
  inner product, 41
  norm, 41, 51, 197–198
error, xxviii, 31, 196
  a priori analysis, *see* a priori error
    estimates
  absolute, 27–28, 54, 166, 283, 293
  accumulation, *see* error propagation
  estimate, 48
    Backward Euler scheme, 83
    Cell-centred FV scheme, 54, 186

Euler schemes, 195
    FE Crank-Nicolson scheme, 204
    Five-point FD scheme, 152, 154
    Forward Euler scheme, 83
    Lax scheme, 85
    modified Three-point FD scheme, 49
    $\mathcal{P}_1$ FEM, 50, 52, 163–164, 217
    right-side quadrature, 173
    Upwind FV scheme, 287
    Upwind scheme, 85
    Vertex-centred FV scheme, 53, 177
    weighted mass scheme, 293
  interpolation, 47–48, 156, 175–176,
    228
  local truncation, 42–43
  mean flux consistency, 185–186
  numerical quadrature, 173
  propagation, xxix, xxxii, 88
  relative, xxviii, xxix, 294
  round-off, xxxi, xxxii, 26
Euler methods, *see* Euler schemes
Euler schemes (first order), 65, 195
  convergence, 81–82, 193–195
existence, 35, 104–105, 133, 165, 181
explicit, 65–66, 88, 290, 296, (quasi) 308
  residual based method, 296
  scheme, 68, 72–75, 87, 194, 205, 290
  time-integration, 70

family of grids, 30, 152
family of meshes, 31, 46, 74, 150, 160–163
FD, 8, (analog) 9, 11, 280
  grid, 7
    uniform, 43, 64, 319
  operator, 88, 151, 243
  stencil, 243–244, 265
FD scheme, 17
  Five-point, 96, 151–154, 193
  modified Three-point, 44, 49
  Seven-point, 242, 244
  Thirteen-point, 264, 266
  Three-point, 10–12, 36, 39
  Upwind, 276–277, 280–282
FDM, xxi, 1, 7–8, 24, 275
  local truncation error, 43
  mimetic, 318

Neumann boundary conditions, 25, 99
residual vector, 43
FE, 171, 224, 246, 250
   advection (techniques), 287
   analog, 42, 245, 267, 280, 306, 308
     Lax scheme, 75
   analysis, 318
   approximate problem, 39, 110, 224, 280
   Bogner-Fox-Schmit, 270–273
   code, 235, 250–251
   counterpart, *see* FE analog
   Crank-Nicolson scheme, 198–204
   Hermite, 266, 268, 270
   interpolate, 267
   isoparametric (FEM), 117, 222
   Lagrange, 212, 216, 224, 266, 317
   linear, *see* $\mathcal{P}_1$ FE
   matrix, 17–18, 125
   mesh, 17, 139, 170, 219, 253
     quasi-uniform family, 185
   piecewise linear, *see* $\mathcal{P}_1$ FE
   piecewise quadratic, 224
   pointwise behaviour, 204
   space, 120, 187, 231, 267, 280
   space semi-discretisation, 75, 198
   subspace, *see* FE space
   three-dimensional computations,
     248–253
FEA, *see* FE analysis
FEM, xxi, 5, 12, 22–25, 31, 73, 226
   accuracy, 224, 299
   conforming, 267–269
   consistency (consistent), 44, 47
   convergence, 161
     order, 47, 170
   Hermite, 212, 266
   implementation, 110, 235–236, 250
   Lagrange, 212
   linear, *see* Piecewise Linear FEM
   lowest-order Raviart-Thomas mixed,
     128, (FE) 133, 186
   matrix, 124, 140
   Navier-Stokes equations, 310
   non-conforming, 267, 268
   Piecewise Linear, *see* $\mathcal{P}_1$ FEM
   piecewise trilinear, 242, 245

$\mathcal{Q}_k$ (k-linear), 224
RT_0 (mixed FEM), *see* lowest order
     Raviart-Thomas mixed FEM
   trilinear, *see* piecewise trilinear FEM
FES, *see* Forward Euler scheme
finite difference, *see* FD
finite difference method, *see* FDM
finite element, *see* FE
finite element method, *see* FEM
finite point method, 316
finite volume element method, 187
finite volume method, *see* FV scheme
finite volume-finite element, 187
first-order PDE, 63
flux, 21, 48, 121, 125, 131, 136, 181,
     285
   approximate, *see* numerical flux
   boundary, 135, 182, 244, 286
   calculation, *see* flux computation
   Cell-centred FV scheme, 137, 178
   computation, 126, 136
   consistency, 185–186
   edge, 127–128, 135–137, 177, 186
   face, 244
   heat, 187, 273
   mean, 128, 131, 134–135, 185
     consistency error, 185–186
   numerical (discrete), 22, 127
     out-going, *see* outward,
     outward, 136, 177, 284–286
   physical quantity, 121
   residual, 186, 286
   shape function, 189
   total, 131
   unknown, 135, 190
   variable, 137, 189, 192
   zero, 20, 137
Forward Euler scheme, 67, 193, 289
   consistency, 81
   error estimate, 83
   local truncation error, 82, 195
   stability condition, 68, 77, 194
Fourier
   analysis, 217
   series, 67, 319
Fredholm integral equation, 314

Friedrichs-Poincaré inequality, 40, 101
    discrete, 174, 179, 285
frontal method, 138
FV scheme
    Cell-centred, 20–23, 25, 27, 178, 244
        convergence, 177
        error estimate, 54, 186
        stability, 181, 184
    flux (computation), 126, 136
        rectangle, 131–137, 178
    order of consistency, 48
    triangle-centred, 128, 187
    Upwind, 73–74, 283–284
        error estimate, 287
        stability, 285
    Vertex-centred, 17–20, 23–24, 93, 122
        convergence, 169
        error estimate, 53, 177
        Poisson equation, 124–125
        right side quadrature, 170–172
FVM, *see* FV scheme

Galerkin, 287–288
    FEM, 317
    formulation, 4, 13, 100, 103, 130, 198,
        213, 236, 245, 267–268, 288, 314
    variational form, *see* Galerkin
        (variational) formulation
Galerkin Least-squares, *see* GLS
Gauss-Seidel, xxx, 26, 146, 319
Gaussian
    elimination, xxxi, xxxii, 26, 138
    mid-point rule, 18, 20
    quadrature formula, 115, 222, 239, 252
global truncation error, 47
GLS, 282, 308
Godunov scheme, 90
golden number, 56, 191–192
gradient, 95 (operator), 221, 300
    symmetric, 118
gradient recovery method, 296
Green's function, 313
Green's identity, 102, 197, 245, 297, 313
grid, 9
    coarse, 68
    curved orthogonal, 241

equally spaced, *see* uniform grid
    FD, 7, 65, 100, 205
    fine, 69, 319–320
    nested, 49, 55, 146, 319
    non-uniform, 11, 19, 152–154, 194, 243
    point, 7, 36, 49, 64, 67, 85, 265
        fictitious, 85, 154
        numbering, 98
    quasi-uniform family, 30, 152
    rectangular, 95–96, 151
    refinement, 154, 195
    size, 30, 48, 68, 83, 88, 151
        intermediate, 11
        minimum, 194
    space-time, 79
    spacing, 193, 264
    squared, 266
    time-step, 83
    uniform, 9, 25, 43, 70, 193–194

h-p, *h-* and *p-* version, 300
heat equation, 62, 67, 197–198
    initial boundary conditions, 62
    maximum principle, 76
    two-dimensional, 193
Helmholz equation, 313
Hermite
    cubics, 273
    FEM, 212, 266
    interpolation, 267–268, 271
    shape functions, 269–270
hessian, 158
homogeneous boundary conditions, 24
Hooke's law, 119
hybrid method, 317
hyperbolic equations, xx, 62
    first-order, 63
    second-order, 62–63, 88

IGA, 317
ill-conditioned, xxviii
implicit
    FD scheme, 66, 71, 193
    time discretisation, 65
incidence matrix, 111, 121, 220, 237, 250
incompressible fluid (flow), 92, 120,
        262–263, 306

inf-sup condition, 189
inhomogeneous boundary conditions, 24
initial conditions, xix, 62–64
inner product, 33
inscribed circle, 161, 223
interior penalty, 315
inverse inequality, 208, 228
inverse problems, 299
irrotational flow, 92
isogeometric analysis, *see* IGA
isoparametric (FE), 117, 218, 222
    mappings, 218–220
        bilinear (Q_1), 218
      P_2 (quadratic), 231
    transformation, 218, 220, 230
iteration divergence, 303

jacobian matrix, 303

Kawahara scheme, 75, 287, 293
Krylov spaces, xxxi, 319

Lagrange
    FE (FEM), 212, 216, 224, 266, 317
    interpolation, 212, 300–301, 317
Lagrange multiplier, 133, 317, 319
Lamé coefficients, 118
Laplace equation, 92, 95, 152
Laplacian (operator), 93–95, 242, 263
Lax scheme, 72, 79, 85
Lax-Milgram Theorem, 104
Lax-Richtmyer equivalence theorem, 31, 84
Lax-Wendroff scheme, 72, 88
linear differential equations, xx
linear elasticity, 120
linear elements, *see* P_1 FE
linear form, 103, 226, 250
linearised form, xxi, 301
Liouville form, 3
Lipschitz domains, 93
local truncation error, 42–43, 82, 98
    Backward Euler scheme, 195
    Cell-centred FV scheme, 48, 181
    centred FD, 70, 276
    Euler schemes, 81, 82

Five-point FD scheme, 152
FD sense, 54
FDM (uniform grid), 43
Forward Euler scheme, 195
Lax scheme, 84
Thirteen-point FD scheme, 266
Upwind FD scheme, 84
Upwind FV scheme, 284
lower bound, 160, 162, 194, 223, 296, 298
lumping (mass), 17, 75, 308–309

mass matrix, 288, 290–291, 293, 315
master
    square, 218–219
    triangle, 116, 230
master element, 115, 220, 230, 235
matrix
    band, 121
    banded, xxxii, 26, 98, 140, 310
    bandwidth, 138
    BEM, 314
    block, 132, 189
        diagonal, 315
        tridiagonal, 98
    condition number, xxviii
    diagonal, 98, 133, 134
    diagonally dominant, 140
        strictly, 26
    factorisation, xxxii, xxxiii, 308, 319
    FD, 320
    FE, 17, 114
    ill-conditioned, xxviii
    inverse, xxviii
    invertible, xxix, xxxi
    jacobian, 171, 303, 305, 306
    lower triangular, xxx, xxxii
    mass, 288, 315
        consistent, 290, 293, 306
        lumped, 290, 293, 298
        weighted, 290, 293
    norm, xxviii
    null, xxviii, 132
    permutation, xxxii
    positive-definite, xxvii, 26, 105
    skyline, 139

matrix (*continued*)
  sparse, 26
  spectral radius, xxix
  stiffness, 288, 316
  symmetric, xxvii, 10, 15, 18, 105
  transformation, 171
  tridiagonal, 10, 26, 66, 98, 303
  upper triangular, xxx, xxxii
maximum principle, 42, 68, 76, 193
maximum-minimum principle, 152
mean flux consistency error, 185–186
mesh, 12, 18, 22, 139, 238, 270, 284, 299
  acute type, 123, 126, 140, 169, 172–175
  adaptivity, 296–300
  admissible, 121, 246
  Cartesian, 77, 177, 184, 205
  dual, 124
  independent constant, 41, 163, 199,
      298
  nodes, 12, 17, 106, 120, 220, 236, 249
  non-uniform, 28, 53–54, 245, 283, 287
  points, *see* mesh nodes
  primal, 124
  quadrilateral, 220
  quasi-uniform family, 31, 162, 226, 228
  rectangle, 129, 132, 135, 138, 244, 246
  refinement 224, 297, 300
  regular family, 161, 163, 174, 177, 225
  size, *see* mesh step size
  step, *see* mesh step size
  step size, 12, 27, 31, 42, 156, 280, 293
  structured, 205, 246, 271
  tetrahedral, 246–247, 249–252
  triangular, 105–107, 122, 129, 132, 139
  uniform, 12, 16, 53, 74–75, 135, 139
  unstructured, 205
  variable, *see* non-uniform mesh
meshless methods, 316
mimetic FDM, 318
minimum principle, 76
mixed Dirichlet-Neumann problem, 95, 217
mixed FEM, 93, 128, 131
modulus, 32
mortar methods, 319
multigrid methods, 319
multi-scale, 279

natural boundary conditions, 42, 264
Navier-Stokes equations, 273, 287, 306
Neumann problem, 212, 217
Newton's iterations, 303
nodes, 12, 106, 214, 219, 224, 236
non-uniform grid, 11–12, 96, 151, 243
non-uniform rational B-splines, *see* NURBS
norm, 31
normal
  derivative, 98, 154, 264, 271, 297, 315
    outer, 94, 102, 124, 127, 135, 227
normal flux, *see* flux
normed vector space, 32
nonzero
  boundary conditions *see* inhomogeneous
      boundary conditions
  entries, 16, 26, 98, 134, 138–139, 315
null element, 31
numerical
  analysis, xxi
  coefficient, 315
  diffusion, 277, 281–282
  fluxes, 22, 135, 185, 286
  integration, *see* numerical quadrature
  mathematics, *see* numerical analysis
  mean fluxes, 22, 136
  method, xix, 31, 274, 301, 307, 315, 317
    consistent, 42
    convergence, 31, 48
    stable, 35
  model, 76
  parameter, 307
  procedure, *see* numerical method
  quadrature, 16–17, 155, 231, 240, 280
    formula, 115, 120, 133, 170, 222
  scheme, 17, 24, 36, 76, 193
  simulation, xx
  solution, xix, xxi, 31
  technique, *see* numerical method
  time integration, 88
NURBS, 317

ODE, xix
  linear system, 64
operator
  difference, 88, 96, 151, 243

differential, 52, 100, 263, 302
gradient, 95
Helmholz, 313
identity, 305
Laplacian, 93, 103, 165
order, xxiii
ordinary differential equation, *see* ODE
orthogonal, 110, 124, 182, 246–248, 263
   grids (boundary fitted), 241
   projection, 298–299

p-version, 300
$\mathcal{P}_1$ FE, 14, 50, 115, 224, 236, 288
   approximation, 213
   GLS formulation, 310
   interpolate (error estimate), 47
   linear elasticity (analog), 120
   representation, 308
   scheme, 14, 19, 38, 277, 287
      consistency, 44, 46
      convergence, 197
      SUPG, 280
      weighted mass, 293
   shape function, 120, 306
   solution, 172, 186, 230
   space (subspace), 296, 298
      semi-discretisation, 75
   test function, 170
$\mathcal{P}_1$ FEM, 12, 17, 105, 122, 213, 296, 306
   convergence (gent) 49, 51, 154, 223
      order, 163, 172
      non-smooth solutions, 166
   curved domains, 226
   element matrix and vector, 112, 124
   error estimate, 50, 52, 163–164
      right side quadrature, 173
   implementation, 26, 110, 121, 237, 250
   numerical quadrature error, 173
   parabolic problems, 205
   Poisson equation, 100, 105
   shape function, 12–14, 106–108, 249
   solution gradient, 131
   stability inequality
      energy norm, 41

   mean-square, 41, 155
   tetrahedron-based, 247
   three-dimensional space, 249
$\mathcal{P}_2$ FEM (FE), 224–226, 230
   curved domain, 230
   implementation, 236–237, 240
   isoparametric, 231
   shape function, 214–215
   variational approximation, 225
$\mathcal{P}_3$ shape functions, 216
$\mathcal{P}_k$ FEM, 212, 217, 224, 237
   Lagrange, 216
   space, 214
parabolic
   equations, xx, 62, 100, 197–198, 293
   edge, 230
parallel computing, xxi, 318, 321
parallelogram law, 34
partial differential equation, *see* PDE
partition, *see* mesh
partition of unity, 73, 76–77, 115, 165–166
patch-test, 268
PDE, xix
   elliptic, *see* elliptic equations
   fourth-order, 263
   hyperbolic, *see* hyperbolic equations
   non-linear, 301, 306, 310
   parabolic, *see* parabolic equations
Péclet number, 274, 288
   element, 282
   high, 277, 279, 315
perpendicular bisector, 123, 125, 127
Petrov-Galerkin
   formulation, 278–279
   streamline upwind, *see* SUPG
   terms, 282
piecewise
   cubic shape function, 214, 270
   linear (function), 23, 169, 187
   linear elements, *see* $\mathcal{P}_1$ FE
   linear FEM, *see* $\mathcal{P}_1$ FEM
   quadratic, 214, 225, 235
   quadratic FE, *see* $\mathcal{P}_2$ FEM (FE)
   trilinear FEM, *see* $\mathcal{Q}_1$ FEM (3D)
pivot, xxxii
pivoting, xxxi, 26, 138–139

plate bending, 118
  clamped, 263–264, 268
  simply supported, 264
Poisson equation, 92–94, 99, 130, 242
  a posteriori error estimation, 298
  FD scheme, 96, 151–152, 243–244
  FEM, 100, 140, 163, 218–224, 247
  FV scheme, 122, 244, 246
    Cell-centred (rectangle), 135, 137
    triangle-centred, 128, 187
    Vertex-centred, 124
  maximum-minimum principle, 152
  mixed formulation, 93, 129–131, 187
  Neumann boundary condition, 99, 165
  Raviart-Thomas analog, 131
  three-dimensional, 242, 246, 249
  variational approximation, 222, 236
Poisson problem, *see* Poisson equation
Poisson's ratio, 118, 263
polygon, 93, 105, 124, 170, 183, 218, 224
  convex, 155, 164, 186, 226
polygonal, 318
  approximate curved domain, 230
  cell, 121
  domain, 192, 218, 231, 235
  line, 242, 250
  mesh, 186
polyhedron (hedral), 242, 246–247, 318
polynomial degree, 212–218, 222–226,
    239–240, 314–316
positive-definite, xxvii, 26, 105, 140
pre-conditioning, 140, 308, 319
primal hybrid method, 317
primal mesh, 124
projection, 189, 198–200, 205, 230,
    298

$\mathcal{Q}_1$ FEM of the, 218, 222–223
  three-dimensional (3D), 242
    rectangular elements, 245
quadrangulation, 224
quadratic elements, *see* $\mathcal{P}_2$ FEM
quadrature, *see* numerical quadrature
  rule, *see* quadrature formula
  formula, 19–21, 23–25, 178, 202,
    239–241, 252

quadrilateral, 183, 218–223
quasi-uniform, 30, 31, 162, 228

radial basis functions, 316
rank condition, 133
Raviart-Thomas mixed FEM of the
    lowest order, 128, 133, 137, 186
regular family
  meshes, 161–163, 174, 177, 224–225
regularity, 49, 51, 93, 155, 185, 224, 263
  assumption, 3, 54, 104, 201, 203
  constant, 162
  datum, 5
remeshing, 296, 316
representative point (CV),20, 54, 182, 246
residual, 42, 54, 81–84, 152, 186, 195, 297
  based method (explicit), 296–298
  function, 44, 46, 281
  numerical flux, 286
  restriction (multigrid), 320
  variational, 160
  vector, 43–44, 140, 152, 320
Reynolds number, 307
Ritz, 216, 316
  method, 104–105, 107, 111, 120
  projection, 198–200, 205
Robin problem, 212, 217, 235–236
RT_0, *see* Raviart-Thomas mixed FEM (of
    the) lowest order

semi-dual mesh, 178
semi-norm, 33
sensitivity analysis, 55, 144
shape function, 218, 236, 240–241
  $\mathcal{P}_1$ FEM, 13, 106–108, 111, 115, 249
  $\mathcal{P}_2$ FEM, 214–215, 225, 230
  piecewise cubic, 216
  piecewise quadratic, *see* $\mathcal{P}_2$ FEM shape
    function
shape regular, 223
shock wave, 302
Simpson's rule, 56, 240–241
skyline matrix, 139
SLAE, xxiv, xxviii, xxix, 216, 220, 296, 298
  BEM, 314
  BES, 65, 194

Crank-Nicolson scheme, 66, 198
Five-point FD scheme, 98
$\mathcal{P}_1$ FE, 15, 110
   elasticity system, 120–121
Raviart-Thomas mixed FEM, 132, 189
Ritz method, 105
solving, xxix, 26–27, 66, 138–140, 314
   direct method, xxxi, 26, 138, 319
   error propagation, xxix
   iterative method, xxix, 26, 139, 320
Three-point FD scheme, 10, 12, 44
Vertex-centred FV scheme, 169–171
Sobolev
   embedding theorem, 155, 230
   space, 39, 45, 101–102, 155
sparse matrix, 10, 26, 98, 314
spectral
   methods, 316
   norm, xxviii
   radius, xxix
splines, 270, 317
stability, xxiii, 35
   in norm, 31
   Von Neumann, 70
stability condition
   Forward Euler scheme, 68, 77, 194
   Lax scheme, *see* CFL condition
   weighted mass scheme, 291–292
stability inequality, 44, 50, 278
   Backward Euler scheme, 78, 194
   Crank-Nicolson scheme, 199, 204
   Five-point FD scheme, 153
   Forward Euler scheme, 77, 194
   Lax scheme, 80
   mean-square sense
      Cell-centred FV scheme, 181, 184
      SUPG FE scheme, 281
      Upwind FV scheme, 284–285
      $\mathcal{P}_1$ FEM, 41, 155
   energy norm, 41
   Three-point FD scheme, 36, 39
   Upwind FD scheme (pointwise), 277
   weighted mass scheme, 292
stable, 31, 35, 78–79, 84, 86, 279, 291
   unconditionally, 66, 71, 199
standard Galerkin, *see* Galerkin

stiffness (body), 2, 41, 62, 117–118
storage (matrix), 26, 138–139, 308, 314
strain rate tensor, 307
strain tensor (small deformations), 118
stream function, 263
streamline upwind Petrov-Galerkin, *see*
   SUPG
strict acute type, 126–127, 169
strictly diagonally dominant, xxxi, 26
strictly positive order, 31, 49
strong form, 1, 3
structured mesh, 205, 246, 271
Sturm-Liouville problem, 3
subscales, 279
successive over-relaxation, xxx, 26
supraconvergence, *see* superconvergence
superconvergence, 54, 58, 137
SUPG, 279
   optimal coefficient, 282
   $\mathcal{P}_1$ FE scheme, 280
support, 290
symmetric gradient, 118
symmetry conditions, 192
system of linear algebraic equations, *see*
   SLAE

Taylor-Couette flow, 308
test function, 4, 187, 279, 315
tetrahedrisation, 247
tetrahedron
   barycentric coordinates, 247–250
   centroid, 248, 251–252
   volume, 247
thin plate bending, 262
time marching, 66, 315
   two level scheme, 88
total internal energy, 41
total potential energy, 42
traces, 103, 220, 225, 231, 249, 271
transient problems, xxiv
transition points, 95, 99, 106, 111, 236
transmissivity property, 182
transport equation, 63–64, 76, 279, 287
   non-linear, 301
   physical meaning, 79
   schemes, 70, 79, 84

trapezoidal quadrature formula, *see* trapezoidal rule
trapezoidal rule, 16–19, 133, 144, 280, 298
trial function, 187, 279
triangle, 105–117, 122–128
    barycentric coordinates, 108–109
    centroid, 110, 115, 137, 170, 239–240
    circumcentre, 122, 124, 127, 170
triangle inequality, 32
triangular meshes, 93, 122, 182, 205
triangulation, 105, 122, 154, 161–163, 299
    Delaunay, 126, 247
tridiagonal matrix, 10, 15, 26, 66, 98, 303
trilinear elements, 245
truncation error, *see* local truncation error

unconditionally stable, 66, 71, 78, 199
uniform grid, 9, 61, 97, 154, 193, 264
uniformity constant, 30–31, 46, 151, 162
uniqueness, 35, 104–105, 133, 218, 236
unit vector, 94, 102, 158, 171, 176, 185
unstable, 68, 277, 308
unstructured mesh, 205
upper bound, 297
upwind, 72–74, 81, 84–85
    CV, 73
    FV scheme, 74, 285

V-cycle, 319
variational form, 39, 100, 267, 296, 306
vectorisation, xxi
velocity, 301, 306
    angular, 308
    characteristic, 307
    components, 309

convective, 288
fluid, 273
initial, 63, 89, 301
Vertex-centred FV scheme, 17, 53, 74
    Poisson equation, 124, 169, 174, 177
vibrating string
    equation, 3, 62, 85
        time marching scheme, 88
    first-order system, 63, 85
    initial condition, 63
virtual element method (VEM), 318
viscous
    fluid, 307–308
    flow, 262
Von Neumann stability, 90
    analysis, 88
    criterion, 70
Voronoi diagram, 247
VS, *see* vibrating string

wave equation, 64
weak form, 1, 4, 103, 130
weighted mass
    scheme, 290, 293–294
    matrix, 291
well-centred tetrahedrisations, 246
WMS, *see* weighted mass scheme
work (of external forces), 42

XFEM, 169, 226

Young
    inequality, 40, 199, 285
    modulus, 118, 263

zero boundary conditions, *see* homogeneous boundary conditions